Quantification of Climate Variability, Adaptation and Mitigation for Agricultural Sustainability

Mukhtar Ahmed • Claudio O. Stockle
Editors

Quantification of Climate Variability, Adaptation and Mitigation for Agricultural Sustainability

Springer

Editors
Mukhtar Ahmed
Department of Agronomy
Pir Mehr Ali Shah Arid Agriculture University
Rawalpindi, Pakistan

Department of Biological Systems Engineering
Washington State University
Pullman, WA, USA

Claudio O. Stockle
Biological System Engineering
Washington State University
Pullman, WA, USA

ISBN 978-3-319-32057-1 ISBN 978-3-319-32059-5 (eBook)
DOI 10.1007/978-3-319-32059-5

Library of Congress Control Number: 2016947061

Printed on acid-free paper

This Springer imprint is published by Springer Nature
The registered company is Springer International Publishing AG Switzerland

Preface

Climate change is one of the burning issues in all fields of life starting from social sciences and going to the applied sciences. Climate vulnerability threatens global climatic cycles and world food production systems, thus affecting the life of people. Most of the world is exposed to the effects of climatic change due to extreme variability in temperature and rainfall. Climate change is the defining issue of time and the extreme case scenario of climate change is very horrible. Future generations might curse us if we will not address this issue in the appropriate way. The Paris Climate Agreement (COP21) is one of the efforts to mitigate climate change in which an agreement has been signed to bring global temperature increase well below 2 °C (3.6 °F) and to pursue efforts to limit to 1.5 °C, but this requires its accurate implementation. Similarly, balance between sources and sinks of greenhouse gases (GHGs) is necessary to reduce the risks related to climate change. Risk reduction interventions represent a major avenue for responding to both existing rise in temperature, carbon dioxide, greenhouse gases, and flood and drought hazards and the increases likely to emerge as a consequence of climate change. However, despite the big risk of climate change, the world has done practically nothing to address this risk. The only reason is that climate change threat is a threat to future generations, so today's actions will benefit only the future generations and not us. Similarly, most of the adaptation actions which can reduce emissions are very expensive. Since climate change impacts are irreversible, therefore, we have to take actions to avoid serious climate change.

The economic development of countries depends upon the climate-sensitive sector (CSS) that is agriculture which is the backbone of most of the developing countries. Similarly, agriculture is the main sector which might help to reduce poverty since it was earlier reported that a proportion of people living less than $1.25/day had dropped. Therefore, to eradicate hunger and poverty, it is imperative to focus concentration toward agriculture sector especially in the context of climate change. The world is ecologically more fragile due to multiple climate stresses, and their effects are more on the nature-dependent sector, i.e., agriculture; therefore, the need for mitigation and adaptation is necessary for this sector. This sector has direct link with the poorest peoples; thus, their vulnerability to future climatic extremes would be more open. The developing countries agriculture would be affected by severe desertification, floods, drought, rising temperature, and extreme events as reported by the Intergovernmental Panel for Climate Change. Therefore, climate change and population growth may threaten food security which would necessitate coordinated efforts to ensure food security on long-term basis. Since agriculture impacts more on the world compared to anything else, therefore, transformation in agriculture is essential to ensure yield sustainability, to reduce the impacts of climate extremes and to build a resilient system according to the changing climate. This resilient system will ultimately reduce the impact of climate change on agriculture. Agriculture depends upon the calamities of nature; if climate is favorable, it would lead to good crop yield thus ensuring food security. However, in the context of climate change, the issue of food security will be more highlighted because of the dependency of a maximum population on agriculture. Since climate change is affecting the agriculture sector at maximum, therefore, adaptation approaches need to be considered for the survival of the agriculture sector. These approaches include empirical (use of past data to study the impact of climate change), mechanistic crop modeling approaches (use of crop models like APSIM, DSSAT, EPIC, etc., to build climate scenarios (temperature, rainfall and CO_2, and different crop responses under these climatic factors)), and niche-based approaches or agroecological zoning approaches (use of global models like GCM to study climatic parameters of climatic adaptations).

In this book, we tried to present the impact of climate variability on different agricultural crops using different approaches which can help to redesign our agricultural management operations and cropping systems. Crop responses like accelerated life cycle, skipping of phenological stages, reduced leaf area and duration, inhibition of metabolism (photosynthesis and respiration), and impaired reproductive growth might be seen under different kinds of climatic stresses. The design of new adaptive genotypes in response to these climatic stresses might include the study of QTL (quantitative trait loci) traits and physiological and genetic options. Multilocation testing approaches using empirical models could be used to study the response of genotypes under contrasting environments (genotypes x environments interactions) which could be helpful for breeders and researchers. The dissection of yield into its physiological components and understanding of stress-adaptive traits (deeper roots, canopy cooling, transpiration efficiency, and delayed senescence) may be the best options to adapt under the changing climate. Therefore, this book is helping to give understanding about the impact of climate variability and further adaptations and mitigation strategies. The first two chapters of the book focus on GHG emissions from different sectors across the globe. Chapter 1 suggested mitigation techniques that include the use of bioenergy crops, fertilizer and manure management, conservation tillage, crop rotations, cover crops and cropping intensity, irrigation, erosion control, management of drained wetlands, lime amendments, residue management, biochar, and biotechnology. Chapter 2 explains the source of livestock-related emissions. Chapter 3 covers the impact of climate variability on crop production in sub-Saharan Africa. Being a region with high climate vulnerability, the quantification and understanding of the extent and rate of impact of climate variability on crop productivity are highly essential. In Chap. 4, the fate of N was discussed for wheat crop using the Agricultural Production Systems Simulator (APSIM) cropping system model. It is shown how the APSIM model successfully explains the nitrogen use efficiency in wheat crop. Climate variability impact on rice production is covered in Chap. 5, and phenotype relationships through QTL analysis in a recombinant inbred population are all discussed in Chap. 6. Chapter 7 explains the crop water productivity (CWP) using the soil and water assessment tool (SWAT) model. Chapter 8 discusses the effects of abiotic stress in crop production since abiotic stresses already represent one of the key factors limiting worldwide crop production. Chapter 9 covers the impacts of drought on cereal crops under the changing climate. This chapter summarizes different aspects of crop breeding for drought tolerance and analyzes how conventional breeding, genetics, biotechnology tools, microarrays, MAS, QTL, bioinformatics and transgenic crops as well as mineral nutrients, and plant growth regulations can participate to advancing the emancipation of drought-resistant rice and maize cultivars. Chapter 10 covers wheat physiological response under drought. Drought was considered responsible for the enhanced production of proline and epicuticular wax, reduced stomatal conductance, high stomatal resistance, and low photosynthetic and transpiration rate in genotypes as a mechanism to bear the harsh conditions. Chapter 11 discusses the silvopastoral systems as the best agroecological practice for resilient production systems under dryland and drought conditions. Climate change impacts on wheat production in Europe are discussed in Chap. 12. The authors suggested that the identification of the best adoption strategy to the wide variation in future climate will be a vital option to sustain crop productivity.

Chapter 13 presented climate change impacts and adaptation options for coping with future climate by individual farm fields in the Wami River sub-basin in Tanzania. Climatic variability impact on wheat-based cropping systems of South Asia is discussed in Chap. 14. Chapter 13 provides the fate of phosphorus under the changing climate and its dynamics study using modeling approaches, since global climate change and its impact on crop production are a major issue. Therefore, future climate change impacts on wheat yield in Pakistan, especially in the rain-fed region of Potohar, are discussed in Chap. 16. Finally, Chap. 17 provides an overview of bioinformatics as an interdisciplinary science emerging from the interaction of computer, statistics, biology, and mathematics to analyze genome arrangement and contents and biological sequence data and predict the structure and function of macromolecules that are used in interpreting and decoding plant genome. Overall, it should be possible to cover about one chapter in 2–3 h of lecture. Therefore, it is the appropriate book for universities and public libraries to develop understanding about climate variability. At last, we are immensely grateful to the contributing authors and acknowledge and appreciate the comments by Stewart Higgins in the preparation of the manuscript.

Pullman, Washington, DC, USA Mukhtar Ahmed

Contents

Chapter 1
Greenhouse Gas Emissions and Climate Variability: An Overview

Mukhtar Ahmed

Contents

Abstract A comprehensive overview of greenhouse gas (GHG) emissions of from different sectors across the globe is provide in this chapter. Particular attention is given to agriculture, forestry, and other land use (AFOLU). Since agricultural activities (cultivation of crops, management activities and rearing of livestock) result in production and emissions of GHG, quantification of GHG and its mitigation is addressed in this chapter. The suggested mitigation techniques include the use of bioenergy crops, fertilizer and manure management, conservation tillage, crop rotations, cover crops and cropping intensity, irrigation, erosion control, management of drained wetlands, lime amendments, residue management, biochar and biotechnology. Furthermore, quantification of GHG emissions is discussed using different process based models. These models could further be used as decision support tools under different scenarios to mitigate GHG emissions if calibrated and validated effectively.

Keywords Greenhouse gas emissions • Climate variability • AFOLU • Mitigation

---

M. Ahmed (✉)
Department of Agronomy, Pir Mehr Ali Shah Arid Agriculture University, Rawalpindi, Pakistan

Department of Biological Systems Engineering, Washington State University, Pullman, WA, USA
e-mail: ahmadmukhtar@uaar.edu.pk; mukhtar.ahmed@wsu.edu

M. Ahmed, C.O. Stockle (eds.), *Quantification of Climate Variability, Adaptation and Mitigation for Agricultural Sustainability*,
DOI 10.1007/978-3-319-32059-5_1

1.1 Introduction

Combustion and extensive use of fossil fuels results in the emission of greenhouse gases (GHGs) which contribute to the greenhouse effect. The fundamental phenomenon of greenhouse effect is based upon absorption and transmission of energy, depending upon its wavelength. High temperature bodies such as sun generally emit radiation which is of short wavelength and cooler bodies like earth emit long wavelength radiation. Longer wavelength radiation is called infrared radiation. Infrared radiation is not as harmful according to Planks Quantum theory of radiation energy is inversely proportional to wavelength (λ) and directly proportional to frequency (v) i.e. $E=hv$ where $v=c/\lambda$. However, short wavelength radiation easily passes through glass then after striking colder bodies such it is transmitted back at a longer wavelength, which is blocked by the glass resulting in an increased temperature under the glass. This phenomenon is largely used in the greenhouse industry to let solar radiation in and block longer wavelength radiation to increase inside temperature for plant growth even if the outside temperature is too low to grow plants. Some atmospheric gases have the same property and maintain earth's temperature at a certain level. These gasses are called GHGs, and they include carbon dioxide (CO_2), methane (CH_4), water vapor and oxides of nitrogen (NO_X). However, due to intensive use of fossil fuels, industrialization, deforestation and mechanization in agriculture the amount of these GHGs, particularly CO_2, has increased significantly resulting in global warming. The Global Warming Potential (GWP) is used as a measure of the global warming impacts of different GHGs. It is measure of how much energy the emission of one ton of gas will absorb in a particular time period in comparison to one ton of CO_2. The larger the GWP, the greater will be the impact of that gas in comparison to CO_2 over a given time period, i.e., 100 years. GWP allows policy makers to compare emissions and design reduction strategies. Since CO_2 is used as reference it has GWP of 1 while methane (CH_4) GWP is 28–36, nitrous Oxide (N_2O) has a GWP 265–298. High GWP gases, called fluorinated gases, have GWPs in range of the thousands or tens of thousands.

Carbon dioxide is the chief GHG emitted through human activities. The emission of CO_2 has increased significantly due to deforestation which resulted in an alteration of the carbon cycle. Since forests are a main sink for CO_2, their destruction results in increased atmospheric CO_2 (NRC 2010). The increase of carbon dioxide in the atmosphere is due to the burning of fossil fuels. Methane (CH_4) is the second dominant GHG emitted by human activities. The main source of methane is raising of livestock, rice paddies and bacterial action on landfills and wastes. The petrochemical industry and coal mines are also big contributors of methane. In general 35 % of the methane emissions are natural, and 65 % are due to human activities. Nitrous oxide (N_2O), another GHG, is naturally present in the atmosphere due to the N-cycle but it also comes from human activities such as agriculture, transportation, and industry (EPA 2010). Nitrous oxide is the main precursor of ozone depletion. Nitrous oxide emissions from natural lands is 55 % of global N_2O emissions. Kim et al. (2013) concluded that nitrous oxide emissions from natural land is lowerer than from agricultural land. Fluorinated gases are the longest lasting and most potent GHGs destroying ozone layer. GHG emission is now a critical topic due to its devastating effect on different sectors of life, which also results in global warming (Kennedy et al. 2009).

The countries that emited the highest amount of GHG include China (23 %), USA (19 %), the European Union (13 %), India (6 %), the Russian Federation (6 %), Japan (4 %), and Canada (2 %) while other countries produced 28 % (IPCC 2007). Global GHG emissions and sinks are related mainly to land use change. The maximum emission of CO_2 globally is due to deforestation, particularly in Africa, Asia, and South America. According to Houghton et al. (2012) net flux of carbon from land use and land cover change (LULCC) accounted for 12.5 % of anthropogenic carbon emissions. Hergoualc'h and Verchot (2014) studied land use change in Southeast Asia where tropical peat swamp forests are located. These forests act as global carbon stores but due to their intensive degradation and conversion to agricultural lands GHG emission in the region have increased significantly. The major driver of environmental change and increased GHG emissions is land use change(LUC) (Turner et al. 2007; Lambin and Meyfroidt 2011; IPCC 2013). Similarly, it leads to alteration in soil organic carbon and changes in biodiversity (Sala et al. 2000). Therefore, there is a dire need to mitigate the impact of LUC through utilization of renewable energy technologies. Similarly, in order to minimize GHG emissions from land use change, quantification of the direct impact of land use change on GHG emissions is important in order to design adaptation strategies. Meta-analysis is a robust statistical method of identifying trends and patterns in the effects of LUC on GHG emissions. Similarly, different approaches like basic estimation equations, models, field measurements, inference and a hybrid equation approach could be used used to estimate GHG emission (IPCC 2013). Harris et al. 2015. used meta-analysis to quantify the impact of LUC on GHG emissions. Greenhouse gas (GHG) emission factors for iLUC are proposed for inclusion into carbon footprints (CF) of biofuels (NRC 2010). LCA is a good tool for quantifying environmental impacts throughout the life cycle of a product. LCA, when applied to agriculture or forestry products, can include upstream (extraction and production of material inputs e.g. fuels, fertilizers) and downstream impacts (use and disposal by the end consumer). If we consider the LCA for a grain product it will include emissions from synthetic fertilizer production and N_2O emissions from fertilizer application (upstream impacts) and emissions from grain transportation, storage, processing, use, and disposal (downstream impacts) (Kennedy et al. 2009). Greenhouse gas fluxes from a managed ecosystem were elucidated by Paustian et al. (2006). The main processes involved are photosynthesis, respiration, decomposition, nitrification, denitrification, enteric fermentation and combustion. These processes govern the carbon and nitrogen dynamics in soil which could be affected by physical and biological processes. The biological processes include microbial as well as animal and plant activity while physical process include combustion, leaching and runoff. (Fig. 1.1)

Davies-Barnard et al. (2014) concluded that land cover has a significant impact on climate and it is significantly affected by agricultural land use. Agricultural and forestry activities and land-use change are responsible for in one third of GHG

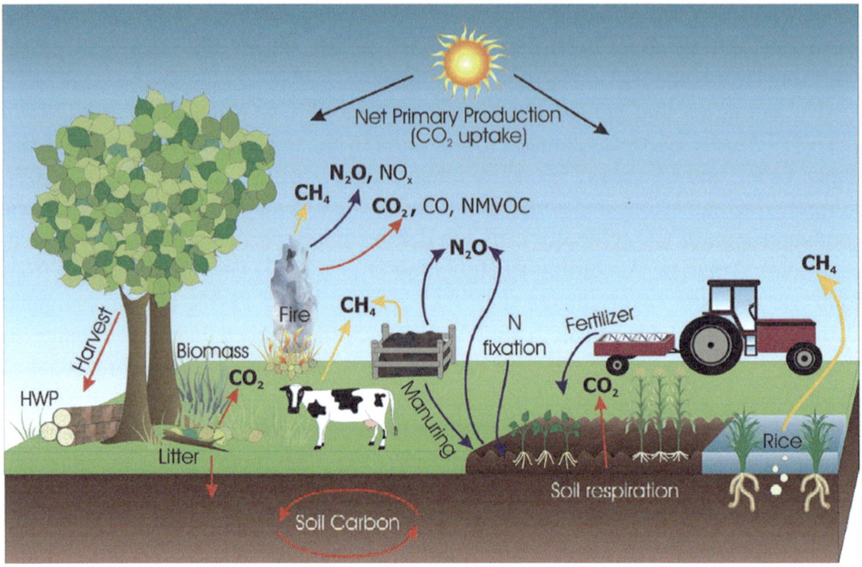

Fig. 1.1 Greenhouse gas emission sources/removals and processes in managed ecosystems (where NMVOC; non-methane volatile organic compounds) (Source: Paustian et al. (2006)

emissions. Agriculture is the dominant land use activity and contributes 5.1–6.1 $GtCO_2$-eq/year (10–12 % of total global anthropogenic emissions of GHGs). N_2O and CH_4 contributions from agriculture are 60 and 50 % respectively. However, these agricultural emissions can be linked to particular crop or animal products (IPCC 2013). The emissions produced by agriculture do not take place at the field level only. There can be spatial dislocation of emissions in which products of agriculture can be transported to another place and utilized there. Similarly, temporal dislocation is the decaying of crop residues over a longer period of time and its later utilization as fuel. The other important source of GHG emissions is the energy sector. The generation and usa of energy results in large emissions of GHGs. Generally more attention of GHG emissions from the energy sector has been given to energy production rather than energy utilization as household electric and electronic equipment (e-products).

Climate change is a major threat to agriculture and food security. GHG emissions from agriculture continue to rise. In order to identify opportunities for reducing emissions while addressing food security, collection of emissions data is necessary to design resilience and rural development goals. FAOSTAT emissions database could be used to estimate GHG emissions from a target regions as it is the most comprehensive knowledge base regarding agricultural greenhouse gas emissions. According to FAOSTAT, (2015) GHG emission (CO_2 equivalent) is continually increasing across the globe (Fig. 1.2). The highest emission is from the agriculture sector followed by land use change. Among continents, Asia is at top with reference to GHG emissions from agriculture followed by America (Fig. 1.2). Greenhouse gase emissions (CO_2 equivalent) from agriculture in Annex I, non-

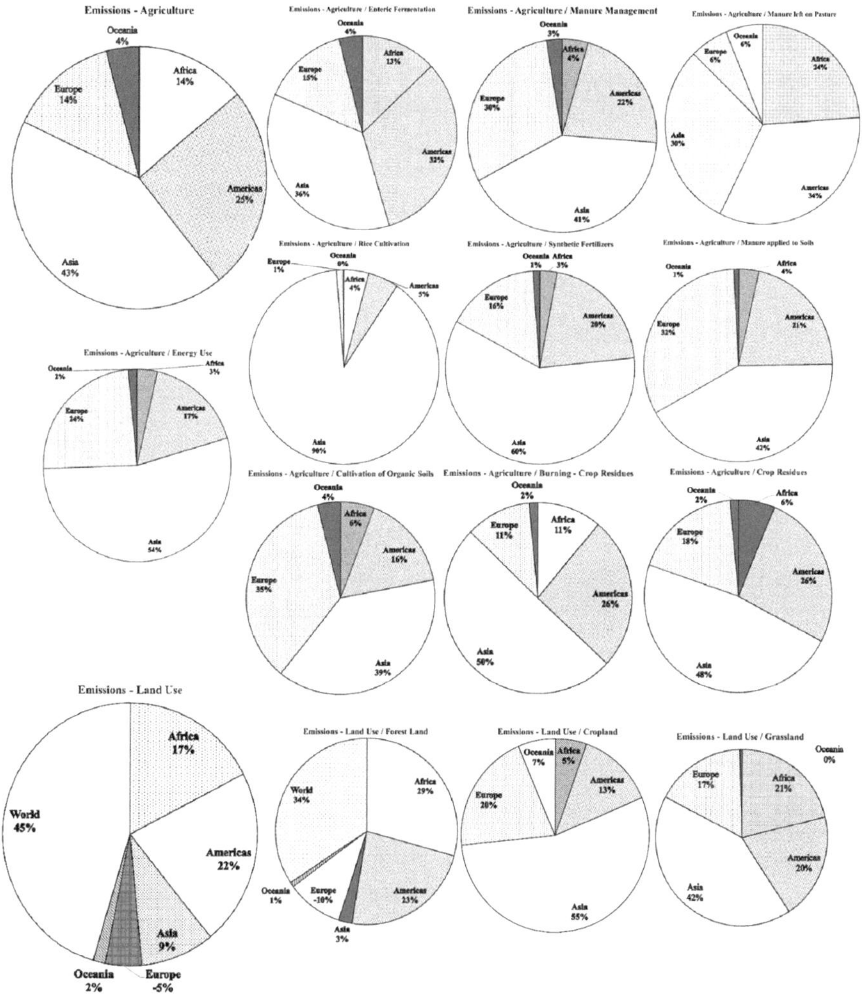

Fig. 1.2 Greenhouse gase (GHG) emissions from agriculture and land use change across the globe

Annex I countries and across the globe provide different pictures in different field of agriculture (Figs. 1.3a and 1.3b). Generally, non-Annex I countries are higher producers of GHGs compared to Annex I countries. Similarly, GHG emissions by sectors involved in agriculture revealed that enteric fermentation contributes the most (40.0 %) to GHG emission while the lowest emissions reported were due to burning crop residues (0.5 %) (FAOSTAT, 2015) (Figs. 1.2, 1.3a and 1.3b). China is the top GHG emitter followed by India. The top ten GHG emitters have been shown in Figs. 1.4a, 1.4b, 1.4c and 1.4d based upon different sectors in agriculture and land use change. FAOSTAT divided GHG emissions under two categories which include agriculture and land use.

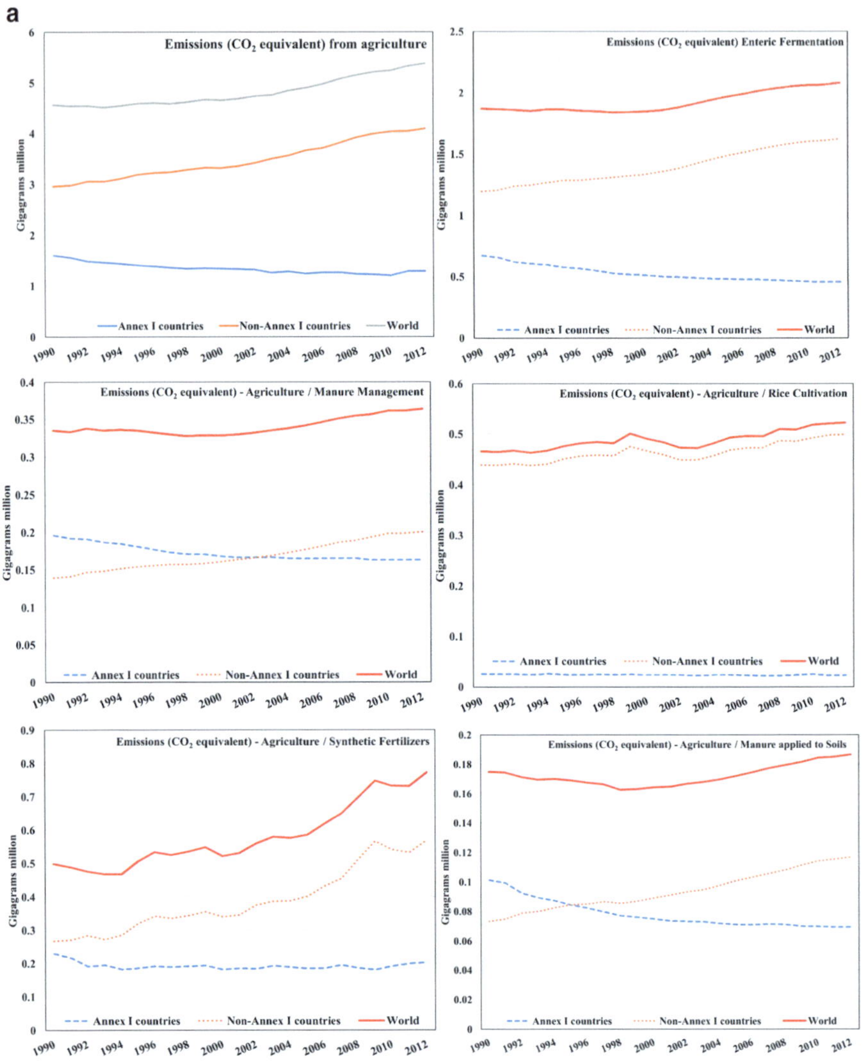

Fig. 1.3a Greenhouse gase (GHG) emissions (CO_2 equivalent) from agriculture

GHG emissions could be controlled or minimized by using different techniques including biofuel, fertilizer and manure, conservation tillage, rotations of crops, cover crops, cropping intensity, irrigation, erosion control, drained wetland management, lime amendments, residue management, biochar and biotechnology. Similarly, GHG emissions from rice based cropping systems could be minimized by water and residue management, organic amendments, ratoon cropping, fallow management, use of nitrification and urease inhibitors and by using different fertilizer placement methods and sulfur products. In case of animal production GHGs emissions is mainly because of enteric fermentation, housing and manure management.

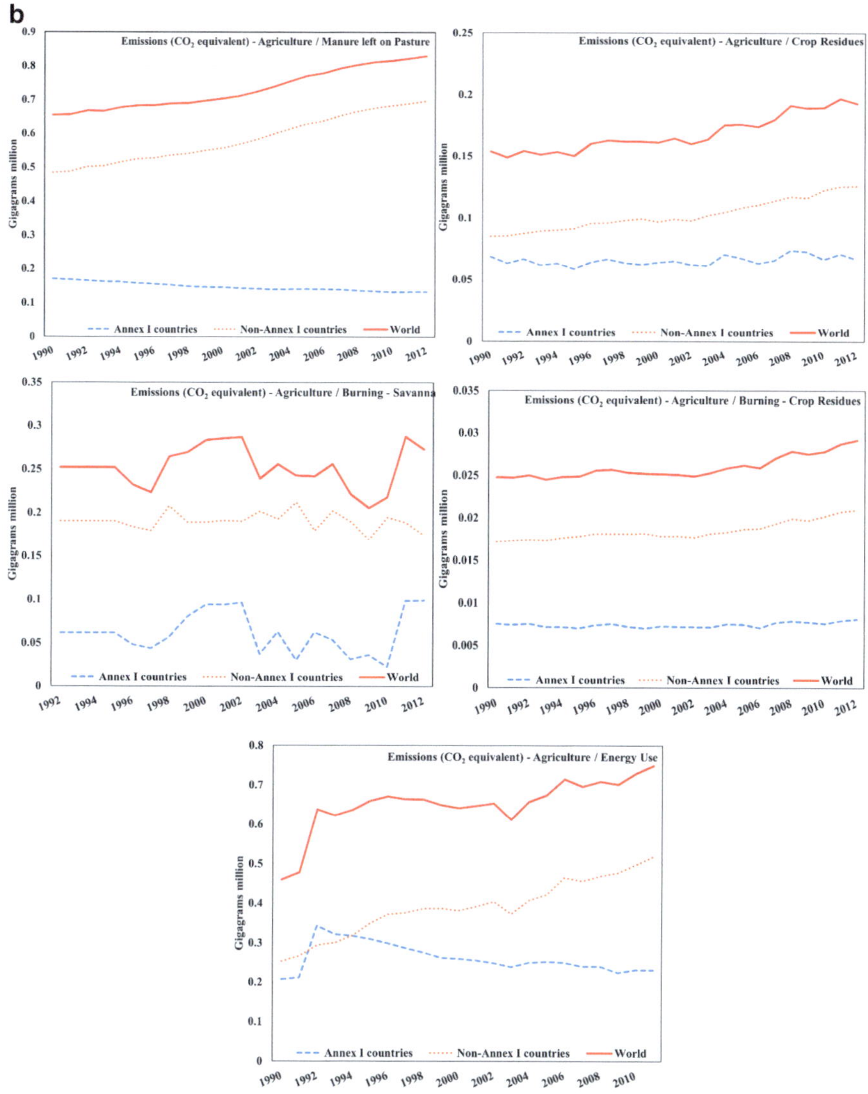

Fig. 1.3b Greenhouse gas (GHG) emissions (CO_2 equivalent) from agriculture

GHG emissions from enteric fermentation and housing could be modified by using different methods. It includes management in the feed and use of different microorganism products. However, in case of manure management techniques like anaerobic digestion, liquid manure storage and treatment practices could be used to minimize or modify GHG emissions.

Forestry has considerable potential to mitigate GHG emissions through the sequestration and storage of forest carbon stocks. Various forestry activities have potential to reduce GHG emissions. According to Morgan et al. (2010) agroforestry

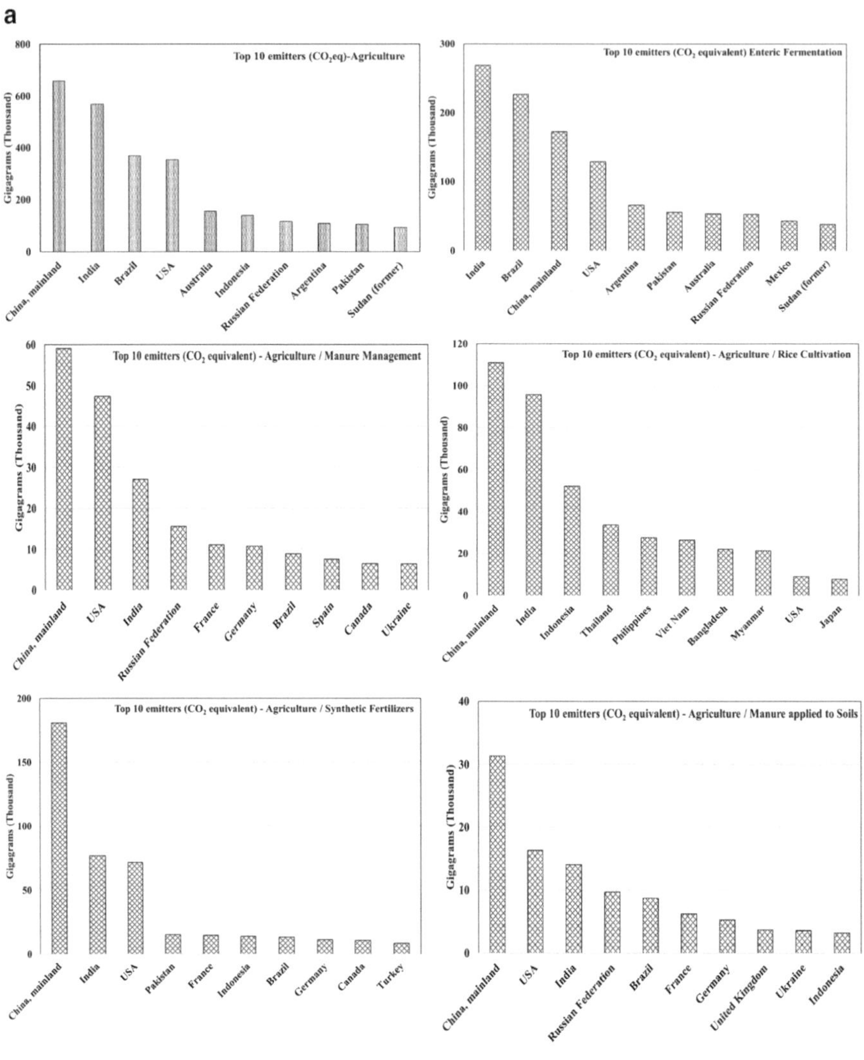

Fig. 1.4a Top 10 greenhouse gase (GHG) emitters (CO_2 equivalent) from agriculture

could contribute to carbon sequestration, GHG mitigation, and adaptation to shifting climate. Land use change is the main contributor to GHG emissions, therefore, it needs to be managed effectively. Land use change mainly includes three directional processes – afforestation, reforestation and deforestation. The balance among these three processes is important to manage GHG flux. Different methods could be used to estimate GHG fluxes from LUC. The GHG flux linked with LUC is the sum of the GHG fluxes from previous land use categories plus the sum of the GHG fluxes related to the current land use (IPCC 2007). Equations 1 and 2 could be used to study annual carbon stock changes for LUC estimates as the sum of changes in all land use categories (Dokoohaki et al. 2016).

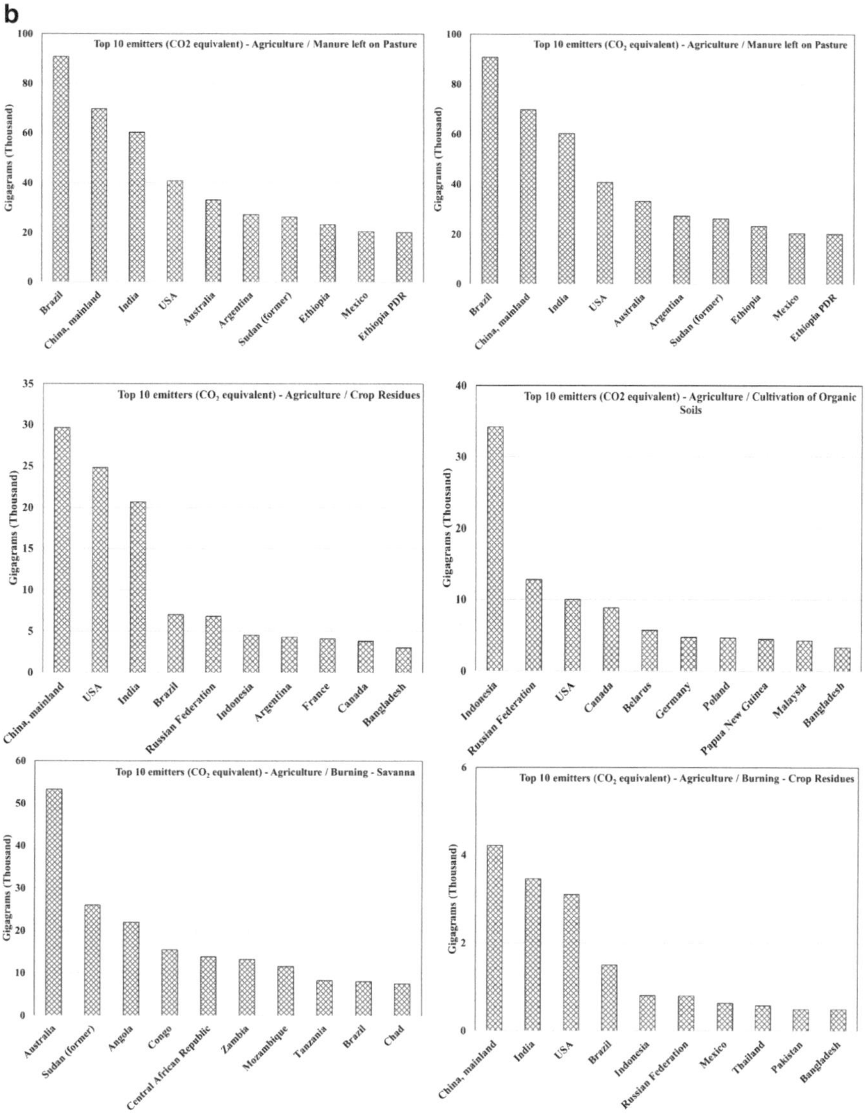

Fig. 1.4b Top 10 greenhouse gase (GHG) emitters (CO_2 equivalent) from agriculture

$$\Delta C_{luc} = \Delta C_{luco} + \Delta C_{lucn} \tag{1}$$

$$\Delta C_{luc} = \Delta C_{luc\,fl} + \Delta C_{lu\,cl} + \Delta C_{lu\,gl} + \Delta C_{lu\,wl} \tag{2}$$

where ΔC; carbon stock change (metric tons CO_2-eq ha^{-1} $year^{-1}$), luc; land use change, o; old land use, n; new land use, fl; forest land, cl; crop land, gl; grazing land and wl; wetlands.

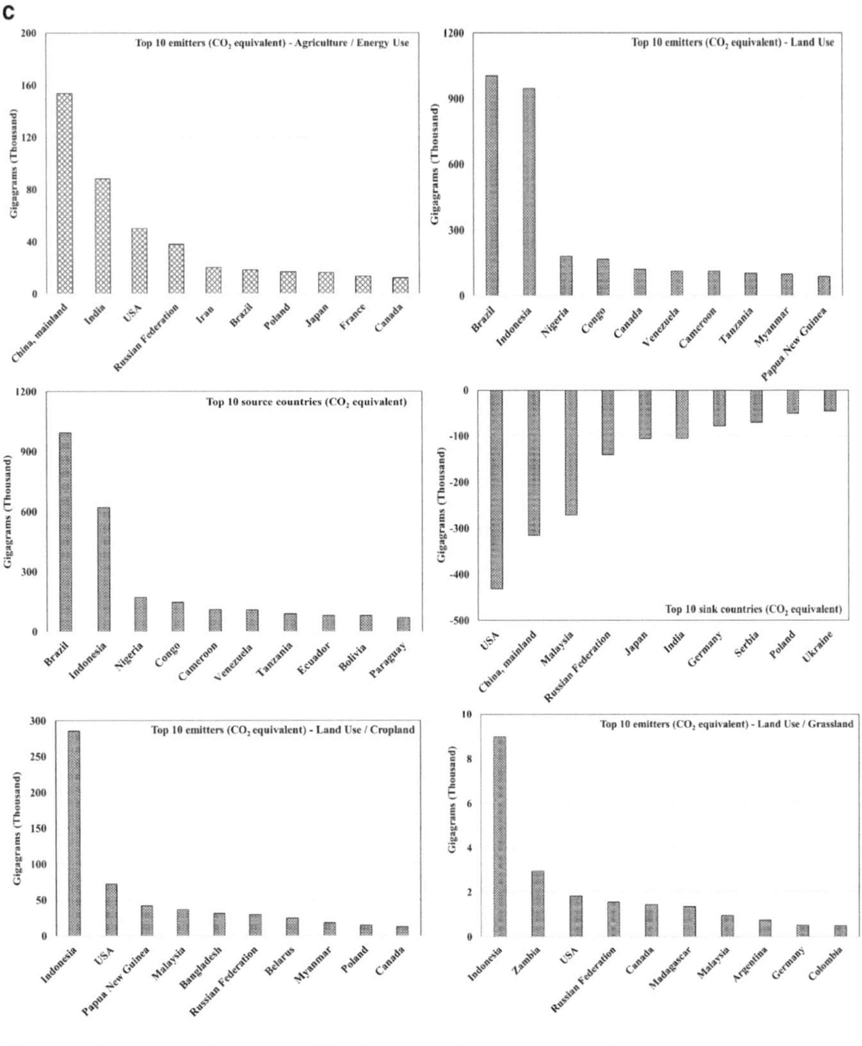

Fig. 1.4c Top 10 greenhouse gase (GHG) emitters (CO_2 equivalent) from agriculture and land use change

The annual carbon stock exchange for a particular section e.g. management regime could be calculated by the following equation

$$C_{luc} = \sum_{i}^{n} \Delta Cluci \tag{3}$$

where ΔCluc; carbon stock changes for a land use change and i denotes a specific division

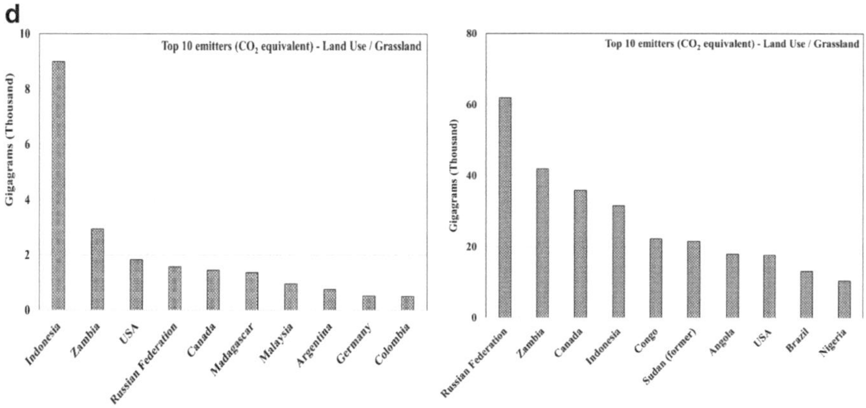

Fig. 1.4d Top 10 greenhouse gase (GHG) emitters (CO_2 equivalent) from agriculture and land use change

Many forest and agricultural lands have live/dead biomass carbon stocks (LDBCS) and soil organic carbon which acts as a good carbon store. The following equation (Dokoohaki et al. 2016) could be used to estimate the annual change in carbon stocks in dead wood due to land conversion.

$$\Delta C_{dom} = (C_n - C_o) \times A_{on} \div T_{on} \tag{4}$$

where ΔC_{dom} = annual change in carbon stocks in dead wood or litter (metric tons C year^{-1}), C_o = dead wood/litter stock, under the old land-use category (metric tons C ha^{-1}); C_n = dead wood/litter stock, under the new land-use category (metric tons C ha^{-1}), A_{on} = area undergoing conversion from old to new land-use category (ha), T_{on} = time period of the transition from old to new land-use category (year) (The default is 20 years for carbon stock increases and 1 year for carbon losses.)

Soil organic carbon stock (SOCS) is also influenced by land use change. The significant change in SOCS occurs due to conversion of land to crop land (Six et al. 2000). Aalde et al. (2006) proposed a method to estimate changes in SOCS from mineral soils.

$$\Delta C_{mineral} = \left[(SOC_f - SOC_i) \times CO_2MW \right] \div D \tag{5}$$

where $\Delta C_{mineral}$ = annual change in mineral SOCS (metric tons CO_2-eq year^{-1}), SOC_f = soil organic carbon stock at the end of year 5 (metric tons C), SOC_i = soil organic carbon stock at the beginning of year 1 (metric tons C), CO_2MW = ratio of molecular weight of CO_2 to C (44/12 dimensionless) and D = time dependence of stock change factors (20 years).

Simialrly, SOCS from mineral soils could be calculated by using the following equation (Aalde et al. 2006)

$$SOCS = SOC_{ref} \times F_{lu} \times F_{mg} \times F_i \times A \tag{6}$$

where SOCS = soil organic carbon stock at the beginning ($SOCS_i$) and end of the 5 years ($SOCS_f$) (metric tons C), SOC_{ref} = reference soil organic carbon stock (metric tons C ha^{-1}), F_{lu} = stock change factor for land use (dimensionless), F_{mg} = stock change factor for management (dimensionless), F_i = stock change factor for input (dimensionless) and A = area of land-use change (ha).

Uncertainty analysis is an important technique to quantify the uncertainty of greenhouse gas (GHG) emissions from different sectors. It can help policy makers and farmers decide management options to minimize GHG emissions based upon an uncertainty range. If uncertainty for an estimate is low farmers can invest in that management practices as it has high probability of GHG emission reduction. A Monte Carlo approach is a comprehensive, sound method that could be used for estimating the uncertainty. *Greenhouse Gas Emissions and Climate Variability: An Overview* covers the GHG emission status by different sectors and how it could be mitigated by using different practices in agriculture and land use sectors. This chapter reviews available methods for studying/quantifying GHG emission for accurate design of strategies to address the issue of climate variability.

1.2 Greenhouse Gas Emission and Climate Variability

Climate variability is one of the burning issues in all fields from social sciences to the applied sciences. Climate vulnerability threatens global climatic cycles and world food production systems, thus affecting the lives of all people. Most of the world is exposed to the effects of climatic change due to extreme variability in temperature and rainfall. Risk reduction represents a major avenue for responding to existing rise in temperature, carbon dioxide, GHGs, flood and drought hazards. Global warming is the greatest environmental challenge of the twenty-first century as it results in increased average air temperature (Gnansounou et al. 2004). Wu et al. (2010) concluded that cities act as heat islands and since large areas of grassland and forest were converted to barren land resulted in greater climate variability. The guiding principle to reduce climate risks is to minimize GHG emission. In recent decades significant changes in the atmospheric temperature have been observed. The global mean annual temperature at the end of the twentieth century was almost 0.7 °C and it is likely to increase further by 1.8–6.4 °C by the end of this century (IPCC 2007). The warmest decade in the last 300 years was 1990–2000 with the increase of 0.5 °C in comparison to the baseline temperature of 1961–1990. A variety of models ranging from simple models to complex earth system models were used to project future warming under different representative concentration pathways (RCPs). The RCP includes RCP 2.6, RCP 4.5, RCP 6.0 and RCP 8.5 (The numbers refer to the rate of energy increase per unit area at the surface of the earth, in watts per square meter). RCP 2.6 is the normal scenario in which a guideline was established to limit global warming to 2 °C (3.6 °F) above the level that existed before industrial times. All other scenarios reflect severe warming due to increasing rates of GHG emission. The scenario RCP 8.5 reflects "business as usual" in which

Table 1.1 Changes in global mean surface temperature in °C and global mean sea level rise in m (bottom) for the two time periods shown, referenced to the baseline period 1986–2005 (The "likely range" gives confidence limits for a 5–95 % interval)

Climate variable	RCP scenario	2046–2065		2081–2100	
		Mean	Range	Mean	Range
Mean temperature change (°C)	RCP2.6	1	0.4–1.6	1	0.3–1.7
	RCP4.5	1.4	09–2.0	1.8	1.1–2.6
	RCP6.0	1.3	0.8–1.8	2.2	1.4–3.1
	RCP8.5	2	1.4–2.6	3.7	2.6–4.8
		Mean	**Range**	**Mean**	**Range**
Mean Sea Level Rise (m)	RCP2.6	0.24	0.17–0.32	0.4	0.26–0.55
	RCP4.5	0.26	0.19–0.33	0.47	0.32–0.63
	RCP6.0	0.25	0.18–0.32	0.48	0.33–0.63
	RCP8.5	0.3	0.22–0.38	0.63	0.45–0.82

no policies are implemented to limit GHG emission. The projected increase in mean temperature and rise in sea level in comparison to baseline (1986–2005) are presented in Table 1.1 (Harris et al. 2015). Climate variability resulted in a change in the intensity and frequency of rainfall which increased flooding and soil erosion.

Crop phenology and productivity will be affected by warmer climates. Craufurd and Wheeler (2009) reported earlier flowering and maturity due to a rise in temperature. Moreover, increased temperature resulted in reproductive failure and yield reductions in many crops. Lobell et al. (2011) reported a 1.7 % reduction in maize yield due to exposure of maize to degree days above 30 °C. Increased night temperature is another effect of GHG which could reduce crop yield. Serious effects have been reported for rice where an increase in night temperature from 27 °C to 32 °C caused 90 % yield reduction (Mohammed and Tarpley 2009). Climate variability can also modify grain quality since high temperature during grain filling affects the protein content of wheat (Hurkman et al. 2009). Pittock (2003) concluded in their findings that the frequency of extreme events will increase due to global warming. Plant processes like photosynthesis will be affected by high temperature which could lead to reduction in growth and yield (Calderini and Reynolds 2000; Talukder et al. 2014; 2013; Wang et al. 2011) (Table 1.2).

A panel of the National Research Council (United States) (2010) on advancing the science of climate change concluded that world mean temperature was 0.8 °C higher during the first decade of twenty-first century compared to first decade of twentieth century. Moreover, they reported that most of the warming was related to CO_2 and other GHGs which can trap heat. The energy sector is the largest contributor to climate change as it involves burning of fossil fuels (coal, oil, and natural gas). Similarly, the panel identified agriculture, forest clearing, and certain industrial activities as big contributors to climate change due to emission of GHGs. Kang and Banga (2013) found that climate change is a well-recognized man made global environmental challenge and that agriculture is significantly influenced by it. Food and Agriculture Organization (FAO) experts reported that each 1 °C rise in temperature would cause annual wheat yield loss of about 6 million tons. However, when

Table 1.2 Impacts of climatic variables on crops with recommended adaptation strategies

Climatic impact	Effect on crop	Adaptation	Reference
Increased temperature (0.67, 0.53, and 0.38 °C decade − 1)	Change in crop life cycle and decreased yield	Adjusting the sowing date, converting tillage system and adopting water-saving technologies	Zhang et al. (2015)
Heat stress	Decreased in number of days to mature (1.8 days for 2025 and by 2.3 days for 2050)	Shift the planting date	Bao et al. (2015)
Increases in precipitation and CO_2 concentration	Soybean projected yield increase from 6 to 22 % for 2025 and 8 to 35 % for 2050 for rainfed conditions.		
El Niño–Southern Oscillation (ENSO)	Might influence growth, maturity, and yield of winter wheat	Shift planting date and cultivar selection	Woli et al. 2014
Temperature	Modification in flowering time of wheat	Use longer-season wheat varieties and varieties with increased heat-stress resistance	Wang et al. 2015
Climate extremes (temperature and precipitation)	Change in rainfed crop yields	Irrigation	Troy et al. 2015
Heat stress	Reproductive growing duration (RGD) and yield	Shifts in cultivars	Tao et al. 2015
Heat stress	Yield losses due to increased frequency and magnitude of heat stress	Heat-tolerant ideotypes	Stratonovitch and Semenov 2015
Elevated temperature	Alteration in the phenology of crops	Agronomic and breeding solutions	Sadras et al. 2015
Reduction of annual precipitation and an increase of air temperature	Shortening of growing season	Supplemental irrigation	Saadi et al. 2015
Higher temperatures	Shortening of the grain filling period, reduce crop yields		Rezaei et al. 2015

losses of all other crops were taken into consideration it might cause loss of US$ 20 billion each year (Swaminathan and Kesavan 2012). Climate variability can reduce crop duration, disturb source sink relationships, increase crop respiration, affect survival and distribution of pest populations, accelerate nutrient mineralization and decrease nutrient use efficiency. It can also lead to changes in the frequency and intensity of drought and floods (Sharma and Chauhan 2011). Overall agricultural production will be significantly affected by climate variability which will influence food security.

1.3 Greenhouse Gas Mitigation and Climate Change Adaptation

Climate change is one of the complex burning issues currently faced by the world. Greenhouse gases are trapping heat energy which results in global warming. It has been reported earlier that if GHGs are stopped completely, climate change will still affect future generations. Therefore, we need to show a high level of commitment to tackle the issue of climate change. Mitigation and adaptation are two approaches used to respond to climate change. Mitigation involves reducing and stabilizing the levels of GHGs while adaptation is adapting to climate change using different techniques. Mitigation is possible by finding ways by which we can increase sinks for GHGs. Mainly the sinks includes forests, soil and oceans, therefore it is necessary to manage those resources which can absorb GHGs. According to Calvin et al. (2015) around 40 % of GHG emissions are from agriculture, forestry, and other land use (AFOLU). Their work further elaborated that implementation of climate policy is necessary to minimize GHG emissions. A multi-model comparison approach was used to study the future trajectory of AFOLU GHG emissions with and without mitigation. a similar approach about the role of land for the mitigation of AFOLU GHG emissions was earlier reported which includes the use of bioenergy crops (Calvin et al. 2013). The models used were Applied Dynamic Analysis of the Global Economy (ADAGE) (Ross 2009.); MIT Emissions Prediction and Policy Analysis (EPPA) (Paltsev et al. 2005); GCAM (Global Change Assessment Model) (Calvin et al. 2011) and TIAM-WORLD (Loulou 2008). The results indicated larger uncertainties in both present and future emissions with and without climate policy.

Bioenergy crops are the biggest potential source that could be used to minimize GHG emissions. Hudiburg et al. (2015) proposed perennial grasses as effective bioenergy crops on marginal lands. They evaluated the DayCent biogeochemical model in their studies and concluded that the model predicted yield and GHG fluxes with good accuracy. They found that with the replacement of traditional corn-soybean rotation with native prairie, switchgrass, and Miscanthus resulted in net GHG reductions of 0.5, 1.0 and 2.0 Mg C ha^{-1} $year^{-1}$ respectively. Since bioenergy crops have the potential to mitigate climate change impacts, they have been under consideration for the past decade. However, these bioenergy crops could only be grown on marginal lands as most of the world land is occupied by major food crops. Albanito

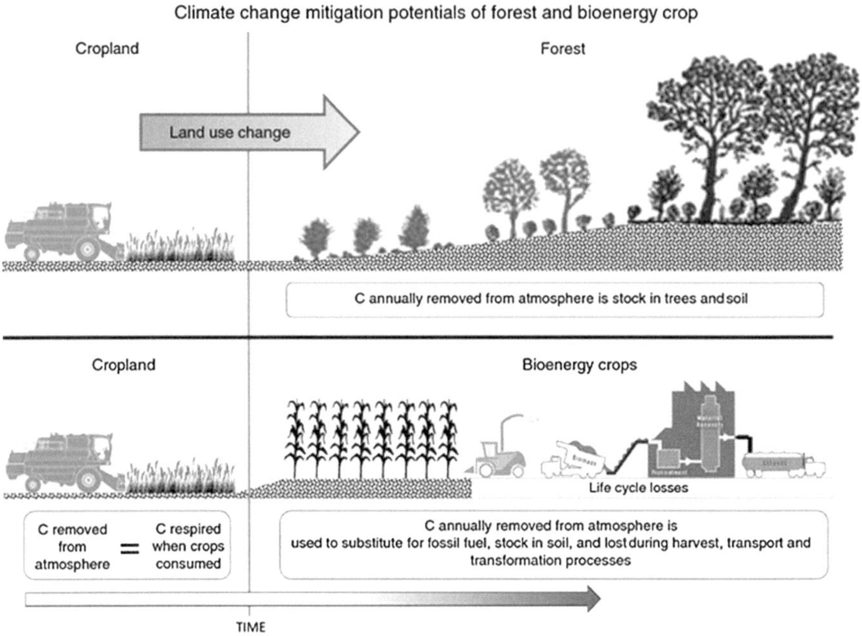

Fig. 1.5 Carbon implications of converting cropland to forest or bioenergy crops for climate mitigation: a global assessment (Source: Albanito et al. 2016)

et al. (2016) reported C4 grasses (Miscanthus and switch-grass) as the potential bioenergy crops with the highest climate mitigation potential. These crops would displace 58.1 Pg of fossil fuel C equivalent (C_{eq} oil) if the proposed land use change took place. Similarly, woody energy crops (poplar, willow and Eucalyptus species) could displace 0.9 Pg C_{eq} oil under proposed land use change. The best climate mitigation option is the afforestation of suggested cropland which would sequester 5.8 Pg C in biomass in the 20-year-old forest and 2.7 Pg C in soil. Croplands could not accumulate carbon for more than a year therefore, in order to mitigate climate change, agricultural lands should either be converted to forest land or bioenergy production (Fig. 1.5). Food security will be a big challenge in the future as the world population will be 9–10 billion by 2050. Therefore, bioenergy crops could not come at the expense of food crops. Earlier researchers accepted the potential of biomass energy production but according to them it was not enough to replace just a few percent of current fossil fuel usage. Increasing biomass energy production beyond a certain level might imperil food security and worsen condition of climate change (Field et al. 2008). However, biomass proponents are recommending the use of grasslands and marginal crop lands as potential sites for bioenergy crops (Qin et al. 2015; Slade et al. 2014). Furthermore, Qin et al. (2015) suggested Miscanthus as the best potential crop to mitigate GHGs emissions on marginal lands compared to switchgrass. Biomass and ethanol yield were higher in Miscanthus. Coyle (2007) concluded that different crops, e.g. corn, sugarcane, rapeseed and soybean could be

Table 1.3 Energy potential from biofuel crops using current technologies and future cellulosic technologies

FT	FM (Mt year^{-1})	GBC (GJ/ton)	GBE (EJ year^{-1})	NEBR (Output/ Input)	NBE (EJ year^{-1})	Refs
Corn kernel	696	8	5.8	1.25	1.2	Hill et al. 2006
Sugar cane	1324	2	2.8	8	2.4	IEA 2004
Cellulosic biomass	–	6	–	5.44	–	Farrell et al. 2006
Soy oil	35	30	1	1.93	0.5	Worldwatch 2006
Palm oil	36	30	1.1	9	1	Worldwatch 2006
Rape oil	17	30	0.5	2.5	0.3	IEA 2004

Source: Field et al. (2008)
Where *FT* Feedstock type, *FM* Feedstock mass, *GBC* Gross biofuel conversion (Useful biofuel energy per ton of crop for conversion into biofuel (1GJ = 10^9 J)), *GBE* Gross biofuel energy (Product of feedstock mass and gross biofuel conversion (1EJ = 10^{18} J)), *NEBR* Net energy balance ratio (Ratio of the energy captured in biomass fuel to the fossil energy input) and *NBE* Net biofuel energy (Energy yield above the fossil energy invested in growing, transporting and manufacturing, calculated as gross biofuel energy × (net energy balance ratio −1)/net energy balance ratio)

used to produce biofuel. Energy potential from different feedstocks have been presented in Table 1.3.

Chum et al. (2011) considered bioenergy as a good renewable source for energy. Bioenergy can easily replace fossil fuels and minimize GHG emissions (Dornburg and Faaij 2005; Dornburg et al. 2008; 2010). Implementation of all these techniques requires identification of terrestrial ecosystems which could contribute to climate mitigation. Many countries have announced different targets to substitute fossil fuels with biofuels (Ravindranath et al. 2008). Table 1.4 shows that C4 bioenergy crops have higher cumulative carbon mitigation potential than SRCW. However, this mitigation potential changes across the continents as in Oceania, SRCW produced higher C savings than energy crops. According to Albanito et al. (2016) cumulative carbon strength due to reforestation is highest in Asia, followed by Africa, North and Central America, South America, Oceania and Europe. However, on a per hectare basis C sequestration strength is higher in South America followed by North and Central America, Oceania, Asia, Africa and Europe. Among climatic regions, warm-dry climates could save 44.7 % of the C in forest followed by warm-moist (42.6 %), cool-dry (11.3 %) and cool-moist (1.4 %) regions (Table 1.5).

Biochar also has potential to mitigate climate change by sequestering carbon. Biochar use improved soil fertility, reduced fertilizer inputs, GHG emissions, and emissions from feedstock, enhanced soil microbial life and energy generation. Its use also increased crop yield. (Woolf et al. 2010) reported that biochar use could minimize GHG emissions by 12 %. The concept of sustainable use of biochar is presented in Fig. 1.6 as proposed by (Woolf et al. 2010). Photosynthesis is a carbon reduction processes in which plants produce biomass by using atmospheric CO_2. Residues from crops and forests were subjected to the process of pyrolysis which produced bio-oil, syngas, process heat and biochar (output). These outputs serve as a good source of energy which could minimize GHG emissions. Furthermore, bio-

Table 1.4 Land use change C mitigation potential

Land use	CR	TCM	CMB	CSSS	ALD
C4 Bioenergy crops	Asia	27.62	24.06	3.56	66.07
	Africa	8.58	7.69	0.89	61.23
	Europe	10.86	7.74	3.12	123.21
	North America	10.49	8.89	1.6	74.34
	South America	10.71	9.58	1.13	58.08
	Oceania	0.19	0.16	0.03	1.98
Forest	Asia	3.84	2.73	1.11	94.52
	Africa	1.56	1.11	0.44	42.31
	Europe	0.31	0.17	0.15	9.97
	North America	1.47	0.96	0.5	24.67
	South America	0.74	0.41	0.34	6.27
	Oceania	0.51	0.39	0.12	8.7
Short Rotation Coppice Woody (SRCW) crops	Asia	0.48	0.2	0.28	10.49
	Africa	0.0045	0.0019	0.0026	0.35
	Europe	0.92	0.52	0.41	12.54
	North America	0.18	0.1	0.07	2.38
	South America	0.03	0.03	0.01	0.46
	Oceania	0.01	0.01	0	0.13

Source: Albanito et al. (2016)

Where *CR* Continental region, *TCM* Total C mitigated, *CMB* C mitigated from biomass use/increment (Pg C forest and Pg Ceq oil for bioenergy crops), *CSSS* C stock sequestered in soil (Pg C) and ALD; Agricultural land displaced (Mha)

char could also be used to improve agricultural soils (Fig. 1.6). Roberts et al. (2009) suggested biochar (biomass pyrolysis) as a good source to mitigate climate change and minimize fossil fuel consumption. They used life cycle assessment (LCA) to estimate the impact of biochar on energy and climate change and concluded that biochar resulted in negative net GHG emissions. However, the economic viability of biochar production depends upon the cost of feedstocks. Similarly, a well-to-wheel (WTW) LCA model was developed to assess the environmental profile of liquid fuels through pyrolysis (Kimball 2011). Bruckman et al. 2014 reported biochar as a potential geoengineering method to mitigate climate change and design adaptation strategies. Biochar as a soil amendment can sequester C and it is a useful option to mitigate climate change (Hudiburg et al. 2015). Biochar stability and decomposition are the best criteria to evaluate its contribution to carbon (C) sequestration and climate change mitigation. (Macleod et al. 2015) reported that around 97 % of biochar contributes directly to C sequestration in soil. Similarly, the biochar effect on soil organic matter (SOM) dynamics depends upon characteristics of biochar and

Table 1.5 Land use change C mitigation potential across global climatic regions

Land use	CR	TCM	CMB	CSSS	ALD
C4 Bioenergy crops	CD	1.69	0.84	0.85	32.47
	CM	18.06	13.94	4.12	176.74
	WD	0.1	0.08	0.02	2.2
	WM	48.59	43.26	5.33	273.49
Forest	CD	0.95	0.59	0.36	47.38
	CM	0.12	0.04	0.08	2.67
	WD	3.77	2.84	0.93	90.96
	WM	3.6	2.3	1.3	45.44
Short Rotation Coppice Woody (SRCW) crops	CD	0.42	0.06	0.36	13.53
	CM	1.05	0.68	0.38	10.37
	WD	0.01	0.01	0.01	0.85
	WM	0.14	0.12	0.02	1.6

Source: Albanito et al. (2016)
Where *CR* Climate region, *TCM* Total C mitigated, *CMB* C mitigated from biomass use/increment (Pg C forest and Pg Ceq oil for bioenergy crops), *CSSS* C stock sequestered in soil (Pg C), *ALD* Agricultural land displaced (Mha), *CD* Cool-Dry, *CM* Cool-Wet, *WD* Warm-Dry and *WM* Warm moist

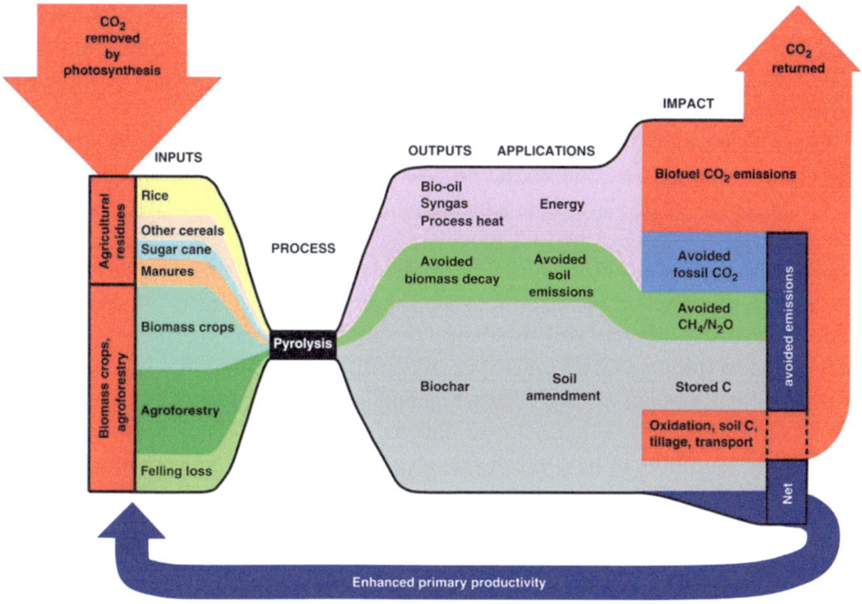

Fig. 1.6 Sustainable biochar concept (Source: Woolf et al. 2010)

soils. Nitrous oxide (N_2O), which is the main GHG coming from agricultural soil, could be minimized by biochar amendment (Ignaciuk 2015). All of these findings support the potential of biochar to be used as a climate change mitigation strategy.

1.4 Modeling and Simulation (Models Used in GHGE Studies)

Quantification of GHG emissions is made possible by the use of process based models. These models could further be used as decision support tools under different scenarios to mitigate GHG emissions. Models should complement field trials in order to have realistic assessments of bioenergy crop production on GHG emissions. The results obtained from simulation studies are sufficiently authentic when validated with field data. Used properly, models could be used to quantify GHG emissions. Models used in GHG studies include CENTURY (Bennetzen et al. 2015), RothC (Albanito et al. 2016), EPIC (Williams 1995; Izaurralde et al. 2006), SOCRATES (Smith 2015), C-Farm (Calvin et al. 2015), ECOSSE (Hill et al. 2006) CropSyst (Stöckle et al. 2003), ALMANAC (Agricultural Land Management Alternatives with Numerical Assessment Criteria) (Nayak et al. 2015), DNDC (DeNitrification DeComposition) (Zhang et al. 2016), ECOSYS (Frank et al. 2015), HOLOS (Wei et al. 2015), DSSAT (Jones et al. 2003) and STICS (Brisson et al. 1998; 2003). The DAYCENT process based model estimates soil organic carbon on daily basis. It also has the potential to simulate GHG fluxes (N_2O, NO_x, and CH_4) for terrestrial ecosystems (Kalafatis et al. 2015) (Table 1.6).

Quantification of GHG emissions is the first step to design mitigation strategies for climate change. Beside these process based models, different publically-accessible tools are also available which could be used to quantify GHG emissions. The calculators include Agri-LCI models, C-PLAN, Carbon Footprint Calculator, DNDC calculator, FarmGAS, Fieldprint Calculator and HOLOS. Similarly, among process based models it is essential to use those models that have low uncertainty. Ogle et al. (2007) reported that simulation modelling is useful to estimate C sequestration and to mitigate GHG emissions under different agricultural managements. However, these models are not accurate enough to simulate C dynamics under different agroecosystems which leads to uncertainty in the results. Quantification of uncertainty is important to confirm the accuracy of models. Uncertainty analysis either uses Monte Carlo Analyses or linear mixed-effects models (empirically based methods). Knightes and Cyterski (2005) suggested comparison between observed and simulated values as good criteria for the evaluation of model performance. This empirically based method was considered a robust estimate of uncertainty (Fig. 1.7a). It is in contrast with error propagation methods (Monte Carlo approach) in which uncertainty is quantified by probability distribution functions. It requires multiple results to obtain approximate confidence intervals for a model estimate (Fig.1.7b). Monte Carlo Analysis could not be used on CENTURY which has too many parameters (Ogle et al. 2007). Webster et al. (2002) concluded that evaluation of uncertainty was important to have accurate prediction from models.

Table 1.6 Use of processes based models in Greenhouse Gases (GHG) studies

Models	Application	References
APSIM	N_2O fluxes	Chum et al. (2011)
CropSyst	Climate change, Estimate long-term soil organic carbon, annual N_2O soil emissions and N balance	Farrell et al. 2006; Ogle et al. 2003; Ipcc 2011
CENTURY	Soil organic carbon (SOC) dynamics in wheat-corn cropping systems	Qin et al. 2015
ECOSSE	Simulate soil C dynamics and GHG emissions	Hill et al. 2006
DSSAT	Irrigation and GHG emissions, GHG emissions reduction potentials	Cardozo et al. 2015; Reddy 2015
DNDC and DayCent	Estimation of nitrous oxide	Calvin et al. 2013; Loulou 2008; Ravindranath et al. 2008
DayCent	Estimation of soil GHG using inverse modelling parameter estimation software (PEST)	Daioglou et al. 2015
DayCent	Estimation of potential of switchgrass (*Panicum virgatum* L.) as bioenergy crop	Paltsev et al. 2005
DayCent	Calibration of model using inverse modeling approach	Ross 2009
DayCent	Studying GHG emissions under different cropping systems	Calvin et al. 2011
EPIC	C dynamics	Izaurralde et al. 2006
RothC	SOC sequestration with the introduction of cover crops	IEA 2004
STICS	Nitrate leaching, N and water dynamics	Dornburg et al. 2008; Constantin et al. 2015

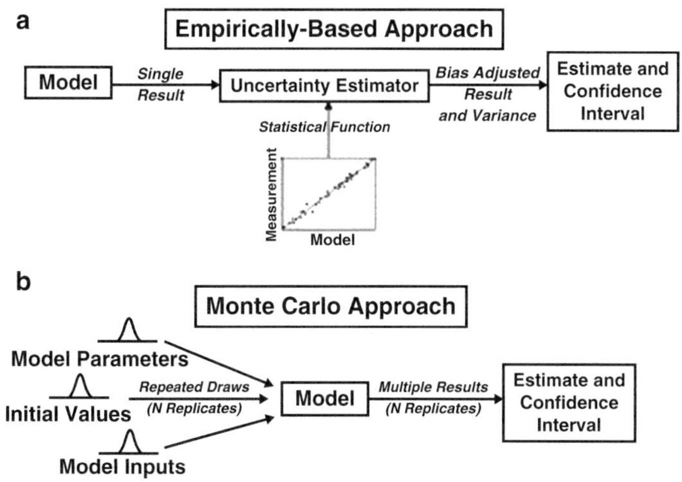

Fig. 1.7 Conceptual diagram with the key components for an uncertainty analysis using (**a**) an empirically based vs. (**b**) Monte Carlo approach. (Source: Ogle et al. 2007)

1.5 Conclusion

Greenhouse Gas Emissions and Climate Variability: An overview elaborated that in order to minimize GHG emissions we need to first accurately estimate GHG emission using different approaches like LCA. Second, after quantification, different mitigation approaches, like changes in land use, should be adopted by considering different factors. Albanito et al. (2016)) suggested bioenergy cropping as the best mitigation strategy under a changing climate. Meanwhile biochar also has potential to mitigate climate change by sequestering carbon. Woolf et al. (2010) proposed the concept of sustainable use of biochar. Different process based models could be used to accurately estimate GHG emissions in response to different land management. These models could be finally used as decision support tools to mitigate climate change. However, in order to utilize these models as effective decision support tools, the use of uncertainty analysis is of utmost importance.

References

Aalde, H., P. Gonzalez, M. Gytarsky, T. Krug, W. Kurz, R.D. Lasco, D. Martino, B. Mcconkey, S. Ogle, and K. Paustian. 2006. Generic methodologies applicable to multiple land-use categories. *IPCC Guidelines for National Greenhouse Gas Inventories,* 4. Institute for Global Environmental Strategies (IGES), Hayama, Japan.

Albanito, F., T. Beringer, R. Corstanje, B. Poulter, A. Stephenson, J. Zawadzka, and P. Smith. 2016. Carbon implications of converting cropland to bioenergy crops or forest for climate mitigation: A global assessment. *GCB Bioenergy* 8: 81–95.

Bennetzen, E.H., P. Smith, and J.R. Porter. 2015. Decoupling of greenhouse gas emissions from global agricultural production: 1970–2050. *Global Change Biology* 22: 763–781.

Brisson, N., C. Gary, E. Justes, R. Roche, B. Mary, D. Ripoche, D. Zimmer, J. Sierra, P. Bertuzzi, P. Burger, F. Bussière, Y.M. Cabidoche, P. Cellier, P. Debaeke, J.P. Gaudillère, C. Hénault, F. Maraux, B. Seguin, and H. Sinoquet. 2003. An overview of the crop model stics. *European Journal of Agronomy* 18: 309–332.

Brisson, N., B. Mary, D. Ripoche, M.H. Jeuffroy, F. Ruget, B. Nicoullaud, P. Gate, F. Devienne-Barret, R. Antonioletti, C. Durr, G. Richard, N. Beaudoin, S. Recous, X. Tayot, D. Plenet, P. Cellier, J.-M. Machet, J.M. Meynard, and R. Delécolle. 1998. STICS: A generic model for the simulation of crops and their water and nitrogen balances. I. Theory and parameterization applied to wheat and corn. *Agronomie* 18: 311–346.

Bruckman, V.J., M. Klinglmüller, B.B. Uzun, and E. Apaydin-Varol. 2014. *Potentials to mitigate climate change using biochar–the Austrian perspective*. Potentials to mitigate climate change using biochar. IUFRO Occasional Papers 27.

Calderini, D.F., and M.P. Reynolds. 2000. Changes in grain weight as a consequence of de-graining treatments at pre- and post-anthesis in synthetic hexaploid lines of wheat (<i>Triticum durum</i>x<i>T. tauschii). *Functional Plant Biology* 27: 183–191.

Calvin, K., L. Clarke, J. Edmonds, J. Eom, M. Hejazi, S. Kim, P. Kyle, R. Link, P. Luckow and P. Patel. 2011. *GCAM Wiki documentation* [Online]. Available: https://wiki.umd.edu/gcam/index.php/Main_Page [Accessed].

Calvin, K., M. Wise, D. Klein, D. Mccollum, M. Tavoni, B. Van Der Zwaan, and D.P. Van Vuuren. 2013. A multi-model analysis of the regional and sectoral roles of bioenergy in near- and long-term CO2 emissions reduction. *Climate Change Economics* 04: 1340014.

Calvin, K. V., R. Beach, A. Gurgel, M. Labriet and A.M. Loboguerrero Rodriguez. 2015. Agriculture, forestry, and other land-use emissions in Latin America. *Energy Economics* 56, May 2016: 615–624.

Cardozo, N.P., R. De Oliveira Bordonal, and N. La Scala Jr. 2015. Greenhouse gas emission estimate in sugarcane irrigation in Brazil: Is it possible to reduce it, and still increase crop yield? *Journal of Cleaner Production* 112: 3988–3997.

Chum, H., A. Faaij, and J. Moreira. 2011. Bioenergy. In *IPCC special report on renewable energy sources and climate change mitigation*, ed. O. Edenhofer, R. Pichs-Madruga, Y. Sokona, K. Seyboth, P. Matschoss, S. Kadner, T. Zwickel, P. Eickemeier, G. Hansen, S. Schlömer, and C. Von Stechow. Cambridge: Cambridge University Press.

Constantin, J., M. Willaume, C. Murgue, B. Lacroix, and O. Therond. 2015. The soil-crop models STICS and AqYield predict yield and soil water content for irrigated crops equally well with limited data. *Agricultural and Forest Meteorology* 206: 55–68.

Coyle, W. 2007. The future of biofuels: A global perspective. *Amber Waves* 5: 24–29.

Craufurd, P.Q., and T.R. Wheeler. 2009. Climate change and the flowering time of annual crops. *Journal of Experimental Botany* 60: 2529–2539.

Daioglou, V., B. Wicke, A.P.C. Faaij, and D.P. Van Vuuren. 2015. Competing uses of biomass for energy and chemicals: implications for long-term global CO2 mitigation potential. *GCB Bioenergy* 7: 1321–1334.

Davies-Barnard, T., P.J. Valdes, J.S. Singarayer, and C.D. Jones. 2014. Climatic impacts of land-use change due to crop yield increases and a universal carbon tax from a scenario model. *Journal of Climate* 27(4): 1413–1424. doi:10.1175/JCLI-D-13-00154.1.

Dokoohaki, H., M. Gheysari, S.-F. Mousavi, S. Zand-Parsa, F.E. Miguez, S.V. Archontoulis, and G. Hoogenboom. 2016. Coupling and testing a new soil water module in DSSAT CERES-Maize model for maize production under semi-arid condition. *Agricultural Water Management* 163: 90–99.

Dornburg, V., and A.C. Faaij. 2005. Cost and Co2-emission reduction of biomass cascading: Methodological aspects and case study of SRF poplar. *Climatic Change* 71: 373–408.

Dornburg, V., B.G. Hermann, and M.K. Patel. 2008. Scenario projections for future market potentials of biobased bulk chemicals. *Environmental Science & Technology* 42: 2261–2267.

Dornburg, V., D. Van Vuuren, G. Van De Ven, H. Langeveld, M. Meeusen, M. Banse, M. Van Oorschot, J. Ros, G. Jan Van Den Born, H. Aiking, M. Londo, H. Mozaffarian, P. Verweij, E. Lysen, and A. Faaij. 2010. Bioenergy revisited: Key factors in global potentials of bioenergy. *Energy & Environmental Science* 3: 258–267.

EPA. 2010. *Methane and nitrous oxide emissions from natural sources* . Washington, DC: U.S. Environmental Protection Agency.

FAOSTAT. 2015. *Food and Agriculture Organization of the United Nations Statistic Division*. Rome: FAO.

Farrell, A.E., R.J. Plevin, B.T. Turner, A.D. Jones, M. O'hare, and D.M. Kammen. 2006. Ethanol can contribute to energy and environmental goals. *Science* 311: 506–508.

Field, C.B., J.E. Campbell, and D.B. Lobell. 2008. Biomass energy: The scale of the potential resource. *Trends in Ecology & Evolution* 23: 65–72.

Frank, S., E. Schmid, P. Havlík, U.A. Schneider, H. Böttcher, J. Balkovič, and M. Obersteiner. 2015. The dynamic soil organic carbon mitigation potential of European cropland. *Global Environmental Change* 35: 269–278.

Gnansounou, E., J. Dong, and D. Bedniaguine. 2004. The strategic technology options for mitigating CO2 emissions in power sector: Assessment of Shanghai electricity-generating system. *Ecological Economics* 50: 117–133.

Harris, Z.M., R. Spake, and G. Taylor. 2015. Land use change to bioenergy: A meta-analysis of soil carbon and GHG emissions. *Biomass and Bioenergy* 82: 27–39.

Hergoualc'h, K., and L.V. Verchot. 2014. Greenhouse gas emission factors for land use and land-use change in Southeast Asian peatlands. *Mitigation and Adaptation Strategies for Global Change* 19: 789–807.

Hill, J., E. Nelson, D. Tilman, S. Polasky, and D. Tiffany. 2006. Environmental, economic, and energetic costs and benefits of biodiesel and ethanol biofuels. *Proceedings of the National Academy of Sciences* 103: 11206–11210.

Houghton, R.A., J.I. House, J. Pongratz, G.R. Van Der Werf, R.S. Defries, M.C. Hansen, C. Le Quéré, and N. Ramankutty. 2012. Carbon emissions from land use and land-cover change. *Biogeosciences* 9: 5125–5142.

Hudiburg, T.W., S.C. Davis, W. Parton, and E.H. Delucia. 2015. Bioenergy crop greenhouse gas mitigation potential under a range of management practices. *GCB Bioenergy* 7: 366–374.

Hurkman, W.J., W.H. Vensel, C.K. Tanaka, L. Whitehand, and S.B. Altenbach. 2009. Effect of high temperature on albumin and globulin accumulation in the endosperm proteome of the developing wheat grain. *Journal of Cereal Science* 49: 12–23.

IEA. 2004. *Biofuels for transport: An international perspective*. Paris: International Energy Agency.

Ignaciuk, A. 2015. *Adapting agriculture to climate change*. France: OECD Publishing.

IPCC. 2007. *Climate change 2007: Mitigation of climate change. .Exit EPA disclaimer contribution of working group III to the fourth assessment report of the intergovernmental panel on climate change,* ed. B. Metz, O.R. Davidson, P.R. Bosch, R. Dave, L.A. Meyer., Cambridge, United Kingdom and New York, NY, USA: Cambridge University Press.

IPCC. 2011. *Special report on renewable energy sources and climate change mitigation*. United Kingdom and New York: NY, USA, Cambridge University Press.

IPCC. 2013. *IPCC Climate change 2013: The Physical science Basis. Contribution of working group I to the Fifth assessment report of the Intergovernmental Panel on climate change,* ed. Stocker, T.F., D. Qin, G.-K. Plattner, M. Tignor, S.K. Allen, J. Boschung, A. Nauels, Y. Xia, V. Bex and P.M. Midgley. Cambridge, United Kingdom and New York, NY, USA, Cambridge University Press.

Izaurralde, R.C., J.R. Williams, W.B. Mcgill, N.J. Rosenberg, and M.C.Q. Jakas. 2006. Simulating soil C dynamics with EPIC: Model description and testing against long-term data. *Ecological Modelling* 192: 362–384.

Jones, J.W., G. Hoogenboom, C.H. Porter, K.J. Boote, W.D. Batchelor, L.A. Hunt, P.W. Wilkens, U. Singh, A.J. Gijsman, and J.T. Ritchie. 2003. The DSSAT cropping system model. *European Journal of Agronomy* 18: 235–265.

Kalafatis, S.E., M.C. Lemos, Y.-J. Lo, and K.A. Frank. 2015. Increasing information usability for climate adaptation: The role of knowledge networks and communities of practice. *Global Environmental Change* 32: 30–39.

Kang, M.S., and S.S. Banga. 2013. Global agriculture and climate change. *Journal of Crop Improvement* 27: 667–692.

Kennedy, C., J. Steinberger, B. Gasson, Y. Hansen, T. Hillman, M. Havránek, D. Pataki, A. Phdungsilp, A. Ramaswami, and G.V. Mendez. 2009. Greenhouse gas emissions from global cities. *Environmental Science & Technology* 43: 7297–7302.

Kim, D.-G., D. Giltrap, and G. Hernandez-Ramirez. 2013. Background nitrous oxide emissions in agricultural and natural lands: A meta-analysis. *Plant and Soil* 373(1): 17–30. doi:10.1007/s11104-013-1762-5.

Kimball, B.A. 2011. Lessons from FACE: CO2 effects and interactions with water, nitrogen, and temperature. In *Handbook of climate change and agroecosystems: Impacts, adaptation, and mitigation*, ed. D. Hillel and C. Rosenzweig. London: Imperial College Press.

Knightes, C.D., and M. Cyterski. 2005. Evaluating predictive errors of a complex environmental model using a general linear model and least square means. *Ecological Modelling* 186: 366–374.

Lambin, E.F., and P. Meyfroidt. 2011. Global land use change, economic globalization, and the looming land scarcity. *Proceedings of the National Academy of Sciences* 108: 3465–3472.

Lobell, D.B., M. Banziger, C. Magorokosho, and B. Vivek. 2011. Nonlinear heat effects on African maize as evidenced by historical yield trials. *Nature Climate Change* 1: 42–45.

Loulou, R. 2008. ETSAP-TIAM: The TIMES integrated assessment model. Part II: Mathematical formulation. *Computational Management Science* 5: 41–66.

Macleod, M., et al. 2015. *Cost-effectiveness of greenhouse gas mitigation measures for agriculture*. Paris: OECD Publishing.

Mohammed, A.R., and L. Tarpley. 2009. High nighttime temperatures affect rice productivity through altered pollen germination and spikelet fertility. *Agricultural and Forest Meteorology* 149: 999–1008.

Morgan, J.A., R.F. Follett, L.H. Allen, S. Del Grosso, J.D. Derner, F. Dijkstra, A. Franzluebbers, R. Fry, K. Paustian, and M.M. Schoeneberger. 2010. Carbon sequestration in agricultural lands of the United States. *Journal of Soil and Water Conservation* 65: 6A–13A.

Nayak, D., E. Saetnan, K. Cheng, W. Wang, F. Koslowski, Y.-F. Cheng, W.Y. Zhu, J.-K. Wang, J.-X. Liu, D. Moran, X. Yan, L. Cardenas, J. Newbold, G. Pan, Y. Lu, and P. Smith. 2015. Management opportunities to mitigate greenhouse gas emissions from Chinese agriculture. *Agriculture, Ecosystems & Environment* 209: 108–124.

NRC. 2010. *Advancing the science of climate change*. Washington, DC: National Research Council. The National Academies Press.

Ogle, S.M., F.J. Breidt, M. Easter, S. Williams, and K. Paustian. 2007. An empirically based approach for estimating uncertainty associated with modelling carbon sequestration in soils. *Ecological Modelling* 205: 453–463.

Ogle, S.M., F. Jay Breidt, M.D. Eve, and K. Paustian. 2003. Uncertainty in estimating land use and management impacts on soil organic carbon storage for US agricultural lands between 1982 and 1997. *Global Change Biology* 9: 1521–1542.

Paltsev, S., J. Reilly, H. Jacoby, R. Eckaus, J. Mcfarland, M. Sarofim, M. Asadoorian, and M. Babiker. 2005. *The MIT emissions prediction and policy analysis (EPPA) model: Version 4*, MIT joint program on the science and policy of global change. Cambridge, MA: MIT.

Paustian, K., N.H. Ravindranath, A. Amstel, and M. Gytarsky. 2006. Chapter 1 - Introduction. In *2006 IPCC Guidelines for National Greenhouse Gas Inventories*, Prepared by the National Greenhouse Gas Inventories Programme, ed. H. S. Eggleston, L. Buendia, K. Miwa, T. Ngara and K. Tanabe. Japan: IGES.

Pittock, B. 2003. *Climate change - An Australian guide to the science and potential impacts*, pp. 239. Report compiled for the Australian Greenhouse Office, Canberra, Australia. Available at http://www.greenhouse.gov.au/science/guide/index.html

Qin, Z., Q. Zhuang, and X. Cai. 2015. Bioenergy crop productivity and potential climate change mitigation from marginal lands in the United States: An ecosystem modeling perspective. *GCB Bioenergy* 7: 1211–1221.

Ravindranath, N., R. Manuvie, J. Fargione, J. Canadell, G. Berndes, J. Woods, H. Watson, and J. Sathaye (eds.). 2008. *Greenhouse gas implications of land use and land conversion to biofuel crops*. New York: Island Press.

Reddy, P.P. 2015. Agriculture as a source of GHGs. In *Climate resilient agriculture for ensuring food security*. India: Springer.

Rezaei, E.E., S. Siebert, and F. Ewert. 2015. Intensity of heat stress in winter wheat—phenology compensates for the adverse effect of global warming. *Environmental Research Letters* 10: 024012.

Roberts, K.G., B.A. Gloy, S. Joseph, N.R. Scott, and J. Lehmann. 2009. Life cycle assessment of biochar systems: Estimating the energetic, economic, and climate change potential. *Environmental Science & Technology* 44: 827–833.

Ross, M. 2009. Documentation of the Applied Dynamic Analysis of the Global Economy (ADAGE) model. Working Paper 09_01, April 2009,.RTI International, Research Triangle Park, NC.

Saadi, S., M. Todorovic, L. Tanasijevic, L.S. Pereira, C. Pizzigalli, and P. Lionello. 2015. Climate change and Mediterranean agriculture: Impacts on winter wheat and tomato crop evapotranspiration, irrigation requirements and yield. *Agricultural Water Management* 147: 103–115.

Sadras, V.O., V. Vadez, R. Purushothaman, L. Lake, and H. Marrou. 2015. Unscrambling confounded effects of sowing date trials to screen for crop adaptation to high temperature. *Field Crops Research* 177: 1–8.

Sala, O.E., F. Stuart Chapin III, J.J. Armesto, E. Berlow, J. Bloomfield, R. Dirzo, E. Huber-Sanwald, L.F. Huenneke, R.B. Jackson, A. Kinzig, R. Leemans, D.M. Lodge, H.A. Mooney, M.N. Oesterheld, N.L. Poff, M.T. Sykes, B.H. Walker, M. Walker, and D.H. WALL. 2000. Global biodiversity scenarios for the year 2100. *Science* 287: 1770–1774.

Sharma, S.K., and R. Chauhan. 2011. Climate change research initiative: Indian network for climate change assessment. *Current Science (Bangalore)* 101: 308–311.

Six, J., E.T. Elliott, and K. Paustian. 2000. Soil macroaggregate turnover and microaggregate formation: A mechanism for C sequestration under no-tillage agriculture. *Soil Biology and Biochemistry* 32: 2099–2103.

Slade, R., A. Bauen, and R. Gross. 2014. Global bioenergy resources. *Nature Climate Change* 4: 99–105.

Smith, P. 2015. Addressing the joint challenges of climate change and food security. 5: 61.
Stöckle, C.O., M. Donatelli, and R. Nelson. 2003. CropSyst, A cropping systems simulation model. *European Journal of Agronomy* 18: 289–307.
Stratonovitch, P., and M.A. Semenov. 2015. Heat tolerance around flowering in wheat identified as a key trait for increased yield potential in Europe under climate change. *Journal of Experimental Botany* 66: 3599–3609.
Swaminathan, M.S., and P.C. Kesavan. 2012. Agricultural research in an era of climate change. *Agricultural Research* 1: 3–11.
Talukder, A.S.M.H.M., G.K. Mcdonald, and G.S. Gill. 2013. Effect of short-term heat stress prior to flowering and at early grain set on the utilization of water-soluble carbohydrate by wheat genotypes. *Field Crops Research* 147: 1–11.
Talukder, A.S.M.H.M., G.K. Mcdonald, and G.S. Gill. 2014. Effect of short-term heat stress prior to flowering and early grain set on the grain yield of wheat. *Field Crops Research* 160: 54–63.
Tao, F., Z. Zhang, S. Zhang, and R.P. Rötter. 2015. Heat stress impacts on wheat growth and yield were reduced in the Huang-Huai-Hai Plain of China in the past three decades. *European Journal of Agronomy* 71: 44–52.
Troy, T.J., C. Kipgen, and I. Pal. 2015. The impact of climate extremes and irrigation on US crop yields. *Environmental Research Letters* 10: 054013.
Turner, B.L., E.F. Lambin, and A. Reenberg. 2007. The emergence of land change science for global environmental change and sustainability. *Proceedings of the National Academy of Sciences* 104: 20666–20671.
Wang, B., D.L. Liu, S. Asseng, I. Macadam, and Q. Yu. 2015. Impact of climate change on wheat flowering time in eastern Australia. *Agricultural and Forest Meteorology* 209–210: 11–21.
Wang, X., J. Cai, D. Jiang, F. Liu, T. Dai, and W. Cao. 2011. Pre-anthesis high-temperature acclimation alleviates damage to the flag leaf caused by post-anthesis heat stress in wheat. *Journal of Plant Physiology* 168: 585–593.
Webster, M.D., M. Babiker, M. Mayer, J.M. Reilly, J. Harnisch, R. Hyman, M.C. Sarofim, and C. Wang. 2002. Uncertainty in emissions projections for climate models. *Atmospheric Environment* 36: 3659–3670.
Wei, D., N. Dormady, and A. Rose. 2015. Development of reduced-form models to evaluate macroeconomic impacts of greenhouse gas mitigation. *Journal of Sustainable Energy Engineering* 2: 377–397.
Williams, J.R. 1995. The EPIC model. In *Computer models of watershed hydrology*, ed. V.P. Singh. Colorado: Water Resources Publications.
Woli, P., B.V. Ortiz, D. Buntin, and K. Flanders. 2014. El Niño-Southern Oscillation (ENSO) effects on Hessian Fly (Diptera: Cecidomyiidae) infestation in the Southeastern United States. *Environmental Entomology* 43: 1641–1649.
Woolf, D., J.E. Amonette, F.A. Street-Perrott, J. Lehmann, and S. Joseph. 2010. Sustainable biochar to mitigate global climate change. *Nature Communications* 1: 56.
Worldwatch. 2006. Worldwatch biofuels for transportation. Washington DC.
Wu, Z., H. Zhang, C. Krause, and N. Cobb. 2010. Climate change and human activities: a case study in Xinjiang, China. *Climatic Change* 99: 457–472.
Bao, Y., G. Hoogenboom, R. McClendon, and P. Urich. 2015. Soybean production in 2025 and 2050 in the southeastern USA based on the SimCLIM and the CSM-CROPGRO-Soybean models. *Climate Research* 63: 73–89.
Zhang, D., G. Pan, G. Wu, G.W. Kibue, L. Li, X. Zhang, J. Zheng, J. Zheng, K. Cheng, S. Joseph, and X. Liu. 2016. Biochar helps enhance maize productivity and reduce greenhouse gas emissions under balanced fertilization in a rainfed low fertility inceptisol. *Chemosphere* 142: 106–113.
Zhang, H.-L., X. Zhao, X.-G. Yin, S.-L. Liu, J.-F. Xue, M. Wang, C. Pu, R. Lal, and F. Chen. 2015. Challenges and adaptations of farming to climate change in the North China Plain. *Climatic Change* 129: 213–224.

Chapter 2
Greenhouse Gas Emissions Due to Meat Production in the Last Fifty Years

Dario Caro, Steven J. Davis, Simone Bastianoni, and Ken Caldeira

Contents

Abstract We estimate greenhouse gas emissions due to the production of beef cattle, pork and chickens for the period 1961–2010, following IPCC guidelines (IPCC. 2006 IPCC guideline for national greenhouse gas inventories. In: Eggleston H S, Buendia L, Miwa K, Ngara T, Tanabe K (eds) IGES, Japan. Available at: http://www.ipcc-nggip.iges.or.jp/public/2006gl/index.html, 2006). We find that during the last 50 years, global greenhouse gas (GHG) emissions released from beef cattle, pork and chickens increased by 59 %, 89 % and 461 % respectively. In 2010, GHG emissions caused by beef cattle contributed 54 % of total livestock emissions; pork and chickens contribute to 5–1 %, respectively. In the same year, the methane

D. Caro (✉)
Department of Animal Science, University of California, Davis, CA 95616, USA

Ecodynamics group, Department of Earth, Environmental and Physical Sciences, University of Siena, Pian dei Mantellini 44, 53100 Siena, Italy
e-mail: darcaro@ucdavis.edu

S.J. Davis
Department of Earth System Science, University of California (Irvine), Irvine, 3232 Croul Hall, CA 92697, USA

S. Bastianoni
Ecodynamics group, Department of Earth, Environmental and Physical Sciences, University of Siena, Pian dei Mantellini 44, 53100 Siena, Italy

K. Caldeira
Department of Global Ecology, Carnegie Institution for Science, 260 Panama Street, Stanford, CA 94305, USA

M. Ahmed, C.O. Stockle (eds.), *Quantification of Climate Variability, Adaptation and Mitigation for Agricultural Sustainability*,
DOI 10.1007/978-3-319-32059-5_2

emissions released from beef cattle represent about 69 % of total emissions associated with that livestock category, in particular due to methane emissions from enteric fermentation. Although beef cattle and pork emissions increased during the period 1961–2010, their respective per capita emissions decreased over the time (−29 % and −15 %, respectively). Beef cattle, especially in developing countries, represent the largest source of livestock-related emissions. For this reason, dietary substitution of pork and chicken products with beef products might be an option for mitigating livestock emissions. However, this is in contrast to the global trend towards increased reliance on cattle.

Keywords Greenhouse gas • Beef • Cattle • Pork • Chicken • Diet

2.1 Introduction

The intensification of livestock sector contributes to climate change (Stehfest et al. 2012; De Vries and De Boer 2010; Naylor et al. 2005). Direct non-CO_2 emissions associated with livestock sector resulting from methane (CH_4) and nitrous oxide (N_2O) emissions. The formers are released by enteric fermentation and manure management whereas the latters are released by manure management and manure left on pasture. Although, compared with carbon dioxide emissions, CH_4 and N_2O emissions represent a lower percentage of global GHG emissions (IPCC 2006), their global warming potentials are 21 and 310 times higher than CO_2 (EPA 2011). In other words one methane and nitrous oxide molecule contribute to climate change, 21 and 310 times more than one carbon dioxide molecule.

Nowadays, greenhouse gas emissions (GHG) released from livestock are an emerging problem for several reasons. Firstly, direct non-CO_2 emissions from livestock represent about 10 % of global GHG emissions (Tubiello et al. 2013) and when indirect CO_2 emissions (including fossil fuel emissions and environmental impact associated with the feed) are taken in account, livestock emissions cover about 18 % of global emissions (F.A.O et al. 2006). Secondly, a recent study highlighted that GHG emissions from livestock have increased by about 50 % in the last five decades (E.P.A 2006). In particular, emissions from livestock in developing countries steadily increased over the time (Caro et al. 2014a). Thirdly, the global population is expected to increase over the time as well as, consequently, the food demand (Godber and Wall 2014). Several papers show the growth in demand for livestock products, especially meat and milk and its implications in terms of environmental impact (Alexandratos and Bruinsma 2012; Reay et al. 2012; Bustamante et al. 2012; Valin et al. 2013). Meat production increased by about 300 % from 1961 to 2010, mainly due to beef cattle, pork and chicken meat demand (F.A.O 2005; F.A.O et al. 2006). As a consequence of this rising demand, livestock production is expected to double by 2050 (Garnett 2009; Godfray et al. 2010).

The Intergovernmental Panel on Climate Change (IPCC), has developed guidelines for national greenhouse gas inventories (IPCC 2006). Guidelines are the tool for estimating emissions on a regional (and global) level. For each country, four sectors are assessed: Energy, Industry, Waste and AFOLU (such as Agriculture, Forestry and Other Land Use). Livestock sector is included in AFOLU sector and a recent study highlighted that GHG emissions released from livestock sector represent more than 50 % of total GHG emissions associated with AFOLU sector (Caro et al. 2014a). Although, in scientific literature, IPCC approach has been subjected to some criticisms (Bastianoni et al. 2004; Caro et al. 2014b; Davis and Caldeira 2010), IPCC guidelines provide a robust and simple accounting method for evaluating GHG emissions yearly released from regional systems. IPCC guidelines provide three tiers (levels of detail) for estimating GHG emissions on the basis of data availability. Because global analysis requires a large amount of data that may be unavailable for each country, tier 1 (basic method) is considered the more appropriate for global analysis (IPCC 2006). For livestock sector, in the last detailed IPCC tier 1 method, default emission factors are recommended for different livestock categories. Emissions factors vary significantly among these different types of livestock, and also depend on characteristics such as mean air annual temperature, geographic location, and level of economic development (IPCC 2006). Here, by using a Tier 1 method, we estimate the non-CO_2 emissions (such as methane and nitrous oxide) due to production of beef cattle, pork and chickens for 237 countries in historical series, during the period 1961 and 2010. We assess the total impact of these three livestock categories, taking into account variation in mean annual air temperature during this time period as well as the level of economic development and geographic location of each country. In particular, we focus on trend of developed and developing countries during the last 50 years. In discussing the results, we compare the different trends and we show where the livestock emissions as well as per capita livestock emissions increased or decreased over the time.

2.2 Methodology

Tier 1 method requires data on the average number of animals in each livestock category. The number of beef cattle, pork and chicken for this analysis was provided by new FAO statistic database (F.A.O 2015).

We take into account three different livestock populations: beef cattle, pigs (market and breeding) and chickens. The equations and emission factors used in this paper to estimate livestock emissions in each emission category were provided by IPCC guidelines (IPCC 2006). We assume that IPCC guidelines, developed for the period 1990–2010, fit also for the three previous decades.

According to IPCC guidelines, emissions due to beef cattle, pork and chicken are estimated by multiplying appropriate specific emission factors and activity data (IPCC 2006). In our analysis we take in account beef cattle, pork and chicken emissions due to enteric fermentation, manure management and manure left on pasture. The three emissions sources are described below.

2.2.1 Enteric Fermentation

The process of enteric fermentation produces methane emissions (JRC 2010). We apply the default emission factors presented in the IPCC guidelines for each of the recommended population subgroup. We use specific emission factors of enteric fermentation provided by IPCC (2006).

2.2.2 Manure Management

Organic material and water are the primarily elements of livestock manure. Methane emissions from manure are mediated by anaerobic and facultative bacteria that decompose the organic material under anaerobic conditions (Bouwman 1996). The methane production potential of manure is also due to the ambient temperature and its management (E.P.A 2006). Again following the Tier 1 methodology, our analysis uses population data for each animal category (F.A.O 2015), mean annual temperature in each nation (NOAA 2014). We use specific emission factors associated with manure management provided by IPCC (2006)). Urine and dung are included in our analysis.

N_2O emissions are released from nitrification and denitrification of nitrogen included in animal excretions (Barton and Atwater 2002). The production of direct N_2O emissions occurs when other nitrites interact in aerobic conditions. According to Tier 1 (IPCC 2006), direct N_2O emissions depend on total amount of N excretion from animals and the IPCC default factors associated with the type of manure management system. N_2O emissions are not dependent on mean air temperature (Klein et al. 2006). We sum estimated N across all manure management systems and multiply by the appropriate default emission factor (IPCC 2006).

Indirect N_2O emissions are due to volatile nitrogen losses (NH_3 and NO_x). The calculation of N volatilization from manure management systems (IPCC 2006) is obtained by multiplying the quantity of nitrogen released and a specific default factor that represent the fraction of volatilized nitrogen. Nitrogen is also released from leaching/runoff into soils that produce indirect N_2O emissions (Meyer et al. 2002). We use a Tier 2 (IPCC 2006) method to estimate this source of emission.

2.2.3 Manure Left on Pasture

Nitrous oxide is released from soils due to the nitrification and denitrification processes (Bateman and Baggs 2005). In this paper we take in account the contribute of N_2O direct and N_2O indirect due to volatilization and leaching/runoff from managed soils (Vogeler et al. 2011). The emissions due to manure applied to soils is not included in this paper, because those emissions are attributed to the fertilized crops and not the livestock which produced the manure.

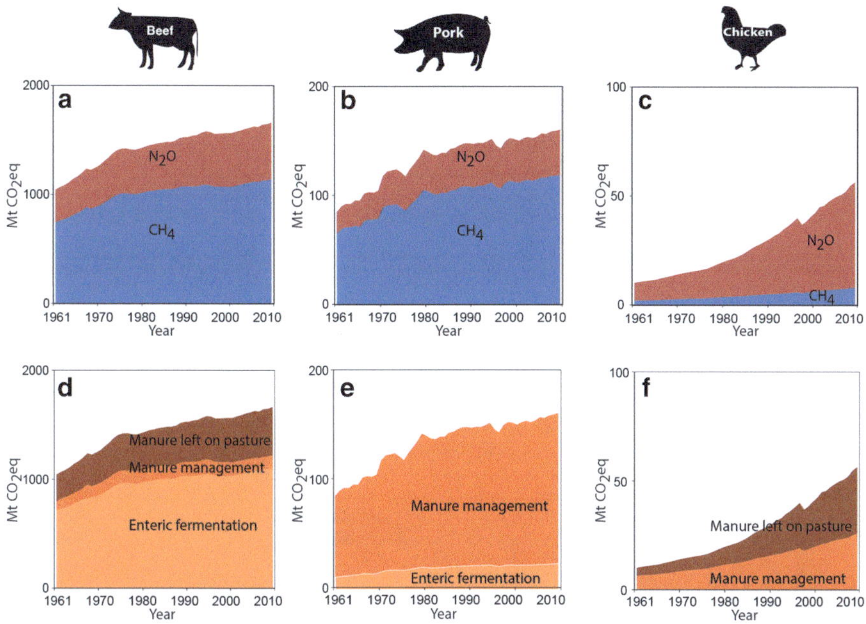

Fig. 2.1 Trend of global beef cattle (**a**, **d**), pork (**b**, **e**) and chickens (**c**, **f**) greenhouse gas (expressed as equivalent CO_2) emissions during the period 1961–2010. (**a**, **b**, **c**) Emissions are subdivided into methane (CH_4) and nitrous oxide (N_2O) and expressed in terms of CO_2 equivalents using 100-year GWP measures. (**d**, **e**, **f**) CO_2 equivalent emissions are subdivided into enteric fermentation, manure management, and manure left on pasture. In general, greenhouse gas emissions from livestock have been growing over time

2.3 Results

Figure 2.1 shows the livestock emissions trend for the principal GHGs directly produced by livestock (CH_4 and N_2O) in different emissions categories (enteric fermentation, manure management, manure left on pasture) for cattle, pork and chickens.

Our analysis estimates that in 2010, the GHG emissions due to beef cattle, pork and chickens are 1660, 160 and 56 Mt CO_{2e}q respectively. In the same year, according to Caro et al. (2014b)), we estimate that in overall, beef cattle, pork and chicken represent about 60 % of total livestock emissions.

We find that between 1961 and 2010, global GHG emissions released from cattle, pork and chickens increased by 59 %, 89 % and 461 % respectively (Fig. 2.1). In 2010, the methane emissions released from beef cattle represent about 69 % of total beef cattle emissions, in particular due to enteric fermentation (66 % of total beef cattle emissions). Even for pigs methane emissions make up the largest emission source, representing about 74 % of total pork emissions. However, for pork, methane is mainly due to manure management (85 % of total pork emissions). Chickens, being not ruminant, do not produce enteric fermentation as well as emissions due to this process. In fact, in 2010, nitrous oxide was the main greenhouse gas released

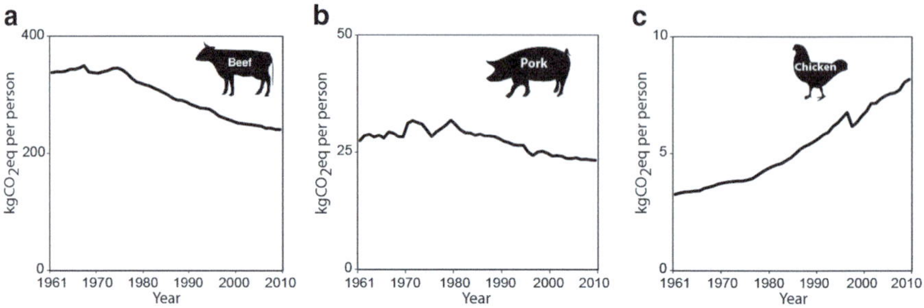

Fig. 2.2 Global beef cattle (**a**), pork (**b**) and chicken (**c**) emissions per capita during the period 1961–2010. Emissions are expressed as kilograms of equivalent CO_2 per person. While beef cattle and pork emissions per capita decrease over the time, chickens emissions per capita strongly increase from 1961 to 2010

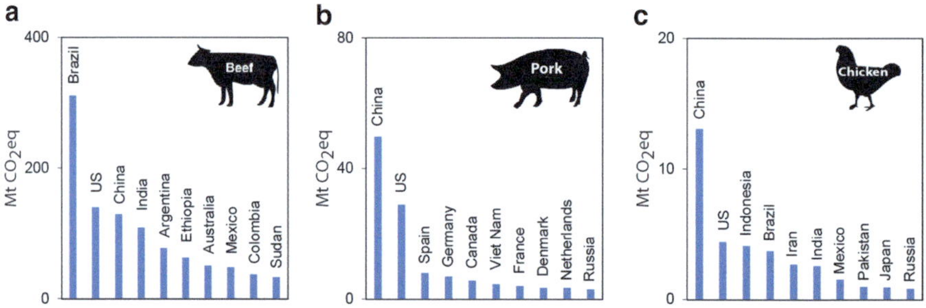

Fig. 2.3 Largest ten emitters of beef cattle (**a**), pork (**b**) and chickens (**c**) greenhouse gas emissions in 2010 (expressed as equivalent CO_2). In general, Brazil, US and China represent the greatest emitters of livestock emissions in the world

from chicken (85 % of total chicken emissions), and CH_4 was responsible for the remainder (15 %). Manure management and manure left on pasture are the only emissions sources associated with chicken production (45 % and 55 % respectively in 2010).

Although, beef cattle and pork emissions increased during the period 1961–2010 (Fig. 2.1), the respective per capita emissions decreased over the time (Fig. 2.2). In particular, beef cattle and pork emissions per capita decreased by 29 % and 15 % respectively. However Fig. 2.2 shows that per capita emissions due to chicken strongly increased over the time (151 % during the period 1961–2010). In general, each person globally, releases more GHG emissions per kg of CO_2eq due to beef cattle than pork and chicken ones: we observe that in 2010, about 240, 23 and 8 kg CO_2 eq per capita are due to beef cattle, pork and chickens respectively (Fig. 2.2).

Figure 2.3 shows that, in 2010, six countries produced 50 % of the global emissions due to beef cattle including (in order) Brazil (311 Mt CO_2eq/y, 19 %), US (140 Mt CO_2eq/y, 8 %), China (129 Mt CO_2eq/y, 8 %), India (109 Mt CO_2eq/y, 7 %), Argentina (77 Mt CO_2eq/y, 5.0 %) and Ethiopia (52 Mt CO_2eq/y, 4 %). In the same year, three countries produced 54 % of the global emissions due to pork including

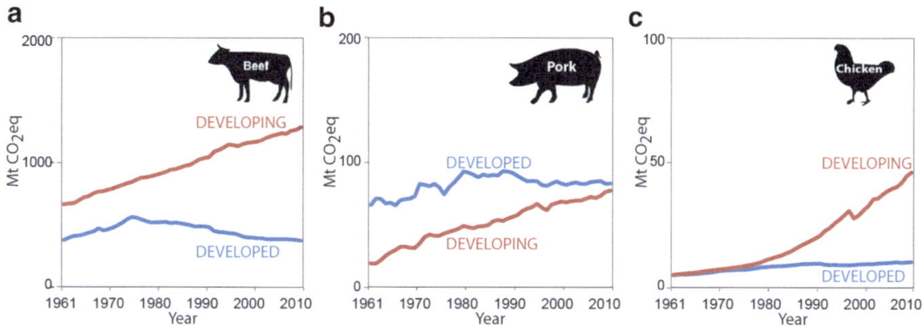

Fig. 2.4 Trend of beef cattle (**a**), pork (**b**) and chickens (**c**) greenhouse gas (expressed as equivalent CO_2) emissions during the period 1961–2010 in Developed and Developing countries. Global greenhouse gas emissions from livestock have been growing over time, primarily due to increased emissions from Developing countries in each type of livestock analyzed

(in order) China (50 Mt CO_2eq/y, 31 %), US (29 Mt CO_2eq/y, 18 %) and Spain (8 Mt CO_2eq/y, 5 %). For chickens, six countries produced 54 % of the global emissions due to beef cattle including (in order) China (13 Mt CO_2eq/y, 23 %), US (4 Mt CO_2eq/y, 8 %), Indonesia (4 Mt CO_2eq/y, 7 %), Brazil (3 Mt CO_2eq/y, 7 %), Iran (2 Mt CO_2eq/y, 5 %) and India (2 Mt CO_2eq/y, 5 %).

We find that, by percentage, beef cattle emissions over the period of 1961–2010 increased the most in Congo (+929 %), Saudi Arabia (+894 %) and Gabon (+765 %), and decreased most in Tunisia (−89 %), Lebanon (−80 %) and Bulgaria (−73 %).

We find that, according to Caro et al. (2014b), on the average, emissions caused by beef cattle contributed 54 % of total livestock emissions. Pork and chickens contribute to 5–1 % respectively.

Therefore, emissions released by beef cattle are substantially higher than other livestock categories. However, the increase was almost entirely in developing nations, where beef cattle emissions increased by 94 % between 1961 and 2010 (Fig. 2.4). In contrast, beef cattle emissions in developed countries were stable over the period, decreasing by just 1 %. The large difference in growth of emissions from developed and developing countries is also observed in pork and chickens. Because of this rapid growth, developing countries produced more chicken emissions than developed countries in 2010, which was not the case in 1961 (Fig. 2.3b, d). Pork emissions (Fig. 2.4) in 2010 were about the same in developed and developing countries. Another noteworthy trend is a marked decrease in livestock emissions in transition economies between 1992 and 2010. Over this 18-year period, beef cattle and pork emissions in transition economies decreased by 47 % and 83 % (4.4 % per year) respectively.

Figure 2.5 decomposes the intensity of livestock emissions by primary livestock products, using FAO data on the masses of beef, pork and chicken meat produced 1961–2010 (F.A.O 2015).

Global emission intensities generally decrease for all products over the time. Although emissions intensities in developing countries are consistently higher than in developed countries (with the noteworthy exception of pork in the last two

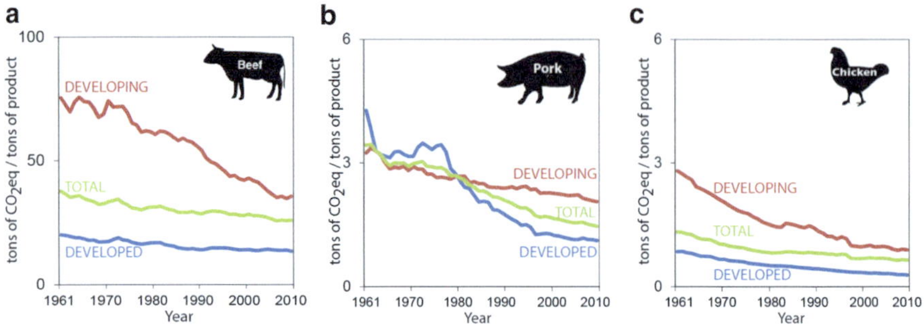

Fig. 2.5 Beef cattle, pork and chicken emissions per ton of beef cattle (**a**), pork (**b**) and chicken (**c**) meat produced during the period 1961–2010 in total (*green*), developed (*blue*) and developing (*red*) countries. Emission intensity in developing countries is higher than developed countries (except for pig meat in the last two decades). Global emission intensity decreases over the time. Beef meat is the product associated to highest intensity. Data on production of livestock products is from FAOstat database (F.A.O 2015)

decades), the developing country intensities are decreasing over the time while intensities in developed countries have changed very little. Between 1961 and 2010, the emissions intensities of beef cattle, pork and chicken meat produced in developing countries have decreased by 52 %, 73 % and 67 %, respectively (Fig. 2.5). Beef production is the most emissions intensive; in 2010, the emissions intensity of beef ranged from 11 tCO_2eq per ton of beef produced in the U.S. and 101 tCO_2eq per ton of beef produced in India.

2.4 Discussion

We find that the GHG emissions due to beef cattle, pork and chickens are increasing globally (Fig. 2.1), mostly due to emissions growth in developing countries (Fig. 2.4). In contrast, GHG emissions in developed countries decrease during the analyzed period. Emissions per capita for beef cattle and pork tend to decrease over the time, however for chicken they strongly increased during the period 1961–2010 (Fig. 2.2). For beef cattle and pork the decrease of both total and per capita GHG emissions from livestock sector reveals a decreased livestock production in developed countries during the analyzed period. Oppositely, our analysis shows that an increased livestock production in developing countries occurs, mainly due to increased population and consequent demand for livestock products.

Beef cattle, especially in developing countries, represent the largest source of livestock-related emissions (Fig. 2.3 according to Caro et al. 2014b). This is due to both the abundance of these animals and the fact that emissions per animal are substantially higher than for other livestock categories. Looking at results obtained from our analysis, we conclude that dietary choices can be a strong driver of livestock emissions. In particular we show that beef cattle meat releases more GHG emissions per calorie than pork and chickens, and much more than vegetables (Figs. 2.1 and 2.4; Engstrom et al. 2007). For this reason, substitution of pork and chickens for beef is an option for mitigating livestock emissions. This is in contrast to the global trend towards increased reliance on cattle (F.A.O 2015). However, pigs and poultry are dependent on other products (in particular grain and soy) that are additional sources of GHG emissions.

The decade of most rapid change was from 1991 to 2000 with substantial increases in emissions from developing countries and decreased emissions in developed countries. Since 1989, a large number of developing countries have liberalized their economic policies shifting from import- to export-oriented regimes (Narula and Dunning 2000). These economic processes also affected the livestock sector in developing countries, encouraging an increase of production and export of livestock products resulting in a contemporaneous growth of livestock emissions from these countries (Moran and Wall 2011). Moreover Fig. 2.5 points out that in spite of the livestock production is less efficient in developing countries, their emission intensity relative to the main type of meat is decreasing more rapidly than developed countries. Thus, international trade that exports livestock products from developing countries and imports these into developed countries may have two effects: first, to increase greenhouse gas emissions and second, to increase global economic productivity (Caro et al. 2014c; Bastianoni et al. 2014).

This study presents Tier 1 method calculations of livestock emissions from 237 countries. However higher Tier inventory for livestock emissions are available for few single countries. Comparing two methods Caro et al. (2014a)) showed that different results may occur. They concluded that unlike tier 1, higher levels of detail reflect the efficiency of farming practices. However, tier 1 can capture the increased demand of livestock products. Moreover, Tier 1, by using generalized parameter values, allows a more appropriate comparison among countries because every nation is treated equally so that emission data can be adequately compared whereas higher Tiers use more specific national values (IPCC 2006). Our study shows that beef cattle, pork and chicken emissions have been growing rapidly in developing countries that have no binding commitments to reduce or limit these livestock emissions. Future emissions might be reduced by the enacting measures aimed at increasing the efficiency of livestock production globally and discouraging the delocalization of livestock production from areas with low emission intensity to areas with high emission intensity (Bastianoni et al. 2014). Our analysis could be extended in two different directions. First, an analysis could consider the total life cycle emissions. Such an analysis would consider factors such as GHG emissions associated with the transportation of livestock products to market as well as land use change emissions. Second, an accounting could adopt a consumption-based perspective in which the GHG emissions are associated with the consumer of a product rather than the producer of a product (Caro et al. 2014c). Another aspect that is worthy of future study is a potential feedback loop involving livestock emissions and increasing global temperatures. Increasing temperatures provoke GHG emissions from livestock to increase, in particular, methane emissions from manure. Thus, livestock emissions both contribute to and are increased by global warming.

References

Alexandratos, N., and J. Bruinsma. 2012. *World agriculture towards 2030/2050: The 2012 revision*, ESA working paper No. 12–03. Rome: FAO.

Barton, P.K., and J.W. Atwater. 2002. Nitrous oxide emissions and the anthropogenic nitrogen in wastewater and solid waste. *Journal of Environmental Engineering* 128: 137–150.

Bastianoni, S., F.M. Pulselli, and E. Tiezzi. 2004. The problem of assigning responsibility for greenhouse gas emissions. *Ecological Economics* 49: 253–257.

Bastianoni, S., D. Caro, S. Borghesi, and F.M. Pulselli. 2014. The effect of a consumption-based accounting method in national GHG inventories: A trilateral application at macro and micro scale. *Frontiers in Energy Systems and Policy* 2: 1–8.

Bateman, E.J., and E.M. Baggs. 2005. Contributions of nitrification and denitrification to N_2O emissions from soils at different water-filled pore space. *Biology and Fertility of Soils* 41: 379–388.

Bouwman, T. 1996. Direct emission of nitrous oxide from agricultural soils. *Nutrient Cycling in Agroecosystems* 46: 53–70.

Bustamante, M.M.C., C.A. Nobre, R. Smeraldi, A.P.D. Aguiar, L.G. Barioni, L.G. Ferreira, K. Longo, and J.P.H.B. Ometto. 2012. Estimating greenhouse gas emissions from beef cattle raising in Brazil. *Climatic Change* 115: 559–577.

Caro, D., S.J. Davis, S. Bastianoni, and K. Caldeira. 2014a. Global and regional trends in greenhouse gas emissions from livestock. *Climatic Change*. doi:10.1007/s10584-014-1197-x.

Caro, D., S. Bastianoni, S. Borghesi, and F.M. Pulselli. 2014b. On the feasibility of a consumer-based allocation method in national GHG inventories. *Ecological Indicators* 36: 640–643.

Caro, D., A. Lopresti, S.J. Davis, S. Bastianoni, and K. Caldeira. 2014c. CH4 and N2O embodied in international trade of meat. *Environmental Research Letters*. doi:10.1088/1748-9326/9/11/114005.

Davis, S., and K. Caldeira. 2010. Consumption-based accounting of CO_2 emissions. *Proceedings of the National Academy of Sciences of the United States of America* 12: 5687–5692.

De Vries, M., and I.J.M. De Boer. 2010. Comparing environmental impacts for livestock products: A review of life cycle assessments. *Livestock Science* 128: 1–11.

Engstrom, R., A. Wadeskog, and G. Finnveden. 2007. Environmental assessment of Swedish agriculture. *Ecological Economics* 60: 550–563.

E.P.A. 2006. *Global anthropogenic non-CO_2 greenhouse gas emissions: 1990–2020*. United States Environmental Protection Agency, EPA 430-R-06-003, June 2006. Washington, DC, USA. www.epa.gov/nonCO2/econ-inv/dow.

E.P.A. 2011. *Global anthropogenic non-CO_2 greenhouse gas emissions: 1990–2030*. Final report 182 pg. Avaliable at: http://www.epa.gov/climatechange/EPAactivities/economics/nonco2projections.html.

F.A.O. 2005. *Livestock policy brief 02* (FAO, Rome, 2005). Food and agriculture organization of the United Nations. Avaliable at: http://www.fao.org/ag/againfo/resources/en/pubssap.html.

F.A.O, Steinfeld H, Gerber P, Wassenaar T, Castel V, Rosales M, De Haan C. 2006. *Livestock's long shadow: Environmental issues and options*. Food and Agriculture Organization of the United Nations (FAO), Rome, Italy. 408 Pp. Available at: http://www.fao.org/docrep/010/a0701e/a0701e00.HTM.

F.A.O. 2015. *FAOSTAT online database*. Available at http://faostat.fao.org/. Accessed Dec 2014.

Garnett, T. 2009. Livestock-related greenhouse gas emissions: Impacts and options for policy makers. *Environmental Science and Policy* 12: 491–503.

Godber, O.F., and R. Wall. 2014. Livestock and food security: Vulnerability to population growth and climate change. *Global Change Biology* 20: 3092–3102.

Godfray, H.C.J., J.R. Beddington, I.R. Crute, L. Haddad, D. Lawrence, J.F. Muir, J. Pretty, S. Robinson, S.M. Thomas, and C. Toulmin. 2010. Food security: The challenge of feeding 9 billion people. *Science* 327: 812–818.

IPCC. 2006. *2006 IPCC guideline for national greenhouse gas inventories*, ed. Eggleston H S, Buendia L, Miwa K, Ngara T, Tanabe K. IGES, Japan. Available at: http://www.ipcc-nggip.iges.or.jp/public/2006gl/index.html.

JRC (Joint Research Centre). 2010. *Evaluation of the livestock sector's contribution to the EU greenhouse gas emissions (GGELS)—Final report*, ed Leip A, F. Weiss, T. Wassenaar, I. Perez, T. Fellmann, P. Loudjani, F. Tubiello, D. Grandgirard, S. Monni and K. Biala, 323. Ispra: European Commission, Joint Research Centre. Available at: http://ec.europa.eu/agriculture/analysis/external/livestock-gas/.

Klein, C.A.M., R.S.A. Novoa, S. Ogle, K.A. Smith, P. Rochette, T. Wirth, B. Mcconkey, A. Mosier, K. Rypdal, M. Walsh and S.A. Williams. 2006. *N_2O emissions from managed soils, and CO_2 emissions from lime and urea application*. Chapter 11. 2006 Intergovernmental Panel on Climate Change guidelines for national greenhouse gas inventories.

Meyer, R.L., T. Kjær, and N.P. Revsbech. 2002. Nitrification and denitrification near a soil-manure interface studied with a nitrate-nitrite biosensor. *Soil Science Society of America Journal* 66: 498–506.

Moran, D., and E. Wall. 2011. Livestock production and greenhouse gas emissions: Defining the problem and specifying solutions. *Animal Frontiers* 1: 19–25.

Narula, R., and J.H. Dunning. 2000. Industrial development, globalization and multinational enterprises: New realities for developing countries. *Oxford Development Studies* 28: 141–167.

Naylor, R., H. Steinfield, W. Falcon, J. Galloway, and V. Smil. 2005. Losing the links between livestock and land. *Science* 310: 1621–1622.

NOAA. 2014. *National climatic data center*. Earth System Research Laboratory of the University of Delware. National oceanic and atmosphere administration. Available at: http://www.esrl.noaa.gov/psd/data/gridded/data.UDel_AirT_Precip.html. Accessed Feb 2014.

Reay, D.S., E.A. Davidson, K.A. Smith, P. Smith, J.M. Melillo, F. Dentener, and P.J. Crutzen. 2012. Global agriculture and nitrous oxide emissions. *Nature Climate Change* 2: 410–416.

Stehfest, E., M. Van Den Berg, and G. Woltjer. 2012. Options to reduce the environmental effects of livestock production – Comparison of two economic models. *Agricultural Systems* 114: 38–53.

Tubiello, F.N., M. Salvatore, S. Rossi, A. Ferrara, N. Fitton, and P. Smith. 2013. The FAOSTAT database of greenhouse gas emissions from agriculture. *Environmental Research Letters* 8: 1–10.

Valin, H., P. Havlik, A. Mosnier, M. Herrero, E. Schimd, and M. Obersteiner. 2013. Agricultural productivity and greenhouse gas emissions: Trade-offs or synergies between mitigation and food security? *Environmental Research and Letters*. doi:10.1088/1748-9326/8/3/035019.

Vogeler, I., D. Donna Giltrap, F. Frank.LI, and V. Snow. 2011. *Comparison of models for predicting nitrification, denitrification and nitrous oxide emissions in pastoral systems*. 19th International Congress on Modelling and Simulation, Perth, Australia, 12–16 December. Available at: http://mssanz.org.au/modsim2011.

Chapter 3
Modeling the Impact of Climate Variability on Crops in Sub-Saharan Africa

Ephraim Sekyi-Annan, Ernest Nti Acheampong, and Nicholas Ozor

Contents

Abstract The impact of climate variability is expected to have significant impacts on crop production in Sub-Saharan Africa. Being a region with high climate vulnerability, the quantification and understanding of the extent and rate of impact of climate variability on crop productivity is highly essential. Crop models have been used to analyze such impacts by predicting crop yields, conditions of growth, and suitable crop types under both current and future climatic conditions. The chapter examines the impact of climate variability on crop production system and analyzes the contribution of crop models in defining appropriate crop management strategies against the threat of high climate risk and uncertainty in sub-Saharan Africa. The region faces a range of climate risk that could have far-reaching implications on future cropping system. Uncertainties in the changing patterns of rainfall and temperature pose a threat to crop production and contribute to increasing rural

---

E. Sekyi-Annan
Soil Research institute, Kumasi, Ghana

E.N. Acheampong (✉)
Stellenbosch University, Stellenbosch, South Africa
e-mail: pope2gee@gmail.com

N. Ozor
African Technology Policy Studies Network, Nairobi, Kenya

M. Ahmed, C.O. Stockle (eds.), *Quantification of Climate Variability, Adaptation and Mitigation for Agricultural Sustainability*,
DOI 10.1007/978-3-319-32059-5_3

vulnerability and poverty in sub-Saharan Africa. Variation in inter-seasonal and intra-seasonal rainfall variability is considered as highly crucial in shaping the outcome of cropping systems during the season.

Keywords Vulnerability • Crop models • Cropping systems • Sub-Saharan Africa • Climate variability

3.1 Introduction

There have been several projections of the impact of climate variability on crop productions in different parts of the world with Africa seen as the continent to bear the brunt (FAO 2011). The impacts of projected climate change during the first half of this century will severely affect the development in sub-Saharan Africa (SSA), and will worsen conditions of poor and vulnerable countries (Scholes and Biggs 2004; IPCC 2007a). The agricultural sector will be hit hardest by the impact of climate change and variability in SSA. Nevertheless, it will continue to play a crucial role of providing an indispensable platform for wider economic growth to reduce poverty (DFID 2005).

Most parts of sub-Saharan Africa still remain under-developed with over 60 % of the population largely dependent on rainfed agricultural production for their socio-economic livelihood (FAO 2003; Yegbemey et al. 2014). Changes in rainfall patterns (i.e. late onset of rainy season, seasonal and intra-seasonal variability in terms of amount and duration, and non-uniform distribution in terms of space and time) and other relevant climatic variables (e.g. temperature, carbon dioxide etc.) threaten crop production as well as vulnerable communities (especially the poor rural communities) (Boko et al. 2007; Graef and Haigis 2001). Yield losses of up to 50 % from rain-fed agriculture occur in many parts of Africa (Fig. 3.1) due to changing climate which seriously affect food security and exacerbate malnutrition (IPCC 2007a, b).

Climate variability will also impact heavily on irrigated agriculture as storage dams and river channels receive low rainfall quantity (Knox et al. 2010). The Intergovernmental Panel on Climate Change (IPCC) has warned that, climate

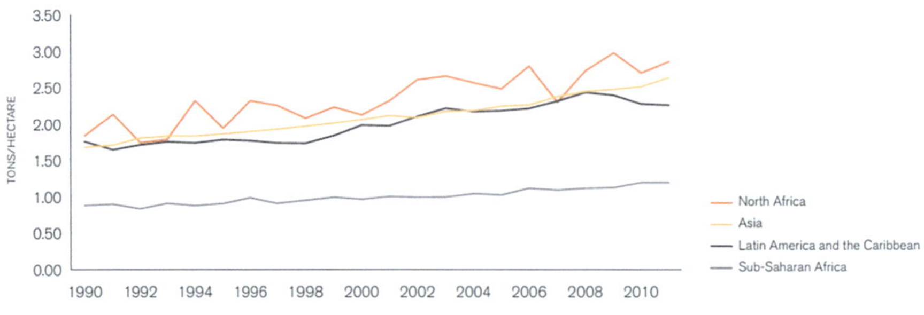

Fig. 3.1 Yields of cereal crops by region in major food-deficit countries (Source: ERS (2013), adapted from FAO)

change and variability will put more pressure on available water, access to water accessibility and demand which will contribute to expansion of arid and semi-arid lands in Africa by 5–8 % by 2080. Evidence of climate impact showed a decline in mean annual rainfall for the period 1960–1998 of about 4 %, 3 % and 2 % in the tropical rain-forest zones of West Africa, North Congo and South Congo, respectively. However, Fabusoro et al. (2014) in a recent study revealed that, there has been a steady increment of mean monthly rainfall by 65 mm per month per decade from 1982 to 2010 in the sub-humid parts of Nigeria. The study also showed a similar pattern of rainfall and temperature in the study area with temperature rising at about 0.4 °C per month per decade in southwest Nigeria.

Recent climatic data shows that 2010 and 2013 recorded the warmest years in Africa. In extreme cases, the hottest temperature of 47.3 °C was recorded on the March 4, 2013 in Vioolsdrif, South Africa. In a related development in West Africa, the warmest temperature (43 %) ever recorded in Navrongo, Ghana occurred on March 6 (AGRA 2014). Future projection indicates a likely decrease in mean annual rainfall in most areas of Northern Sahara region, Mediterranean Africa, and Southern Africa, while a possible increase is expect in East Africa (Christensen et al. 2007; IPCC 2007a, b). The IPCC forth assessment report depicts a warming experience greater than the global annual mean in the entire season across Africa and predicts a temperature increment of 3.3 °C by close of the twenty-first Century (Christensen et al. 2007).

A review by Hertel and Lobell (2014) highlighted that, the capacity of farmers, agri-businesses and economies to adapt to changes and variability in climate will determine their resilience to the negative impacts these changes and/or variability are likely to bring. The effect of climate variability such as low variation in rainfall distribution and increasing temperature provides a direct causal relation between agricultural production and food security. Studies indicate that such climatic variables has the potential to impact severely on yields of crop such as maize, wheat, rice and other food crops in semi-arid areas across the world (Lobell et al. 2009). Rainfall quantity and distribution is single out as the most important factor which does not only affect crop yield among smallholder farmers in SSA, but directly impact on farm sizes, crop enterprises, sowing dates, incidence of pests, diseases and weeds (Yengoh et al. 2010).

Different scientists, in an effort to model the impact of climate variability on crop production, have used a variety of models and scales of analysis depending on their interests and scenarios which resulted in heterogeneity of the projections (Boko et al. 2007; Cooper et al. 2008). The chapter examines the impact of climate variability on crop production system and analyzes the contribution of crop models in defining appropriate crop management strategies against the threat of high climate risk and uncertainty in Sub-Saharan Africa.

3.1.1 Cropping System of Sub-Saharan Africa

A cropping system consists of a cultivation of diverse plants, over time and land, for specific purposes including grain, fodder, fibre, oil or other raw material, income and/or ecosystem services (eg soil cover to fight soil erosion). Diverse crops cultivation characterizes SSA agriculture, however cereal crops and tubers account for more than 60 % of the population's total energy intake (Diao et al. 2012).

3.1.2 Sequential Cropping Systems

Sequential cropping entails cultivating two different crops on the same farmland in turns in the same year. This may involve one being grown during long rains and the other during short rains. In Mali, the introduction of sorghum cultivar for short season in sequential cropping with other short duration groundnut and cowpea cultivars produced substantial sorghum and legumes yields (Sedogo and Shetty 1991). A research by Sivakumar (1988) suggested that sequential cropping can potential improve soil productivity in Sahelian zone. Waha et al. (2013) noted that, maize forms the fundamental crop in all sequential cropping systems in Eastern Africa while maize – wheat systems are common in Southern Africa. In Western Africa, Ghana and Cameroon have the highest crop diversity in sequential systems of cropping with groundnut as the primary crop (Fig. 3.2).

The most common cropping systems in SSA include intercropping, sequential cropping and crop rotation. Intercropping is regarded as the ancient and widely practiced system of cropping in sub-Saharan Africa. Sequential cropping and crop rotation (Table 3.1) are also common indigenous management practices of agricultural production (Waha et al. 2013).

3.1.3 Intercropping System

Intercropping is one of the traditional systems in sub-Saharan Africa. According to Steiner (1984) it covers over three-quarters of cultivated areas in the semi-arid tropics. Norman (1974) notes that the main importance of intercropping to farmers includes; resource mobilization, soil conservation and maintenance, risk minimization, flexibility, profit and weed control. In Sudano-Sahelian zone, growing of many crops in association as intercrops or mixtures is the common system of cropping (Bationo et al. 2003). According to Bationo et al. (2003), a sorghum-based cropping system is common in the Sudanian zone with groundnuts, maize, pearl millet, and cowpea as the main components. In the Sahelian region, millet-based is the cropping system is common and the system has millet-cowpea and millet-groundnut as the cropping pattern. According to Swinton et al. (1984), over 85 % of millet

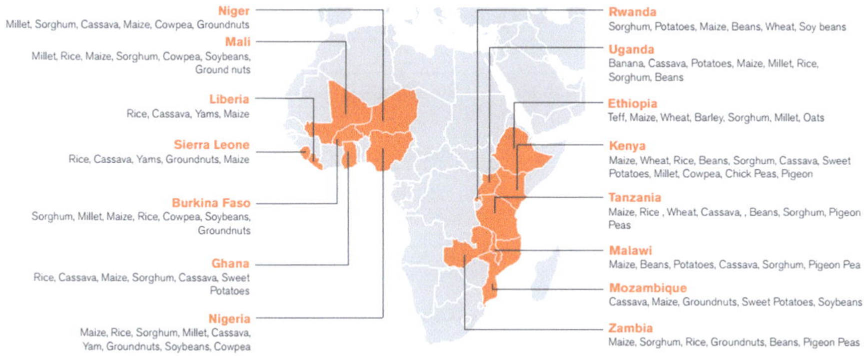

Fig. 3.2 Major crops in selected SSA countries (Source: AGRA 2014)

Table 3.1 Sequential cropping systems in some countries in sub-Saharan Africa

Country	Sequential cropping system
Burkina Faso	Maize – rice; Rice – rice
Cameroon	Wheat – maize; Maize – wheat; Maize – maize; Cassava – maize
Ethiopia	Cassava – cowpea
Ghana	Cassava – cowpea
Kenya	Wheat – maize; Rice – rice; Maize – maize; Cassava – maize Cassava – cowpea; Ground – cassava; Groundnut – groundnut
South Africa	Wheat – maize; Maize – wheat; Cassava – maize; Cassava – cowpea
Zimbabwe	Wheat – maize

Source: Waha et al. (2013); www.fao.org

cultivated area in Niger is intercropped and about 50 % of sorghum planted area in northern Nigeria is also intercropped (Norman 1974).

The common intercropping combinations include cereal/groundnut, cereal/cowpea and cereal/cereal like millet/sorghum/maize and millet/sorghum/cowpea. Intercropping system is an advantageous system since it allows the exploitation of the time differences between crops and reduces the potential of competition during growth periods (Baker 1979). For example, in the millet-cowpea system of intercropping, the planting of cowpea is always done in 3–4 weeks after millet (i.e. relay intercropping). This allows millet to effectively maximize nutrients and moisture use thereby significantly reducing the chances of crop yield since there is longer growth duration in legumes.

3.1.4 Crop Rotation

Crop rotation has a long history as one of the most productive agricultural practice in SSA countries. It is a practice of growing different crops on the same agricultural land in a regular recurring sequence season after season. Bationo et al. (1998)

Table 3.2 Examples of crop rotation in parts Africa

Country	Year 1	Year 2	Year 3
Swaziland	Maize – cowpea/lablab	Sorghum – lablab	Maize – cowpea
Cameroon	Sorghum – cowpea	Cotton	Sorghum – cowpea
Kenya	Maize – lablab	Maize – cowpea/bean	Maize – lablab
Tanzania	Maize – lablab	Sorghum – cowpea	Maize – lablab

Source: www.fao.org

observed increased cereal yields as a result of cereal-legume rotation in Sahelian region of West Africa. For example, there was an improvement in total dry matter production of about 3 tons/hectares for millet-cowpea rotation (Bagayoko et al. 1996; Bationo et al. 1998). Bationo and Mokwunye (1991) noted that in Niger, crop rotation system resulted in Effective Cation Exchange Capacity (ECEC), reduce soil PH and increased saturation of the base. According to the FAO, crop rotation is regarded as key strategy for conservation agriculture in Swaziland, Cameroon, Kenya and Tanzania. Table 3.2 above presents some common examples of crop rotation in these countries.

3.2 Quantification of Cimate Variability Impact on Cropping System

Overview of Responses to High Temperature, Drought, or [eCO_2] The impact of climate variables including temperature, rainfall (drought) and eCO_2 on cropping systems in Africa varies in different degrees and extents (Challinor et al. 2007; Sivakumar and Hatfield 1990). In Southern Mali, an annual minimum temperature increase of 0.5 °C per decade was recorded over the period 1965–2005 (Traore et al. 2013), which is higher than a forecast on a global scale of 0.3 °C per decade (Abrol and Ingram 1997). Predicted increase in eCO_2 concentration will yield positive results by increasing crop yield since rising atmospheric CO_2 levels act like a natural carbon fertilization thereby enhancing crop growth in general terms (Drake *et al* 1997). However, climate variability is expected to have a significant impact on agriculture especially in semi-arid region where crop growth, yields and production are very sensitive to the climate and other environmental factors. Increasing temperatures will hasten crop ontogenetic development and probably have some negative impacts on photosynthesis, thus obstructing the production of biomass (Kersebaum et al. 2009). Shortage of rainfall will require irrigation in today's rain-fed agricultural production systems or at least lead to prolong dry spells with decreasing yields. The combination of these climatic factors may also not only influence crop yield but also on crop quality, with the potential of affecting predominant smallholder farmer's income more severely in sub-Saharan Africa.

(a) ***High Temperature and Drought Interaction***

High temperature and drought often occur simultaneously. In the tropics, episodes of drought are always aggravated both by high temperature and high solar radiation and thus drought is always viewed as multi-dimensional stress (DaMatta 2003). In this paper, we view drought as an environmental stress factor that leads to deficit of water in plants which is triggered when low osmotic potential develops and turgor of the cell goes below its maximum value. Drought impact is not seen to occur suddenly but slowly develops and the longer it lasts the greater the intensity. The consequences of drought are severely pronounced at high temperature than low temperature. High temperature has the potential to rapidly slow down plant photosynthetic rate and development of leaf area, decrease shoot and grain quantity, and reduce water-use efficiency. In a study of the interaction of high temperature and drought on wheat during photosynthesis and grain-filling by Shah and Paulsen, (2003), they observed that the synergistic interactions showed a reduced productivity more by the combined stresses than by either stress alone, and that much of the effect is on photosynthetic processes.

(i) ***Impact Assessment on cropping system***

In Africa, heat shock (high temperature) and drought are the common stress factors that results into reduction of crop yield by 50 % more (Larkindale and Knight 2002; Macar and Ekmekci 2009; Tayyar 2010). According to Utrillas and Alegre (1997), the reduction on crop yield always depends on severity and duration of the stress. When plants are subjected to drought and high temperature, it results in changes in plants' osmotic potential and this leads to unavoidable stress for the functioning and structure of mitochondria and chloroplast. For example, in sorghum the induced drought impacts on chloroplast are; stroma distortion, reduction in amount of starch in chloroplast, increased swelling of chloroplast outer membrane and lipid droplet accumulation (Giles et al. 1976; Olmos et al. 2007; Vassileva et al. 2011). High temperature and drought lead to plant metabolism disturbance and this is due to increased acceleration of reaction kinetics, loosening of macromolecular bonds and increased fluidity of lipid layers from bio-membranes. Further, excessive temperature results in denaturation and cellular protein aggregation, ROS overproduction and normal transcription and translation inhibition (Larcher 1995; Krishna 2004). All these have implication on cropping system and food security at large. A research by Aranjuelo et al. (2007) also shows a great decrease in plant production as a result of elevated drought and temperature. For instance a combination of elevated drought and temperature negatively affects leaves, shoots and roots dry matter.

(ii) ***Impact on Water Use Efficiency***

High temperature and drought is likely to cause a reduction is water use efficiency, which is the ratio of crop yield to crop water requirement, since an already water stress situation will be exacerbated by an increase in evapotranspiration (Hertel and Lobell 2014) as a result of increase in temperature. There is increase

Table 3.3 Projected reduction in crop yields in Africa by sub-regions using GCM

Sub-region	Mean yield reduction (%)	Significant crop yield variation	Non-significant Crops yield variation
Southern Africa	11	Maize	Wheat; Sorghum; Sugarcane
Central Africa	15	Maize	Wheat
West Africa	13	Maize	Wheat; Sorghum; Cassava
Sahel	11	Maize; Millet	Sorghum

Source: Adapted from Knox et al. (2012)

demand for crop water requirements during the dry and windy conditions than in humid and calm climatic conditions (Brouwer and Heibloem 1986). The limiting factor to agricultural production in most parts of sub-Saharan Africa (particularly in the Sahelian part of West Africa and parts of southern Africa) is inadequate soil moisture which contributes to reduced crop productivity (Schlesinger et al. 1990). When full crop water needs are unmet, plants respond to water deficit through stomatal closure to reduce further loss of water and stress. Plant stomatal closure also contributes to reduction in other plant growth and phenological processes such as CO_2 uptake, photosynthesis and biomass production. The extent of water deficits in crop can thus seriously derail crop growth, crop development and yields (Kassam and Smith 2001). Also influencing crop water needs is the combined effect of other factors including humidity, wind speed and solar radiation.

(iii) ***Impact on Agronomic Efficiency: Yield and Yield Components***

The plant growth and yield processes rely to a large extent on temperature. Crop can potentially benefit or suffer from increase in temperature in terms of yield and crop water requirements. Studies have shown that increasing temperature contribute to the crop prematuration by hastening the rate of crop development. This leads to shortened crop life cycle and gain filling period resulting reduced crop yield and poor quality of grains (Adams et al. 1998).

High temperature and drought affect crop production in different ways as crops' response to these effects of climate variables differ (Hassan and Nhemachena 2008; Hartwell et al. 1997). High temperatures usually cause a reduction in crop yield due to the fact that, they normally occur in conjunction with drought (Fisher et al. 1997). Crops are more sensitive to water scarcity (drought) than to high temperature (Kang et al. 2009). A review by Knox et al. (2012) using General Circulation Models (GCM) to predict the impact of two main climate variables- temperature and rainfall showed a projected yield reduction of about 8 % in all crops across Africa. From their work, Knox et al. (2012)) indicated mean yield reductions of 17 %, 5 %, 15 % and 10 % across Africa in wheat, maize, sorghum and millet, respectively. An increase in maximum temperature (by 0.08 °C) and dry spell (drought) during the rainy season negatively affects cotton production in Southern Mali and this effect is the most important feature of climate variability in the area (Traore et al., 2013). Table 3.3 gives the details of the mean yield reductions by sub-regions in Africa as reported by Knox et al. (2012).

Knox et al. (2012) also observe some increment in crop yield in other regions of Africa including Eastern Africa (0.4 %) and North Africa (0.8 %) in both wheat and maize but this was considered not significant. Information on yield changes for rice, cassava and sugarcane were, however, found to be inconclusive, absent or contradictory (Knox et al. 2012). According to Lobell et al. (2011) temperature increase during the growing season is likely contribute to a decrease in maize yield by 3 % in Eastern and Southern Africa.

(b) **High Temperature and [eCO_2] Interaction**

Keeling and Whort (1991) noted that there is an increase in concentration of global atmospheric carbon dioxide by 30 %. Further increase in CO_2 concentration and other greenhouse gases could cause rise in global average atmospheric temperature by about 3–5 °C due to present day doubling of CO_2 (Grotch 1988; Adams et al. 1990). Therefore, combine interaction of CO_2 and high temperature leads to partial closure of stomata and increased resistance of stomata resulting in decreased transpiration per unit leaf area which increase temperature of the leaf (Jones et al. 1985; Idso et al. 1987).

(i) ***Impact on Water Use Efficiency***

Due to the fact that, an increase in CO_2 concentration results in a partial closure of stomata, doubling of CO_2 results in about 40 % decrease of water vapor through stomatal conductance. Also a decrease in stomatal conductance impacts plants by resulting into decreased leaves transpiration (Allen and Prasad 2004). Allen and Prasad (2004) further indicated that although there is a slight decrease in crop transpiration under elevated CO_2; rise in temperature will increase water use. An average daily range of temperature of 20–40 °C could lead to increased water use by about fourfold (Allen and Prasad 2004). Elevated CO_2 concentration causes the stomatal effect which reduces water losses through transpiration thereby increasing water use efficiency in both C_3 and C_4 plants whereas high temperature leads to increase in evapotranspiration and lower soil moisture (Hertel and Lobell 2014). However, according to Hertel and Lobell (2014), the combined effect of high temperature and elevated CO_2 levels on soil moisture and water stress is unknown.

(ii) ***Impact on Agronomic Efficiency: Yield and Yield Components***

High temperature results in shortened growing period due to the fact that, crops attain maturity faster with increase in temperature than usual in the year leading to potential yield decrease (Waha et al. 2013). Acoording to Hertel and Lobell (2014), the rate of crop development increases linearly with temperature especially in the range 0–30 °C. Increase in atmospheric CO_2 concentration, however, can cause an increase in the productivity of plants (especially C_3 plants) (Long et al. 2006).

(c) **Overview of responses to biotic stresses**

Increasing temperature and elevated CO_2 concentration in the atmosphere are likely to make crops more vulnerable to biotic stress such as weeds invasion and pests and diseases damage because, weeds for example are more responsive than crops to

Table 3.4 Contribution of climate variability to biotic stress in crop production

High temperature	Elevated CO_2 concentration
Invasive weeds often more climate tolerant; also more responsive to changes in temperature due to short juvenile period, long-distance dispersal	Invasive weeds more responsive to changes in elevated CO_2 concentrations due to short juvenile period, long-distance dispersal
Reduced frequency of frost will expand range of pests and diseases	Elevated CO_2 can make weed management more difficult

Source: Adapted from Hertel and Lobell (2014)

elevated CO_2 (Ziska et al. 2011 in Hertel and Lobell 2014). Ziska et al. (2011) stated in their work 'evasive species and climate change: an agronomic perspective' that, rust caused yield reductions in soybean and increased cost of production in Africa and other parts of the world including Asia, Australia and South America. They further noted that increasing climate variability will probably aggravate the spread of biotic stresses on crop production. Currently an epidemic of stem rust associated with a strain of wheat is spreading in Africa, Asia and the Middle East. According to Ziska et al. (2011), increases in precipitation and wetter environments are conducive to stem rust establishment and therefore the incidence of drought could curb this spread but would also take a toll on crop yield. Ward et al. (1999) also highlighted the issue of Grey leaf spot (*Cercospora zeae-maydis*) in corn production becoming an epidemic in parts of Africa. Table 3.4 shows the response of biotic stress (pests and disease damage) to high temperature and elevated CO_2 concentration.

3.3 Modeling and Simulation

3.3.1 Cropping Systems Models to Understand Climatic Variability

Crop production and food security can be affected by the impact of climate variability. Using scientific approach to understand the extent and rate of climate variability impact on crop productivity has gained recognition significantly. Crop models have been used to analyze such impacts by predicting crop yields, conditions of growth, suitable crop types, etc. under current and future climatic conditions (Donatelli et al. 2002; Tripathy et al. 2011).

Crop models have been used as principal tools for the assessment of climate change impact on crop productivity, simulating the robustness of context-specific adaptation strategies, simulating the effects of key drivers on adaptation strategies and describing how cropping systems respond to key drivers (Webber et al. 2014; UNFCCC 2012).

Webber et al. (2014) note that, when crop models are combined with some information about water availability they could be useful for spatial targeting of irrigation possibilities. Due to the fact that large scale crop models are calibrated in industrialized countries (and with parameters from that area, for that matter), it is important to calibrate crop growth parameters to local conditions in order to produce accurate and reliable location-specific results (Folberth et al. 2012). According to Webber et al. (2014), to ensure that predictions and simulated results of crop models are reliable and context-specific, crop models should be integrated with farmers' knowledge and knowledge from outside the field of modelers and agronomists. They further noted that, farmers' are more likely to adopt resultant adaptation options if they are involve from the outset and their input incorporated. Crop modeling studies conducted in most part of SSA spelt out increase in mean temperature, elevated CO_2 concentration and increased frequency of drought and floods to be the key drivers of future impacts (Webber et al. 2014).

In current conditions of high climate variability in several parts of Sub-Saharan Africa, crop simulation models offer important contribution by predicting possible scenarios in the future so that effective agricultural management options can be exploited (van Ittersum et al. 2003; Hoogenboom 2000). As a decision support tool, models can be used to assess optimum management practices, either strategic or tactic, such as cultivar selection, planting dates, fertilization and pesticides usage for making seasonal or within-season decisions (Boote et al. 1996).

There have been a lot of studies in sub-Saharan Africa (SSA) using crop models with different characteristics to determine the impact of climate variability and change on cropping systems which resulted in a variety of outcomes (Schlenker and Lobell 2010; Müller 2009 in Webber et al. 2014). Webber et al. (2014) noted that, what one should look out for in the choice of a crop model for effective predictions and simulation of adaptation options for a particular location are, (i) it should, at least, be responsive to the key drivers (climatic variables) expected to impact on the cropping systems of the region (ii) it should be able to model the principal crops, cropping systems and management strategies. Table 3.5 shows the characteristics of different cropping systems models suitable for adaptation studies in SSA.

Simulation models have not fully been applied in all regions. For example in Kenya it has not been applied in cropping systems on large scale as a determination, prediction and forecasting tool of cropping systems properties/behavior like crop productivity and crop growth. According to Staggenborg and Vanderlip (2005), the use of crop simulation models helps in efficient resource-use by scientists through giving an insight in responses on potential plants in cropping system alteration. Alva et al. (2010) indicated that crop simulation models assist as decision tools in improving the efficiency of cropping system input management and environmental negative impacts minimization. Crop simulation models are developed to provide alternative options or solution in the following areas (Murthy 2004) such as;

(a) **Policy management**

This is one area where crop simulation models have been very useful. Thornton et al. (1997) noted that in Burkina Faso there is the use of crop simulation models using ground-based and satellite data to estimate production of millet for early warning of famine. This gives policy makers time to act on effect of food shortages on vulnerable population in rural and urban. They further noted that crop models can assist policy makers to understand climate change effects such as effects of

Table 3.5 Characteristics of cropping systems models suitable for adaptation studies in SSA

Model	Where has it been tested in West Africa?	Which crops relevant for West Africa?	Which processes are responsive to temperature?	Sensitivity to high emperature at key growth stages?	Sensitivity to atmospheric CO2?	Sensitivity to water stress?	Which management options?	Other comments
DSSAT – CERES (Ritchie and S. Otter 1985; Jones et al. 1986)	Sudan-Savanna in Ghana (Maccarthy et al. 2010); Sudan Savanna in Burkina Faso (Tojo Soler et al. 2011); sub-humid Ghana Maccarthy et al. 2012); humid Togo (Dzotsi et al. 2003)	Maize (Maccarthy et al. 2012; Tojo, Soler et al. 2011; Dzotsi et al. 2003); rice (Xiong et al. 2009); sorghum (Maccarthy et al. 2010; Tojo, Soler et al. 2011); millet (Sharma et al. 2010; Thornton et al. 1997)	Phenology, photosynthesis (RUE), kernel # per plant (via impact on the duration of the grain setting phase), kernel growth rate, evapotranspiration	No – kernel number appears to be influenced by temperature other than as it affects carbohydrate supply check if affects translocation	Yes (tested with FACE experiments for wheat under N and water stress using ZWQM2 for soil water processes – Ko et al. 2010)	Flooding and drought	Nitrogen fertilization (MacCarthy et al. 2010, 2012; Tojo Soler et al. 2011); crop rotations (Tojo Soler et al. Tojo, Soler et al. 2011); residue management (SOM with Century in Tojo Soler et al. 2011); intercropping (maize/ wheat – Knörzer et al. 2011); tillage; irrigation	

DSSAAT – CROPGRO (Jones et al. 2003)	Sudan-Savanna of Burkina Faso (Tojo Soler et al. 2011); Guinean Savanna zone of Ghana (Naab et al. 2004)	Cowpea (Bastos et al. 2002); cotton (Tojo Soler et al. 2011); groundnut (Tojo Soler et al. 2011; Naab et al. 2004); tomato (Ventrella et al. 2012)	Phenology, leaf level photosynthesis, respiration, seed lipid concentration, nitrogen fixation, specific leaf area, leaf expansion rate, internode elongation, evapotranspiration	Yes, in PODS. for, relative # of pods set in a day decreases past threshold temperature to 0 at a second threshold	Yes (tested at leaf scale for soybean – Alagarswamy et al. 2006)	Flooding and drought	Nitrogen fertilization (Tojo Soler et al. 2011); crop rotations (Tojo Soler et al. 2011); residue management (SOM with Century in Tojo Soler et al. 2011); irrigation	

(continued)

Table 3.5 (continued)

Model	Where has it been tested in West Africa?	Which crops relevant for West Africa?	Which processes are responsive to temperature?	Sensitivity to high emperature at key growth stages?	Sensitivity to atmospheric CO2?	Sensitivity to water stress?	Which management options?	Other comments
APSIM (Keating et al. 2003)	Sudan-Savanna in Ghana (Maccarthy et al. 2009); subhumid Ghana (Fosu-Mensah et al. 2012); Sahel in Niger (Akponikpè et al. 2010)	Maize (Fosu-Mensah et al. 2012); sorghum (Maccarthy et al. 2009); millet (Akponikpè et al. 2010); chickpea (Robertson et al. 2001b); cowpea (Adiku et al. 1993); cotton; pigeon pea (Robertson et al. 2001a); mucuna; groundnut (Robertson et al. 2001a)	Phenology, photosynthesis (RUE), leaf area, N content, rate of senescence, speeds grain filling rate (but duration decreases with temperature – therefore final kernel weight reduced), grain N-content, rate of rooting depth advance, evapotranspiration	No – cotton, legumes Yes – sorghum, millet and maize – potential kernel number is reduced past a threshold value during flowering	Yes (tested with FACE experiments for wheat under N and water stress – Asseng et al. 2004)	Drought (Song et al. 2010)	Nitrogen fertilization (Maccarthy et al. 2009); P-fertilization; residue management (SOM and erosion); intercropping; tillage; irrigation	Phenology responsive to N and P deficiency; accounts for acidity

EPIC (Williams et al. 1989)	Sub-humid Guinean zone Bénin (Gaiser et al. 2010a; Srivastava et al. 2012; Worou 2012); semi-arid Sudan-Savanna in Bénin (Gaiser et al. 2010a; Srivastava et al. 2012); Nigeria (National scale – Adejuwon 2006)	Maize (Gaiser et al. 2010a; Adejuwon 2006); yam (Srivastava et al. 2012); rice (upland and lowland – Worou 2012); millet (Adejuwon 2006); sorghum (Adejuwon 2006); cassava (Adejuwon 2006)	Phenology, reduces biomass accumulation (RUE), harvest index, evapotranspiration	No	Yes	Drought (Gaiser et al. 2010a)	Nitrogen fertilization; P fertilization; intercrops; crop rotations; residue management (with CENTURY); fallow availability (Gaiser et al. 2010b; Srivastava et al. 2012); irrigation	Considers acidity, but poor performance on highly acidic tropical soils (aluminum saturation >35 %) (Gaiser et al. 2010a); considers erosion; does not consider iron toxicity for rice (Worou 2012)

(continued)

Table 3.5 (continued)

Model	Where has it been tested in West Africa?	Which crops relevant for West Africa?	Which processes are responsive to temperature?	Sensitivity to high emperature at key growth stages?	Sensitivity to atmospheric CO2?	Sensitivity to water stress?	Which management options?	Other comments
CROPSYST (Stöckle et al. 2003)	Burkina Faso (National scale – Badini et al. 1997); Cameroon (Tingem et al. 2009)	Bambara nut (Tingem and Rivington 2009); maize (Tingem et al. 2009); sorghum (Tingem et al. 2009); millet (Badini et al. 1997); cotton (Sommer et al. 2008)	Phenology, biomass accumulation (RUE), rate of senescence, evapotranspiration	No	Yes	Drought	Flooded rice management (Confalonieri and Bocchi 2006); crop rotations; residue management (effect on SOM); fallow availability; tillage (Donatelli et al. 1997); irrigation	
STICS (Brisson et al. 2003)	Mali (Folliard et al. 2004)	Sorghum (Folliard et al. 2004); maize (Brisson et al. 1998); legumes (Corre-Hellou et al. 2009)	Phenology, biomass accumulation (RUE), rate of senescence, grain filling, energy balance to determine canopy temperature, evapotranspiration	No	Yes	Drought	Intercropping (Brisson et al. 2004); irrigation (Mailhol et al. 2001); N fertilization (Mailhol et al. 2001); residue management (Scopel et al. 2004)	

SIMPLACE/APES (Adam et al. 2012; Rötter et al. 2012)		Maize (Adam et al. 2012; Belhouchette et al. 2009; Therond et al. 2011)	Phenology, leaf area growth rate, biomass accumulation (RUE), evapotranspiration, root penetration and growth	No	Yes	Drought changes rooting patterns, reduces LUE and LAI and increases leaf death rate	N(PK) fertilization (Van Ittersum et al. 2008; Belhouchette et al. 2011; Adam et al. 2012), crop rotation; residue management; irrigation (Belhouchette et al. 2011)	Modular framework which can combine different model components into a site-specific, problem and data oriented model solution
SARRA-H (Dingkuhn et al. 2003; Berg et al. 2011)	Sahel and Sudan savanna of Senegal, Mali, Burkina Faso, Niger and Chad (Berg et al. 2011; Dingkuhn et al. 2003)	Millet (Berg et al. 2011); sorghum and maize	Biomass accumulation, phenology, maintenance respiration, senescence, evapotranspiration	No	Yes	Drought	No	

Source: Webber et al. (2014)

elevated CO_2 and temperature and rainfall variability on development, growth and yield of crops.

(b) **Understanding of research**

According to Penning de Vries (1977), simulation models leads to our understanding of real systems and thus assist to bridge areas and knowledge levels. Penning de Vries (1977) further noted that due to the diverse nature of crop simulation modelling there is increased efficacy and improved direction of research through direct feedback. And as a result Wit and Goudriaan (1978) came up with BAsic CROp growth Simulator (BACROS) to be used as model reference for other models development.

(c) **Agronomic management and farm decision-making**

Crop simulation models give chance for evaluation of available one or more options with regard to other decision options of agronomic management such as-determining optimum planting date; weather risk evaluation and determining best choice of cultivars. Crop models also assist to predict performance of crops in areas where crops has never been grown or grown but not under optimal required conditions. Crop models also adds value in developing countries with regards to agricultural planning and regional development (Van Keulen and Wolf 1986).

3.3.2 Adaptation/Mitigation to Climate Variability

As climate variability is unprecedented it is incumbent on everybody to adapt their lifestyle and social systems to the variability in order to benefit from concomitant positive impacts (e.g. increased rainfall, elevated CO_2), alleviate negative impacts (e.g. drought, high temperature) that are likely to occur and cope with them thereby increasing the resilience of the ecosystem (IPCC 2001). Adaptation strategies range from behavioral changes (e.g. avoiding cultivation along river banks) through institutional arrangements (e.g. buffer zone policy) (Kurukulasuriya and Mendelsohn 2006) to building of physical structures (e.g. water-harvesting structures) or bringing ecological changes (e.g. planting of trees for catchment protection). Although matters of adaption to climate variability are context-specific and thus it is mostly handled locally, it is sometimes necessary for an integrated and coordinated action at various levels (Paavola and Adger 2005). Adger et al. (2005) noted that climate change variability takes place at different scales and thus successful adaptation only depends on actions applied at various levels (Paavola and Adger 2005). For example, a national level strategy may include; development of policy of climate change directed to vulnerable sectors with an aim of reducing poverty and sustaining food security. According to Downing et al. (1997), warrant of adaptation may be in a situation where climatic hazards and mean climate changes are frequent and the consequences to the vulnerable populations are significant. They further noted that such situations call for adaptive strategies which include;

Table 3.6 Adaptation strategies in response to climatic and non-climatic drivers of climate change

Strategies for climatic drivers	Strategies for non-climatic drivers
Formation of farmers associations which enabled a range of other changes and experimentation (risk taking)	Formation of farmers associations
Diversified crop production	Diversified crop production
Investments in labor and irrigation	Investments in labor and irrigation
Soil conservation practices (contour tillage and mulching)	Soil conservation practices (contour tillage and mulching)
Shifting production between cropping and livestock keeping	Shifting production between cropping and livestock keeping
Reapportioning areas between crops and livestock	Reapportioning area between crops and livestock
Collective actions such as livestock holdings and commercialization	Collective actions such as livestock holdings and
Use of resilient varieties	Commercialization (esp. Vegetables)
Water harvesting	Increased commercial production
Using shorter season varieties	Water harvesting

Anticipatory Adaptation This entails variety of strategies for agricultural improvement such as irrigation schemes planning for areas where water availability and supply is uncertain. Also long life projects like construction of reservoirs, in areas where marginal adaptation cost is less to bring protection against extreme weather events and reduce irreversible impacts.

Research and Education In a situation of limited adaptation to climate variability, research and education are recommended to allow development of new solutions to accommodate the change of climate.

Capacity Building Development Assistance This entails efficient use of information on climate and resources to improve production, monitor water resources and adapt to risk of climate and reducing vulnerability through development, sustainability, mitigation of drought and preparedness and integration of regional economies. Hassan and Nhemachena (2008) conducted a study on the determinants of African farmers' strategies for adapting to climate change and highlighted some key strategies to insure farmers against climate variability. These strategies include diversifying crops with drought tolerant and/or stress resistant crops; improve efficient use of available water; and promoting crop variety of same plot or different plots to reduce chances of complete failure. Webber et al. (2014) also assessment of crop models for climate change adaptation decisions in sub-Saharan Africa catalogued current adaptation strategies by farmers in the region as measures to climatic and non-climatic drivers as shown in the Table 3.6.

Institutional and Regulatory Adaptation Institutional and regulatory adaptation is applied in a situation where a developmental project (e.g. coastal development) leads to vulnerability/unable to guard the vulnerable. For instance irrigation water resources may be unreliable and reduce potential of agriculture due to soil salinization. Example of such a situation is the irrigation project of South Chad in 1970s (Kolawole 1987).

3.3.3 Introduction of Legume in the Cropping System

The intercropping of legumes with staples is increasing gaining popularity in sub-Saharan Africa (SSA) due to their ability to conserve soil moisture by preventing direct sunshine on soil surface, protect the soil against erosion and fix nitrogen resulting in soil fertility improvement. In Malawi, a research into adapting cropping systems to climate variability by intercropping maize with legumes was conducted. In a matter of 10 years after the research, this cropping system gained widespread adoption by 70 % of farmers in the study area due to the resultant yield increase, soil quality improvement, protection against crop losses as a result of drought and unreliable rainfall patterns (IDRC 2001). Woomer et al. (2004) conducted an on-farm experiment in West Kenya where they sought to modify the conventional maize-legume intercropping into a new system called the MBILI maize-legume intercropping where every other maize row is staggered by 25 cm and legumes planted in the resultant wider row and also maintaining a fix maize population of 44,444 ha-1 and legumes population of 88,888 ha-1. When MBILI was compared to conventional intercropping it was realized that MBILI increased in Land Equivalent Ratio (LER) more than conventional intercropping. LER is a measure of the efficiency of an intercrop. When LER is unity (=1), there is no additional production advantage of mixed culture; when LER is less than unity, there is disadvantage; and when LER is more than unity, there is advantage (de Wit and van den Bergh 1965; Willey and Osiru 1972). Thus the adoption of legume intercrop has the potential of being a key adaptation technology to climate variability for subsistence farming in SSA.

- **Alteration of the agricultural calendar**

Rainfed agriculture which is the main source of livelihood in most part of Africa (especially sub-Saharan Africa) is predicted to be much affected by current and future climate variability and change (Müller et al. 2011). For most farmers in Africa the only evidence of climate variability/change are the changes occurring in rainfall pattern and increase in temperature (Hassan and Nhemachena 2008) and therefore, they are forced to alter their agricultural calendar (which primarily depends on the sowing date) to suit the prevalent rainfall pattern (Yegbemey et al. 2013). Several other studies (Nhemachena and Hassan 2007; Gnanglè et al. 2012) indicate that famers are changing the sowing dates. A study conducted by Yegbemey et al. (2014) in Northern Benin on managing the agricultural calendar as coping mechanism to climate variability using maize farming as a case study revealed that, 84 % of respondents have adjusted their agricultural calendar to jibe with the rainy season. This change was mainly in terms of changes in dates of land preparation and sowing since all the remaining farming activities including weeding (or herbicide application), fertilizing and harvesting depend on the sowing date (Yegbemey et al.

2014). According to Van Duivenbooden et al. (2000) in Waha et al. (2013), adjustment of the sowing dates to match the actual commencement of the rainy season guarantees optimal growing conditions (i.e. enough soil moisture) and eschew crops from the risk of drought (dry spell) at crucial crop development stages resulting thereby enhancing yield. The change in sowing dates, however, lengthens the cropping season since after land preparation sowing which relies on the onset of the rainy season might take a while longer (Yegbemey et al. 2014). In a related work, Thornton et al. (2006) found from their work 'Mapping climate vulnerability and poverty in Africa' that, highland of sub-Saharan Africa will experience longer cropping season due to rainfall variability but on the contrary, a larger part of the region is likely to experience shortening of the growing season.

In addition to changing the sowing dates, farmers in Northern Benin adopted double sowing (i.e. a first sowing at the onset of a major rainfall and a second sowing in case of the occurrence of a dry spell after the rainfall) to safeguard crop production (Yegbemey et al. 2014). These adaptation strategies are adopted in Tanzania, South Africa and semi-arid West Africa (O'Brien et al. 2000; Benhin 2006).

- **Adaptation through the choice of cropping system**

Shifting agro-ecological zones in sub-Saharan Africa due to variability in climate has necessitated the adoption of cropping systems which are better adapted to climate variability and, thus, farmers' decision on an appropriate cropping system and crop variety might be a significant adaption strategy (Waha et al. 2013). Single cropping in a year is likely to suffer more from unreliable rainfall patterns than multiple cropping systems since the later eschews the risk of complete crop failure and stabilizes crop production (Francis 1986a). In the event of low yields in the first season, the cropping that follows (i.e. the second cropping) is likely to benefit from some soil nutrients like nitrogen (in case of leguminous cropping in the first season) and phosphorus from deep-rooted crops (Sisworo et al. 1990 and Francis 1986a in Waha et al. 2013). In sub-Saharan Africa, cereals which include maize, sorghum, millet and wheat are normally rotated with legumes in multiple cropping systems (Van Duivenbooden et al. 2000). Francis (1986b) observed in humid East and West Africa that, cassava and maize-based mixed cropping systems are prevalent while millet-based mixed cropping is notable common in dry areas of East and West Africa.

- **Rainfall harvesting as an adaptation strategy**

In consequence of the observed unreliable rainfall patterns and predicted decline of future precipitation, rainwater harvesting for supplementary irrigation is gaining roots in most parts of Africa as coping mechanism against recurrent drought (Christensen *et al* 2007). The storage of rainwater comes is different forms including farm ponds, small reservoirs, dugouts, tanks, water pans etc. (Ngigi 2009). Farm ponds owned by households are widely used in Kenya and have resulted in remarkable improvement in crop production, diversification and enhancement of farmers' income (Blank et al. 2007 and Malesu et al. 2006). Ngigi (2009) reports on a high adoption rate of rainwater harvesting with farm pond lined with ultra-violet-resistant plastic (geo-membrane) in Ethiopia, where over 7600 farmers adopted over a period of 5 years. Zai pits are also use for water conservation in most part of Africa including the Sahel region, West Africa (e.g. Burkina Faso), East Africa (e.g. Kenya) and Southern Africa (Ngigi 2003). Furthermore, as indicated above, rainwater harvesting is widely practiced as a key water management adaptation strategy in response to both climatic and non-climatic drivers (Webber et al. 2014)

- **Organic agriculture as adaptation strategy**

Organic agriculture is a management system of production which enhances and promotes agro-ecosystem health which includes biodiversity, activity of soil biological and biological cycles. Through organic agriculture there is no exploitation of nutrients and hence organic matter content of the soil increases. Also there is increased capturing and storage of water in soils under organic agriculture (Niggli et al. 2008). Therefore according to IPCC (2007a, b), production systems under organic agriculture is less subjected to adverse weather condition like water logging, flooding and drought.

Furthermore, there is increased adaptation by farmers to climate variability under organic agriculture. First, the highly diverse systems of farming under organic agriculture increase the diversity of sources of income and the flexibility to adapt to effects of adverse change of climate and variability. As a result there is greater ecological and economic stability through optimized balance of ecology and risk-spreading. Secondly, organic agriculture is a farming strategy with low risk and thus reduces the cost of inputs to farmers. This therefore, reduces risks due to adverse weather events or change of climate and variability in case of partial or total failure of crops (El-Hage Scialabba and Hattam 2002).

3.4 Conclusion

Future change in climate is expected to have profound impacts on agricultural production in the semi-arid region, particularly the combined impact of high rainfall variation culminating into increased probability of droughts and reduced crop-water availability, and elevated temperatures. The expectation is that climate change impact will modify the rate of evaporation and soil moisture storage. We set out to do a review on modeling the impact of climate variability on crops in Africa to ascertain current knowledge on cropping systems in the region and discover possible areas which need further research thereby contributing knowledge aimed at optimizing crop production to improve food security in sub-Saharan Africa. It was clear that, the evidence of climate variability and predicted positive and negative (mostly) impacts on crop production in Africa are well established. Evidently, changes in global climate phenomenon and the resultant changes in climate conditions will drastically affect crop growth. Being the region to be most affected by current and future climate change and variability, Africa's preparedness in devising adaptation and mitigation strategies on the premises of evidence-based information is highly essential. For this reason, crop models and climate models should be parameterized with local data in conjunction with farmers' input to develop feasible, context-specific adaptation and mitigation strategies against climate change and variability. Several adaptation options are available to farmers in this region, but it was found that the widely adopted strategies (for instance, changing of agricultural calendar) are largely dependent on rainfall pattern. Although a lot of research has been done to ascertain the impacts of the most prominent agricultural-related climate variables (such as temperature, precipitation and CO_2 concentration) on crop production in Africa (especially sub-Saharan Africa), there exist a gap with regards to the combined effects of these variables (especially temperature-CO2 interaction) on crops in the region. High temperature and low rainfall (drought) are the most important climate variables affecting crop production in Africa (especially sub-Saharan Africa).

References

Abrol, Y.P., and K.T. Ingram. 1997. Chapitre 6: Les effets de la hausse des températures diurnes et Nocturnes sur la croissance et les rendements de certaines plantes cultivées. In: Changements du climat et production agricole, Agriculture mondiale: Horizon 2010, ed. FAO, Polytechnica, p. 472. Rome: FAO/POLYTECHNICA.

Adam, M., H. Belhouchette, M. Corbeels, F. Ewert, A. Perrin, E. Casellas, F. Celette, and J. Wery. 2012. Protocol to support model selection and evaluation in a modular crop modeling framework: An application for simulating crop response to nitrogen supply. *Computers and Electronics in Agriculture* 86: 43–54.

Adams, R.M., B.H. Hurd, S. Lenhart, and N. Leary. 1998. Effects of global climate change on agriculture: An interpretative review. *Climate Research* 11: 19–30.

Adams, R.M., C. Rosenzweig, R.M. Peart, J.T. Ritchie, B.A. McCart, J.D. Glyer, R.B. Curry, J.W. Jones, K.J. Boote, and L.H. Allen Jr. 1990. Global climate change and U.S. agriculture. *Nature* 345: 219–224.

Adejuwon, J.A. 2006. Food crop production in Nigeria. II. Potential effects of climate change. *Climate Research* 32: 229–245.

Adger, W.N., N.W. Arnell, and E.L. Tompkins. 2005. Successful adaptation to climate change across scales. *Global Environmental Change* 15: 77–86.

Adiku, S.K., P.S. Carberry, C.W. Rose, R.L. Mccown, R. Braddock. 1993. *Assessing the performance of maize (Zea mays/cowpea (Vigna unguiculata)) intercrop under variable soil and climate conditions in the tropics*. In: Proceedings of the 7th Australian society of agronomy conference, September, p. 382. South Australia: Adelaide.

Akponikpè, P.B.I., B. Gérard, K. Michels, and C. Bielders. 2010. Use of the APSIM model in long term simulation to support decision making regarding nitrogen management for pearl millet in the Sahel. *European Journal of Agronomy* 32: 144–154.

Alagarswamy, G., K.J. Boote, L.H. Allen, and J.W. Jones. 2006. Evaluating the CROPGRO – Soybean model ability to simulate photosynthesis response to carbon dioxide levels. *Agronomy Journal* 98: 34–42.

Allen, L.H., and P.V. Prasad. 2004. Crop responses to elevated carbon dioxide. In *Encyclopedia of plant and crop science*, 346. New York: Marcel Dekker.

Alliance for a Green Revolution in Africa (AGRA) Database. 2014. https://agra.basecamphq.com/projects/12313519-datasharing-portal/files/cat/127115925

Alva, A.K., J. Marcos, C. Stockle, V.R. Reddy, and D. Timlin. 2010. A crop simulation model for predicting yield and fate of nitrogen in irrigated potato rotation cropping system. *Journal of Crop Improvement* 24: 142–152.
Aranjuelo, I., J.J. Irigoyen, and M. Sánchez-Díaz. 2007. Effect of elevated temperature and water availability on CO2 exchange and nitrogen fixation of nodulated alfalfa plants. *Environmental and Experimental Botany* 59(2): 99–108.
Asseng, S., P.D. Jamieson, B. Kimball, P. PINTER, K. Sayre, J.W. Bowden, and A.M. Howden. 2004. Simulated wheat growth affected by rising temperature, increased water deficit and elevated atmospheric CO2. *Field Crops Research* 85: 85–102.
Badini, O., C.O. Stöckle, and E.H. Franz. 1997. Application of crop simulation modeling and GIS to agroclimatic assessment in Burkina Faso. *Agriculture Ecosystem and Environment* 64: 233–244.
Bagayoko, M., S.C. Mason, S. Traore, and K.M. Eskridge. 1996. Pearl millet/cowpea cropping systems yield and soil nutrient levels. *African Crop Science Journal* 4: 453–462.
Baker, E.F.I. 1979. Mixed cropping in northern Nigeria. III, mixtures of cereals. *Experimental Agriculture* 15: 41–48.
Bastos, E.A., M.V. Folegatti, R. Faria, Aderson Soares De Andrade Jr., and M.J. Cardoso. 2002. Simulation of growth and development of irrigated cowpea in Piauí State by CROPGRO model. *Pesquisa Agropecuária Brasileira* 37: 1381–1387.
Bationo, A., F. Lompo, and S. Koala. 1998. Research on nutrient flows and balances in West Africa: State-of-the-art. *Agriculture Ecosystem and Environment* 71(19): 35.
Bationo, A., and A.U. Mokwunye. 1991. Alleviating soil fertility constraints to increased crop production in West Africa. The experience in the Sahel. *Fertilizer Research* 29: 95–115.
Bationo, A., Z. Traore, J. Kimetu, M. Bagayoko, J. Kihara, V. Bado and S. Koala. 2003. *Cropping systems in the Sudano-sahelian zone: Implications on soil fertility management*, 14. Nairobi: The Tropical Soil Biology and Fertility Institute of CIAT
Belhouchette, H., M. Adam, E. Casellas, F. Celette, M. Corbeels, and J. Wery. 2009. Performances of two crop models in various conditions: The importance of underlying assumptions. In *AgSAP conference 2009*, ed. M.K. Van Ittersum, J. Wolf, and H.H. Van Laar, 196–197. Egmondaan Zee: Wageningen University and Research Centre.
Belhouchette, H., K. Louhichi, O. Therond, I. Mouratiadou, J. Wery, M.V. Ittersum, and G. Flichman. 2011. Assessing the impact of the nitrate directive on farming systems using a bio economic modelling chain. *Agriculture Systems* 104: 135–145.
Benhin, J.K.A. 2006. *Climate change and South African agriculture: Impacts and adaptation options*. Pretoria: Centre for Environmental Economics and Policy in Africa.
Berg, A., B. Sulta, and N. Noblet-Ducoudré. 2011. Including tropical croplands in a terrestrial biosphere model: Application to West Africa. *Climate Change* 104: 755–782.
Blank, H.G., S.N. Ngigi and I.V. Sijali. 2007. Sustainable agricultural water management systems in Kenya. Paper to be presented at International Commission on Irrigation and Drainage (ICID) conference in Sacramento, California, USA. http://www.icid2007.org/07prog.PDF.
Boko, M., I. Niang, A. Nyong, C. Vogel, A. Githeko, M. Medany, B. Osman-Elasha, R. Tabo, and P. Yanda. 2007. Africa. In *Climate change 2007: Impacts, adaptation and vulnerability. Contribution of working group II to the fourth assessment report of the intergovernmental panel on climate change*, ed. M.L. Parry et al., 433–467. Cambridge UK: Cambridge University Press.
Boote, K.J., J.W. Jones, and N.B. Pickering. 1996. Potential uses and limitations of crop models. *Agronomy Journal* 88: 704–716.
Brisson, N., F. Bussière, H. Ozier-Lafontaine, R. Tournebize, and H. Sinoquet. 2004. Adaptation of the crop model STICS to intercropping. Theoretical basis and parameterisation. *Agronomie* 24: 409–421.
Brisson, N., C. Gary, E. Justes, R. Roche, and B. Mary. 2003. An overview of the crop model STICS. *European Journal of Agronomy* 18: 309–332.
Brisson, N., B. Mary, D. Ripoche, M.H. Jeuffroy, F. Ruget, B. Nicoullaud, P. Gate, F. Devienne-Barret, R. Antonioletti, C. Durr, G. Richard, N. Beaudoin, S. Recous, X. Tayot, D. Plenet,

P. Cellier, J.M. Machet, J.M. Meynard, and R. Delécolle. 1998. STICS: A generic model for the simulation of crops and their water and nitrogen balances I. Theory and parameterization applied to wheat and corn. *Agronomie* 18: 311–346.
Brouwer, C., and M. Heibloem. 1986. Irrigation water management: irrigation water needs. Training manual 3.
Challinor, A., T. Wheeler, C. Garforth, P. Craufurd, and A. Kassam. 2007. Assessing the vulnerability of food crop systems in Africa to climate change. *Climatic Change* 83: 381–399.
Christensen, J.H., B. Hewitson, A. Busuioc, A. Chen, X. Gao, R. Held, and K. Dethloff. 2007. Regional climate projections. In *Climate change, 2007: The physical science basis. Contribution of working group I to the fourth assessment report of the intergovernmental panel on climate change*, 847–940. Cambridge: Cambridge University Press.
Confalonieri, R., and S. Bocchi. 2006. Evaluation of CropSyst for simulating the yield of flooded rice in Northern Italy. *European Journal of Agronomy* 23: 315–326.
Cooper, P.J.M., J. Dimes, K.P.C. Rao, B. Shapiro, B. Shiferaw, and S. Twomlow. 2008. Coping better with current climatic variability in the rain-fed farming systems of sub-Saharan Africa: An essential first step in adapting to future climate change? *Agriculture Ecosystem and Environment* 126: 24–35.
Corre-Hellou, G., M. Faure, M. Launay, N. Brisson, and Y. Crozat. 2009. Adaptation of the STICS intercrop model to simulate crop growth and N accumulation in pea–barley intercrops. *Field Crops Research* 113: 72–81.
Damatta, F.M. 2003. Drought as a multidimensional stress affecting photosynthesis in tropical tree crops. In *Advances in plant physiology*, vol. 5, ed. A. Hemantaranjan, 227–265. Jodhpur: Scientific Publishers.
De Wit, C.T., and J.P. Van Den Bergh. 1965. Competition among herbage plants. Neth. *Agriculture Science* 13: 212–221.
De Wit, C.T., and J. Goudriaan. 1978. *Simulation of assimilation, respiration and transpiration of crops*, Simulation monograph. Wageningen: PUDOC.
Department for International Development (DFID). 2005. DFID's Draft Strategy for Research on Sustainable Agriculture (SRSA) 2006–2016. 15 pp. Online at http://www.dfid.gov.uk/research/srsa-consultaton.pdf.
Diao, X., J. Thurlow, S. Benin, and S. Fan (eds.). 2012. *Strategies and priorities for African agriculture: Economywide perspectives from country studies*. Washington, D.C: Intl Food Policy Res Inst.
Dingkuhn, M., C. Baron, V. Bonnal, F. Maraux, B. Sarr, B. Sultan, A. Clopes, and F. Forest. 2003. Decision support tools for rain-fed crops in the Sahel at the plot and regional scale. In *A practical guide to decision-support tools for agricultural productivity and soil fertility enhancement in Sub-Saharan Africa*, ed. T.E. Struif-Bontkes and M.C.S. Wopereis, 127–139. Muscle Shoals: IFDC, CTA.
Donatelli, M., C.O. Stöckle, and E. Ceotto. 1997. Evaluation of CropSyst for cropping systems at two locations of northern and southern Italy. *European Journal of Agronomy* 6: 35–45.
Donatelli, M., F.N. Tubiello, U. Peruch, and C. Rosenzweig. 2002. Impacts of climate change and elevated CO2 on sugar beet production in Northern and Central Italy. *Italian Journal of Agronomy* 6: 133–142.
Downing, T.E., L. Ringius, M. Hulme, and D. Waughray. 1997. Adapting to climate change in Africa. *Mitigation and adaptation strategies for global change* 2(1): 19–44.
Drake, B.G., M.A. Gonzalez-Meler, and S.P. Long. 1997. More efficient plants? a consequence of rising atmospheric CO? *Annual Revolution in Plant Physiology and Plant Molecular Biology* 48: 609–639.
Dzotsi, K., A. Agboh-Noameshie, T.E. Struif Bontkes, U. Singh, and P. Dejean. 2003. Using DSSAT to derive optimum combinations of cultivar and sowing date for maize in southern Togo. In *Decision support tools for smallholder agriculture in Sub-Saharan Africa: A practical guide*, ed. T.E. Struif-Bontkes and M.C.S. Wopereis, 100–113. The Netherlands: ACP-EU Technical Centere for Agricultural and Rural Cooperation.

Economic Research Services (ERS). 2013. *Sub-Saharan African cereal yields show signs of growth*. http://www.ers.usda.gov/data-products/chart-gallery/detail.aspx? chartId=37396&ref=collection#.U4da_fldV8E
El-Hage, Scialabba N., and C. Hattam (ed). 2002. Organic agriculture, environment, and food security. Environment and natural resources service, sustainable development department, food and agriculture organization of the United Nations (FAO). http://www.fao.org/docrep/005/y4137e/y4137e00.htm.
Fabusoro, E., C.I. Sodiya, M. Fasona, J. Oyedepo. 2014. Vulnerability of settled Fulani Agro-Pastoralists' livelihoods to climate change and emerging innovations for adaptation and land accessibility in Southwest Nigeria, a global environmental change research project sponsored by global system for analysis, research and training (START), subaward grant 2013–02
Fisher, G., K. Frouhberg, M. Parry, C. Rosenzweig. 1997. Les effets potentiels du Changement de climat sur la production et la sécurité alimentaires mondiales. In Changements du climat et production agricole, Agriculture mondiale: Horizon 2010, ed. FAO, Polytechnica, 472. Rome: FAO/Polytechnica.
Folberth, C., T. Gaiser, K.C. Abbaspour, R. Schulin, and H. Yang. 2012. Regionalization of a large-scale crop growth model for sub-Saharan Africa: Model setup, evaluation, and estimation of maize yields. *Agriculture, Ecosystem and Environment* 151: 21–33.
Folliard, A., P.C.S. Traoré, M. Vaksmann, and M. Kouressy. 2004. Modeling of sorghum response to photoperiod: A threshold–hyperbolic approach. *Field Crops Research* 89: 59–70.
Food and Agriculture Organization (FAO). 2011. Climate change and food systems resilience in sub-Saharan Africa. In Conference on ecological agriculture: Mitigating climate change, providing food security and self-reliance for rural livelihoods in Africa. Addis Ababa, Ethiopia: African Union Headquarters.
Food, and Agriculture Organization (FAO). 2003. *Responding to agricultural and food insecurity challenges mobilising Africa to implement nepad programmes FAO*. Maputo: FAO.
Fosu-Mensah, B.Y., D.S. Maccarthy, P.L.G. Vlek, and E.Y. Safo. 2012. Simulating impact of seasonal climatic variation on the response of maize (Zea mays L.) to inorganic fertilizer in sub-humid Ghana. *Nutrient Cycling in Agroecosystems* 94: 255–271.
Francis, C.A. 1986a. Future perspectives of multiple cropping. In *Multiple cropping system*, ed. A. Francis, 351–369. New York: MacMillan Publishing Co.
Francis, C.A. 1986b. *Multiple cropping systems*. New York: Macmillan Publishing Company.
Gaiser, T., I. DE Barros, F. Sereke, and F.-M. Lange. 2010a. Validation and reliability of the EPIC module to simulate maize production in small-holder farming systems in tropical sub-humid West Africa and semi-arid Brazil. *Agriculture Ecosystem and Environment* 135: 318–327.
Gaiser, T., M. Judex, C. Hiepe, and A. Kuhn. 2010b. Regional simulation of maize production in tropical savanna fallow systems as affected by fallow availability. *Agriculture Systems* 103: 656–665.
Giles, K.L., D. Cohen, and M.F. Beardsell. 1976. Effects of water stress on the ultrastructure of leaf cells of Sorghum bicolor. *Plant Physiology* 57: 11–14.
Gnanglè, P.C., J.A. Yabi, N.R. Yegbemey, L.R. Glèlè Kakaï, and N. Sokpon. 2012. Rentabilité économique des systèmes de production des parcs à Karité dans le contexte de l'adaptation au changement climatique du Nord-Bénin. *African Crop Science Journal* 20(2): 589–602.
Graef, F., and J. Haigis. 2001. Spatial and temporal rainfall variability in the Sahel and its effects on farmers' management strategies. *Journal of Arid Environment* 48: 221–231.
Grotch, S.L. 1988. *Regional intercomparison of general circulation model prediction and historical climate data*, USDE and USDAARS TR041. Washington, DC: Printing Office.
Hartwell, L., J.R. Allen, J.T. Baker and J.K. Boote. 1997. L'effet fertilisant du CO2: Production et rétention accrues d'hydrates de carbone en termes de rendement en biomasse et en grain. In ed, FAO, Polytechnica, *Changements du climat et production agricole, Agriculture mondaile: Horizon 2010*, 472. Rome: FAO/Polytechnica.
Hassan, R., and C. Nhemachena. 2008. Determinants of african farmers' strategies for adapting to climate change: Multinomial choice analysis. *African Journal of Agricultural and Resource Economics* 2: 83–104.

Hertel, T.W., and D.B. Lobell. 2014. Agricultural adaptation to climate change in rich and poor countries: Current modeling practice and potential for empirical contributions. Energy Economics http://dx.doi.org/10.1016/j.eneco.2014.04.014.

Hoogenboom, G. 2000. Contribution of agrometeorology to the simulation of crop production and its applications. *Agricultural and Forest Meteorology* 103: 137–157.

Idso, S.B., B.A. Kimball, and J.R. Mauney. 1987. Atmospheric carbon dioxide enrichment effects on cotton midday foliage temperature: Implications for plant water use and crop yield. *Agronomy Journal* 79: 667–672.

Intergovernmental Panel on Climate Change (IPCC). 2007a. Climate change 2007: Impacts, adaptation and vulnerability. Summary for policy makers. Online at http://www.ipcc.cg/SPM13apr07.pdf.

IPCC. 2007b. Summary for policy makers. In IPCC fourth assessment report, Working group II report: Impacts, adaptation, and vulnerability. See specifically on adaptation, chapter 17; on inter-relationships between adaptation and mitigation, chapter 18; on vulnerability, chapter 19.

IPCC. 2001. IPCC 3rd assessment report – Climate change 2001: Working group II: Impacts, adaptation and vulnerability. http://www.ipcc.ch/ipccreports/index.htm.

International Development Research Center. 2001. *Adapting to climate change in Malawi by improving soils and nutrition*. Canada: Ecosystems and Health.

Jones, C.A., J.T. Ritchie, J.R. Kiniry, and D.C. Godwin. 1986. Subroutine structure. In *CERES-Maize: A simulation model of maize growth and development*, ed. C.A. Jones and J.R. Kiniry, 49–111. College Station: Texas A&M University Press.

Jones, J.W., G. Hoogenboom, C.H. Porter, K.J. Boote, W.D. Batchelor, L.A. Hunt, P.W. Wilkens, U. Singh, A.J. Gijsman, and J.T. Ritchie. 2003. The DSSAT cropping system model. *European Journal of Agronomy* 18: 235–265.

Jones, P., L.H.J.R. Allen, and J.W. Jones. 1985. Responses of soybean canopy photosynthesis and transpiration to whole-day temperature changes in different CO2 environments. *Agronomy Journal* 77: 242–249.

Kang, Y., S. Khan, and X. Ma. 2009. Climate change impacts on crop yield, crop water productivity and food security – A review. *Progress in Natural Science* 19: 1665–1674.

Kassam, A., and M. Smith. 2001. *FAO methodologies on crop water use and crop water productivity*. Expert meeting on crop water productivity, 3–5 December 2001, Paper No. CWP-M07, Rome.

Keating, B.A., P.S. Carberry, G.L. Hammer, M.E. Probert, M.J. Robertson, D. Holzworth, N.I. Huth, J.N.G. Hargreaves, H. Meinke, Z. Hochman, G. Mclean, K. Verburg, V. Snow, J.P. Dimes, M. Silburn, E. Wang, S. Brown, K.L. Bristow, S. Asseng, S. Chapman, R.L. Mccown, D.M. Freebairn, and C.J. Smith. 2003. An overview of APSIM, a model designed for farming systems simulation. *European Journal of Agronomy* 18: 267–288.

Keeling, C.D., and T.P. Whort. 1991. Atmospheric CO2 – A modem record, Manna Loa. In *Trends'91: A compendium of data on global change, ORNL/CDIAC-46*, ed. T.A. Boden, R.J. Sepanski, and F.W. Stoss, 12–15. Oak Ridge: Carbon Dioxide Information Analysis Center, Oak Ridge National Laboratory.

KERSEBAUM, K.C., W. MIRSCHELA, K.O. WENKELA, R. MANDERSCHEIDB, H.J. WEIGEL, and C. NENDEL. 2009. Modelling climate change impacts on crop growth and management in Germany. In *Climate variability, modeling tools*, 703–708. New York: Nova Science Publishers, Inc. ISBN 978-1-60692-703-8.

Knörzer, H., H. Grözinger, S. Graeff-Hönninger, K. Hartung, P. Piepho, and W. Claupein. 2011. Integrating a simple shading algorithm into CERES-wheat and CERES-maize with particular regard to a changing microclimate with in a relay intercropping system. *Field Crops Research* 121: 274–285.

Knox, J.W., J.A. Rodriguez-Diaz, D.J. Nixon, and M. Mkhwanazi. 2010. A preliminary assessment of climate change impacts on sugarcane in Swaziland. *Agriculture Systems* 103: 63–72.

Knox, J., H. Tim, A. Daccache, and T. Wheeler. 2012. Climate change impacts on crop productivity in Africa and South Asia. *Environment Research Letters* 7: 034032.

Ko, J., L.R. Ahuja, B. Kimball, S.A. Saseendran, L. Ma, T.R. Green, A.C. Ruane, G.W. Wall, P. Pinter, and D.A. Bader. 2010. Simulation of free air CO2 enriched wheat growth and interactions with water, nitrogen, and temperature. *Agriculture and Forest Meteorology* 150: 1331–1346.

Kolawole, A. 1987. Environmental change and the South Chad irrigation project (Nigeria). *Journal of Arid Environments* 13: 169–176.

Krishna, P. 2004. Plant responses to heat stress. In *Plant responses to abiotic stress*, Topics in Current Genetics, ed. H. Hirt and K. Shinozaki, 72–101. Berlin: Springer.

Kurukulasuriya, P., and R. Mendelsohn. 2006. *A Ricardian analysis of the impact of climate change on African cropland*. CEEPA discussion paper No. 8 (2006). Centre for Environmental Economics and Policy in Africa. University of Pretoria.

Larcher, W. 1995. *Physiological plant ecology*, 3rd ed. Berlin: Springer.

Larkindale, J., and M.R. Knight. 2002. Protection against heat stress-induced oxidative damage in Arabidopsis involves calcium, abscisic acid, ethylene, and salicylic acid. *Plant Physiology* 128: 682–695.

Lobell, D.B., K.G. Cassman, and C.B. Field. 2009. Crop yield gaps: Their importance, magnitudes, and causes. *Annual Review of Environment and Resources* 34: 179–204.

Lobell, D.B., W. Schlenker and J. Costa-Roberts. 2011. Climate trends and global crop production since 1980. Science, http://dx.doi.org/10.1126/science.1204531:1-9.

Long, S.P., E.A. Ainsworth, A.D.B. Leakey, J. Noesberger, and D.R. Ort. 2006. Food for thought: Lower-than-expected crop yield stimulation with rising CO2 concentrations. *Science* 312: 1918–1921.

Macar, K.T., and Y. Ekmekçi. 2009. Alterations in photochemical and physiological activities of chickpea (Cicer arietinum L.) cultivars under drought stress. *Journal of Agronomy and Crop Science* 195(5): 335–346.

Maccarthy, D.S., R. Sommer, and P.L.G. Vlek. 2009. Modeling the impacts of contrasting nutrient and residue management practices on grain yield of sorghum (Sorghum bicolor (L.) Moench) in a semi-arid region of Ghana using APSIM. *Field Crops Research* 113: 105–115.

Maccarthy, D.S., P.L.G. Vlek, A. Bationo, R. Tabo, and B.Y. Fosu-Mensah. 2010. Modeling nutrient and water productivity of sorghum in smallholder farming systems in a semiarid region of Ghana. *Field Crops Research* 118: 251–258.

Maccarthy, D.S., P.L.G. Vlek, and B.Y. Fosu-Mensah. 2012. The response of maize to N fertilization in a sub-humid region of Ghana: Understanding the processes using a crop simulation model. In *Improving soil fertility recommendations in Africa using the Decision Support System for Agrotechnology Transfer (DSSAT)*, 61–75. Dordrecht/New York: Springer.

Mailhol, J., P. Ruelle, and I. Nemeth. 2001. Impact of fertilisation practices on nitrogen leaching under irrigation. *Irrigation Science* 20: 139–147.

Malesu, M.M., J.K. Sang, J.O. Odhiambo, A.R. Oduor, and M. Nyabenge. 2006. *Rainwater harvesting innovations in response to water scarcity: The lare experience*, Technical manual No. 5. Nairobi: SearNet Secretariat, RELMA-in-ICRAF, World Agroforestry Centre.

Müller, C. 2009. Climate change impact on Sub-Saharan Africa: An overview and analysis of scenarios and models. DIE – (Discussion Paper/Deutsches Institut für Entwicklungspolitik; 3/2009), Bonn.

Müller, C., W. Cramer, W.L. Hare, and H. Lotze-Campen. 2011. Climate change risks for African agriculture. *Proceedings of the National Academy of Sciences. USA* 108: 4313–4315.

Murthy, V.R.K. 2004. Crop growth modeling and its applications in agricultural meteorology. In *Satellite remote sensing and GIS applications in agricultural meteorology*, ed. Sivakumar, MV K, 235–261. Dehra Dun, India: WMO.

Naab, J.B., P. Singh, K.J. Boote, J.W. Jones, and K.O. Marfo. 2004. Using the Cropgro – peanut model to quantify yield gaps of peanut in the Guinean Savanna Zone of Ghana. *Agronomy Journal* 96: 1231–1242.

Ngigi, S.N. 2003. *Rainwater harvesting for improved food security: Promising technologies in the Greater Horn of Africa*. Nairobi: Greater Horn of Africa Rainwater Partnership (GHARP), Kenya Rainwater Association (KRA). 266 pp.

Ngigi, S.N. 2009. *Climate change adaptation strategies: Water resources management options for small-holder farming systems in sub-Saharan Africa*, The MDG centre for East and Southern Africa. New York: The Earth Institute at Columbia University. 189p.

Nhemachena, C., and R. Hassan. 2007. *Micro-level analysis of farmers' adaptation to climate change in Southern Africa*, IFPRI discussion paper No. 00714. Washington, DC: International Food Policy Research Institute.

Niggli, U., A. Fliessbach, P. Hepperly, and N. Scialabba. 2008. *Low greenhouse gas agriculture: Mitigation and adaptation potential of sustainable farming systems*. Rome: FAO.

Norman, D.W. 1974. Rationalizing mixed cropping under indigenous conditions: The example of northern Nigeria. *Journal of Development Studies* 11: 3–21.

O'brien, K., L. Synga, L.O. Naess, R. Kingamkono, and B. Hochobeb. 2000. *Is information enough? User response to seasonal climate forecasts in Southern Africa*, CICERO report. Oslo: Center for International Climate and Environmental Research.

Olmos, E., M.J. Sanchez-Blanco, T. Ferrandez, and N.J.J. Alarco. 2007. Subcellular effects of drought stress in Rosmarinus officinalis. *Plant Biology* 9: 77–84.

Paavola, J., and T.W. Adger. 2005. Analysis of fair adaptation to climate change. *Ecological Economics* 56: 594–609.

Penning de Vries, F.W.T. 1977. Evaluation of simulation models in agriculture and biology: Conclusions of a workshop. *Agricultural Systems* 2: 99–105.

Ritchie, J.T., and S. Otter. 1985. Description and performance of CERES-wheat. In *A user oriented wheat yield model. USDA wheat yield model*, ed. W.D. Willis, vol. 38, 159–175. US Department of Agriculture – ARS.

Robertson, M.J., P.S. Carberry, Y.S. Chauhan, R. Ranganathan, and G.J. O'leary. 2001a. Predicting growth and development of pigeonpea: A simulation model. *Field Crops Research* 71: 195–210.

Robertson, M.J., P.S. Carberry, N.I. Huth, J.E. Turpin, M.E. Probert, P.L. Poulton, M. Bell, G.C. Wright, S.J. Yeates, and R.B. Brinsmead. 2001b. Simulation of growth and development of diverse legume species in APSIM. *Australian Journal of Agricultural Research* 53: 429–446.

Rötter, R.P., T. Palosuo, K.C. Kersebaum, C. Angulo, M. Bindi, F. Ewert, R. Ferrise, P. Hlavinka, M. Moriondo, C. Nendel, J.E. Olesen, R.H. Patil, F. Ruget, J. Takáčˇ, and M. Trnka. 2012. Simulation of spring barley yield in different climatic zones of Northern and Central Europe: A comparison of nine crop models. *Field Crops Research* 133: 23–36.

Schlenker, W., and D.B. Lobell. 2010. Robust negative impacts of climate change on African agriculture. *Environmental Research Letters* 5: 1–8.

Schlesinger, W.H., J.F. Reynolds, G.L. Cunningham, L. Huenneke, W. Jarrell, R. Virginia, and W. Whitford. 1990. Biological feedbacks in global desertification. *Science* 247: 1043–1048.

Scholes, R.J., and R. Biggs. (ed). 2004. Ecosystem services in southern Africa: A regional assessment. Millennium Ecosystem Assessment. Online at http://www.millenniumassessment.org.

Scopel, E., F.A.M.D. Silva, M. Corbeels, F. Affholder, and F. Maraux. 2004. Modelling crop residue mulching effects on water use and production of maize under semi-arid and humid tropical conditions. *Agronomie* 24: 383–395.

Sedogo, D., and S.V.R. Shetty. 1991. Progres de la recherché vers une technologie amelioree de production du sorgho en culture pluviale. Paper presented at the inter-network conference in food grain improvement, 7–14, March 1991, Niamey, Niger (proc, in Press).

Sharma, K., O.P. Bishnoi, R. Niwas, and M.L. Khichar. 2010. Impact assessment of climatic variability on wheat and pearl millet productivity using CERES models in arid zone of Haryana. *Journal of Agrometeorology* 12: 123–127.

Sisworo, W.H., M.M. Mitrosuhardjo, H. Rasjid, and R.J.K. Myers. 1990. The relative roles of N fixation, fertilizer, crop residues and soil in supplying N in multiple cropping systems in a humid, tropical upland cropping system. *Plant and Soil* 121: 73–82.

Sivakumar, M.V.K. 1988. Predicting rainy season potential from the onset of rains in Southern Sahelian and Sudanian climatic zones of West Africa. *Agricultural and Forest Meteorology* 42: 295–305.

Sivakumar, M.V.K., and J.L. Hatfield. 1990. Spatial variability of rainfall at an experimental station in Niger, West Africa. *Theoretical and Applied Climatology* 42: 33–39.
Sommer, R., K. Kienzler, C. Conrad, N. Ibragimov, J. Lamers, C. Martius, and P. Vlek. 2008. Evaluation of the CropSyst model for simulating the potential yield of cotton. *Agronomy and Sustainable Development* 28: 345–354.
Song, Y., C. Birch, S. Qu, A. Doherty, and J. Hanan. 2010. Analysis and modelling of the effects of water stress on maize growth and yield in dryland conditions. *Plant Production Science* 13: 199–208.
Srivastava, A.K., T. Gaiser, D. Cornet, and F. Ewert. 2012. Estimation of effective fallow availability for the prediction of yam productivity at the regional scale using. *Field Crops Research* 131: 32–39.
Staggenborg, A.S., and L.R. Vanderlip. 2005. Crop simulation models can be used as dryland cropping systems research tools. *Agronomy Journal* 97: 378–384.
Steiner, K.G. 1984. *Intercropping in tropical smalholder agriculture with special reference to West Africa*. West Germany: Stein. 304pp.
Stöckle, C.O., M. Donatelli, and R. Nelson. 2003. CropSyst, a cropping systems simulation model. *European Journal of Agronomy* 18: 289–307.
Stoop, W.A., and J.P.V. Staveren. 1981. Effects of cowpea in cereal rotations on subsequent crop yields under semi-arid conditions in Upper-Volta. In *Biological nitrogen fixation technology for tropical agriculture*, ed. P.C. Graham and S.C. Harris. Cali: Centro International de Agricultura Tropical.
Swinton, S.M., G. Numa and L.A. Samba. 1984. Les cultures associees en milieu paysan dans deux regions du Niger: Filingue et Madarounfa. Pages 183–194 In Proceedings of the regional workshop on intercropping in the Sahelian and Sahelo-sudanian zones of West Africa, 7–10 Nov 1984, Niamey, Niger, Bamako, Mali: Institute du Sahe.
Tayyar, S. 2010. Variation in grain yield and quality of Romanian bread wheat varieties compared to local varieties in northwestern turkey. *Romanian Biotechnological Letters* 15(2): 5189–5196.
Therond, O., H. Hengsdijk, E. Casellas, D. Wallach, M. Adam, H. Belhouchette, R. Oomen, G. Russell, F. Ewert, J.E. Bergez, S. Janssen, J. Wery, and M.K. Van Ittersum. 2011. Using a cropping system model at regional scale: Low-data approaches for crop management information and model calibration. *Agriculture Ecosystem and Environment* 142: 85–94.
Thornton, K., P.G. Jones, T.M. Owiyo, R.L. Kruska, M. Herero, P. Kristjanson, A. Notenbaert, and N. Bekele. 2006. *Mapping climate vulnerability and poverty in Africa. Report to the Department for International Development*. Nairobi: ILRI.
Tingem, M., and M. Rivington. 2009. Adaptation for crop agriculture to climate change in Cameroon: Turning on the heat. *Mitigation and Adaptation Strategies Global Change* 14: 153–168.
Tingem, M., M. Rivington, G. Bellocchi, and J.J. Colls. 2009. Crop yield model validation for Cameroon. *Theoretical and Applied Climatology* 96: 275–280.
Tojo Soler, C.M., V.B. Bado, K. Traore, W. Mcnair Bostik, J.W. Jones, and G. Hoogenboom. 2011. Soil organic carbon dynamics and crop yield for different crop rotations in a degraded ferruginous tropical soil in a semi-arid region: A simulation approach. *Journal of Agriculture Science* 149: 579–593.
Traore, B., M. Corbeels, M.T. van Wijk, M.C. Rufino, and K.E. Giller. 2013. Effects of climate variability and climate change on crop production in southern Mali. *European Journal of Agronomy* 49: 115–125.
Tripathy, R., S.S. Ray, H. Kaur, S.K. Jalota, S.K. Bal, and S. Panigrahy. 2011. Understanding spatial variability of cropping system response to climate change in Punjab state of India using remote sensing data and simulation model. *ISPRS-International Archives of the Photogrammetry, Remote Sensing and Spatial Information Sciences* 3820: 29–33.
Shah, N.H., and G.M. Paulsen. 2003. Interaction of drought and high temperature on photosynthesis and grain-filling of wheat. *Plant and Soil* 257(1): 219–226.

UNFCCC. 2012. Compendium on methods and tools to evaluate impacts of, and vulnerability and adaptation to climate change <http://unfccc.int/adaptationnairobi_workprogramme/knowledge_resources_and_publications/items/5457.php>.

Utrillas, M.J., and L. Alegre. 1997. Impact of water stress on leaf anatomy and ultrastructure in Cynodcon dactylon L. Pers. under natural conditions. *International Journal Plant Science* 158(3): 313–324.

Van Duivenbooden, N., M. Pala, C. Studer, C.L. Bielders, and D.J. Beukes. 2000. Cropping systems and crop complementarity in dryland agriculture to increase soil water use efficiency: A review. *NJAS – Wageningen Journal of Life Sciences* 48: 213–236.

van Ittersum, M.K., P.A. Leffelaar, H. van Keulen, M.J. Kropff, L. Bastiaans, and J. Goudriaan. 2003. On approaches and applications of the Wageningen crop models. *European Journal of Agronomy* 18: 201–234.

Van Ittersum, M.K., F. Ewert, T. Heckelei, J. Wery, J. Alkan- Olsson, E. Andersen, I. Bezlepkina, F. Brouwer, M. DONATELLI, G. Flichman, L. Olsson, A.E. Rizzoli, T. Van Der Wal, J.E. Wien, and J. Wolf. 2008. Integrated assessment of agricultural systems – A component based framework for the European Union (SEAMLESS). *Agriculture System* 96: 150–165.

Van Keulen, H., and J. Wolf (eds.). 1986. *Modelling of agricultural production: Weather, soils and crops*, Simulation monographs. Wageningen: PUDOC.

Vassileva, V., C. Signarbieux, I. Anders, and U. Feller. 2011. Genotypic variation in drought stress response and subsequent recovery of wheat (*Triticum aestivum* L.). *Journal of Plant Research* 124(1): 147–154.

Ventrella, D., M. Charfeddine, M. Moriondo, M. Rinaldi, and M. Bindi. 2012. Agronomic adaptation strategies under climate change for winter durum wheat and tomato in southern Italy: Irrigation and nitrogen fertilization. *Regional Environment Change* 12: 407–419.

Waha, K., C. Muller, A. Bondeau, J.P. Dietrich, P. Kurukulasuriya, J. Heinke, and H. Lotze Campen. 2013. Adaptation to climate change through the choice of cropping system and sowing date in sub-Saharan Africa. *Global Environmental Change* 23: 130–143.

Ward, J.M.J., E.L. Stromberg, D.C. Nowell, and F.W.J.R. Nutter. 1999. Gray leaf spot: A disease of global importance in maize production. *Plant Diseases* 83: 884–895.

Webber, H., T. Gaiser, and F. Ewert. 2014. What role can crop models play in supporting climate change adaptation decisions to enhance food security in Sub-Saharan Africa? *Agricultural Systems* 127: 161–177.

Willey, R.W., and D.S.O. Osiru. 1972. Studies on mixtures of maize and beans (Phaseolus vulgaris) with particular reference to plant population. *Journal of Agricultural Science* 79: 519–529.

Williams, J.R., C.A. Jones, J.R. Kiniry, and D.A. Spanel. 1989. The EPIC crop growth model. *Transactions of ASAE* 32: 497–511.

Woomer, P., M. Lan'gat, and J. Tungani. 2004. Innovative maize-legume intercropping results in above-and below-ground competitive advantages for understorey legumes. *West African Journal of Applied Ecology* 6(1): 85–94.

Worou, O.N. 2012. *Experimental analysis and modeling of the rainfed rice cropping systems in West Africa*. Ph.D. Thesis, Universität Bonn, Bonn, Germany, 126 p.

Xiong, W., D. Conway, E. Lin, and I. Holman. 2009. Potential impacts of climate change and climate variability on China's rice yield and production. *Climate Research* 40: 23–35.

Yegbemey, R.N., H. Kabir, O.H.R. Awoye, J.A. Yabi, and A.A. Paraïso. 2014. Managing the agricultural calendar as coping mechanism to climate variability: A case study of maize farming in northern Benin, West Africa. *Climate Risk Management* 3: 13–23.

Yegbemey, R.N., J.A. Yabi, D.S. Tovignan, G. Gantoli, and S.E.H. Kokoye. 2013. Farmers' decisions to adapt to climate change under various property rights: A case study of maize farming in Northern Benin (West Africa). *Land Use Policy* 34: 168–175. http://dx.doi.org/10.1016/j.landusepol.2013.03.001.

Yengoh, G.T., F.A. Armah, E.E. Onumah, and J.O. Odoi. 2010. Trends in agriculturally-relevant rainfall characteristics for small-scale agriculture in Northern Ghana. *Journal of Agricultural Science* 2: 3.

Ziska, L.H., D.M. Blumenthal, G.B. Runion, E.R. Hunt, and H. Diaz-Soltero. 2011. Invasive species and climate change: An agronomic perspective. *Climate Change* 105(1): 13–42.

Chapter 4
Modeling Nitrogen Use Efficiency Under Changing Climate

Muhammad Aqeel Aslam, Mukhtar Ahmed, Fayyaz-ul-Hassan, and Riffat Hayat

Contents

M.A. Aslam (✉) • Fayyaz-ul-Hassan
Department of Agronomy, Pir Mehr Ali Shah Arid Agriculture University, Rawalpindi, Pakistan
e-mail: sawan_malik@yahoo.com; drsahi63@gmail.com

M. Ahmed
Department of Agronomy, Pir Mehr Ali Shah Arid Agriculture University, Rawalpindi, Pakistan

Department of Biological Systems Engineering, Washington State University, Pullman, WA, USA
e-mail: ahmadmukhtar@uaar.edu.pk; mukhtar.ahmed@wsu.edu

R. Hayat
Department of Soil Science and SWC, Pir Mehr Ali Shah Arid Agriculture University, Rawalpindi, Pakistan
e-mail: hayat@uaar.edu.pk

M. Ahmed, C.O. Stockle (eds.), *Quantification of Climate Variability, Adaptation and Mitigation for Agricultural Sustainability*,
DOI 10.1007/978-3-319-32059-5_4

Abstract Nitrogen is the most limiting element in the production of cereal crops after water hence leads plant nutrition. Since, nitrogen uptake and supply directly depends upon soil physical conditions, climate and plant genetic features, so N requirement could be varied by place to place. Crop simulation models can be complementary tools in field experiments to develop innovative crop management systems under continuous varying nitrogen regime. Data regarding total nitrogen, nitrogen uptake efficiency, nitrogen utilization efficiency and nitrogen utilization efficiency, drymatter accumulation at three phenological stages (Three leaf, Anthesis and Maturity), and yield parameters (Number of tillers, Biological yield, Thousand grain weight, Grain yield and Harvest index) were recorded. The present study revealed that different nitrogen rates and application methods have significant impact upon crop growth and development. Wheat crop responded well to nitrogen fertilizer. Maximum grain yield obtained for N_{100} when nitrogen was applied as split dose. Similarly, genotypes responded significantly to nitrogen fertilizers for grain production. Genotype NARC-2009 performed well under different nitrogen regime of rainfed zone of pothwar. APSIM model was parameterized using different agronomic parameters (days after sowing, biomass total nitrogen, root total nitrogen, grain yield and grain total nitrogen). The modeled nitrogen was satisfactory compared to observed nitrogen. The analysis of the modeling results depicted the strong dependency of the mineral nitrogen content upon plant nitrogen uptake and growth. By concluding APSIM model performed well under rainfed conditions of pothwar for modeling nitrogen use efficiency. Modeling approaches should be adopted by farmers and policy makers to get maximum crop production and eliminate extra nitrogen losses.

Keywords Wheat • APSIM • Nitrogen use efficiency • Nitrogen uptake efficiency

4.1 Introduction

The anthropogenic emission of greenhouse gases is projected to affect crop production over the globe due to climate change. A shift in cropping pattern is induced by Climate change. This shift might eliminate one crop while creating good growing environment for the other crops. Crop production depends different nutrients among which carbon, hydrogen and oxygen are taken from the soil and atmosphere while in most cases nitrogen, phosphorus and potassium are artificially applied to the crops at farm level. The applied nitrogen could flows (a) into the system, (b) out of the system with products (grains, stem, leaves and roots), and (c) could be balanced and the system would be safe with respect to nitrogen (Oenema et al. 2003), whereas any excessive import of nitrogen may lead to nitrogen accumulation and/or towards gaseous nitrogen (e.g. N_2, NO, N_2O and NH_3) losses into the environment and to the hydrosphere as nitrate. Nitrogen is the most limiting element in the production of cereal crops after water hence leads plant nutrition (Russell 2010). Main grain crops (wheat, paddy and maize) utilize 1 kg of nitrogen to produce 68, 44 and 49 kg of paddy, wheat, and maize grain, respectively (Pathak et al. 2003; Janssen et al. 1990). At present a huge amount of N is being used by the world population, almost 83 million metric tons, which is almost a 100-fold increase over the last century. For the production of world's three major cereals: rice, wheat, and maize on average 60 % of the nitrogen fertilizer is used. Nitrogen availability regulates numerous aspects of plant growth. The resource capturing tissues (meristematic activity and cell extension) are expanded by the presence of nitrogen, as well as in their photosynthetic activity. It is assumed that by the end of 2050, 50–70 % grains from cereal crops will be required to fulfill the food requirement of a huge population of 9.3 billion (Smil 2005).

Nitrogen use efficiency (NUE) can be defined as yield harvested per unit of nitrogen applied. NUE idea delivers a numerical measure of the usefulness of plants to absorb and transform available N into potential yield under different cropping systems. N fertilizer is among the central inputs for cereal production (Giller 2004). Over the globe (NUE) for grain crops is nearly 33 % including wheat (Raun and Johnson 1999). Suitable N application rates and timing are precarious for fulfilling plant requirements and enhancing NUE. Wheat is among the crops which are the most fertilized. Nitrogen is the most important fertilizer for wheat crop. High use of fertilizer is a great threat to ecological pollution (Abril et al. 2007). Crop rotation, soil edaphic features, temperature, soil water, N fertilizer rates and crop types affect NUE (Halvorson et al. 2002). Mahler et al. (1994) specified necessity of efforts to enhance plant NUE and productivity in semi-arid situation and to improve sustainable farming systems in reply to continually increasing financial and ecological stresses. Losses of N have been endorsed to the mutual effects of de-nitrification, volatilization and leaching. Ground-water toxification and other severe climatic problems are the result of adding a huge amount of nitrogen to the environment (Chen et al. 2010). Nitrate leaching into soil could be lessening by reducing rate of N application (Power et al. 2000). Urea-N when applied to soil undergoes three nitrogen transformation processes i.e. rapid hydrolysis to NH_4^+ followed by ammonia volatilization (Praveen and Aggarwal 1998).

Creating new plans and conclusion making in crop production gradually makes implementation of numerous model-based decision support tools especially in the context of changing climatic issues. Simulation models which are used to simulate crop growth are generally mechanistic, i.e. these models not only try to explain relationship between simulated variables and parameters but also the appliance of the designated methods (Challinor et al. 2009; Porter and Semenov 2005). Although many crop growth simulation models (crop models) are established and assessed at the field scale, and the only problem was there that they were not made to simulate huge areas. Now a day it is a common practice to use these dynamic models in evaluation of agricultural impacts and alteration to climate from a field to the national level (Parry et al. 2005; Rosenberg 2010).

Nitrogen use efficiency (NUE) was assessed in 25 wheat varieties for 2 years where nitrogen uptake efficiency accounted for 54 % of the genotypic variation in NUE for yield and 72 % of the genotypic variation in NUE for protein. Nitrogen uptake efficiency had direct effect on NUE revealed by a path coefficient analysis (Van Sanford and MacKown 1986). In the early developmental stages of plant, NUE can be improved by providing mineral nitrogen to fulfill the partial requisite of crop and saving nitrogen for later stages before the onset of prompt crop growth (Sowers et al. 1994; López-Bellido et al. 2005). The nitrogen balance is mostly estimated by linking various N inputs and outputs in soil-crop systems by considering variations of soil mineral nitrogen (Sogbedji et al. 2000). Nitrogen loss by nitrate leakage from cultivated fields is an emerging concern as raised nitrate levels were observed in soil water in numerous countries (Diez et al. 2000).

Soils low in nitrogen content requires N management for beneficial and sustainable wheat Bakht et al. (2009) reported that in low nitrogen soils crop production can be enhanced by contribution of legumes, residues and by applying N fertilizers. Dhungana et al. (2006) conducted experiment to develop a strategy beneficial in ascertaining crop technologies for future climatic conditions.

In a field experiment on clayey soil different treatments of nitrogen fertilizers and irrigation on wheat crop was performed to observe and simulate plant growth and development, N uptake and mineral nitrogen division among roots, leaves, shoot and grains. SOILN-CROP model was used to simulate crop growth. This model is run by a hydrological model and it has its bases on the experiential allometric functions and the light intervention concept. Growth can be reduced by fluctuation in the mineral nitrogen in the soil and is the principle driving force of N uptake. SOILN-CROP model components describing the fraction of soil mineral N accessible for plant uptake had a firm stimulus on model behavior (Liang et al. 2016).

4.2 Methodology

An experiment was carried out to parameterize and evaluate APSIM model for nitrogen use efficiency of two wheat genotypes at research area of PMAS, Arid Agriculture University, Rawalpindi during 2010–2011 and 2011–2012. Experiment was laid out in accordance with four way factorial randomized complete block design (RCBD) with four replication. Phosphorous was applied @ 50 kg ha^{-1} in the form of single super phosphate (SSP). Individual plot size for each treatment was 6 m × 4 m for each genotype with row spacing of 25 cm. To isolate treatments, plot to plot distance was maintained at 1 m and the experiment was repeated for 2 years.

Treatments applied were four nitrogen rates [T1 = Control (N0), T2 = 50 kg N (N50), T3 = 100 kg N (N100) and T4 = 150 kg N (N150)], two application methods (AM1 = Full dose of nitrogen at sowing and AM2 = Three equal doses (1/3rd of each treatment) of nitrogen at sowing, tillering and at flag leaf stage), two genotypes (G1 = NARC-2009 and G2 = Chakwal-50) and two environments (Y1 = 2010–2011, Y2 = 2011–2012). Amount of Nitrogen was determined at Zadok's growth stages i.e. Zadoks et al. (1974) (Three leaf, Anthesis and at Maturity) from roots and grains from a randomly selected area of 0.25 m^2 from each plot. Total nitrogen contents were determined calorimetrically as prescribed by Anderson and Ingram (1993).

Then nitrogen use efficiency, nitrogen uptake efficiency, nitrogen utilization efficiency were calculated. Yield and yield parameters like number of fertile tillers per unit area, thousand grain weight, grain yield, biological yield and harvest index recorded under changing climate and varying nitrogen regimes. The data collected from the field experiments was used for model evaluation. *A*gricultural *P*roduction *S*ystems *S*imulator (APSIM) model was used to parameterize and evaluate to increase the nitrogen use efficiency based upon crop plants and soil data. APSIM-wheat is a dynamic model, software through which agricultural systems are simulated (McCown et al. 1996). Wheat crop simulate LAI, plant biomass, grain yield, grain nitrogen and nitrogen uptake by wheat crop (Wang et al. 2003).

4.3 Results and Discussion

4.3.1 Nitrogen Estimation

4.3.1.1 Total Nitrogen

Nitrogen contents at three leaf stage were calculated to determine nitrogen uptake by wheat crop. Significant variation was observed among wheat genotypes at varying nitrogen rates and application methods for both years. Highest total nitrogen was calculated for treatment N_{150} (5.33 kg ha^{-1}) while lowest was calculated at treatment N_0 (3.71 kg ha^{-1}) (Table 4.1). There was 44 % difference between highest and lowest value of total nitrogen. The nitrogen application methods varied significantly at three leaf stage for nitrogen uptake. Total nitrogen calculated in plant biomass was higher (4.77 kg ha^{-1}) in split dose compared to full dose (4.36 kg ha^{-1}) of nitrogen application. Split nitrogen dose application accumulated about 9 % more nitrogen than full dose. There was a considerable difference between growing years (2010–2011 and 2011–2012) for total nitrogen at three leaf stage. Higher total nitrogen was calculated during 2010–2011 (4.84 kg ha^{-1}) while lower amount of nitrogen (4.26 kg ha^{-1}) was calculated during 2011–2012. During 2010–2011, about 13 % more total nitrogen was calculated at three leaf stage than 2011–2012. There was significant difference among both genotypes during both years. Genotype NARC-2009 has taken more nitrogen (4.79 kg/ha) than Chakwal-50 (4.34 kg/ha). The interaction among treatments, doses, genotypes and years were non-significant at three leaf stage. Considerable variation observed among wheat genotypes at varying nitrogen rates and application methods for both years at anthesis stage. Total nitrogen differed considerably at different nitrogen rates at anthesis stage. Treatment N_0 accumulated minimum nitrogen (14.76 kg ha^{-1}) while N_{150} accrued maximum nitrogen (55.55 kg ha^{-1}). In split doses higher total nitrogen (40.39 kg ha^{-1}) was measured than full dose nitrogen application method (35.25 $kgha^{-1}$). Split dose application accumulated about 15 % high total nitrogen than full dose nitrogen application method. Considerable difference for total nitrogen observed among years at anthesis stage. During 2010–2011 highest nitrogen was calculated

Table 4.1 Dry matter nitrogen, nitrogen uptake efficiency, nitrogen utilization efficiency and nitrogen use efficiency for varying nitrogen rates and application methods among two wheat genotypes during 2010–2011 and 2011–2012

Treatments	TN Z-13	TN Z-60	TN Z-92	NUpE	NUtE	NUE
Nitrogen Rate (NR)						
N_0	3.71^d	14.77^d	18.94^d	0.53^a	203.19^a	108.49^a
N_{50}	4.18^c	28.15^c	36.07^c	0.42^c	121.65^b	51.35^b
N_{100}	5.03^b	52.83^b	67.73^b	0.50^b	80.52^c	40.12^c
N_{150}	5.33^a	55.55^a	71.20^a	0.38^d	66.76^d	25.47^d
Application Methods (AM)						
Split	4.36^b	40.39^a	51.78^a	0.49^a	120.63^{NS}	58.95^a
Full	4.77^a	35.25^b	45.19^b	0.43^b	115.43^{NS}	53.77^b
Years (Y)						
Y1	4.84^a	39.06^a	50.07^a	0.46^a	125.56^a	60.55^a
Y2	4.29^b	36.58^b	46.90^b	0.45^b	110.5^b	52.17^b
Genotypes (G)						
NARC-2009	4.79^a	39.39^a	50.49^a	0.48^a	125.13^a	62.19^a
CHAKWAL-50	4.34^b	36.26^b	46.49^b	0.44^b	110.93^b	50.52^b
Interactions						
NR×AM	NS	***	***	NS	NS	***
NR×AM	NS	***	***	NS	***	***
NR×Y	NS	***	***	***	***	**
AM×G	NS	NS	NS	NS	NS	NS
AM×Y	NS	***	***	***	*	NS
G×Y	NS	NS	NS	NS	*	**
NR×AM×G	NS	NS	NS	NS	NS	NS
NR×AM×Y	NS	**	**	NS	NS	NS
NR×G×Y	NS	NS	NS	NS	NS	NS
AM×G×Y	NS	NS	NS	NS	*	NS
NR×AM×G×Y	NS	NS	NS	NS	NS	NS

Abbreviations: *TN* Total Nitrogen, *Z-13* Three Leaf Stage, *NUpE* Nitrogen Uptake Efficiency, *Z-60* Anthesis Stage, *NUtE* Nitrogen Utilization Efficiency, *Z-92* Maturity Stage, *NUE* Nitrogen Use Efficiency

Different letters indicate a statistical difference among the treatments at $P < 0.05$

(39.06 kg ha^{-1}) whereas, lowest (36.58 kg ha^{-1}) total nitrogen in plant biomass calculated during 2011–2012. During 2010–2011, about 7 % more nitrogen calculated than 2011–2012 at anthesis stage. Genotype NARC-2009, harvested maximum nitrogen (39.39 kg ha^{-1}) than Chakwal-50 (36.26 kg ha^{-1}). There was about 9 % difference between both genotypes for total nitrogen. The interactive effects NR×AM, NR×G, NR×Y and AM×Y was significant at 1 % significance level whereas, the interaction among NR×AM×Y was significant at 5 % P level (Table 4.1).

Significant variation was observed between wheat genotypes at varying nitrogen rates and application methods for both years at maturity stage. Treatment N_0 accumulated minimum nitrogen (18.94 kg ha^{-1}) while N_{150} accrued maximum nitrogen (71.2 kg ha^{-1}). There was 73 % difference between highest and lowest level of total nitrogen. In split doses higher (51.78 kg ha^{-1}) nitrogen was measured as compared to 45.19 kg ha^{-1} nitrogen was calculated at full dose nitrogen application method. Nitrogen application methods differed by about 13 %. Immense difference for total nitrogen at maturity stage observed during both years. During 2010–2011 higher total nitrogen was calculated (50.07 kg ha^{-1}) whereas, minimum total nitrogen (46.90 kg ha^{-1}) was calculated during 2011–2012. Similarly, for genotype NARC-2009, harvested total nitrogen (50.49 kg ha^{-1}) was higher as compared to Chakwal-50 (46.49 kg ha^{-1}). There was 8 % difference between two genotypes for total nitrogen. The interactions among NR×AM, NR×AM, NR×Y and AM×Y were significant at 1 % significance level, whereas, NR×AM×Y was significant at 5 % significance level.

4.3.1.2 Nitrogen Uptake Efficiency (NUpE)

The results showed the significant difference among different nitrogen rates and application methods on both wheat genotypes for 2 years for nitrogen uptake efficiency. Higher nitrogen uptake efficiency calculated for N_0 (0.53) as compared to (0.38) for N_{150} (Table 4.1). There was about 39 % difference for NUpE from highest to lowest value. Regarding nitrogen application methods higher NUpE was recorded in split doses (0.49) compared to full doses (0.43). Similarly, between years the higher nitrogen uptake efficiency was calculated during 2010–2011 (0.46) as compared to (0.45) during 2011–2012. Meanwhile wheat genotypes differed greatly for nitrogen uptake efficiency. Maximum nitrogen uptake efficiency (0.48) calculated for genotype NARC-2009 compared to Chakwal-50 which calculated minimum NUpE (0.44). The interactive effect of NR×Y and AM×Y illustrated significant differences at $P < 1\ \%$.

Nitrogen uptake efficiency is the measure how much nitrogen is taken up by the wheat crop. It was suggested that to increase NUE, nitrogen uptake must be increased (Raun and Johnson 1999). The results of present study depicted that NUpE is affected by nitrogen application rates and methods for both the years between two wheat genotypes. Highest nitrogen uptake efficiency was calculated for control (0.53) nitrogen rate while lowest (0.38) NUpE calculated for N_{150}. Our findings were in accordance with Rahimizadeh et al. (2010) who stated that nitrogen uptake efficiency decreased by increase in nitrogen rates.

4.3.1.3 Nitrogen Utilization Efficiency (NUtE)

Nitrogen rates varied considerably in depicting nitrogen utilization efficiency. Higher nitrogen utilization efficiency was calculated for N_0 (203.19 kg kg^{-1}) compared to (66.76 kg kg^{-1}) N_{150}. But the nitrogen application methods viz. split and full dose nitrogen application did not varied significantly for nitrogen utilization efficiency (Table 4.1). Whereas, varied nitrogen utilization efficiency calculated during both years. Maximum nitrogen utilization efficiency calculated during 2010–2011 (125.56 kg kg^{-1}) and minimum NUtE recorded during 2011–2012 (110.5 kg kg^{-1}).

Wheat genotypes due to their genetic make-up responded differently to nitrogen utilization efficiency. NARC-2009 responded well to give highest NUtE (125.13 kg kg^{-1}) compared to Chakwal-50 (110.93 kg kg^{-1}). The interactive effect of NR × AM and NR × Y depicted significant differences at $P < 1\%$ while AM × Y, G × Y and AM × G × Y interaction was significant at $P < 5\%$. Nitrogen rates and application methods had profound effect on grain yield. Nitrogen utilization efficiency represents the capability of a plant to convert up-taken nitrogen into grain (Delogu et al. 1998). The response of split dose and full dose application methods were same for NUtE. In the present study nitrogen utilization efficiency decreased with the increase in nitrogen rates. The same findings were elaborated by Delogu et al. (1998) who stated that nitrogen utilization efficiency reduced with enhancing nitrogen fertilizer rates.

4.3.1.4 Nitrogen Use Efficiency (NUE)

Nitrogen use efficiency was calculated to determine the ability of wheat to respond upon the application of nitrogen fertilizers. Nitrogen rates differed significantly for showing nitrogen use efficiency. Maximum NUE calculated (108.49 kg kg^{-1}) for N_0 while minimum nitrogen use efficiency (25.47 kg kg^{-1}) calculated for N_{150} (Table 4.1). Similarly, nitrogen application methods varied significantly for NUE. Split nitrogen doses considerably gave higher nitrogen use efficiency (58.95 kg kg^{-1}) compared with (53.77 kg kg^{-1}) for full dose nitrogen application method. In the same way, both the years differed considerably for nitrogen use efficiency. More nitrogen use efficiency calculated during 2010–2011 (60.55 kg kg^{-1}) and less NUE calculated during 2011–2012 (52.17 kg kg^{-1}). Both genotypes differed significantly in showing nitrogen use efficiency. Genotype NARC-2009 depicted higher nitrogen use efficiency (62.19 kg kg^{-1}) compared with Chakwal-50 who gave lower NUE (50.52 kg kg^{-1}).

The results depicted that NUE of wheat affected by nitrogen fertilizer rates and application methods. The nitrogen use efficiency for split dose application was more than full dose nitrogen application method. Reduction in wheat NUE during 2011–2012 was due to lower grain yield than 2010–2011. In the present study results depicted that NUE reduced with increasing nitrogen rates.

Nitrogen use efficiency is actually the measure of how much grain yield produced by applying one unit of fertilized nitrogen. In control treatment no fertilizer nitrogen applied in the field but grain yield produced due to nitrogen present in the soil profile so, maximum nitrogen use efficiency calculated for control nitrogen. Our results were in line with Zhao et al. (2006) who stated that nitrogen use efficiency diminished with increase N rates. Likewise, Timsina et al. (2001) were of the view that nitrogen use efficiency declined by enhancing nitrogen fertilizers.

4.3.2 *Yield Parameters*

4.3.2.1 Number of Tillers

Number of tillers per meter square varied significantly for varying nitrogen rates. Maximum number of tillers calculated for N_{100} (211) while minimum number of tillers (202) counted for N_0. Percentage difference for number of tillers per meter square among nitrogen rates was 2.5. N_{50} and N_{150} gave same results (206) for number of tillers per meter square. However, nitrogen application methods differed considerably. Higher number of tillers (208) calculated for split dose nitrogen application method whereas, lower number of tillers (205) counted for full dose nitrogen application method. There was 1 % difference among both nitrogen application methods. Similarly, both wheat genotypes differed considerably for number of tillers per square meter. More number of tillers (219.74) counted for NARC-2009 than Chakwal-50 (192.72). Among years there was huge variation among number of tillers per square meter. During 2010–2011, 251 tillers ware calculated from the field while during 2011–2012 only 161.03 tillers were calculated (Table 4.2). Fertile tillers hold a vital place to depict crop productivity as they are major constituent of crop yield. Higher number of fertile tillers often attributed to more yield. With the increase in nitrogen level number of tillers increased upto an optimum level. López-Bellido and López-Bellido (2001) reported that with the enhancement of nitrogen fertilizer the number of tillers and grain yield increased.

4.3.2.2 Thousand Grain Weight (g)

Significant variations were observed for thousand grain weight among different nitrogen rates and application methods for wheat genotypes during both years under present study. Nitrogen rates did not varied for thousand grains weight. However, nitrogen application methods gave distinct variations for thousand grain weight. Split dose nitrogen application method gave maximum thousand grain weight (43.44 g) while minimum thousand grain weight (40.42 g) was produced by full dose nitrogen application method (Table 4.2). Same as previous, 7 % variation was calculated among nitrogen application methods for thousand grain weight. In the same way, thousand grain weight varied during both growing years. During 2010–2011 higher thousand grain weight (48.02 g) was calculated while less thousand grain weight (35.84 g) was produced during 2011–2012. Percentage difference of 25 % recorded among both years. Meanwhile, wheat genotypes due to their genetic make behaved differently for thousand grain weight. Highest thousand grain weight (43.28 g) was calculated by NARC-2009 while lowest thousand grain weight (40.58 g) was calculated by Chakwal-50. Chakwal-50 accumulated 6 % less thousand grain weight compared to NARC-2009. The interactive effect of Y × G was highly significant at 1 % significance level. All the other interactions were not significant at 5 % significance level. Thousand grain weight is a very crucial varietal character contributing towards final yield. This variation might be due to increase in temperature and moisture stress during later growth stages of wheat crop and ultimately it had

Table 4.2 Yield and yield parameters under different nitrogen rates and application methods among two wheat genotypes during 2010–2011 and 2011–2012

Treatments	No. of tillers	Thousand grain weight	Biological yield	Grain yield	Harvest index
Nitrogen Rate (NR)					
N_0	202[c]	41.63[NS]	11,813[c]	3864[d]	33.76[b]
N_{50}	206[b]	41.88	12,388[bc]	4395[c]	35.56[ab]
N_{100}	211[a]	42.31	14,764[a]	5441[a]	37.38[a]
N_{150}	206[b]	41.91	12,999[b]	4727[b]	37.49[a]
Application Methods (AM)					
Split	208[a]	43.44[a]	12,862[NS]	4752[a]	38.27[a]
Full	205[b]	40.42[b]	13,119[NS]	4461[b]	33.83[b]
Year (Y)					
Y1	251[a]	48.02[a]	14,820[a]	5027[a]	34.39[b]
Y2	161[b]	35.84[b]	11,161[b]	4186[b]	37.71[a]
Genotypes (G)					
NARC-2009	220[a]	43.28[a]	13,756[a]	5005[a]	37.15[NS]
CHAKWAL-50	193[b]	40.58[b]	12,225[b]	4208[b]	34.94
Interactions					
NR×AM	NS	NS	NS	NS	NS
NR×AM	NS	NS	NS	***	***
NR×Y	NS	NS	NS	NS	NS
AM×G	NS	NS	NS	NS	NS
AM×Y	NS	NS	***	**	*
G×Y	**	***	**	**	NS
NR×AM×G	NS	NS	NS	NS	NS
NR×AM×Y	NS	NS	NS	NS	NS
NR×G×Y	NS	NS	NS	NS	NS
AM×G×Y	NS	NS	NS	NS	**
NR×AM×G×Y	NS	NS	NS	NS	NS

Different letters indicate a statistical difference among the treatments at $P < 0.05$

marked influence on grain yield of crop. Results were in the line with the findings of (Singh and Agrawal 2005) who stated that nitrogen application levels change thousand grain weights in wheat. The interaction among years, genotypes, N application methods and treatments was highly significant (Table 4.2). Our results were in accordance with Jun-Hua et al. (2010) who stated that nitrogen directly influence kernel weights of wheat crop by increasing thousand grain weight. (Yang et al. 2000) also reported higher thousand grain weight for high nitrogen levels. Our findings were in line with Nakano and Morita (2009) who were of the point of view that grain weight increased when N was applied in split doses.

4.3.2.3 Biological Yield (kg ha^{-1})

At maturity maximum drymatter was produced for N_{100} (14,764 kg/ha) while minimum biomass was calculated at control (11,813 kg/ha). N_{100} produced about 20 % more biomass than N_0 (Table 4.2). There was no significant variation for drymatter production among nitrogen application methods at maturity. Similarly, during 2010–2011 more biomass was harvested (14,820 kg/ha) than 2011–2012 (11,161 kg/ha). A variation of about 25 % was calculated among both years for biomass production at anthesis stage. While in case of genotypes more biomass was produced by NARC-2009 (13,756 kg/ha) than Chakwal-50 (12,225 kg/ha). In the same way, about 11 % variation was calculated among genotypes for drymatter production. The interactive effect of AM×Y was significant at 1 % P level and G×Y was significant at 5 % P level.

The drymatter accumulation enhanced at post anthesis stages. Same to our findings Jun-Hua et al. (2010) reported that with the addition of nitrogen at late growth stages like anthesis the dry matter accumulation is enhanced. Dry matter translocation efficiency (12.15–28.25 %) was not affected by N treatments, but it was affected by the cultivars and the growing period. The values reported in the study were higher than the values reported in other studies on cereals (Dordas and Sioulas 2009). The change in drymatter production was due to variation in soil moisture status. Similar to our findings, White and Wilson (2006) testified that crop drymatter was expressively affected due to change in environments. (Khayatnezhad and Gholamin 2012) depicted that drymatter production is increased by increasing nitrogen levels for wheat crop. Our results were also in accordance with Marino et al. (2011) who stated that nitrogen had principle role in dry matter accumulation and enhancing grain yield in wheat crop.

4.3.2.4 Grain Yield (kg ha^{-1})

Outcomes of the current study highlighted that genotypes behaved differently at different nitrogen rates and application methods during both years for grain yield. Grain yield differed significantly in response to different nitrogen rates (Table 4.2). Maximum grain yield (5441 kg ha^{-1}) was harvested for N_{100} whereas minimum grain yield (3864 kg ha^{-1}) obtained for N_0. The percentage difference for grain yield among highest and lowest nitrogen rates was about 29. Nitrogen application methods differed greatly for grain yield in present study. Higher grain yield (4752 kg ha^{-1}) was obtained by split dose nitrogen application as compared to (4461 kg ha^{-1}) obtained for full dose. A variation of about 7 % calculated between nitrogen application methods. In the same way, significant variation in grain yield recorded during both years. Higher grain yield (5027 kg ha^{-1}) was recorded during 2010–2011 against 2011–2012 (4186 kg ha^{-1}). During 2011–2012, about 20 % less grain yield obtained than 2010–2011. Both genotypes varied considerably for grain yield production. NARC-2009 produced higher grain yield (5005 kg ha^{-1}) than Chakwal-50 (4208 kg ha^{-1}). A percentage difference of about 19 % calculated among both wheat genotypes. The interactive effect of NR×G was significant at 1 % significance level while AM×Y and G×Y were significant at 5 % P level. Other interactions were non-significant.

Yield is the final outcome of the research depends upon the fertility of soil. Nitrogen rates and application methods varied grain yield for both wheat genotypes during 2010–2011 and 2011–2012. Muurinen et al. (2007) reported higher grain yield with high nitrogen rates in cereals. Bakht et al. (2009) reported that in low nitrogen soils by adding crop residue and nitrogen fertilizer, grain yield could be increased.

4.3.2.5 Harvest Index

The results revealed significant difference for harvest index among different genotypes at varying nitrogen rates and application methods for both years. Results depicted that there were no significant difference among nitrogen rates (Table 4.2). Similarly split and full dose nitrogen application methods did not differ significantly. Meanwhile, no variation among years and wheat genotypes were calculated during present study. The harvest index is important crop parameter that is obtained by dividing economically valuable part of crop (grain yield) with the above ground biomass (biological yield). Nitrogen treatments did not provide much difference in harvest index. The harvest index was not affected by the N level, as the proportion of change of the total biomass and grain yield was similar. Similar results were reported for other crop species, such as winter wheat, and safflower (Dordas 2009; Dordas and Sioulas 2009) where N application did not affect the HI.

4.3.3 Model Parameterization and Evaluation

Model testing consists of two main activities (i) establishing the source codes representing the models performance as intended, and (ii) confirming that simulation models accurately reproduce empirical data (Meinke 1996). These two activities were referred as model verification and validation (V & V) (Kleijnen 1995). Model verification and validation against an independent data set is an essential step in model development. APSIM model was parameterized and evaluated for nitrogen dynamics in wheat. In the present study the APSIM model was evaluated for simulation of days after sowing, dry matter accumulation (biological yield), grain yield, biomass nitrogen, root nitrogen, grain total nitrogen as these were the major constituent of optimal crop productivity.

4.3.3.1 Days After Sowing

There was a close association among observed and simulated days after sowing from for Zadok's scale (Three leaf, Anthesis and Maturity). Figure 4.1 represents observed and simulated days after sowing of two wheat genotypes at different nitrogen rates and application methods for both years. Observed days after sowing (DAS) (32) was higher at three leaf stage than the APSIM simulated yield (26). APSIM simulated days after sowing with acute accuracy for nitrogen application methods. Higher days after sowing (33) was accumulated for split dose nitrogen

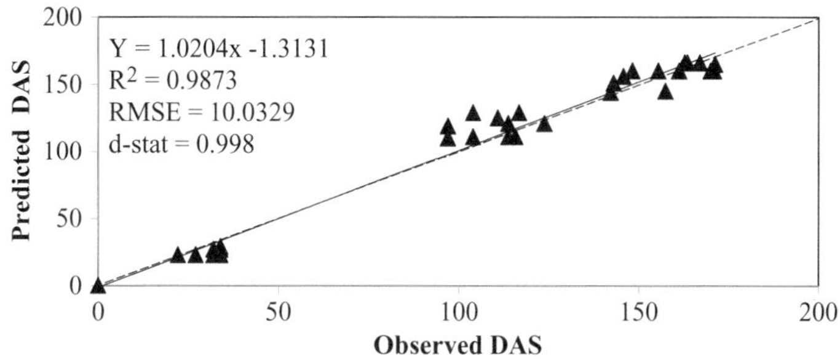

Fig. 4.1 Observed and simulated days after sowing (DAY) by APSIM of two wheat genotypes for different nitrogen application method during 2010–11 and 2011–2012

application method than full dose method (28) and it was close to observed days after sowing recorded from full dose application method. APSIM simulated higher days after sowing (27.75) for NARC-2009 at three leaf stage than Chakwal-50 (27). Similarly during 2010–2011 simulated days after sowing was higher (33.5) than 2011–2012 (31.78). While, at anthesis APSIM simulated days after sowing (116.93) was lower than the observed yield (119.68). APSIM simulated days after sowing with acute accuracy for nitrogen application methods. Higher days after sowing (115.37) was accumulated for split dose nitrogen application method than full dose method (118.8). APSIM simulated higher days after sowing (119) for NARC-2009 at three leaf stage than Chakwal-50 (115). Similarly during 2010–2011 simulated days after sowing was higher (120) than 2011–2012 (114). Whereas, APSIM simulated days after sowing (159.46) was close to observed yield (158.68) at maturity. Higher days after sowing (159.38) was accumulated for split dose nitrogen application method than full dose method (158) and it was close to observed biomass total nitrogen recorded from split dose application method. APSIM simulated higher days after sowing (162.25) for NARC-2009 at maturity stage than Chakwal-50 (155). Similarly during 2010–2011 simulated days after sowing was higher (160.2) than 2011–2012 (157). Our results were in accordance to Zhang et al. (2008) who were of the view that yield simulation may be improved if models can simulate a more accurate days after sowing due to variable nutrient conditions. The accurate simulation of DAS by APSIM showed that model can work with good accuracy and can be used to made decisions about crop managements.

4.3.3.2 Biomass Total Nitrogen (g m^{-2})

APSIM model was parameterized to simulate biomass total nitrogen contents under different nitrogen regime and application methods during 2010–2011 and 2011–2012 for two wheat genotypes at three phenological stages (Three leaf, Anthesis and Maturity). Observed biomass total nitrogen (1.06875 g m^{-2}) was higher at three leaf stage than the APSIM simulated biomass total nitrogen (0.55 g m^{-2}). APSIM simulated biomass total nitrogen with acute accuracy for nitrogen application methods. Higher biomass total nitrogen (1.165 g m^{-2}) was accumulated for split dose nitrogen

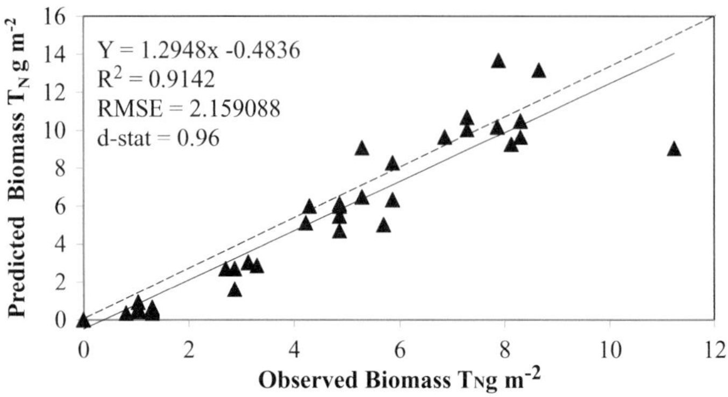

Fig. 4.2 Observed and simulated Biomass total nitrogen by APSIM of two wheat genotypes for different nitrogen application method during 2010–11 and 2011–2012

application method than full dose method (1.06875 g m^{-2}) and it was close to observed biomass total nitrogen recorded from full dose application method. APSIM simulated higher biomass total nitrogen (0.60625) for NARC-2009 at three leaf stage than Chakwal-50. Similarly during 2010–2011 simulated biomass total nitrogen was higher than 2011–2012 (0.4725 g m^{-2}). While, at anthesis APSIM simulated biomass total nitrogen (5.085) was lower than the observed yield (7.524). APSIM simulated biomass total nitrogen with acute accuracy for nitrogen application methods. Higher biomass total nitrogen (5.256 g m^{-2}) was accumulated for split dose nitrogen application method than full dose method (4.9865 g m^{-2}) and it was close to observed biomass total nitrogen recorded from split dose application method. APSIM simulated higher biomass total nitrogen (7.498 g m^{-2}) for NARC-2009 at three leaf stage than Chakwal-50 (6.281 g m^{-2}). Similarly during 2010–2011 simulated biomass total nitrogen was higher (7.080 g m^{-2}) than 2011–2012 (5.960 kg/ha). Whereas, APSIM simulated biomass total nitrogen (3.139) was close to observed yield (3.656 g m^{-2}). APSIM simulated biomass total nitrogen with acute accuracy for nitrogen application methods. Higher biomass total nitrogen (2.868 g m^{-2}) was accumulated for split dose nitrogen application method than full dose method (2.178 g m^{-2}) and it was close to observed biomass total nitrogen recorded from split dose application method. APSIM simulated higher biomass total nitrogen (3.656 g m^{-2}) for NARC- 2009 at maturity stage than Chakwal-50 (1.9419 g m^{-2}). Similarly during 2010–2011 simulated biomass total nitrogen was higher (3.542 g m^{-2}) than 2011–2012 (2.4523 g m^{-2}). This variation might be due to the reason that there was variation in moisture contents at critical growth stages. Figure 4.2 represents observed calculated by APSIM model for both the years. The reduction in simulating grain and simulated days after sowing of two wheat genotypes at different nitrogen rates and application methods for both years. The use of models to simulate biomass total nitrogen was reported with good accuracy in earlier work who concluded that APSIM-wheat module can simulate biomass nitrogen and model was able to explain more than 90 % variation in crop biomass (Chen et al. 2010).

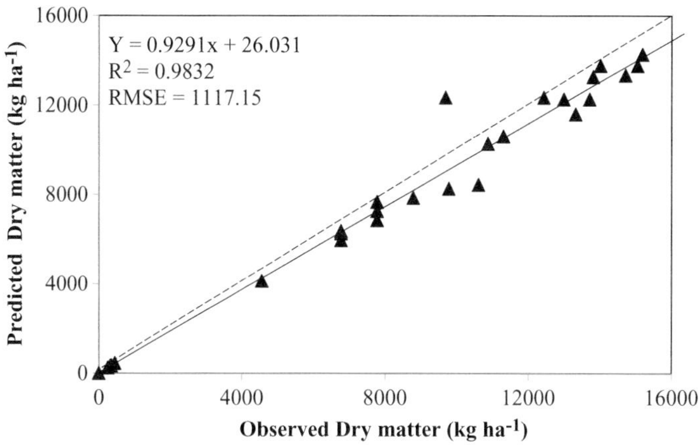

Fig. 4.3 Observed and simulated dry matter (biological yield) by APSIM of two wheat genotypes for different nitrogen application method during 2010–11 and 2011–2012

4.3.3.3 Dry Matter

APSIM model was parameterized to simulate dry matter accumulation under different nitrogen regime and application methods during 2010–2011 and 2011–2012 for two wheat genotypes at three phenological stages (Three leaf, Anthesis and Maturity). APSIM model simulated dry matter contents with great accuracy with observed dry matter contents (biological yield). Figure 4.3 represents observed and simulated dry matter contents of two wheat genotypes at different nitrogen rates and application methods for both years. Observed biological yield (357 kg/ha) was higher at three leaf stage than the APSIM simulated yield (341 kg/ha). APSIM simulated biological yield with acute accuracy for nitrogen application methods. Higher dry matter (351 kg/ha) was accumulated for full dose nitrogen application method than split dose method (330 kg/ha) and it was close to observed dry matter recorded from full dose application method (396 kg/ha). APSIM simulated higher drymatter (357 kg/ha) for NARC-2009 at three leaf stage than Chakwal-50 (330 kg/ha). Similarly during 2010–2011 simulated biological yield was higher (344 kg/ha) than 2011–2012 (338 kg/ha). While, at anthesis APSIM simulated biological yield (7833.625 kg/ha) was higher than the observed yield (7498 kg/ha). APSIM simulated biological yield with acute accuracy for nitrogen application methods. Higher dry matter (7373 kg/ha) was accumulated for split dose nitrogen application method than full dose method (6706 kg/ha) and it was close to observed dry matter recorded from split dose application method. APSIM simulated higher drymatter (7498 kg ha^{-1}) for NARC-2009 at three leaf stage than Chakwal-50 (6581 kg ha^{-1}). Similarly during 2010-2011 simulated biological yield was higher (7180 kg/ha) than 2011–2012 (6900 kg/ha). Whereas, APSIM simulated biological yield (13,139 kg/ha) was close to observed yield (13,556 kg/ha). APSIM simulated biological yield with acute accuracy for nitrogen application methods. Higher dry matter (12,868 kg/ha) was accumulated for split dose nitrogen application method than full dose method (12,180 kg/ha) and it was close to observed dry matter recorded from split dose application method. APSIM simulated higher drymatter (13,556 kg ha^{-1}) for NARC-2009 at three leaf stage than Chakwal-50 (11,419 kg ha^{-1}). Similarly during 2010–2011 simulated biological yield was higher (13,525 kg/ha) than 2011–2012 (11,523 kg/ha).

Meinke (1996) stated that model simulation is dependent upon triangle of climate, soil and plant genetic features. Same like observed biological yield modeled grain yield differed greatly for varying nitrogen rates and application methods among both wheat genotypes during both years. Our results were in line with Kmoch et al. (1957) who stated that with the enhancement of nitrogen fertilizer levels the root weight increase which ultimately increase biological yield. Hocking and Meyer (1991) were of the point of view that control nitrogen treatments had less biological yield than applied nitrogen fertilizers.

4.3.3.4 Grain Yield

APSIM model was parameterized to simulate grain yield under different nitrogen regime and application methods during 2010–2011 and 2011–2012 for two wheat genotypes. Figure 4.4 represents observed and simulated grain yield of two wheat genotypes at different nitrogen rates and application methods for both years. Observed and simulated grain yield were very close to each other. Nitrogen application rates and methods varied potentially for simulating grain yield of wheat crop. A direct relation with nitrogen fertilizing rates calculated in simulating grain yield by APSIM model. At higher nitrogen fertilizer levels (N_{100}) maximum grain yield (5094 kg/ha) simulated whereas, minimum grain yield (3545 kg/ha) simulated for control nitrogen rate (N_0). Similarly, variation for grain yield simulation was less yield during 2011–2012 (4028 kg/ha) than 2010–2011 (4611 kg/ha) was due the less moisture availability. Meinke (1996) stated that model simulation is dependent

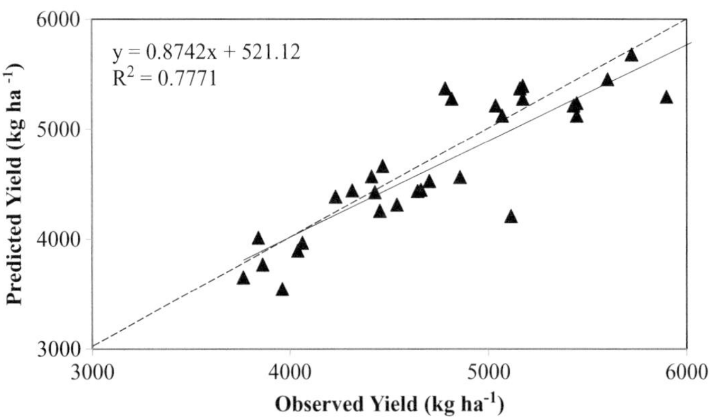

Fig. 4.4 Observed and simulated grain yield by APSIM of two wheat genotypes for different nitrogen application method during 2010–11 and 2011–2012

Table 4.3 Showing work on nitrogen in relation to crop traits and its effect using different techniques

Sr. no.	Findings	References
1	Root Zone Water Quality Model (RZWQM) for winter wheat (Triticum aestivum L.) production and compared with the CERES-Wheat model to assess their potential as N management tools	Saseendran et al. (2004)
2	The potential effects of climate change, NO3-N losses were assessed by using RZWQM2 and concluded that under the future climate, NO3-N loss and flow-weighted average NO3-N concentration increased	Wang et al. (2015)
3	Two ecological models of nitrogen processes using the Modelica modelling and simulation language evaluated and compared and results depicted that MathModelica Model Editor could better predict nitrogen losses in the form of Nitrification/Denitrificatin	Edelfeldt and Fritzson (2008)
4	Pasture Simulation Model (PaSim) and CropSyst models were used to analyze the shift in the ratio of N lost via leaching, denitrification and volatilization	Dueri et al. (2007)
5	Climate is influencing nitrogen cycle so NO3 level is affected by mineralization-immobilization processes in the soil	Melillo et al. (2002)
6	To investigate climate change impacts on drainage and N loss DRAINMOD used under agricultural production systems	Dayyani et al. (2012), Singh et al. (2009)
7	RZWQM2 validated by using 16-year (1989–2004) drainage data to predict NO3-N loss	Qi et al. (2012)

upon triangle of climate, soil and plant genetic features. Same like observed grain yield modeled grain yield differed greatly for varying nitrogen rates and application methods among both wheat genotypes during both years. Maximum grain yield (4496 kg/ha) modeled by APSIM for genotype NARC-2009 during 2010–2011 for nitrogen rate N_{100} when it was applied as split dose. While minimum grain yield (4143 kg/ha) simulated for Chakwal-50 with higher nitrogen application rate (N_{100}). Tadayon (2007) was of the point of view that genotypes vary for grain yield production due to their genetic behavior under different nitrogen regime. Our results were in line with Melaj et al. (2003) who stated that grain yield increases due to increase in applied N. Martre et al. (2006) simulated grain yield with varying nitrogen rates and found direct relation among grain yield and applied nitrogen. In Table 4.3 the recent work on nitrogen modeling and its fate have been elaborated.

4.4 Recommendations

- The study about nitrogen modelling under changing climate found to be highly valuable for predicting the yield for policy makers.
- Split dose application methods should be adopted to increase wheat yield under rainfed agriculture.
- Higher nitrogen applications rates like 100 kgN/ha must be applied to get higher wheat grain yield.

References

Abril, A., D. Baleani, N. Casado-Murillo, and L. Noe. 2007. Effect of wheat crop fertilization on nitrogen dynamics and balance in the Humid Pampas, Argentina. *Agriculture, Ecosystems & Environment* 119: 171–176.

Anderson, J.M., and J.S.I. Ingram. 1993. *Tropical soil biology and fertility: A handbook of methods*. Wallingford: CAB International.

Bakht, J., M. Shafi, M.T. Jan, and Z. Shah. 2009. Influence of crop residue management, cropping system and N fertilizer on soil N and C dynamics and sustainable wheat (Triticum aestivum L.) production. *Soil and Tillage Research* 104: 233–240.

Challinor, A.J., F. Ewert, S. Arnold, E. Simelton, and E. Fraser. 2009. Crops and climate change: Progress, trends, and challenges in simulating impacts and informing adaptation. *Journal of Experimental Botany* 60: 2775–2789.

Chen, C., E. Wang, and Q. Yu. 2010. Modelling the effects of climate variability and water management on crop water productivity and water balance in the North China Plain. *Agricultural Water Management* 97: 1175–1184.

Dayyani, S., S.O. Prasher, A. Madani, and C.A. Madramootoo. 2012. Impact of climate change on the hydrology and nitrogen pollution in a tile-drained agricultural watershed in Eastern Canada. *Transactions of the ASABE* 55: 389–401.

Delogu, G., L. Cattivelli, N. Pecchioni, D. De Falcis, T. Maggiore, and A.M. Stanca. 1998. Uptake and agronomic efficiency of nitrogen in winter barley and winter wheat. *European Journal of Agronomy* 9: 11–20.

Dhungana, P., K.M. Eskridge, A. Weiss, and P.S. Baenziger. 2006. Designing crop technology for a future climate: An example using response surface methodology and the CERES-Wheat model. *Agricultural Systems* 87: 63–79.

Diez, J.A., R. Caballero, A. Tarquis, M.C. Cartagena, and A. Vallejo. 2000. Integrated fertilizer and irrigation management to reduce nitrate leaching in central Spain. *Journal of Environmental Quality* 29: 1539–1547.

Dordas, C. 2009. Dry matter, nitrogen and phosphorus accumulation, partitioning and remobilization as affected by N and P fertilization and source–sink relations. *European Journal of Agronomy* 30: 129–139.

Dordas, C.A., and C. Sioulas. 2009. Dry matter and nitrogen accumulation, partitioning, and retranslocation in safflower (Carthamus tinctorius L.) as affected by nitrogen fertilization. *Field Crops Research* 110: 35–43.

Dueri, S., P.L. Calanca, and J. Fuhrer. 2007. Climate change affects farm nitrogen loss – A Swiss case study with a dynamic farm model. *Agricultural Systems* 93: 191–214.

Edelfeldt, S., and P. Fritzson. 2008. Evaluation and comparison of models and modelling tools simulating nitrogen processes in treatment wetlands. *Simulation Modelling Practice and Theory* 16: 26–49.

Giller, K.E. 2004. Emerging technologies to increase the efficiency of the use of fertilizer nitrogen. In *Agriculture and nitrogen cycle. Scope 65*, ed. A.R. Moiser, J.K. Syers, and J.R. Frency. Washington, DC: Island Press.

Halvorson, A.D., B.J. Wienhold, and A.L. Black. 2002. Tillage, nitrogen, and cropping system effects on soil carbon sequestration. *Soil Science Society of America Journal* 66: 906–912.

Hocking, P.J., and C.P. Meyer. 1991. Effects of CO_2 enrichment and nitrogen stress on growth, and partitioning of dry matter and nitrogen in wheat and maize. *Australian Journal of Plant Physiology* 18: 339–356.

Janssen, B.H., F.C.T. Guiking, D. Van-Der-Eijk, E.M.A. Smaling, J. Wolf, and H. Reuler. 1990. A system for quantitative evaluation of the fertility of tropical soils (QUEFTS). *Geoderma* 46: 299–318.

Jun-Hua, Z., L. Jian-Li, Z. Jia-Bao, Z. Fu-Tao, C. Ya-Nan, and W. Wei-Peng. 2010. Effects of nitrogen application rates on translocation of dry matter and nitrogen utilization in rice and wheat. *Acta Agronomica Sinica* 36: 1736–1742.

Khayatnezhad, M., and R. Gholamin. 2012. Effect of nitrogen fertilizer levels on different planting remobilization of dry matter of durum wheat varieties Seimareh. *African Journal of Microbiology Research* 6: 1534–1539.
Kleijnen, J.P.C. 1995. Verification and validation of simulation models. *European Journal of Operational Research* 82: 145–161.
Kmoch, H.G., R.E. Ramig, R.L. Fox, and F.E. Koehler. 1957. Root development of winter wheat as influenced by soil moisture and nitrogen fertilization. *Agronomy Journal* 49: 20–26.
Liang, H., K. Hu, W.D. Batchelor, Z. Qi, and B. Li. 2016. An integrated soil-crop system model for water and nitrogen management in North China. *Scientific Reports* 6, 25755.
López-Bellido, R.J., and L. López-Bellido. 2001. Efficiency of nitrogen in wheat under Mediterranean conditions: Effect of tillage, crop rotation and N fertilization. *Field Crops Research* 71: 31–46.
López-Bellido, L., R.J. López-Bellido, and R. Redondo. 2005. Nitrogen efficiency in wheat under rainfed Mediterranean conditions as affected by split nitrogen application. *Field Crops Research* 94: 86–97.
Mahler, R.L., F.E. Koehler, and L.K. Lutcher. 1994. Nitrogen source, timing of application, and placement: Effects on winter wheat production. *Agronomy Journal* 86: 637–642.
Marino, S., R. Tognetti, and A. Alvino. 2011. Effects of varying nitrogen fertilization on crop yield and grain quality of emmer grown in a typical Mediterranean environment in central Italy. *European Journal of Agronomy* 34: 172–180.
Martre, P., P.D. Jamieson, M.A. Semenov, R.F. Zyskowski, J.R. Porter, and E. Triboi. 2006. Modelling protein content and composition in relation to crop nitrogen dynamics for wheat. *European Journal of Agronomy* 25: 138–154.
Mccown, R.L., G.L. Hammer, J.N.G. Hargreaves, D.P. Holzworth, and D.M. Freebairn. 1996. APSIM: A novel software system for model development, model testing and simulation in agricultural systems research. *Agricultural Systems* 50: 255–271.
Meinke, K. 1996. Topological methods for algebraic specification. *Theoretical Computer Science* 166: 263–290.
Melaj, M.A., H.E. Echeverría, S.C. López, G. Studdert, F. Andrade, and N.O. Bárbaro. 2003. Timing of nitrogen fertilization in wheat under conventional and no-tillage system. *Agronomy Journal* 95: 1525–1531.
Melillo, J.M., P.A. Steudler, J.D. Aber, K. Newkirk, H. Lux, F.P. Bowles, C. Catricala, A. Magill, T. Ahrens, and S. Morrisseau. 2002. Soil warming and carbon-cycle feedbacks to the climate system. *Science* 298: 2173–2176.
Muurinen, S., J. Kleemola, and P. Peltonen-Sainio. 2007. Accumulation and translocation of nitrogen in spring cereal cultivars differing in nitrogen use efficiency. *Agronomy Journal* 99: 441–449.
Nakano, H., and S. Morita. 2009. Effects of seeding rate and nitrogen application rate on grain yield and protein content of the bread wheat cultivar 'Minaminokaori' in Southwestern Japan. *Plant Protection Science* 12: 109–115.
Oenema, O., H. Kros, and W. De-Vries. 2003. Approaches and uncertainties in nutrient budgets: Implications for nutrient management and environmental policies. *European Journal of Agronomy* 20: 3–16.
Parry, M., C. Rosenzweig, and M. Livermore. 2005. Climate change, global food supply and risk of hunger. *Philosophical Transactions of the Royal Society, B: Biological Sciences* 360: 2125–2138.
Pathak, H., P.K. Aggarwal, R. Roetter, N. Kalra, S.K. Bandyopadhaya, S. Prasad, and H. Van Keulen. 2003. Modelling the quantitative evaluation of soil nutrient supply, nutrient use efficiency, and fertilizer requirements of wheat in India. *Nutrient Cycling in Agroecosystems* 65: 105–113.
Porter, J.R., and M.A. Semenov. 2005. Crop responses to climatic variation. *Philosophical Transactions of the Royal Society B: Biological Sciences* 360: 2021–2035.
Power, J.F., R. Wiese, and D. Flowerday. 2000. Managing nitrogen for water quality—Lessons from management systems evaluation area. *Journal of Environmental Quality* 29: 355–366.
Praveen, K., and R.K. Aggarwal. 1998. Interdependence of ammonia volatilization and nitrification in arid soils. *Nutrient Cycling in Agroecosystems* 51: 201–207.

Qi, Z., L. Ma, M.J. Helmers, L.R. Ahuja, and R.W. Malone. 2012. Simulating nitrate-nitrogen concentration from a subsurface drainage system in response to nitrogen application rates using RZWQM2. *Journal of Environmental Quality* 41: 289–295.
Rahimizadeh, M., A. Kashani, A.Z. Feizabadi, A.R. Kooocheki, and M.N. Mahallati. 2010. Nitrogen use efficiency of wheat as effected by preceding crop, application rate of nitrogen and crop residues. *Australian Journal of Crop Science* 4: 363–368.
Raun, W.R., and G.V. Johnson. 1999. Improving nitrogen use efficiency for cereal production. *Agronomy Journal* 91: 357–363.
Rosenberg, N. 2010. Climate change, agriculture, water resources: What do we tell those that need to know? *Climatic Change* 100: 113–117.
Russell, E.J. 2010. *Soil conditions and plant growth*, 11th ed. London: Longman Publishing.
Saseendran, S.A., D.C. Nielsen, L. Ma, L.R. Ahuja, and A.D. Halvorson. 2004. Modeling nitrogen management effects on winter wheat production using RZWQM and CERES-wheat. *Agronomy Journal* 96: 615–630.
Singh, R., and S.K. Agrawal. 2005. Effect of levels of farmyard manure and nitrogen fertilizer on grain yield and use efficiency of nutrients on wheat (Triticum aestivum). *Indian Journal of Agricultural Sciences* 75: 408–413.
Singh, R., M.J. Helmers, A.L. Kaleita, and E.S. Takl. 2009. Potential impact of climate change on subsurface drainage in Iowa's subsurface drained landscapes. *Journal of Irrigation and Drainage Engineering* 135: 459–466.
Smil, V. 2005. Do we need higher farm yields during the first half of the 21st century? In *Yields of farmed species*, ed. R. Sylvester-Bradley and J. Wiseman, 1–14. Nottingham: Nottingham University Press.
Sogbedji, J.M., H.M. Van Es, C.L. Yang, L.D. Geohring, and F.R. Magdoff. 2000. Nitrate leaching and nitrogen budget as affected by maize nitrogen rate and soil type. *Journal of Environmental Quality* 29: 1813–1820.
Sowers, K.E., B.C. Miller, and W.L. Pan. 1994. Optimizing yield and grain protein in soft white winter wheat with split nitrogen applications. *Agronomy Journal* 86: 1020–1025.
Tadayon, M.R. 2007. *Effects of supplemental irrigation and N fertilization increase of two wheat cultivars under dryland conditions. (In Persion with abstract in English).* PhD, Shiraz University, Shiraz, Iran.
Timsina, J., U. Singh, M. Badaruddin, C. Meisner, and M.R. Amin. 2001. Cultivar, nitrogen, and water effects on productivity, and nitrogen-use efficiency and balance for rice–wheat sequences of Bangladesh. *Field Crops Research* 72: 143–161.
Van Sanford, D.A., and C.T. Mackown. 1986. Variation in nitrogen use efficiency among soft red winter wheat genotypes. *Theoretical and Applied Genetics* 72: 158–163.
Wang, E., Oosterom, E.V., Meinke, H., Asseng, S., Robertson, M., Huth, N., Keating, B., and M. Probert. 2003. The new APSIM-Wheat Model – Performance and future improvements. In *Proceedings of the 11th Australian Agronomy conference.* Geelong, Victoria.
Wang, Z., Z. Qi, L. Xue, M. Bukovsky, and M. Helmers. 2015. Modeling the impacts of climate change on nitrogen losses and crop yield in a subsurface drained field. *Climatic Change* 129: 323–335.
White, E.M., and F.E. Wilson. 2006. Responses of grain yield, biomass and harvest index and their rates of genetic progress to nitrogen availability in ten winter wheat varieties. *Irish Journal of Agricultural and Food Ressearch* 45: 85–101.
Yang, J., J. Zhang, Z. Huang, Q. Zhu, and L. Wang. 2000. Remobilization of carbon reserves is improved by controlled soil-drying during grain filling of wheat. *Crop Science* 40: 1645–1655.
Zadoks, J.C., T.T. Chang, and C.F. Konzak. 1974. A decimal code for the growth stages of cereals. *Weed Research* 14: 415–421.
Zhang, X., S. Chen, H. Sun, D. Pei, and Y. Wang. 2008. Dry matter, harvest index, grain yield and water use efficiency as affected by water supply in winter wheat. *Irrigation Science* 27: 1–10.
Zhao, J.-J., Y.-J. Hua, D.-G. Sun, X.-X. Meng, H.-S. Xiao, and X. Ma. 2006. Genome-wide microRNA profiling in human fetal nervous tissues by oligonucleotide microarray. *Child's Nervous System* 22: 1419–1425.

Chapter 5
Climate Variability Impact on Rice Production: Adaptation and Mitigation Strategies

Mukhtar Ahmed, Fayyaz-ul-Hassan, and Shakeel Ahmad

Contents

M. Ahmed (✉)
Department of Agronomy, Pir Mehr Ali Shah Arid Agriculture University, Rawalpindi, Pakistan

Department of Biological Systems Engineering, Washington State University, Pullman, WA, USA
e-mail: ahmadmukhtar@uaar.edu.pk; mukhtar.ahmed@wsu.edu

Fayyaz-ul-Hassan
Department of Agronomy, Pir Mehr Ali Shah Arid Agriculture University, Rawalpindi, Pakistan
e-mail: drsahi63@gmail.com

S. Ahmad
Faculty of Agricultural Sciences and Technology, Bahauddin Zakariya University, Multan 60800, Pakistan
e-mail: shakeelahmad@bzu.edu.pk

M. Ahmed, C.O. Stockle (eds.), *Quantification of Climate Variability, Adaptation and Mitigation for Agricultural Sustainability*,
DOI 10.1007/978-3-319-32059-5_5

Abstract Rice feeds half of humanity. Global climate change has given rise to food security issues. Changes in temperature and rainfall may affect the yield of rice as its water requirement is higher than other crops. Though rice is adaptable to a variety of environments, seasonal rainfall variability, and even at times complete absence of rainfall, are major issues in rice growing areas. This chapter discusses problems in the rice growing areas and possible solutions. The need of the hour is to find new strategies and ways to exploit the genetic yield potential of rice. Water use efficiency improvement is vital for the crop so that it may be grown under water-limiting conditions. The crop may be improved by selection and breeding techniques as well as molecular and biotechnological techniques. Crop management for enhanced water use efficiency has great significance. Production systems such as the system of rice intensification (SRI), alternate wetting and drying irrigation (AWD), aerobic rice system (ARS), raised beds and ground cover rice production system (GCRPS) to enhance water use efficiency are beneficial. Incorporation of the C4 photosynthetic pathway into rice is another approach to increase rice yield for food security problems in future. The conversion of rice from C_3 to C4 will enhance the yield of the crop. All these techniques can help tackle the problems of water scarcity and food security.

Keywords Rice • Crop management • Water use efficiency • Aerobic rice system

Abbreviations

IRRI	International Rice Research Institute
CIAT	Centro Internacional de Agricultura Tropical (Spanish: International Center for Tropical Agriculture Colombia)
ROS	Reactive Oxygen Species
DNA	Deoxyribose Nucleic Acid
UV	Ultra Violet
ABA	Abscisic Acid
Mha	Million hectares
SRI	System of Rice Intensification
MAS	Marker Assisted Selection

5.1 Introduction

5.1.1 Importance of Rice As a Major Cereal Food Crop

Rice is the most important among cereal grains. Rice feeds over half of mankind particularly in Asia. Rice is the vital grain from the perspective of human nutrition and caloric intake, contributing almost one fifth of the calorie intake by the human beings. Globally, it is the chief nutritional energy source. Rice fulfills 20 % of the world food energy requirement, whereas wheat contributes 19 % and maize 5 %. Rice has a pivotal role in the food security of rural populations. Asia provides 90 % of the total rice around the globe, i.e., about 640 million tonnes while Latin America contributes nearly 25 million tonnes and Sub-Saharan Africa produces 19 million tonnes. Ninety-five percent of the total production of rice comes from developing countries, whereas half of the world production is met by China and India. Variation in rice yields ranges from less than 1 ton/ha to more than 10 ton/ha in very poor rainfed areas and well irrigated temperate systems respectively.

Rice is adapted to a variety of environments from lowland irrigated to lowland and upland rainfed areas. About 80 million hectares of lowland irrigated rice produces 75 % of total world rice production. The irrigated lowland system comprises 56 % of the total area under rice in Asia (Swain and Singh 2005).These areas are the central rice producing areas from the perspective of food security, especially on the Asian continent. The lowland rainfed system has almost a 20 % share of global rice production. These areas include South Asia, parts of Southeast Asia and the African continent. These environments mostly suffer from various abiotic stresses accompanied by unpredictable rainfall. The upland rice system prevails in drylands without irrigation as well as areas where puddling is common. Its contribution in total world rice production is 4 % although it occupies an area of nearly 14 million hectares.

5.1.2 Botany of Rice

Asian rice has been categorized in a single group called the *Oryza sativa* complex (Tateoka 1962). All rice species have a well-developed root system. Rice species have long but somewhat branched adventitious roots however, since rice is a grass, a main root and concealed shoots are absent. In contrast to other crop plants rice has adaptive traits to tolerate submergence. Longitudinal interconnection of gas spaces, known as aerenchyma, is the distinguishing feature of rice. The aerenchyma cells enable internal aeration between shoots and roots (Colmer 2003). Oxygen is supplied by aerenchyma cells (Evans 2003). *Oryza sativa* has rooting nodes that sometimes produce new shoots.

5.1.3 Challenges in Rice Growing Areas

The major problem in rainfed rice areas is an unpredictable or abrupt rainfall pattern that results in several abiotic stress incidents. Drought prevails in an area of almost 27 million hectares. Devastating floods hit about 20 million hectares where deep water remains for a few months. Fields remain flooded with more than 100 cm of water periodically. Degraded soils also affect the crop. Coastal areas face the salinity problem. In lowland rainfed rice environments, poverty is the main issue that affects the reliability of yields because farmers cannot afford fertilizer or improved seed.

Climates in the upland rice environments are extremely unpredictable varying from humid to sub-humid. The soils range from moderately fertile to exceedingly infertile while the topography varies from level to sharply sloping. Limited population and access to markets are additional constraints. Poverty and the traditional cultivation system comprising long fallow periods are the elements that limit the crop potential in these upland ecosystems.

Periodic drought and floods bring abiotic stresses and have the central role in lowering the productivity of rainfed environments. According to a recent prediction of climate change the water deficit is going to worsen in the coming years (Wassmann et al. 2009) and both the concentration and incidence of drought are expected to become severe (Bates et al. 2008). Climate change has caused an increase in the minimum air temperatures in the crop seasons resulting in reduced rice yields mainly in China and the Philippines, and increases are predicted to continue. As a combined effect of abiotic stresses and increasing human population, rice prices have risen with consequences of intensified hunger and famine across the globe.

Rice productivity is primarily limited by drought. Drought disrupts plant water relations and biological membrane structures. From an agricultural perspective, drought is eventually expressed by its effects on yield, because it is the chief issue restricting crop expansion under water deficit conditions (Passioura 2007). Timing is the most important aspect of drought regarding its effect on rice.

Loss of equilibrium between the production and utilization of reactive oxygen species occurs under drought stress conditions (Smirnoff 1998), resulting in reduced production potential of the crop. High reactivity of ROS causes protein destruction, DNA disintegration, lipid peroxidation and, in the end, cell death (Beligni and Lamattina 1999). ROS are mainly produced in mitochondria and chloroplasts (Breusegem et al. 2001). As a consequence of all above effects, yields are reduced in a range of plant species under drought (Abdul-Jaleel et al. 2009). To avoid these damages a variety of primary and secondary metabolites are produced by the plant body as a protective strategy (Zhu 2002; Wahid et al. 2007). Evidence of free proline synthesis under a variety of stresses have been provided (Zhu 2002; Wahid et al. 2007). The influential antioxidant activity of phenolics (tannins, flavonoids, lignins) has been reported under moisture deficit conditions (Wahid 2007).

Ultraviolet and visible radiation disrupt the photosynthetic machinery as soon as they come in contact with it (Garcia-Plazaola and Becerril 2000), whereas phenolics impart resistance towards these harmful effects due to the presence of the benzene ring in their structure (Bilger et al. 2001). The production of highly water soluble anthocyanins saves plant from the devastating effects as they act as UV screens and osmolytes (Wahid et al. 2007). Polyamines like spermidine (Spd), spermine (Spm) and putrescine (Put) are small, universal nitrogenous compounds. They are now considered plant growth regulators and are also believed to be secondary messengers in signaling pathways (Kusano et al. 2008; Davies 2004; Liu et al. 2007). The role of polyamines in the abiotic stress tolerance was first reported by Richards and Coleman in 1952.

5.1.4 Effect of Drought on Rice Plants

If drought prevailed at the earlier stages of rice development, a considerable reduction in the transpiration rate was observed as the earliest response. Initially leaf growth and stem elongation were affected. Then a linear decline was observed as soon as available moisture reached 70% (Lilley and Fukai 1994). Leaf rolling leads to the reduced light interception and transpiration. Loss of water takes place through leaf surfaces even under stomatal closure since rice has low cuticular hydraulic resistance because of inadequate wax deposition. Young plants can tolerate drought by maintaining leaf area in stress conditions and by retaining their capacity to tiller after drought (Lilley and Fukai 1994). If drought occurs mid-season or during flowering it causes reduced grain number as well as yield since spikelet fertility is vulnerable to diminishing water levels.

Delayed flowering takes place under drought stress. The inflorescence may not emerge. Low turgor decreases panicle exertion while plant water potential has a positive correlation with flowering (Pantuwan et al. 2002a). The moisture available in the rhizosphere is utilized in transpiration by the plants. If permanent wilting of plants occurs death ensues.

The main determinants of dry matter production under drought are the potential of moisture extraction of the root system and the water use efficiency of the plant. The soil water extraction ability of rice depends on the root depth, root density and root length. Rice has a greater net root length than maize under normal circumstances but during extreme stress conditions upland rice fails to maintain root growth. There is a distinction between the root distribution patterns of rice and other crops (Kondo et al. 1999). Rice has lower potential to extract water from the deeper soil profile compared to other crops.

Transpiration is the main process to determine the performance of plant under drought (Lawlor and Tezara 2009). Reduced leaf net carbon uptake gives altered patterns of partitioning of photo assimilates that has the effect of an increased root to shoot ratio (Sharp 2002). The reason behind this type of response is hormonal activity, mainly by ethylene, ABA and interactions between them (Wilkinson and Davies 2010). Root/shoot conversion, accumulation of reserves in the stem under water stress (Chaves et al. 2002) with modification in C and N metabolism in various organs might lead to adaptation of crop under limited water as reported by Antonio et al. (2008). Carbohydrates are the main players of assimilation at the plant level that respond to internal and external stimuli (biotic and abiotic stresses). Their main role might be in different types of enzymatic reactions either to act as substrates or modulators in C-related pathways that control gene expression for C, N and lipid metabolism (Rolland et al. 2006).

Rice is the most extensively cultivated of all crops under irrigation around the world. In contrast to other cereal crops it requires 2–3 times as much water for 1 kg grain production (Barker et al. 1998). Presently more than 80% of the freshwater resources of Asia are being utilized for irrigating crops and half of these are consumed in rice production (Dawe et al. 1998). Depletion of water resources is a major threat for the irrigated rice crop giving rise to issues of food security and living of all the people directly or indirectly related to rice (Tuong et al. 2004). Seventeen Mha of irrigated rice land in Asia was susceptible to physical water shortage while 22 Mha may face economic water shortage by 2025 (Tuong and Bouman 2002). The true yield potential of rice might decline under water stress (Tuong et al. 2004). The challenge in the context of future climate change is to develop novel technologies and production systems that could maintain yield in the rice even under stress. Farming systems should be developed that increase or at least maintain production with the declining water availability. Therefore, attention must be paid to aerobic rice cultivation rather than flooded rice cultivation with the development of varieties that have better yield potential in aerobic environments (Castaneda et al. 2003). Techniques like aerobic rice (Bouman 2003), saturated soil culture (Borrell et al. 1997), bed planting or raised beds (Singh et al. 2003), rice intensification systems (Stoop et al. 2002), ground-cover systems (Lin et al. 2003a, b) and alternate wetting and drying (Tabbal et al. 2002) could be used as potential techniques under limited water. At present, research is being conducted in the areas of varietal development through traditional breeding, MAS as well as rendering of biotechnological tools for water-scarce conditions (Atlin and Lafitte 2002).

5.1.5 How Rice Can Be Adapted to Drought?

Secondary traits involved in water retention and yield are selected through conventional breeding techniques and have started to be used to bring genetic improvement for adaptation to water-limited conditions (Farooq et al. 2009). Several studies have highlighted the response of rice to limited water availability and the genetic traits involved like deeper, thicker roots (Yadav et al. 1997), root-pulling resistance (Pantuwan et al. 2002b), greater root penetration (Ali et al. 2000), osmotic adjustment (OA) (Lilley and Ludlow 1996), and membrane stability (Tripathy et al. 2000). Varieties suitable for aerobic rice culture should be medium-statured with fair drought tolerance to resist lodging and ultimately provide an improved harvest index (Atlin et al. 2004, 2006). Average rice yield has increased with the reduction in crop duration due to the development of rice varieties with higher yield and early maturity characteristics. Consequently water productivity has been enhanced three-fold with reference to the inputs (Bouman et al. 2006). In this regard QTL mapping is very helpful to identify and select the important traits for developing new varieties with efficient water use under limited water conditions (Kirigwi et al. 2007).

5.1.6 Selection and Breeding Strategies

Breeding rice has induced earliness in the crop with enhanced water use efficiency and low transpiration (Tuong 1999). Research has shown that reduction in leaf size reduces the transpiration losses. Reduced leaf area index contributes towards water economy during stress periods (Ball et al. 1994). Harvest index is the above-ground biomass accumulated by photosynthetic activity divided by grain yield. Rice yield has increased in the last decades due to improved harvest index. Harvest index is now approaching its hypothetical limits in major crops (Richards et al. 1993). A slight variation was found in the photosynthetic rate of various universally grown rice cultivars. Peng et al. (1999) suggested that tropical japonica rice has 25–30 % higher water use efficiency than indica rice. The rice germplasm showed considerable variability in regarding the photosynthesis to transpiration ratio, which was used as the basis of selection for yield (Atlin and Lafitte 2002).

Subbarao et al. (1995) and Turner et al. (2001) have declared that root features like length, density depth and biomass contribute towards water economy. According to Kavar et al. (2007) extraction of water from lower depths was performed better by deeper and thicker roots. Water losses were minimized by waxy bloom or glaucousness on leaves which helps maintain high tissue water potential. Glaucous leaves was a mandatory character for drought tolerance (Ludlow and Muchow 1990; Richards et al. 1986). Breeding programs should sort out how the plant responds to the transition from anaerobic to aerobic conditions, so the crop may bear irregular drought spells, higher impedance of soil and low air humidity. There is broad genetic variability among rice cultivars as well as in its wild ancestors. Response of root growth in drying soil situations is also variable. Rice cultivars also differ genetically in leaf area development and the number of spikelets in response to either soil or atmospheric water stress and under aerobic environments.

Less water is required for a short duration rice crop. Crop genotype and environment are the main determinants of the crop's potential to mature quickly (Dingkuhn and Asch 1999). Flowering time is the main crop attribute to adjust when water deficit and increased temperatures limit the growth period. Yield losses due to drought can be minimized by the development of early maturing varieties which have the potential to avoid drought spells (Kumar and Abbo 2001).

Osmotic adjustment is another important feature of drought tolerance (Blum 1988). In osmotic adjustment higher turgor potential is maintained at a particular water potential. In rice, the role of osmotic adjustment during drought delays the leaf curling, tissue death and leaf senescence (Hsiao et al. 1984). Zhang et al. (1999) observed that osmotic adjustment boosts the grain yield of rice and other crops under drought conditions.

5.1.7 *Molecular and Biotechnological Approaches*

Yield potential and drought tolerance have been improved by recent progress in the fields of genomics, molecular genetics and genetic engineering. The discovery and consequent manipulation of dogmatic genes controlling the complex responses of rice plants to water deficit on the biochemical and physiological levels will speed up breeding for enhancing water use efficiency. Water use efficiency is enhanced by the expression of stress regulating genes. Efforts are being conducted for crop plant bioengineering (Bahieldina et al. 2005). But growth may be retarded due to the improved expression of the genes which would narrow their applications. For the recognition of key genes responding to drought, genomics and other relevant techniques are used (Bruce et al. 2002).

5.1.8 Water-Use Efficiency and Transpiration Efficiency

The ratio between photosynthesis and transpiration is known as transpiration efficiency (Tuong and Bouman 2003) while the ratio of total biomass or grain production to the total amount of transpired water is water use efficiency. Water use efficiency is mainly determined by transpiration and photosynthetic rate (Tuong and Bouman 2003). These processes and ultimately the water economy of plant are controlled by the stomata. The stomatal density, size and morphology vary from species to species. Photosynthesis and water use efficiency are increased and energy is saved by the rapid opening and closing of stomata (Grantz and Assmann 1991).

Drought tolerance mechanisms are complex and interlinked with the molecular and physiological bases of water storage (Chaves et al. 2009). Water use efficiency can be enhanced by management as well as biology (Giordano et al. 2007). Biological water conservation is a very proficient means to enhance the water use efficiency by utilization of limited input. Increased water use efficiency enhanced yield due to increased use of water during drought. In a breeding strategy, selection for elevated water use efficiency causes reduced or earlier flowering that results in lower water usage along with lower yield capacity (Blum 2005). Hence, it is vital to produce the genotypes having higher water use efficiency as well as higher yields compared to present varieties (Farooq et al. 2009).

Peng and Bouman (2007) reported that it is better to develop rice varieties that perform better by the using water saving techniques like AWD and aerobic cultivation as they may lead to considerable progress and enhancement of lowland irrigated rice production. A comparison of indica varieties and improved tropical japonica lines grown in flooded environments revealed that japonica lines have 25–30 % greater transpiration efficiency than the indica varieties at the single leaf level. Therefore, tropical japonica lines have a lower transpiration rate than the indica varieties with a negligible difference in the photosynthetic rates. But unfortunately the potential for the utilization of this important feature still lacks proper research (Farooq et al. 2009). Enhanced water use efficiency was linked to the non-dwarf growth habit and for that reason its incorporation in the commercial varieties may not prove helpful for increased WUE. To reduce non-stomatal transpiration, increased leaf waxes were proposed, but their impact on water productivity was unclear (Lafitte and Bennett 2002).

Another approach to enhance transpiration efficiency is C_3 to C_4 transformation of rice by using genetic engineering (Farooq et al. 2009). Ku et al. (2000) observed that non-transformed rice plants had 30–35 % lower photosynthesis than the transgenic rice plants. Conversely, enhanced stomatal conductance was related to the increased photosynthetic activity that reduced the transpiration efficiency by the conversion of rice from C_3 to C_4. But still C_4 transition of rice is the current research issue (Farooq et al. 2009).

5.1.9 Crop Management

Selection of a good germplasm and planting site are of prime importance involving land and seedbed preparation, the production system, date of planting, method of planting, plant protection measures and nutrient management strategies from sowing till maturity. For the best rice production, soil type, weed management, irrigation method and land preparation are prime factors (Farooq et al. 2009). Minimizing land preparation time results in reduced evaporation losses and enhanced water use efficiency. Canopy closure after crop establishment is also beneficial in reducing evaporation losses. Early canopy closure is achieved by maintenance of proper plant density as well as by the selection of varieties having better seedling vigor (Tuong et al. 2000). Additionally, harmful weeds are suppressed, transpiration is enhanced and production is improved (Tuong et al. 2000). From the perspective of high water productivity, the following rice production systems are well recognized in various agro-ecological regions of the world:

- Aerobic rice (Bouman et al. 2007)
- Alternate wetting and drying (Cabangon et al. 2001)
- System of rice intensification (Uphoff and Randriamiharisoa 2003)
- Ground cover rice production system (Dittert et al. 2003)

In addition to these production systems, there are physiological techniques that were used for the enhancement of the rice water productivity and they include seed priming (Harris et al. 2002), silicon nutrition and the application of osmoprotectant (Yang et al. 2007).

5.2 Rice Production Systems

5.2.1 Aerobic Rice System

A new production technology involves the cultivation of specially developed rice varieties in non-puddled and non saturated soils, and these varieties have a peculiar feature of aerobic adaptation (Bouman et al. 2007). The main purpose of this production system is the balanced and economical use of water. With this production system, the use of saturated and flooded rice fields is abandoned (Bouman and Tuong 2001). Research showed that in the aerobic rice system, yields range from 4.5–6.5 t ha^{-1} which is 20–30 % less than the traditional lowland varieties grown under saturated and flooded field conditions, but two times higher than the traditional upland cultivars. Water useage was 60 % lower than the lowland rice, net water use efficiency was 1.6–1.9 times greater and total profit to water use was doubled (Farooq et al. 2009).

ARS is less labor intensive compared to lowland rice and may involve a high degree of mechanization (Huaqi et al. 2003). From a yield perspective ARS is the best alternative because it maximizes water utilization and is an appropriate management in water deficient ecosystems. Castaneda et al. (2003) reported rice yields to be 14–40 % lower than the yields taken in flooded fields whereas water productivity was enhanced by 20–40 % compared to flooded fields. ARS lowers water usage by eliminating puddling and flooding, consequently raising the water use efficiency. Irrigation water for land preparation and for crop growth was conserved 73 % and 56 % respectively (Castaneda et al. 2003). In short, ARS is a smart choice and it can be promoted as a water conserving cultivation system while maintaining yield stability.

5.2.2 Alternative Wetting and Drying Irrigation

In Asia most rice is transplanted in puddled soils (Farooq et al. 2009). Puddling is considered essential for a variety of purposes like weed control (Tabbal et al. 2002), facilitation of field leveling and ease of transplantation (Farooq et al. 2009) in addition to reduction of percolation losses (Kukal and Aggarwal 2003). But according to different researchers puddling has no significant effect on growth and yield of rice. Like yields were obtained for direct-seeded or transplanted rice cultivated with and without puddling (Kukal and Aggarwal 2003). High-yielding rice cultivation systems of Australia and California lack puddling (Farooq et al. 2009). Puddling does not essentially reduce the net water application in rice regardless of the reduced percolation rate (Tabbal et al. 2002). Kukal and Aggarwal (2003) concluded that puddling results in high soil bulk density, low permeability in subsurface layers and increased soil strength that limited root growth and development and restricted root activity regarding moisture and nutrient uptake in rice-wheat cropping system (Gajri et al. 1992).

For more than a decade AWD has been in use globally as a water-conserving technology (Cabangon et al. 2001). In the AWD system of rice cultivation, application of irrigation water is done when there are dry conditions after the disappearance of flooded water (Rice and water). Soil dries for a few days between irrigations depending on crop developmental phases (Gani et al. 2003; Lu et al. 2003). Higher water use efficiency was observed in AWD system of rice cultivation compared to a constantly flooded system (Belder et al. 2003). Although some researchers have reported an increase in yield under AWD, recent research indicates it is an exception instead of a rule (Belder et al. 2004; Cabangon et al. 2001; Tabbal et al. 2002). Tuong and Bouman (2002) performed a series of field experiments and concluded that there was a reduction in yield ranging up to 70 % in 92 % of the AWD treatments when compared to the saturated controls. But in all treatments water productivity was enhanced due to the decrease in the water applications, and the water conservation was greater than the yield reduction. They also reported that the high variation in the results was a function of variation in the number of days between irrigations as well as soil and moisture conditions.

With AWD technique different research trials have been conducted in lowland rice region having shallow groundwater tables and heavy soils in China and

Philippines where researchers reported that water applications were reduced almost 15–30 % without appreciably affecting yield (Lampayan et al. 2005). By means of AWD system of rice cultivation, much water can be conserved in addition to increased water use efficiency by extending the duration of dry soil with minimal plant moisture deficit and with minimal yield loss (Bouman and Tuong 2001). The number of days without flooding in AWD system is adjustable in accordance with the soil type, ground water depth and prevailing climate. Alternate wetting and drying is a widespread technique. In China, it has been extensively adopted where it is thought to be the regular practice of lowland rice (Li and Barker 2004). A lot of work is still to be done related to the impacts of AWD technology on water losses by percolation, seepage and evaporation. Evaporation losses were measured to be 2–33 % less than the saturated conditions (Cabangon et al. 2001).

5.2.3 System of Rice Intensification

The system of rice intensification was developed in Madagascar in the 1980s and 1990s. It allowed farmers having a narrow resource base to take approximately 15 t ha^{-1} paddy yields on unproductive soils with low irrigation water applications and it reduced additional inputs (Stoop et al. 2002). This system involves the transplantation of young seedlings singly following square pattern having wide row spacing, manual weed control, organic fertilizer use and maintaining the soil moist throughout the vegetative phase (Stoop et al. 2002). Noteworthy alterations occur not only in form and function but also yield and yield elements of plants subjected to SRI. This system enhanced yields 50–100 % or even more and it entails merely half of the water than the conventional rice system (Uphoff and Randriamiharisoa 2003). With the implementation of SRI technique yield of any variety can be improved while the highest yields have been obtained from the improved high yielding varieties. Synergistic dynamics were explained between SRI techniques by the factorial experiments conducted in Madagascar in which yield increased 100–200 %. SRI methods enhance the gains to labor, water and capital (Uphoff and Randriamiharisoa 2003).

McHugh et al. (2003) concluded after conducting a survey of farmers in Madagascar that farmers have adapted the AWD as a component of SRI in response to the prevailing conditions of soil type, moisture and labor availability. Lack of a consistent water source was the basic hurdle in implementation of AWD as reported by the farmers. Grain production can be enhanced even with reduced irrigation by a combination of SRI and AWD (McHugh et al. 2003). Besides its usage and advantages, the SRI system is hard to practice as it involves extra labor at a time when the farmer is often unable to invest the extra time, and when the whole family is already highly involved in this effort. From the perspective of promoting the water-conserving techniques it is a big challenge of which policy makers should be mindful. Despite all these problems, production can be raised while conserving water, but the adoption of the technology by farmers is still an important question to be answered (Moser and Barrett 2003).

5.2.4 The Ground-Cover Rice Production System

This system was developed for China in 1990 and it involves straw or plastic film mulching for rice cultivation in order to enhance tolerance of low temperatures (Shen et al. 1997). In this type of mulching lowland rice varieties are grown by covering soil with covering materials in order to keep it humid (Kreye et al. 2007). Plastic film, plant or paper mulch can be used to cover the soil surface in order to control the evaporation losses (Lin et al. 2003b). Water is applied up to 80 % of water holding capacity. Water saved by this technique may be 60–85 % of that required in the conventional rice system without any harm to the final yield (Huang et al. 1999). There is, however, experimental evidence that illustrates a fair to severe decline in grain yield (Borrell et al. 1997; Castillo et al. 1992). The elongation of inter-nodes along with number of panicles and overall rate of crop growth are decreased due to the lower soil water potentials when compared with saturated ecosystems. The water requirement of a rice crop was reduced up to 60 % in GCRPS while the yield was 10 % less than the conventional lowland rice system. The yield reduction was linked to difficulty in N-fertilizer management as well as micronutrient deficiency in GCRPS (Lin et al. 2003b).

In one experiment, two ground cover rice production systems were studied by comparing thin plastic film and straw mulch cover with the conventional rice cultivation system. Methane release was higher in the flooded fields and nitrous oxide (N_2O) emission was recorded prior to beginning of panicle growth in the drainage phase. On the other hand negligible methane emission was observed in GCRPS but increased N_2O emission was recorded in GCRPS, apparently associated with the fertilization events (Dittert et al. 2003). In another study three treatments of GCRPS were compared with the traditional lowland rice production system. The three treatments included GCRPS- plastic where soil surface was covered with a plastic film, GCRPS-straw where straw mulch was used and GCRPS-bare where soil was left uncovered. In GCRPS treatments 32–54 % of irrigation water was given in contrast to paddy control. GCRPS had smaller plants with fewer panicles and smaller LAI in comparison with the control. In terms of yield, control had significantly higher yield than the GCRPS-straw and GCRPS-bare whereas GCRPS-plastic had just 8 % lower yield than the control paddy yield. The paddy control had lower water use efficiency than the GCRPS-plastic (Tao et al. 2006).

5.2.5 Raised Beds

In the late 1970s, cultivation of crops other than rice on raised beds was initiated in the heavy clay soils of rice growing areas of Australia (Maynard 1991) and it was introduced in the Indo-Gangetic plains for cultivation of wheat in rice-wheat cropping system (Hossain et al. 2003; Sayre and Hobbs 2004). Raised beds improved soil structure, waterlogging was reduced and drainage was better due to timely mechanical operations. Raised beds are beneficial for weed control and better fertilizer placement (Farooq et al. 2009).

In the Indo-Gangetic plains, different trials have been conducted by researchers and farmers and these trials suggest that 12–60 % of irrigation water can be saved for direct-seeded and transplanted rice on beds. As far as yield is concerned, the same or lesser yields were observed for transplanted rice in comparison with puddle saturated transplanted rice and slightly lower yields were obtained in case of direct seeded rice. Conversely several studies in the northwest Indo-Gangetic plains suggested that rice cultivated on beds had a negligible effect on water use efficiency as less water resulted in poorer yields (Singh et al. 2003). No water saving was observed by Beecher et al. (2006) when rice on raised beds was compared with the rice cultivated conventionally on flat fields. Singh et al. (2003) declared that a variety cultivated on raised beds should have the ability to balance the loss of cropped area by production of additional fertile tillers as there is broad row spacing in the beds.

5.2.6 C4 Conversion of Rice

Asia contains 60 % of the total world's population and at present the land used for rice here supports 27 people per hectare. However, the number of people will change from 27 to 43 per hectare by 2050. Adverse shifts in existing climate patterns are expected to occur due to climate change. Climate extremes will result in water scarcity. Price hikes will occur due to increasing demand for biofuels causing competition between grain for fuel and grain for food. There will be a wide gap between demand and production of food. Consequently inadequate yields of rice will cause food insecurity, environmental destruction, non-sustainable agricultural practices and social conflict. This destructive cycle must be replaced with a more righteous cycle raising productivity (IRRI 2013) (Table 5.1).

All of these egregious practices and factors should be controlled as they are growing at a time when the rice production has slowed and farmers have reached the yield limits. Photosynthetic solar energy conversion efficiency has a great role in this regard. According to recent scientific opinions it is better to modify the biophysical structure of rice plants to fulfill the future rice needs. By changing the structure of rice plants, solar radiation will be used much more efficiently. Rice has the C3 photosynthetic pathway that is less efficient than the C4 pathway which is present in maize. The conversion of rice from C3 to C4 would involve a reorganization of cellular structures inside leaves and proficient manifestation of several enzymes involved in the photosynthetic pathway. All requirements for C4 photosynthesis are present in the rice plant but they lack the C4 structure (IRRI 2013).

Table 5.1 Rice improvement work using different techniques

Sr. #	Effect	References
01	Effect of water stress is ameliorated by the ABA treatments	Majid et al. (2007)
02	Identification of major and consistent QTL regarding yield may enhance the rice yield in drought situations	Bernier et al. (2007), Kumar et al. (2008), Venuprasad et al. (2007) and Vikram et al. (2012)
03	Mechanisms of growth control and of internal aeration are used in rice to tolerate submergence	Nishiuchi et al. (2012)
04	Handling of waterlogging stress by formation of lysigenous aerenchyma along with a barrier to radial O_2 loss (ROL) in roots	Nishiuchi et al. (2012)
05	Yield can be improved under drought by marker assisted recurrent selection	Ribaut and Ragot (2007)
06	Genome-wide selection may improve production under water stress	Bernardo and Yu (2007)
07	Activation of gene TLD1 under drought stress improves adaptation to drought stress	Zhang et al. (2009)
08	Salt and drought tolerance can be induced by the gene OsLE A3-2	Duan and Cai (2012)
09	Effect of drought stress on spikelet infertility and ultimately grain yield minimized by the traits causing the maintenance of high leaf water potential	Karim and Rahman (2015)
10	Tolerance to moisture and saline stress can be achieved by colonizing the rice varieties with Class 2 fungal endophytes	Redman et al. (2011)
11	Improved drought tolerance is accomplished by the improvement of root system architectture of rice	Ahmadi et al. (2014)
12	Early flowering genotypes not only escape the drought stress but also have higher yields	Pantuwan et al. (2002a)

5.3 Science of C4 Rice

C4 plants are more efficient utilizers of carbon dioxide and hence have higher water and nitrogen use efficiency. They also have better adaptation to hot and dry environments. C4 photosynthesis has evolved more than 50 times in nature in a broad range of flowering plants signifying that it may be a rather easy pathway to develop. Its development should first consider how to incorporate Kranz anatomy into C3 plants (IRRI 2013).

The incorporation of C4 pathway in rice would result in 50% yield increase, double water use efficiency and reduce fertilizer use. C4 rice would require less water and has reduced water losses. Stomata in C4 plants would be partially closed in the warmest part of the day. C4 plants compartmentalize carbon dioxide in the bundle sheath cells which is why they are efficient utilizers of carbon dioxide. Nitrogen use efficiency would increase by 30% in C4 rice because lower amounts of Rubisco are utilized for carbon dioxide fixation. By using fewer enzymes C4 rice may attain the same productivity with less nitrogen because enzymes and other proteins have 15 % nitrogen. As far as yield is concerned, models illustrate that increased water and nitrogen use efficiencies and other characteristics would result in yield increases of 30–50 % established on relative studies between rice and maize (IRRI 2013).

References

Abdul-Jaleel, C., P. Manivannan, A. Wahid, M. Farooq, R. Somasundaram, and R. Panneerselvam. 2009. Drought stress in plants: A review on morphological characteristics and pigments composition. *International Journal of Agriculture and Biology* 11: 100–105.

Ahmadi, N., A. Audebert, M.J. Bennett, A. Bishopp, A.C. de Oliveira, B. Courtois, A. Diedhiou, A. Diévart, P. Gantet, A. Ghesquière, E. Guiderdoni, A. Henry, Y. Inukai, L. Kochian, L. Laplaze, M. Lucas, D.T. Luu, B. Manneh, X. Mo, R. Muthurajan, C. Périn, A. Price, S. Robin, H. Sentenac, B. Sine, Y. Uga, A.A. Véry, M. Wissuwa, P. Wu, and J. Xu. 2014. The roots of future rice harvests. *Rice* 7: 1–9.

Ali, M.L., M.S. Pathan, J. Zhang, G. Bai, S. Sarkarung, and H.T. Nguyen. 2000. Mapping QTLs for root traits in a recombinant inbred population from two indica ecotypes in rice. *Theoretical and Applied Genetics* 101: 756–766.

Antonio, C., C. Pinheiro, M.M. Chaves, C.P. Ricardo, M.F. Ortuno, and J. Thomas-Oates. 2008. Analysis of carbohydrates in Lupinus album stems on imposition of water deficit, using porous graphitic carbon liquid chromatography-electrospray ionization mass spectrometry. *Journal of Chromatography* 1187: 111–118.

Atlin, G.N., and H.R. Lafitte. 2002. Marker-assisted breeding versus direct selection for drought tolerance in rice. In *Field screening for drought tolerance in crop plants with emphasis on rice*, eds. N.P. Saxena, and J.C.O'Toole, 208. Proceedings of an international workshop on field screening for drought tolerance in rice, Patancheru, India, 11–14 December 2000. ICRISAT, Patancheru/The Rockefeller Foundation, India/New York.

Atlin, G.N., M. Laza, M. Amante, and H.R. Lafitte. 2004. Agronomic performance of tropical aerobic, irrigated, and traditional upland rice varieties in three hydrological environments at IRRI. In *New directions for a diverse planet*. Proceedings of the 4th international crop science Congress, Brisbane, Australia.

Atlin, G.N., H.R. Lafitte, D. Tao, M. Laza, M. Amante, and B. Courtois. 2006. Developing rice cultivars for high-fertility upland systems in the Asian tropics. *Field Crops Research* 97: 43–52.

Bahieldina, A., H.T. Mahfouz, H.F. Eissa, O.M. Saleh, A.M. Ramadan, I.A. Ahmed, W.E. Dyer, H.A. El-Itriby, and M.A. Madkour. 2005. Field evaluation of transgenic wheat plants stably expressing the HVA1 gene for drought tolerance. *Physiologia Plantarum* 123: 421–427.

Ball, R.A., D.M. Oosterhuis, and A. Mauromoustakos. 1994. Growth dynamics of the cotton plant during water-deficit stress. *Agronomy Journal* 86: 788–795.

Barker, R., D. Dawe, T.P. Tuong, S.I. Bhuiyan, and L.C. Guerra. 1998. The outlook for water resources in the year 2020: Challenges for research on water management in rice production. In *Assessment and orientation towards the 21st century*, 96–109. Proceedings of 19th session of the International Rice Commission, Cairo, Egypt, 7–9 September 1998. FAO.

Bates, B.C., Z.W. Kundzewicz, S. Wu, and J.P. Palutikof, eds. 2008. Climate change and water. In *Technical paper of the intergovernmental panel on climate change*. Geneva: IPCC Secretariat.

Beecher, H.G., B.W. Dunn, J.A. Thompson, E.S. Humphreys, K. Mathews, and J. Timsina. 2006. Effect of raised beds, irrigation and nitrogen management on growth, water use and yield of rice in south-eastern Australia. *Australian Journal of Experimental Agriculture* 46: 1363–1372.

Belder, P., B.A.M. Bouman, J.H.J. Spiertz, L. Guoan, and E.J.P. Quilang. 2003. Water use of alternately submerged and non submerged irrigated lowland rice. In *Water-wise rice production*, eds. B.A.M. Bouman, H. Hengsdijk, B. Hardy, P.S. Bindraban, T.P. Tuong, and J.K. Ladha. Proceedings of a thematic workshop on water-wise rice production, 8–11 April 2002 at IRRI Headquarters in Los Banos, Philippines. International Rice Research Institute, Los Banos, Philippines.

Belder, P., B.A.M. Bouman, J.H.J. Spiertz, R. Cabangon, L. Guoan, E.J.P. Quilang, Li Yuanhua, and T.P. Tuong. 2004. Evect of water and nitrogen management on water use and yield of irrigated rice. *Agricultural Water Management* 65: 193–210.

Beligni, M.V., and L. Lamattina. 1999. Nitric oxide counteracts cytotoxic processes mediated by reactive oxygen species in plant tissues. *Planta* 208: 337–344.

Bernardo, R., and J. Yu. 2007. Prospects for genomewide selection for quantitative traits in maize. *Crop Science* 47: 1082–1090.

Bernier, J., A. Kumar, V. Ramaiah, D. Spaner, and G. Atlin. 2007. A large-effect QTL for grain yield under reproductive-stage drought stress in upland rice. *Crop Science* 47: 507–518. doi:10.2135/cropsci2006.07.0495.

Bilger, W., T. Johnsen, and U. Schreiber. 2001. UV-excited chlorophyll fluorescence as a tool for the assessment of UV-protection by the epidermis of plants. *Journal of Experimental Botany* 52: 2007–2014.

Blum, A. 1988. *Plant breeding for stress environments*. Boca Raton: CRC Press.

Blum, A. 2005. Drought resistance, water-use efficiency, and yield potential: Are they compatible, dissonant, or mutually exclusive? *Australian Journal of Agricultural Research* 56: 1159–1168.

Borrell, A.K., A.L. Garside, and S. Fukai. 1997. Improving efficiency of water use for irrigated rice in a semi-arid tropical environment. *Field Crops Research* 52: 231–248.

Bouman, B.A.M. 2003. Examining the water-shortage problem in rice systems: Water-saving irrigation technologies. In *Rice science: Innovations and impact for livelihood*, eds. T.W. Mew, D.S. Brar, S. Peng, D. Dawe, and B. Hardy. 519–535. Proceedings of the international rice research conference, Beijing, 16–19 September 2002. International Rice Research Institute, Chinese Academy of Engineering and Chinese Academy of Agricultural Sciences, Beijing, China.

Bouman, B.A.M., and T.P. Tuong. 2001. Field water management to save water and increase its productivity in irrigated lowland rice. *Agricultural Water Management* 49: 11–30.

Bouman, B.A.M., X.G. Yang, H. Wang, Z. Wang, J. Zhao, and B. Chen. 2006. Performance of aerobic rice varieties under irrigated conditions in North China. *Field Crops Research* 97: 53–65.

Bouman, B.A.M., R.M. Lampayan, and T.P. Tuong. 2007. *Water management in irrigated rice coping with water scarcity*. Los Banos: International Rice Research Institute.

Breusegem, F.V., E. Vranova, J.F. Dat, and D. Inze. 2001. The role of active oxygen species in plant signal transduction. *Plant Science* 161: 405–414.

Bruce, W.B., G.O. Edmeades, and T.C. Barker. 2002. Molecular and physiological approaches to maize improvement for drought tolerance. *Journal of Experimental Botany* 53: 13–25.

Cabangon, R.J., E.G. Castillo, L.X. Bao, J. Lu, G.H. Wang, Y.L. Cui, T.P. Toung, B.A.M. Bouman, Y.H. Li, C.D. Chen, and J.Z. Wang. 2001. Impact of alternate wetting and drying irrigation on rice growth and resource-use efficiency. In *Water-saving irrigation for rice*, eds. R. Barker, R. Loeve, Y.H. Li, and T.P. Tuong. Proceedings of an international workshop held in Wuhan, China, 23–25 March 2001. International Water Management Institute Colombo, Sri Lanka.

Castaneda, A.R., B.A.M. Bouman, S. Peng, and R.M. Visperas. 2003. The potential of aerobic rice to reduce water use in water-scarce irrigated lowlands in the tropics. In *Water-wise rice production*, eds. B.A.M. Bouman, H. Hengsdijk, B. Hardy, P.S. Bindraban, T.P. Tuong, and J.K. Ladha. Proceedings of a thematic workshop on water-wise rice production, 8–11 April 2002 at IRRI Headquarters in Los Banos, Philippines. International Rice Research Institute, Los Banos, Philippines.

Castillo, E.G., R.J. Buresh, and K.T. Ingram. 1992. Lowland rice yield as affected by timing of water deficit and nitrogen fertilization. *Agronomy Journal* 84: 152–159.

Chaves, M.M., J.S. Pereira, J.P. Maroco, M.L. Rodrigues, C.P.P. Ricardo, M.L. Osorio, I. Carvalho, T. Faria, and C. Pinheiro. 2002. How plants cope with water stress in the field: Photosynthesis and growth. *Annals of Botany* 89: 907–916.

Chaves, M.M., J. Flexas, and C. Pinheiro. 2009. Photosynthesis under drought and salt stress: Regulation mechanisms from whole plant to cell. *Annals of Botany* 103: 551–560.

Colmer, T.D. 2003. Long-distance transport of gases in plants: A perspective on internal aeration and radial oxygen loss from roots. *Plant, Cell and Environment* 26: 17–36.

Davies, P.J. 2004. The plant hormones: Their nature, occurrence and function. In *Plant hormones, biosynthesis, signal transduction, action*, ed. P.J. Davies. Dordrecht: Kluwer.

Dawe, D., R. Barker, and D. Seckler. 1998. Water supply and research for food security in Asia. In *Proceedings of the workshop on increasing water productivity and efficiency in rice-based systems*, July 1998, International Rice Research Institute, Los Banos, Philippines.

Dingkuhn, M., and F. Asch. 1999. Phenological responses of Oryza sativa, O. glaberrina and interspecific rice cultivars on a toposequence in West Africa. *Euphytica* 110: 109–126.

Dittert, K., L. Shan, C. Kreye, Z. Xunhua, X. Yangchun, L. Xuejuan, H. Yao, S. Qirong, F. Xiaolin, and B. Sattelmacher. 2003. Saving water with ground-cover rice production systems (GCRPS) at the price of increased greenhouse gas emissions. In *Water-Wise Rice Production*, eds. B.A.M. Bouman, H. Hengsdijk, B. Hardy, P.S. Bindraban, T.P. Tuong, and J.K. Ladha. Proceedings of a thematic workshop on water-wise rice production, 8–11 April 2002 at IRRI Headquarters in Los Baños, Philippines. International Rice Research Institute, Los Banos, Philippines.

Duan, J., and W. Cai. 2012. OsLEA3-2, an abiotic stress induced gene of rice plays a key role in salt and drought tolerance. *PloS One* 7(9), e45117.

Evans, D.E. 2003. Aerenchyma formation. *New Phytologist* 161: 35–49.

Farooq, M., A. Wahid, N. Kobayashi, D. Fujita, and S.M.A. Basra. 2009. Plant drought stress: Effects, mechanisms and management. *Agronomy for Sustainable Development* 29: 185–212.

Gajri, P.R., V.K. Arora, and S.S. Prihar. 1992. Tillage management for efficient water and nitrogen use in wheat following rice. *Soil and Tillage Research* 24: 167–182.

Gani, A., A. Rahman, D. Rustam, and H. Hengsdijk. 2003. Synopsis of water management experiments in Indonesia. In *Water-wise rice production*, eds. B.A.M. Bouman, H. Hengsdijk, B. Hardy, P.S. Bindraban, T.P. Tuong, and J.K. Ladha. Proceedings of a thematic workshop on water-wise rice production, 8–11 April2002 at IRRI Headquarters in Los Banos, Philippines. International Rice Research Institute, Los Banos, Philippines.

Garcia-Plazaola, J.I., and J.M. Becerril. 2000. Effects of drought on photoprotective mechanisms in European beech (Fagus sylvatica L) seedlings from different provenances. *Trees* 14: 485–490.

Giordano, R., G. Passarella, V.F. Uricchio, and M. Vurro. 2007. Integrating conflict analysis and consensus reaching in a decision support system for water resource management. *Journal of Environmental Management* 84: 213–228.

Grantz, D.A., and S.M. Assmann. 1991. Stomatal response to blue light: Water use efficiency in sugarcane and soybean. *Plant, Cell & Environment* 14: 683–690.

Harris, D., R.S. Tripathi, and A. Joshi. 2002. On-farm seed priming to improve crop establishment and yield in dry direct-seeded rice. In *Direct seeding: Research strategies and opportunities*, ed. S. Pandey, M. Mortimer, L. Wade, T.P. Tuong, K. Lopes, and B. Hardy, 231–240. Los Banos: International Rice Research Institute.

Hossain, M.I., M.A. Sufian, A.B.S. Hossain, C.A. Meisner, J.G. Lauren, and J.M. Duxbury. 2003. *Performance of bed planting and nitrogen fertilizer under rice–wheat–mungbean cropping systems in Bangladesh*, Wheat research centre internal review reports. Dinajpur: Bangladesh Agricultural Research Institute.

Hsiao, T.C., J.C. O'toole, E.B. Yambao, and N.C. Turner. 1984. Influence of osmotic adjustment on leaf rolling and tissue death in rice (Oryza sativa L.). *Plant Physiology* 75: 338–341.

Huang, Y.D., Z.L. Zhang, F.Z. Wei, and J.C. Li. 1999. Ecophysiological effect of dry-cultivated and plastic film-mulched rice planting. *Chinese Journal of Applied Ecology* 10: 305–308.

Huaqi, W., B.A.M. Bouman, D. Zhao, W. Changgui, and P.F. Moya. 2003. Aerobic rice in northern China: Opportunities and challenges. In *Water-wise rice production*, eds. B.A.M. Bouman, H. Hengsdijk, B. Hardy, P.S. Bindraban, T.P. Tuong, and J.K. Ladha. 207–222. Proceedings of a thematic workshop on water-wise rice production, 8–11 April 2002 at IRRI Headquarters in Los Banos, Philippines. International Rice Research Institute, Los Banos, Philippines.

International Rice Research Institute (IRRI). 2013. http://irri.org/resources/publications/annual-reports/annualreport-2013

Karim, M.R., and M.A. Rahman. 2015. Drought risk management for increased cereal production in Asian Least Developed Countries. *Weather and Climate Extremes* 7: 24–35.

Kavar, T., M. Maras, M. Kidric, J. Sustar-Vozlic, and V. Meglic. 2007. Identification of genes involved in the response of leaves of Phaseolus vulgaris to drought stress. *Molecular Breeding* 21: 159–172.

Kirigwi, F.M., M. Van Ginkel, G. Brown-Guedira, B.S. Gill, G.M. Paulsen, and A.K. Fritz. 2007. Markers associated with a QTL for grain yield in wheat under drought. *Molecular Breeding* 20: 401–413.

Kondo, M., M.V.R. Murty, D.V. Aragones, K. Okada, T. Winn, and K.S. Kwak. 1999. Characteristics of the root system and water uptake in upland rice. In Genetic improvement of rice for water limited environments, eds. O. Ito, J. O'Toole, and B. Hardy. 117–131. Proceedings of the workshop on genetic improvement of rice for water-limited environments, 1–3 December 1998, Los Banos, Philippines. Manila (Philippines): International Rice Research Institute.

Kreye, C., K. Dittert, X. Zheng, X. Zhang, S. Lin, H. Tao, and B. Sattelmacher. 2007. Fluxes of methane and nitrous oxide in water-saving rice production in north China. *Nutrient Cycling in Agroecosystems* 77: 293–304.

Ku, M.S.B., D. Cho, U. Ranade, T.P. Hsu, X. Li, D.M. Jiao, J. Ehleringer, M. Miyao, and M. Matsuoka. 2000. Photosynthetic performance of transgenic rice plants overexpressing maize C_4 photosynthesis enzymes. In *Redesigning Rice Photosynthesis to Increase Yield*, eds. J.E. Sheehy, P.L. Mitchell, and B. Hardy, 193–204. Proceedings of the workshop on the quest to reduce hunger held in Los Banos, Philippines, 30 November–3 December 1999. Elsevier, Amsterdam (Studies in Plant Science no. 7).

Kukal, S.S., and G.C. Aggarwal. 2003. Puddling depth and intensity effects in rice–wheat system on a sandy loam soil. I. Development of subsurface compaction. *Soil and Tillage Research* 72: 1–8.

Kumar, J., and S. Abbo. 2001. Genetics of flowering time in chickpea and its bearing on productivity in the semi-arid environments. *Advances in Agronomy* 72: 107–138.

Kumar, A., J. Bernier, S. Verulkar, H.R. Lafitte, and G.N. Atlin. 2008. Breeding for drought tolerance: Direct selection for yield, response to selection and use of drought-tolerant donors in upland and lowland-adapted populations. *Field Crops Research* 107: 221–231.

Kusano, T., T. Berberich, C. Tateda, and Y. Takahashi. 2008. Polyamines: Essential factors for growth and survival. *Planta* 228: 367–381.

Lafitte, H.R., and J. Bennet. 2002. Requirements for aerobic rice: Physiological and molecular considerations. In *Water-wise rice production*, eds. B.A.M. Bouman, H. Hengsdijk, B. Hardy, P.S. Bindraban, T.P. Tuong, and J.K. Ladha. Proceedings of a thematic workshop on water-wise rice production, 8–11 April2002 at IRRI Headquarters in Los Banos, Philippines. International Rice Research Institute, Los Banos, Philippines.

Lampayan, R.M., B.A.M. Bouman, J.L. De Dios, A.T. Lactaoen, A.J. Espiritu, T.M. Norte, E.J.P. Quilang, D.F. Tabbal, L.P. Llorca, J.B. Soriano, A.A. Corpuz, and R.B. Malasa. 2005. Transfer of water saving technologies in rice production in the Philippines. In *Transitions in agriculture for enhancing water productivity*, eds.T.M. Thiyagarajan, H. Hengsdijk, and P. Bindraban, 111–132. Proceedings of the International symposium on transitions in agriculture for enhancing water productivity, September 23–25, 2003, Kilikulam, Tamil Nadu Agricultural University, Tamil Nadu, India.

Lawlor, D.W., and W. Tezara. 2009. Causes of decreased photosynthetic rate and metabolic capacity in water-deficient leaf cells: A critical evaluation of mechanisms and integration of processes. *Annals of Botany*. doi:10.1093/aob/mcn244. Available online at www.aob.oxfordjournals.org.

Li, Y.H., and R. Barker. 2004. Increasing water productivity for paddy irrigation in China. *Paddy and Water Environment* 2: 187–193.

Lilley, J.M., and S. Fukai. 1994. Effect of timing and severity of water deficit on four diverse rice cultivars. III. Phenological development, crop growth and grain yield. *Field Crops Research* 37: 225–234.

Lilley, J.M., and M.M. Ludlow. 1996. Expression of osmotic adjustment and dehydration tolerance in diverse rice lines. *Field Crops Research* 48: 185–197.

Lin, S., H. Tao, K. Dittert, Y. Xu, X. Fan, Q. Shen, and B. Sattelmacher. 2003a. Saving water with the ground cover rice production system in China. In *Technological and institutional innovations for sustainable rural development*. Conference on international agricultural research for development. Deutscher Tropentag, 8–10 October 2003, Gottingen.

Lin, S., K. Dittert, H. Tao, C. Kreye, Y. Xu, Q. Shen, X. Fan, and B. Sattelmacher. 2003b. The ground-cover rice production system (GCRPS): A successful new approach to save water and increase nitrogen fertilizer efficiency. In *Water-wise rice production*, eds. B.A.M. Bouman, H. Hengsdijk, B. Hardy, P. S. Bindraban, T.P. Tuong, and J.K. Ladha. Proceedings of a Thematic Workshop on Water-Wise Rice Production, 8–11 April 2002 at IRRI Headquarters in Los Banos, Philippines. International Rice Research Institute, Los Banos, Philippines.

Liu, J.H., H. Kitashiba, J. Wang, Y. Ban, and T. Moriguchi. 2007. Polyamines and their ability to provide environmental stress tolerance to plants. *Plant Biotechnology Journal* 24: 117–126.

Lu, G., R. Cabangon, T.P. Tuong, P. Belder, B.A.M. Bouman, and E. Castillo. 2003. The effects of irrigation management on yield and water productivity of inbred, hybrid, and aerobic rice varieties. In *Water-wise rice production*, eds. B.A.M. Bouman, H. Hengsdijk, B. Hardy, P.S. Bindraban, T.P. Tuong, and J.K. Ladha. Proceedings of a thematic workshop on water-wise rice production, 8–11 April2002 at IRRI Headquarters in Los Banos, Philippines. International Rice Research Institute, Los Banos, Philippines.

Ludlow, M.M., and R.C. Muchow. 1990. A critical evaluation of traits for improving crop yields in water-limited environments. *Advances in Agronomy* 43: 107–153.

Majid, S.A., R. Asghar, and G. Murtaza. 2007. Yield stability analysis conferring adaptation of wheat to pre- and post-anthesis drought conditions. *Pakistan Journal of Botany* 39: 1623–1637.

Maynard, M. 1991. Permanent beds, their potential role in soil management for the future. *Farmers' Newsletter – Large Area* 137: 14–18.

Mchugh, O.V., T.S. Steenhuis, J. Barison, E.C.M. Fernandes, and N.T. Uphoff. 2003. Farmer implementation of alternate wet–dry and nonflooded irrigation practices in the System of Rice Intensification (SRI). In *Water-wise rice production*, eds. B.A.M. Bouman, H. Hengsdijk, B. Hardy, P.S. Bindraban, T.P. Tuong, and J.K. Ladha. Proceedings of a thematic workshop on water-wise rice production, 8–11 April 2002 at IRRI Headquarters in Los Banos, Philippines. International Rice Research Institute, Los Banos, Philippines.

Moser, C.M., and C.B. Barrett. 2003. The disappointing adoption dynamics of a yield increasing, low external input technology: The case of SRI in Madagascar. *Agricultural Systems* 76: 1085–1100.

Nishiuchi, S., T. Yamauchi, H. Takahashi, L. Kotula, and M. Nakazono. 2012. Mechanisms for coping with submergence and water logging in rice. *Rice* 5: 1–14.

Pantuwan, G., S. Fukai, M. Cooper, S. Rajatasereekul, and J.C. O'toole. 2002a. Yield response of rice (Oryza sativa L.) genotypes to different types of drought under rainfed lowlands. Part 3. Plant factors contributing to drought resistance. *Field Crops Research* 73: 181–200.

Pantuwan, G., S. Fukai, R. Rajatasereekul, and J.C. O'toole. 2002b. Yield responses of rice (*Oryza sativa* L.) genotypes to different types of drought under rainfed lowlands part 1. Grain yield and yield components. *Field Crops Research* 73: 153–168.

Passioura, J. 2007. The drought environment: Physical, biological and agricultural perspectives. *Journal of Experimental Botany* 58: 113–117.

Peng, S., and B.A.M. Bouman. 2007. Prospects for genetic improvement to increase lowland rice yields with less water and nitrogen. In *Scale and complexity in plant systems research: Gene–plant–crop relations*, ed. J.H.J. Spiertz, P.C. Struik, and H.H. van Laar, 251–266. New York: Springer.
Peng, S., K.G. Cassman, S.S. Virmani, J. Sheehy, and G.S. Khush. 1999. Yield potential trends of tropical rice since the release of IR8 and the challenge of increasing rice yield potential. *Crop Science* 39: 1552–1559.
Redman, R.S., Y.O. Kim, C.J.D.A. Woodward, C. Greer, L. Espino, S.L. Doty, and R.J. Rodriguez. 2011. Increased fitness of rice plants to abiotic stress via habitat adapted symbiosis: A strategy for mitigating impacts of climate change. *PloS One* 6, e14823.
Ribaut, J.-M., and M. Ragot. 2007. Marker-assisted selection to improve drought adaptation in maize: The backcross approach, perspectives, limitations, and alternatives. *Journal of Experimental Botany* 58: 351–360.
Richards, R.A., H.M. Rawson, and D.A. Johnson. 1986. Glaucousness in wheat: Its development, and effect on water-use efficiency, gas exchange and photosynthetic tissue temperatures. *Australian Journal of Plant Physiology* 13: 465–473.
Richards, R.A., C. Lopez-Castaneda, H. Gomez-Macpherson, and A.G. Condon. 1993. Improving the efficiency of water use by plant breeding and molecular biology. *Irrigation Science* 14: 93–104.
Rolland, F., E. Baena-Gonzalez, and J. Sheen. 2006. Sugar sensing and signaling in plants: Conserved and novel mechanisms. *Annual Review of Plant Biology* 57: 675–709.
Sayre, K.D., and P.R. Hobbs. 2004. The raised-bed system of cultivation for irrigated production conditions. In *Sustainable agriculture and the international rice–wheat system*, ed. R. Lal, P.R. Hobbs, N. Uphoff, and D.O. Hansen, 337–355. New York: Marcel Dekker.
Sharp, R.E. 2002. Interaction with ethylene: Changing views on the role of abscisic acid in root and shoot growth responses to water stress. *Plant, Cell and Environment* 25: 211–222.
Shen, K.R., X.C. Wang, J. Liu, and X.S. Luo. 1997. Test and demonstration on wet-cultivation with film mulching of rice. *Hebei Agricultural Science* 5: 18–22 (in Chinese).
Singh, A.K., B.U. Choudhury, and B.A.M. Bouman. 2003. Effects of rice establishment methods on crop performance, water use, and mineral nitrogen. In *Water-wise rice production*, eds. B.A.M. Bouman, H. Hengsdijk, B. Hardy, P.S. Bindraban, T.P. Tuong, and J.K. Ladha, 223–235. Proceedings of a thematic workshop on water-wise rice production, 8–11 April 2002 at IRRI Headquarters in Los Banos, Philippines. International Rice Research Institute, Los Banos, Philippines.
Smirnoff, N. 1998. Plant resistance to environmental stress. *Current Opinion in Biotechnology* 9: 214–219.
Stoop, W.A., N. Uphoff, and A. Kassam. 2002. A review of agricultural research issues raised by the system of rice intensification (SRI) from Madagascar: Opportunities for improving farming systems for resource-poor farmers. *Agricultural Systems* 71: 249–274.
Subbarao, G.V., C. Johansen, A.E. Slinkard, R.C.N. Rao, N.P. Saxena, and Y.S. Chauhan. 1995. Strategies and scope for improving drought resistance in grain legumes. *Critical Reviews in Plant Sciences* 14: 469–523.
Swain, S., and D. Singh. 2005. Tall tales from sly dwarves: Novel functions of gibberellins in plant development. *Trends in Plant Science* 10: 123–129.
Tabbal, D.F., B.A.M. Bouman, S.I. Bhuiyan, E.B. Sibayan, and M.A. Sattar. 2002. On-farm strategies for reducing water input in irrigated rice; case studies in the Philippines. *Agricultural Water Management* 56: 93–112.
Tao, H., H. Brueck, K. Dittert, C. Kreye, S. Lin, and B. Sattelmacher. 2006. Growth and yield formation of rice (Oryza sativa L.) in the water-saving ground cover rice production system (GCRPS). *Field Crops Research* 95: 1–12.
Tateoka, T. 1962. Taxonomic studies of Oryza I. O. litifolia complex. *The Botanical Magazine* 75: 418–427.
Tripathy, J.N., J. Zhang, S. Robin, T.T. Nguyen, and H.T. Nguyen. 2000. QTLs for cell membrane stability mapped in rice (*Oryza sativa* L.) under drought stress. *Theoretical and Applied Genetics* 100: 1197–1202.

Tuong, T.P. 1999. Productive water use in rice production: Opportunities and limitations. *Journal of Crop Production* 2: 241–264.
Tuong, T.P., and B.A.M. Bouman. 2002. *Rice production in water-scarce environ-ments*. Paper Presented at the water productivity workshop, 12–14 November 2001.Colombo, Sri Lanka.
Tuong, T.P., and B.A.M. Bouman. 2003. Rice production in water-scarce environments. In *Water productivity in agriculture: Limits and opportunities for improve-ment*, ed. J.W. Kijne, R. Barker, and D. Molden, 53–67. Wallingford: CAB International.
Tuong, T.P., P.P. Pablico, M. Yamauchi, R. Confesor, and K. Moody. 2000. Increasing water productivity and weed suppression of wet-seeded rice: Effect of water management and rice genotypes. *Journal of Experimental Agriculture* 36: 1–19.
Tuong, T.P., B.A.M. Bouman, and M. Mortimer. 2004. More rice, less water-Integrated approaches for increasing water productivity in irrigated rice-based systems in Asia. In *New directions for a diverse planet*. Proceedings of the 4th international crop science congress, 26 September–1 October 2004. Brisbane Australia (published on CDROM).
Turner, N.C., G.C. Wright, and K.H.M. Siddique. 2001. Adaptation of grain legumes (pulses) to water-limited environments. *Advances in Agronomy* 71: 123–231.
Uphoff, N., and R. Randriamiharisoa. 2003. Reducing water use in irrigated rice production with the Madagascar System of Rice Intensification (SRI). In *Water-wise rice production*, eds. B.A.M. Bouman, H. Hengsdijk, B. Hardy, P.S. Bindraban, T.P. Tuong, and J.K. Ladha. Proceedings of a thematic workshop on water-wise rice production, 8–11 April 2002 at IRRI Headquarters in Los Baños, Philippines. International Rice Research Institute, Los Banos, Philippines.
Venuprasad, R., H. Lafitte, and G. Atlin. 2007. Response to direct selection for grain yield under drought stress in rice. *Crop Science* 47: 285–293.
Vikram, P., B.M. Swamy, S. Dixit, H. Ahmed, M.S. Cruz, A.K. Singh, G. Ye, and A. Kumar. 2012. Bulk segregant analysis:"An effective approach for mapping consistent-effect drought grain yield QTLs in rice". *Field Crops Research* 134: 185–192.
Wahid, A. 2007. Physiological implications of metabolite biosynthesis for net assimilation and heat-stress tolerance of sugarcane (*Saccharum officinarum*) sprouts. *Journal of Plant Research* 120: 219–228.
Wahid, A., S. Gelani, M. Ashraf, and M.R. Foolad. 2007. Heat tolerance in plants: An overview. *Environmental and Experimental Botany* 61: 199–223.
Wassmann, R., S.V.K. Jagadish, K. Sumfleth, H. Pathak, G. Howell, A. Ismail, R. Serraj, E. Redona, R.K. Singh, and S. Heuer. 2009. Regional vulnerability of climate change impacts on Asian rice production and scope for adaptation. *Advances in Agronomy* 102: 91–133.
Wilkinson, S., and W.J. Davies. 2010. Drought, ozone, ABA and ethylene: New insights from cell to plant to community. *Plant, Cell and Environment* 33: 510–525.
Yadav, R., B. Courtois, N. Huang, and G. Mclaren. 1997. Mapping genes controlling root morphology and root distribution in a doubled-haploid population of rice. *Theoretical and Applied Genetics* 94: 619–632.
Yang, J., J. Zhang, K. Liu, Z. Wang, and L. Liu. 2007. Involvement of polyamines in the drought resistance of rice. *Journal of Experimental Botany* 58: 1545–1555.
Zhang, J., H.T. Nguyen, and A. Blum. 1999. Genetic analysis of osmotic adjustment in crop plants. *Journal of Experimental Botany* 50: 291–302.
Zhang, S.-W., C.-H. Li, J. Cao, Y.-C. Zhang, S.-Q. Zhang, Y.-F. Xia, D.-Y. Sun, and Y. Sun. 2009. Altered architecture and enhanced drought tolerance in rice via the down-regulation of indole-3-acetic acid by TLD1/OsGH3.13 activation. *Plant Physiology* 151: 1889–1901.
Zhu, J.K. 2002. Salt and drought stress signal transduction in plants. *Annual Review of Plant Biology* 53: 247–273.

Chapter 6
QTL Modelling: An Adaptation Option in Spring Wheat for Drought Stress

Muhammad Umair Aslam, Armghan Shehzad, Mukhtar Ahmed, Muhammad Iqbal, Muhammad Asim, and M. Aslam

Contents

Abstract A project was executed to study genotype to phenotype relationships through QTL analysis in a recombinant inbred population of 77 lines and its integration in crop simulation modeling. RILs were generated from a cross between wheat cultivar Opata and SH-349. At two leaves stage drought was imposed using

M.U. Aslam (✉) • A. Shehzad • M. Iqbal • M. Asim
National Agricultural Research Centre, PARC- Institute of Advanced Studies in Agriculture, Islamabad, Pakistan
e-mail: umairaslamgill@gmail.com; armghan_shehzad@yahoo.com; iqbal2m@yahoo.com; asimnarc@gmail.com

M. Ahmed
Department of Agronomy, Pir Mehr Ali Shah Arid Agriculture University, Rawalpindi, Pakistan

Department of Biological Systems Engineering, Washington State University, Pullman, WA, USA
e-mail: ahmadmukhtar@uaar.edu.pk; mukhtar.ahmed@wsu.edu

M. Aslam
Ministry of Food and Agriculture, Islamabad, Pakistan
e-mail: aslamgill@hotmail.com

M. Ahmed, C.O. Stockle (eds.), *Quantification of Climate Variability, Adaptation and Mitigation for Agricultural Sustainability*,
DOI 10.1007/978-3-319-32059-5_6

gravimetric method for drought maintenance at 40 % of field capacity and control was maintained at 100 % field capacity. At three phenological stages viz. jointing, flag leaf and anthesis; photosynthetic rate, stomatal conductance, transpiration rate, stomatal resistance were determined and chlorophyll content was measured. The RILs under study exhibited high phenotypic variation under drought stress. The physiological and phenological data was used to parameterize and validate Agricultural Production Systems Simulator (APSIM); a crop growth and development modeling tool. It was noted that APSIM predicted the phenology of all the 77 RILs with R^2 value ranging from 0.72 to 0.98. The same mapping population was used for QTL mapping using computational approaches with observed data and simulated data from crop simulation model APSIM. In linkage group 1 a single QTL controlling 13 physiological traits and another QTL controlling a single trait for phenology was found. In linkage group 2 one QTL controlling 7 phenological traits was mapped. The QTLs which were mapped with real data were the same as with simulated data. This indicated that the simulated data with crop models under different environmental scenarios could be efficiently used for QTL mapping reducing the environmental contribution in G x E complex and suggesting the QTLs with more precision. Photosynthetic attributes of these RILs under drought stress at different phenological stages suggests complex physiological aspects critical for coping moisture stress and provides a strong basis for their utilization in wheat cultivar improvement for drought stress adaptation under changing climatic scenarios.

Keywords QTL • Wheat • APSIM • Modeling • RILs • Phenology

6.1 Introduction

Agriculture all around the globe is encountering biophysical limitations that are getting starker with climatic changes. The predicament is leading to more food insecurity and increase in poverty. These limitations include drought, salinity, desertification and new challenges like variation in pest and disease dispersals attributed to climate change. Currently more than 7 billion people need to be nurtured and it is obvious that population will increase to more exploding figures in coming years. Consequently the demand for food will increase and agricultural systems will remain under pressure to meet the targets. Multidisciplinary efforts in agricultural sciences yielded sustainable increase in crop productivity but with present scenario of biophysical and socio-natural constraints it is imperative to gear up the rate of genetic improvement to cope up this perplexity. Breeding improved cultivars and delivering them efficiently in shorter time frames is a key solution by using cutting edge biotechnological tools to augment better adapted crop ideotypes.

Drought is foremost reason of yield loss in spring wheat. Supplemental irrigation can be used to alleviate drought stress but it is not a cost effective and sustainable solution to changing severities of climate. In temperate agro ecological regions, terminal drought stress is getting more prevalent and affecting grain yield since it overlaps the grain filling critical stages of cereal crops including spring wheat.

Productivity losses attributed towards drought stress can be coped sustainably with genetic improvement for better adaptability to drought. Drought tolerance in wheat is a quantitative trait and is controlled by many genes which are spread over the wheat genome each gene contributing only small part to the observed phenotypic variation. The environmental variance resulting from differences in growing conditions further obscures the relation between phenotype and genotype. Furthermore, phenotypic selections and breeders' experience based conventional breeding often results in low breeding efficiency and inaccurate predictions. The genomic sites which house genes linked to pertinent quantitative trait are also called quantitative trait loci (QTLs). QTLs are the Genetic factors that are responsible for a part of the observed phenotypic variation for a quantitative trait. QTL-mapping is a process to analyze linkage between observed trait values and presence/absence of alleles of markers that have been mapped onto a linkage map. A significant correlation is found with minimum environmental variance to claim detection of a QTL. The size of the allelic effect of the detected QTL can also be estimated. A breeder can anticipate QTL occurrences and use this information to his benefit, for example by using in marker assisted breeding (Collard et al. 2005).

Integrating eco-physiological modeling and genetic mapping is critical to predict complex plant or crop traits under variable environmental conditions. Progress in molecular plant breeding is limited by the ability to predict plant phenotype based on its genotype, especially for complex adaptive traits. Efforts are being made around the world to integrate molecular biology and crop simulation models into a useful tool that breeders can use in planning a breeding program towards target environment. \DNA markers have been successfully used to explore plant genetic diversity to investigate variation and similarities at species level. SSR markers; more commonly known as microsatellites are being extensively used in wheat breeding programmes due to their better ability for polymorphism detection, reproducibility, specificity, easiness of use and transferability from one species to another within tribe *Triticeae* (Zhang et al. 2006).

The application of crop eco-physiological models to simulate crop development and management has been extensively researched successfully over many years (Sinclair and Seligman 1996). The utilization of crop modeling and simulation approaches for G-to-P interaction is in its formative stages (Hammer et al. 2002; Hammer and Jordan 2007). For significant Genotype to Phenotype prediction, the factors supportive to the illustrative capability of the crop eco-physiological models should link effectively to the QTLs associated with quantitative trait (Hammer et al. 2006; Chenu et al. 2009; Messina et al. 2009). The juvenile stages in the research and development of this paradigm of eco-physiological modeling resulted in identification of QTLs for several responsive traits of a model that simulate crop yields (Yin et al. 1999). QTLs based aspects were further exploited in model input to estimate yield performance of individuals in the population under study (Yin et al. 2000a). The association between predicted yields using QTLs based determinants and those using with the observed, phenotypic parameters was high. However, the capability of existing crop eco-physiological models is not yet adequate to simulate and predict variation in complex adaptive traits like yield amongst individuals of a segregating population (Yin et al. 2000a, b).

Zadoks et al. (1974) devised a decimal coding system for cereals which has been widely utilized to observe development stages. The Zadoks scale is a non-linear measure based on irregularly spaced phenological events starting from sowing of crop to its maturity. Using a two-digit cipher, the Zadok's system can study comprehensive information about the growth and developmental status of the plant. Crop phenological events direct the patterns plant development and are controlled by complex genetic and environmental determinants.

The Agricultural Production Systems Simulator (APSIM) is a simulation model, designed to cartel precise predictions of economic yield (e.g. grain, biomass, or sugar produce) for various crop species in response to environment and management situations, with forecasts of the long-term consequence of cropping systems on soil physical and chemical aspects (Keating et al. 2003). APSIM integrates a generic crop model (Wang et al. 2002), which uses a library of procedures for simulating crop progression and developmental processes.

A generic cereal template, programmed in object-oriented C++, has been developed based on the generic crop model. It is based on a framework of the physiological determinants of crop growth and development (Charles-Edwards 1982) and is focused at the organ scale. It generates the phenotype of a crop as a consequence of underlying physiological processes, using the concept of supply and demand balances for light, carbon, water, and nitrogen (Hammer et al. 2001).

Asim (2008) reported that simulation modeling using APSIM can be used to understand crop phenological and developmental behavior and its specificity to cultivars and climatic scenarios. Simulation modeling approaches can be successfully used to enhance understanding of crop bio-dynamism, climate and crop management under diverse environmental conditions. This can increase the understanding of underlying crop physiological and phenological processes for complex traits.

Difference between species and genotypes in the template is introduced through differences in input parameter values, rather than through differences in the underpinning crop physiological science for each species. The approach ensures scientific transparency, efficient use of code (Wang et al. 2002), and a more explanatory approach to the modeling of the underlying physiology (Hammer et al. 2006).

A study by Letort et al. (2008) employed a simulation of hypothetical genotypes to examine QTL associations with model parameters versus phenotypic traits. They argued that a functional–structural model was required to achieve satisfactory associations for model parameters. Ahmed (2011), studied climatic resilience of wheat cultivars under changing climate scenarios and suggested planting window adjustment for the target environments using simulation modeling approach to delineate G*M*E interaction to better understand the growth and development of wheat with varying climatic situations. The G-to-P prediction process is often characterized by partitioning into gene-to-trait and trait-to-phenotype components (Messina et al. 2009) in the simulation studies reported to date (Chapman et al. 2003; Hammer et al. 2005; Chenu et al. 2008; Letort et al. 2008). Chenu et al. (2009) have reported the first G-to-P modelling study that derives estimates of the effects on grain yield in target production environments of known quantitative trait loci (QTL) controlling a specific adaptive trait—leaf and silk elongation in maize. Their study highlighted the value of the G-to-P modelling approach in interpreting the genetic control of yield and, hence, its relevance to plant breeding.

Credible simulation of the complex G*M*E crop adaptation landscape can be utilized to add value to plant breeding. Albeit of defining environment types in the target population of environments (Chapman et al. 2000) and using that quantification to weight selection decisions to improve the rate of genetic gain (Podlich et al. 1999), simulated landscapes can be used as a test-bed for statistical techniques for QTL detection (Chapman 2008; Letort et al. 2008) and to aid design of breeding strategies by linking with breeding system simulation capability (Cooper et al. 2002; Cooper et al. 2005; Chapman et al. 2003; Hammer et al. 2005) to support operational molecular breeding (Messina et al. 2009). Bioinformatics in agriculture and use of various crop models and simulation have become accepted tools for agricultural research (Meinke 1996). Modeling crop bio dynamics is widespread and famous research area around the globe. Model testing and validation are the first steps to check the models performance in the wide spectrum. In an ideal case the models needs to be completely validated under range of ecological conditions to verify that its in-built relationships and mathematical language hold for any type of data incorporated into it. The capability of model to claim for reality outputs, complex set of data is required to run the model with lot of experimentation under range of climatic scenarios.

The present study was carried out at National Agricultural Research Centre, Islamabad, Pakistan. The objectives of the study were:

- To study photosynthetic attributes in recombinant inbred lines (RILs) of a wheat mapping population at different phenological stages under drought.
- To analyze genotype-to-phenotype relationships through QTL analysis coupled with an eco-physiological model.

6.2 Methdology

The study was carried out on a recombinant inbred population of 77 lines generated from a cross between wheat cultivar Opata and SH-349 (DrMP5 population). Opata is relatively drought susceptible and SH-349 is drought tolerant parent. A pot experiment was conducted under glass house conditions in CRD with three replications. Pot size used was 36 cm × 15 cm. Germination was carried out under non-stressed condition (Fig. 6.1). At two leaf stage (Z1.2 on Zadock's scale) drought was imposed. Gravimetric method was used for drought maintenance. Moisture content in control was maintained at field capacity and drought was maintained at 40 % of field capacity throughout the course of experiment (Earl 2003). Plant height (cm), spike length (cm), spikelets per spike, seed per spike and hundred grains weight (g) were recorded at maturity Z (92). The data collected from the glass house experiment and from QTL analysis (Sect. 6.3.3) was used for evaluation and parameterization of a crop physiological model named (Agricultural Production Systems Simulator) APSIM V.7.4. Observed data for control (without drought stress) was used to parameterize model and it was validated with observed values for the treatment (drought stress). QTL mapping was performed using same recombinant inbred

Fig. 6.1 Wheat mapping population planted in pots under glass house situated at National Institute of Genomics and Advance Biotechnology (NIGAB)

population (DrMP 5) of 77 lines generated from cross between wheat genotypes Opata and SH-349. Data on a set of markers were taken from the similar studies carried out on the same population at this institute by Karamat (2012). These markers were used; WMC-457, WMC-430, WMC-153, WMC-367, WMC-154, WMC-357, WMC-484, WMC-177, WMC-441, BARC-286, WMC-476, WMC-171, WMC-28, WMC-159, WMC-319, and WMC-160. The marker data and the morpho-physiological and phenological data were used in QTL cartographer V 2.5 for QTL mapping. Composite Interval Mapping model 6 was used was used keeping 5 background markers as control, window size used was 10 cM and walking speed of 2 cM was used. Significance of a QTL being detected was determined by LOD (likelihood of odds ratio) scale. Three hundred permutation tests were conducted to establish a LOD threshold value i.e. 2.0 for declaring a QTL and reducing the chances of false positive detection. The coefficient of determination (R^2) value determined the contribution of that QTL in explaining the total variation in the population for a particular trait. QTL mapping was also performed using simulated data from APSIM wheat module. The parameters used were phenology, plant height and hundred grain weight.

6.3 Results

6.3.1 Phenology of Mapping Population

Phenological development of wheat using Zadock's scale (Z) for 77 RIL's revealed a statistically significant difference (Figs. 6.2a and 6.2b). The result showed that in recombinant inbred population under drought the phenological stages emergence (Z9), three leaf (Z13), tillering (Z20), jointing (Z31), flag leaf (Z47), heading (Z50), anthesis (Z60), milky (Z73), soft dough (Z85) and maturity (Z92) took number of days statistically less than the control showing that during drought (treatment) the time taken by wheat crop to pass from one stage to another was shorter. Crop growth and development is significantly affected by moisture stress. Phenological development of wheat using Zadock's scale (Z) for 77 RIL's revealed a statistically significant difference. The result showed that in recombinant inbred population under treatment the phenological stages emergence (Z9), three leaf (Z13), tillering (Z20),

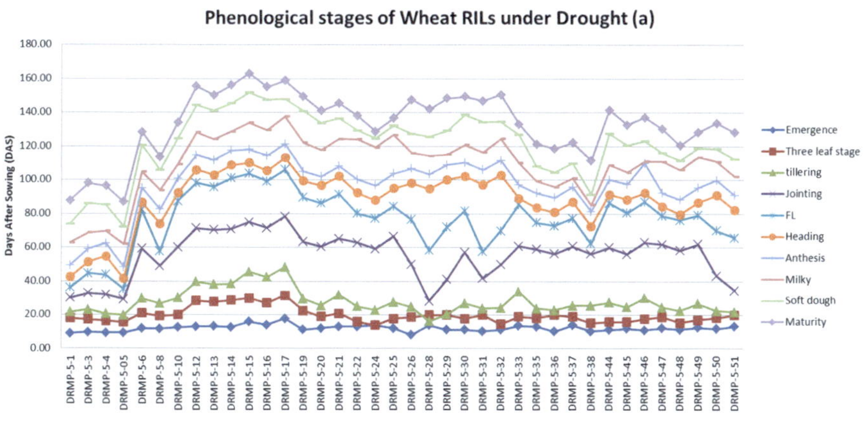

Fig. 6.2a Phenological stages of Wheat Recombinant Inbred Lines under drought. The chart represents the first 38 RILs mapping population

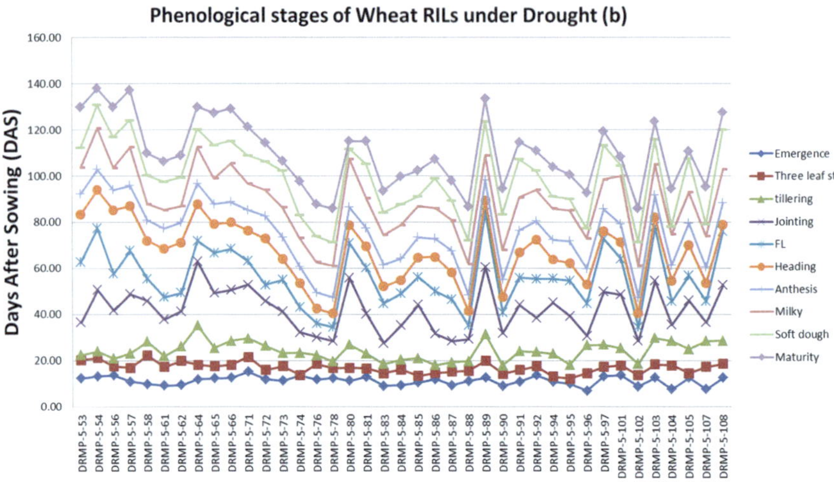

Fig. 6.2b Phenological stages of Wheat Recombinant Inbred Lines under drought. The chart represents other 39 RILs mapping population

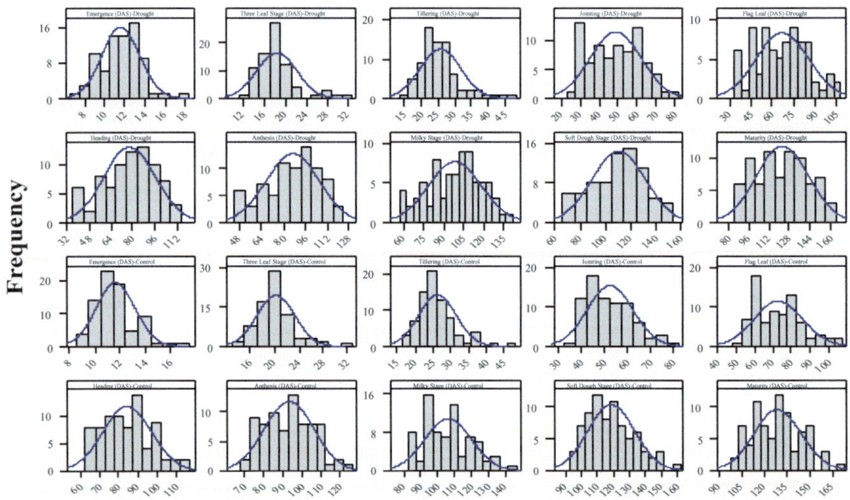

Fig. 6.3 Frequency distribution of phenology (Days after sowing, DAS) for wheat mapping population under study

jointing (Z31), flag leaf (Z47), heading (Z50), anthesis (Z60), milky (Z73), soft dough (Z85) and maturity (Z92) took number of days statistically less than the control showing that during (treatment) the time taken by wheat crop to pass from one stage to another was less (Fig. 6.3). This quick transition of crop from one stage to another may be due to deficient soil moisture to which crop was exposed. However in control, the time taken by wheat crop to pass from one stage was more may be due to availability of optimum moisture to crop during control which may resulted in completion of crop life cycle by utilizing maximum growing degree days (GDD). The study of crop phenology in relation to changing environment is very important application for crop growth modeling. Ludwig and Asseng (2010) reported similar conclusion about shorter growing length due to warmer and drier climates. Since crop growth and development is influenced by soil moisture therefore phenology of crop in this study was significantly affected by moisture stress.

6.3.2 *Yield and Yield Components*

The maximum height recorded in treatment was 82.7 cm while least plant height recorded was 52.9 cm (SD ± 8.50). The mean plant height recorded for stress treatment was 59.93 cm and for control it was 77.35 cm with LSD $\alpha_{0.05}$ 1.298. The treatment results portrayed that maximum spike length recorded was 15.91 cm while minimum spike length recorded was 9.60 cm (SD ± 1.85). The mean value recorded for stress treatment was 11.10 cm and for control was 14.77 cm with LSD $\alpha_{0.05}$ 0.964. The mean maximum spikelets per spike recorded in drought treatment were 16.23 and minimum recorded were 9.94 (SD ± 1.68). The mean value recorded for treatment was of 11.11 and for control was 14.67 with LSD $\alpha_{0.05}$ 0.972 (.6). Mean seed per spike among the mapping population under stress treatment and control revealed that seed per spike differed significantly. Mean maximum seeds per spike in stress treatment were 47.74 while the mean minimum seed per spike were 28.79 (SD ± 5.55). The mean value for treatment was recorded as 33.30 and for control was 44.32 with LSD $\alpha_{0.05}$ 2.894. Hundred grains weight has marked influence on wheat yield. The results for stress treatment revealed that the maximum HGW recorded was 4.21 g while minimum was 2.58 g (SD ± 0.44). The mean HGW under treatment was 2.88 g and control was 3.82 g with LSD $\alpha_{0.05}$.0.252. The significant difference in height may be due to difference in soil moisture and temperature among two environments and the genotypes. Highly significant correlation between environments and plant height was also reported by (Asif et al. 2003). Optimum plant height in control may be due to favorable environmental conditions. Correspondingly optimum plant height range increases yield potential of a crop (Araus et al. 2008). Spike length has significant correlation with grain yield. The significant difference in spike length among two environments may be due to variable temperature and moisture. During control moisture remained optimum which might ultimately helped in the translocation of photosynthate from source to sink. However in treatment there was extreme drought which may lead to decline in source to sink activity. Birsin (2005) concluded similar findings about spike and grain yield of wheat. Source sink relationship at the grain developmental stage had positive effect on spike length and grain formation (Li et al. 2008). Spikelet per spike is an important determinant factor for grain yield. Significant variation for spikelets per spike between two environments may be due to variability in soil moisture and temperature. Since in control there was optimum supply of water as compared to treatment. The optimum moisture conditions led to good spikelets per spike because of good source to sink activity. Significant variation in yield components with drought stress was concluded by earlier researcher (Kilic and Yagbasanlar 2010). Mean seed per spike among mapping population in two environments revealed that seed per spike differed significantly with changes moisture resulting in better source to sink activity for translocation of photosynthate. Hundred grains weight has marked influence on wheat yield. The significant difference in hundred grains weight between two environments may be due to extreme temperature and moisture stress. The highest HGW may be due to optimum temperature at grain development stage of crop while terminal heat and moisture stress prevailed in treatment. Heat and moisture stress at terminal stages caused reduction in grain yield (Nagarajan 2005).

6.3.3 Parameterization of APSIM Model

APSIM wheat module performs through its responsiveness to environmental and management scenarios by a set of generic genetic coefficients which are needed to be cultivar specific. The control data set from experiment was used for the parameterization of model. The generic genetic coefficients that yielded a good match between observed and simulated values were retained as it is, while the others were modified according to cultivar specificity. Simulation performance was evaluated through calculating root mean square error (RMSE) and bias.

Using the modified genetic coefficients in APSIM wheat module simulated values were obtained at coefficient of determination (R^2) of 0.98 which were in close agreement with recorded values for days to anthesis and days to maturity under control and then validated under drought stress conditions. Under drought stress the R^2 value was 0.83 between observed and simulated values indicating that the model was successfully parameterized thus, efficiently simulating phenological stages of wheat mapping population (Figs. 6.4 and 6.5). Parameterization and testing of a simulation model comprises of two processes; firstly, establishment of the source codes for assessment of the model's performance according to the required scenario and secondly validation of simulation modeling tool for accuracy to precisely imitate experimental facts (Meinke 1996). Kleijnen (1995) described these two processes as Validation and Verification of model. Model verification and validation for independent data is significant for the development of model. This study envisages parameterization and validation of APSIM wheat module for phenology of wheat drought mapping population and some yield parameters i.e. plant height and hundred grain weight to assess the predictive accuracy of model for these parameters. Validation and verification of model gives a relative assessment of model performance under a particular environmental conditions and the linear regression states the stability of model across variable climatic scenarios (regression value closer to 1:1 line means better model stability and accuracy) and the ratio of simulated to

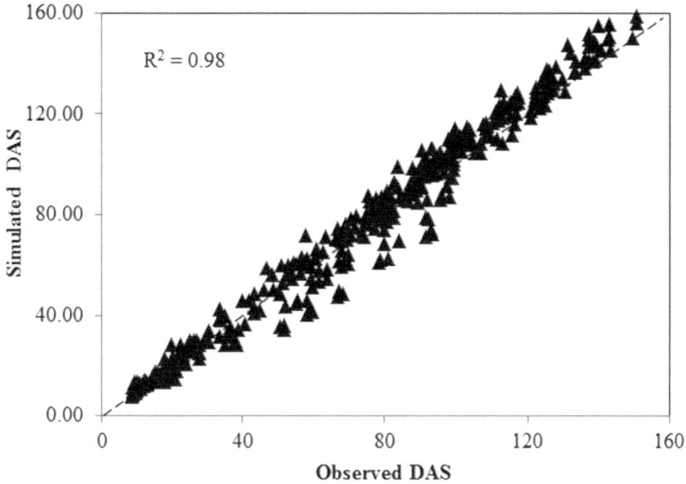

Fig. 6.4 Observed vs. simulated phenology of drought mapping population under control

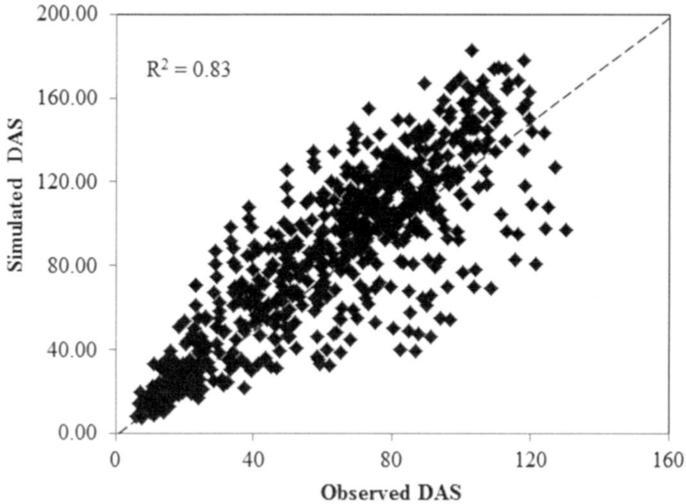

Fig. 6.5 Observed vs. simulated phenology of drought mapping population under drought stress

observed data assesses the predictive performance. In this study the parameterization of model was performed using the actual observations in non-stressed conditions. The generic genetic coefficients were calibrated to cultivar specific coefficients to obtain an appropriately fit agreement between observed and simulated values. The modeling results for comparison of simulated and observed phenological stages of wheat under control and drought stress conditions had a close association.

The results of this study were similar to (Ahmed et al. 2016) who discussed that yield simulation ability can be increased if simulation models can simulate phenology more accurately. The association between observed and simulated plant height under control and drought stress conditions was also appropriately close. Meinke (1996) discussed that simulation model is dependent on the triangle of climatic, edaphic and genetic factors. Hundred grain weight, a yield parameter was also simulated through APSIM wheat module which depicted a close relationship between observed and simulated values under both control and drought stress regimes suggesting optimum model parameterization with climatic and genetic factors that are responsible to trigger under variable moisture regimes.

These results helped in development of an increased level of confidence on the APSIM-wheat module with a much greater reliability and accuracy for simulation of crop growth and development. It increased the application of this bioinformatics tool for cultivar specific modeling of underlying physiological processes of complex genotypic characters responsible in development of drought stress adaptation in wheat.

6.3.4 QTL Mapping

QTL mapping of the wheat mapping population through selected markers along with computational analysis using QTL cartographer revealed that two markers were closely linked to two QTLs for 14 traits on linkage group 1 and a single marker was linked to QTL that controlled 7 traits on linkage group 2 as shown in Figs. 6.6, 6.7 and 6.7a and Tables 6.1 and 6.1. The simulated data for phenology, morphology and yield parameters was used for QTL mapping yielding 1 QTL for phenological stage; three leaf stage under control in linkage group 1 (Fig. 6.8) and 1 QTL for seven phenological traits in linkage group 2 (Fig. 6.9). The data generated from QTL mapping can be utilized to improve significant gene functions and heritability. Mapped position of QTLs gives information about actions of QTL in different plant populations and gene pools (Moose and Mumm 2008). Test of Permutation was carried out to identify the threshold and composite interval mapping was done. Through computational analysis of software package Windows QTL Cartographer V.2.5 QTLs were identified. The mapped QTLs had strong correlation with various important traits under control conditions as well as under drought stress.

Two QTLs for 14 various physiological and phenological traits in linkage group 1 were mapped using composite interval mapping which shows strong reliability of results by using this technique. In linkage group 2 one QTL for phenological stages under drought (Heading, Anthesis, Milky stage, Soft dough, Maturity) and for phenological stages under control (Heading and Anthesis) were mapped with composite interval mapping using observed data All these phenological traits mapped in linkage group 2 were found to be linked with marker WMC-160. The flanking marker for WMC-160 was WMC-319. QTL map position for these traits is 23 cM. The threshold values for these QTLs suggest strong correlation and significant contribution of these QTLs with the trait and percentage of variability conferred by them with an average of 13.67 % of trait variation explained by these

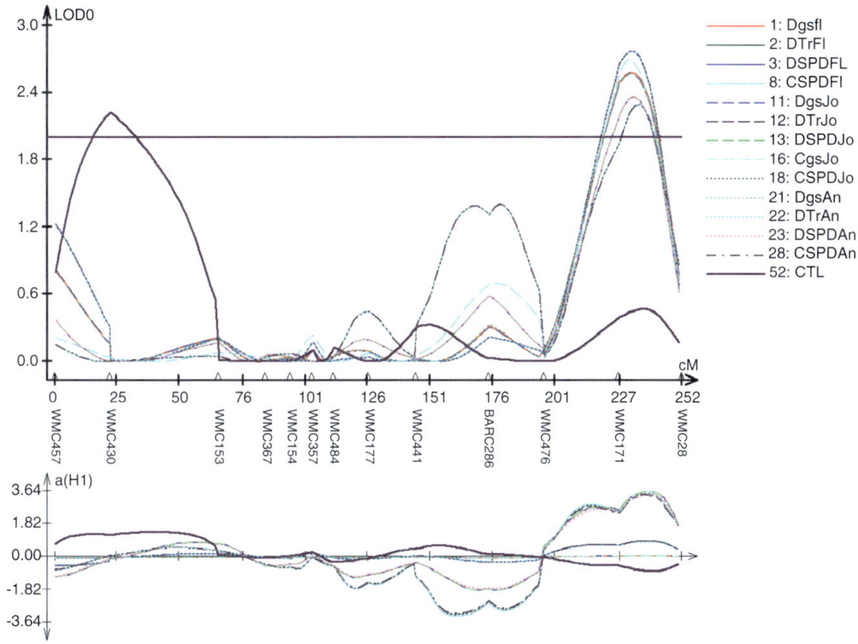

Fig. 6.6 Graph showing QTLs for photosynthetic attributes under drought conditions at different phenological stages (Stomatal conductance at flag leaf-DgsFL, Transpiration rate at flag leaf-DTrFL, Chlorophyll content at flag leaf-DSPDFL, Stomatal Conductance at jointing-DgsJo, Transpiration rate at jointing-DTrJo, Chlorophyll content at jointing-DSPDJo, Stomatal conductance at anthesis-DgsAn, Transpiration rate at anthesis-DTrAn, Chlorophyll content at anthesis-DSPDAn) and QTLs for photosynthetic attributes under control at different phenological stages (Chlorophyll content at flag leaf-CSPDFL, Stomatal Conductance at jointing-CgsJo, Chlorophyll content at jointing-CSPDJo and Chlorophyll content at anthesis-CSPDAn) and QTL for phenological stage under control (Three leaf stage-CTL) on linkage group 1 mapped with composite interval mapping using observed data

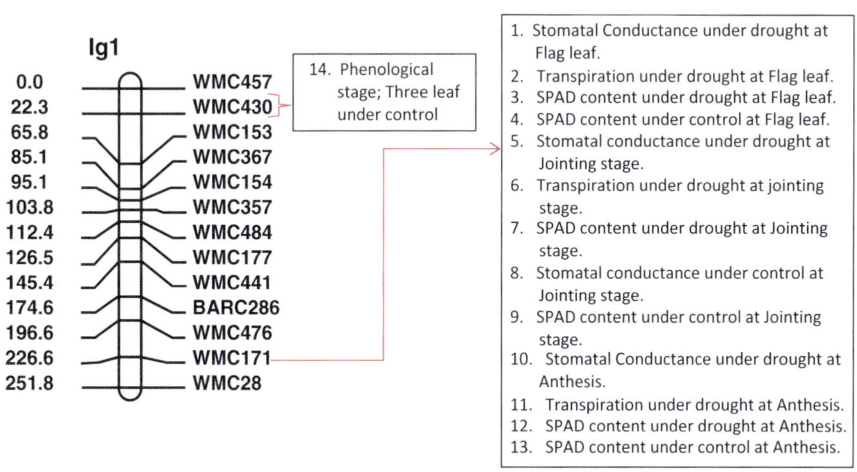

Fig. 6.7 Linkage group 1 and recorded data QTLs mapped on it

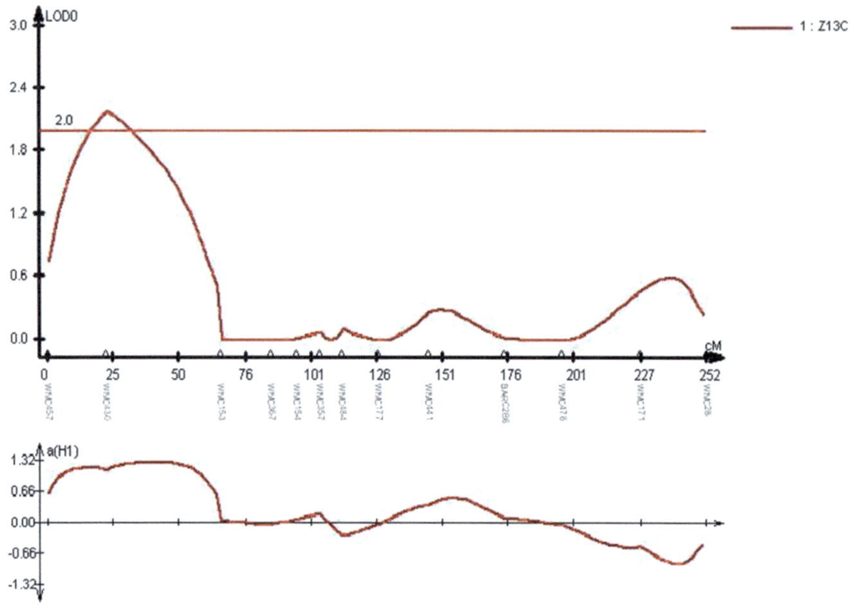

Fig. 6.7a Graph showing QTL for phenological stage under control (Three leaf stage-CTL) on linkage group 1 mapped with composite interval mapping using simulated data. Only one QTL was found for phenology using simulated data for computational analysis for QTL mapping because we could only simulate the phenological stages in simulation tool APSIM. The occurrence of QTL for Three leaf stage under control (CTL) coincides efficiently in mapping through both observed and simulated data

seven QTLs mapped on linkage group 2. QTL for all seven traits determined in linkage group 2 had negative additive effect. (Mason et al. 2010) studied that the marker WMC-160 was linked to leaf length and anthesis and exhibited R^2 value 0.13 in case of leaf length and 0.21 for anthesis. They mapped WMC-160 on chromosome 5B of wheat. The molecular markers and the linked QTLs that were mapped here having strong association in conferring drought tolerance in wheat can be utilized to increase and improve the breeding efficiency in wheat crop improvement against drought stress for vulnerable climatic regimes and agro-ecologies in future research through marker assisted selection (MAS) saving valuable resources.

With the APSIM model parameterized and validated and QTL mapping performed on the basis of observed values for physiology and phenology the next step ahead to bridge the gap between understanding of eco-physiological modeling and genetic factors responsible for phenotypic traits QTL mapping was performed again using the simulated values of phenology, plant height and hundred grain weight. The concept coincides with the idea floated by Hammer and Jordan (2007) of dissecting the complex quantitative trait into its physiological basis and then integrating the underlying genetic factors and physiological determinants with the help of a suitable eco-physiological model to efficiently predict G to P prognosis for target population of environments. It was found that the correlation between the observed values and simulated

Table 6.1 QTLs found in composite interval mapping under controlled and drought conditions in linkage group 1 along with LOD scores and percentage of variation explained by each QTL

Trait	QTLs	Markers associated	LOD	Variation explained (R^2) (%)
Stomatal Conductance under drought at Flag leaf	QTL 1	WMC-171	2.463	19.7
Transpiration under drought at Flag leaf	QTL 1	WMC-171	2.628	21
Chlorophyll content under drought at Flag leaf	QTL 1	WMC-171	2.280	18.3
Chlorophyll content under control at Flag leaf	QTL 1	WMC-171	2.192	19.6
Stomatal conductance under drought at Jointing stage	QTL 1	WMC-171	2.452	19.6
Transpiration rate under drought at Jointing stage	QTL 1	WMC-171	2.561	21.1
Chlorophyll content under drought at Jointing stage	QTL 1	WMC-171	2.280	18.3
Stomatal conductance under control at Jointing stage	QTL 1	WMC-171	2.505	20.5
Chlorophyll content under control at Jointing stage	QTL 1	WMC-171	2.191	19.5
Stomatal Conductance under drought at Anthesis	QTL 1	WMC-171	2.449	19.6
Transpiration under drought at Anthesis	QTL 1	WMC-171	2.559	21.1
Chlorophyll content under drought at Anthesis	QTL 1	WMC-171	2.281	18.3
Chlorophyll content under control at Anthesis	QTL 1	WMC-171	2.191	19.6
Phenological stage; Three leaf under control	QTL 2	WMC-430	2.128	13.8

values with the basis of QTL based factors was significantly high; the capability of eco-physiological crop growth models is not adequate to envisage variation in complex adaptive traits (Yin et al. 2000a, b). Crop simulation modeling has yet contributed very little in QTL analysis of a complex quantitative trait but improvement in crop model to efficiently predict G to P relationships can assist in accurate QTL analysis under diverse environmental conditions. The identified QTLs based on simulated data for target population of environments can be efficiently utilized in increasing efficiency of breeding strategies by predicting and augmenting an appropriate and dynamic crop ideotype, improved for complex adaptive traits for changing climatic scenarios over diverse locations. Modified Genetic Coefficients in APSIM wheat-module for wheat mapping population are presented in Appendix 1. Data showing work on QTL in relation to crop traits and its effect using different techniques is presented in Table 6.2.

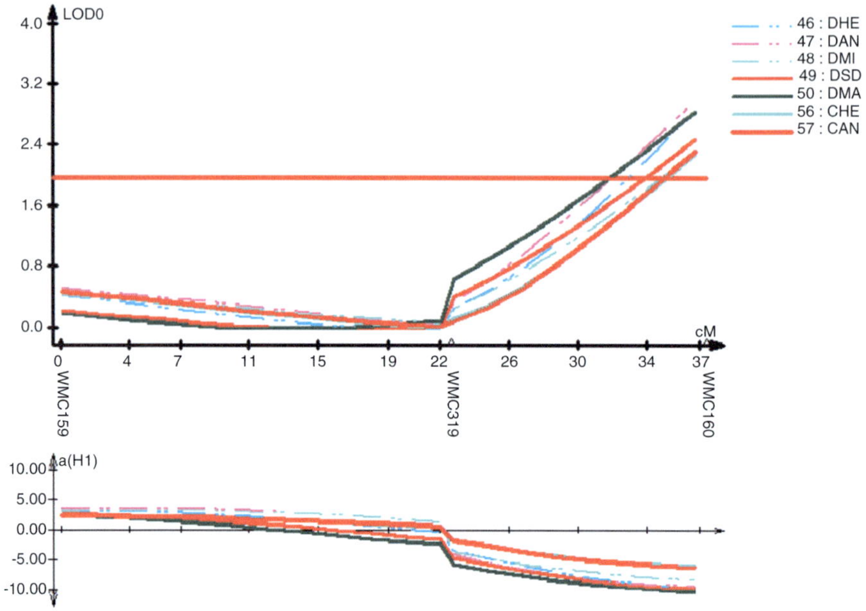

Fig. 6.8 Graph showing QTLs for phenological stages under drought (Heading-DHE, Anthesis-DAN, Milky stage-DMI, Soft dough-DSD, Maturity- DMA) and QTLs for phenological stages under control (Heading-CHE and Anthesis-CAN) on linkage group 2 mapped with composite interval mapping using observed data

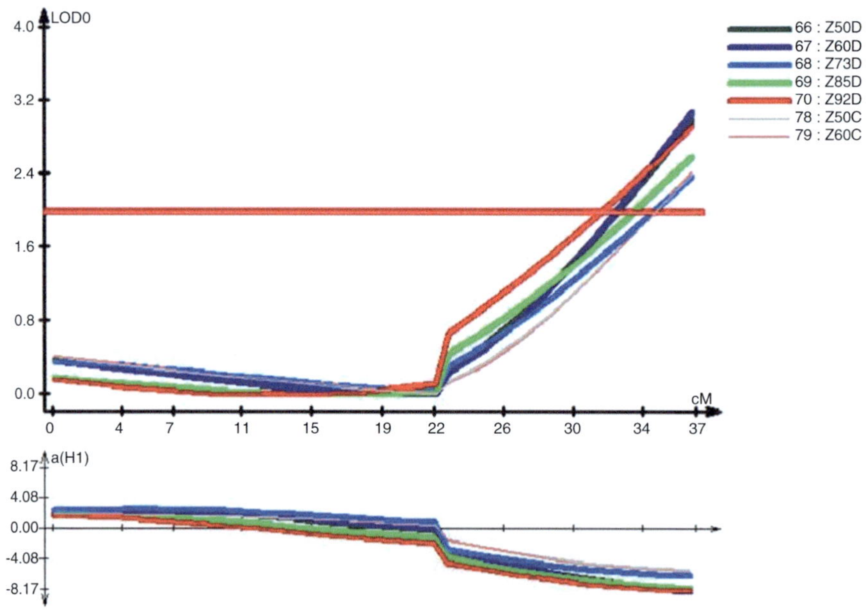

Fig. 6.9 Graph showing QTLs for phenological stages under drought (Heading-Z50D, Anthesis-Z60D, Milky stage-Z73D, Soft dough-Z85D, Maturity- Z92D) and QTLs for phenological stages under control (Heading-Z50C and Anthesis-Z60C) on linkage group 2 mapped with composite interval mapping using simulated data. The occurrence of these QTLs fully coincides in mapping through both observed and simulated data

Table 6.2 QTL found in composite interval mapping under controlled and drought conditions in linkage group 2 along with LOD scores and percentage of variation explained by each QTL

Trait	QTLs	Associated markers	LOD	Variation explained (R^2) (%)
Days to Heading under Drought	QTL 3	WMC-160	2.897	15.4
Days to Anthesis under Drought	QTL 3	WMC-160	3.028	16.1
Days to Milky stage under Drought	QTL 3	WMC-160	2.335	11.9
Days to Soft dough stage under drought	QTL 3	WMC-160	2.502	12.6
Days to Maturity under Drought	QTL 3	WMC-160	2.867	14.6
Days to Heading under Control	QTL 3	WMC-160	2.302	12.4
Days to Anthesis under Control	QTL 3	WMC-160	2.338	12.7

Table 6.3 Data showing work on QTL in relation to crop traits and its effect using different techniques

Sr. No.	Effect	Traits	References
1	The QTL clusters coinciding with the marker wmc41 were associated mainly with grain-size traits, which led to decreases in GW and TGW and to increases in GLW	Grain length (GL), Grain Width (GW), GL/GW ratio (GRW), Thousand Grain Weight (TGW)	Zhang et al. (2015)
2	A set of recombinant inbred lines (RILs) were used to map quantitative trait loci (QTLs) for source–sink size and heading date. Thirty QTLs consistently detected in at least two trials and generally located in the clusters. Using a set of BC4F2 lines, the QTL cluster in C5-1–C5-2 on chromosome 5 was validated to be a major QTL affecting heading date, source size (flag leaf area)	Source Size, Heading Traits, Sink size	Zhang et al. (2015)
3	A new genetic map with high density was constructed and used to detect the QTLs for heading date, kernel width, spike length, spikelet number, and thousand kernel weight. Thirteen QTLs were located on D genomes of SHW-L1, six of them showed positive effect on agronomic traits	Spike length, Heading date, Spikelet number, Kernel width, and Thousand kernel weight	Yu et al. (2014)
4	A total of 14 QTLs, comprising eight for seed yield, five for vegetative traits and one for cold tolerance, were detected on nine linkage groups: seven linkage groups of the E142 genetic map and two linkage groups of the E022 genetic map. Each of these QTLs explained 8.71–23.61 % of the phenotypic variation	seed yield, vegetative traits and cold tolerance	Nezhadahmadi et al. (2013)
5	Three QTL for A-type starch granule content were mapped on chromosomes 1DL, 7BL and 4AL, explaining 5.6 %, 5.2 % and 3.8 % of the phenotypic variation, respectively	Starch Quality, Phenotypic variation	Feng et al. (2013)

(continued)

Table 6.3 (continued)

Sr. No.	Effect	Traits	References
6	Identification of quantitative trait loci (QTL) associated with heat susceptibility index (HSI) of yield components in response to a short-term heat shock during early grain-filling in wheat	Heat susceptibility index (HSI), Yield components	Mason et al. (2010)
7	The QTL Cartographer software was used to study QTL detection of simulated plant traits. Virtual QTL detection was compared in the case of phenotypic traits – such as cob weight – and when traits were model parameters, and was found to be more accurate in the latter case	G × E interaction, Modeling and Virtual QTL detection	Letort et al. (2008)
8	Identification of quantitative trait loci (QTL) controlling grain yield and yield components under reduced moisture. An SSR/EST-STS marker map was constructed and a grain yield QTL on the proximal region of chromosome 4AL was found to have a significant impact on performance under reduced water	Grain Yield	Narjesi et al. (2015)

6.4 Summary

Drought is one of the main limitation in wheat production thus drought tolerance a main objective of wheat crop improvement programs in the country. Supplementing to the numerous yet hopeful efforts of ongoing breeding strategies in different national institutes, this study was carried out at National Agricultural Research Centre (NARC) on recombinent inbred lines (RILs) of wheat mapping population based on a cross between Opata and SH-349. The study was designed to analyze photosynthetic attributes of these recombinant inbred lines at different phenological stages under drought stress and secondly to investigate genotype-to-phenotype relationships through QTL analysis coupled with an eco-physiological model.

- Photosynthetic attributes were studied at three critical phenological stages; flag leaf, jointing and anthesis. The analysis of photosynthetic attributes of these RILs revealed significant physiological determinants responsible for key role in developing physiological adaptation to drought stress regimes. The RILs mapping population was found to be highly variable for these photosynthetic attributes, phenology and yield parameters.
- Phenological, morphological and physiological parameters and yield attributes of mapping population were studied and the variation among the mapping population for these characteristics was found to be statistically highly significant.
- The physiological and phenological data generated from the study was used to parameterize and validate an ecophysiological crop growth and development modelling tool APSIM. Crop phenology, morphology and yield was predicted with adequate accuracy and close corelation was found among observed and simulated values.

- The recombinant inbred lines mapping population were also evaluated with SSR molecular markers for QTL mapping. The markers that were used in the study were WMC-457, WMC-430, WMC-153, WMC-367, WMC-154, WMC-357, WMC-484, WMC-177, WMC-441, BARC-286, WMC-476, WMC-171, WMC-28, WMC-159 WMC-319, and WMC-160.
- With composite interval mapping using observed data, two QTLs were mapped in linkage group 1 for a total of 14 traits and 1 QTL was mapped in linkage group 2 for a total of 7 traits.
- The simulated data from the model was then subjected to QTL mapping and with composite interval mapping technique a single QTL for phenological development was found in likage group 1 and 1 QTL for phenological development controlling 7 traits was mapped in linkage group 2.
- QTL mapping with simulated data as a cross referance to the mapping based on observed values, minimizes the environmental variance and increases the reiability of the identified QTLs. It can also be used for QTL mapping for diverse target population of environments.
- This integrated modelling-molecular (QTL mapping) approach through connecting these QTLs to physiological determinants and modeling the interacion with variable climatic scenarios can help in predicting phenotype for complex quantitative traits under diverse environmental conditions.

6.5 Conclusions and Recommendations

- The drought mapping population under study was found to be highly variable for the phenological, physiological (particularly photosynthetic attributes) and yield traits.
- Parameterized and validated APSIM (crop simulation model) on drought mapping population under control conditions. The simulated data under stressed conditions for phenology, plant height and HGW suggested a close proximity of recorded and simulated data with R^2 values ranging from 0.72 to 0.89.
- QTLs mapped with recorded data and QTLs mapped with simulated data from APSIM for phenology, HGW and plant height were the same suggesting a minimal environmental interaction with the associated QTLs.
- The data in this study can be used for increased ability to predict plant phenotype based on its genotype for complex adaptive traits using suitably constructed crop growth model to bridge the predictability gap (G to P interaction).
- The integration of QTL mapping and eco-physiological modeling can effectively simulate crop attributes throughout the crop life cycle for variable climatic regimes to predict a potential crop ideotype based on a stronger genotype to phenotype prognosis with minimal environmental variance.

Appendix 1: Modified Genetic Coefficients in APSIM Wheat-Module for Wheat Mapping Population

Parameters	Vernalization sensitivity	Photothermal sensitivity	Thermal time for grain filling	Radiation use efficiency at floral initiation/flowering
Default Value in APSIM-Wheat Module	1.5	3	610	1.24/1.24
DRMP-5-1	0	3.45	635	3.00/2.24
DRMP-5-3	0	3.5	644	3.00/2.24
DRMP-5-4	0	3.46	636	3.00/2.24
DRMP-5-05	0	3.44	633	3.00/2.24
DRMP-5-6	0	3.45	635	3.00/2.24
DRMP-5-8	0	3.5	644	3.00/2.24
DRMP-5-10	0	3.6	662	3.00/2.24
DRMP-5-12	0	3.5	644	3.00/2.24
DRMP-5-13	0	3.3	624	3.00/2.24
DRMP-5-14	0	3.4	626	3.00/2.24
DRMP-5-15	0	3.34	625	3.00/2.24
DRMP-5-16	0	3.45	635	3.00/2.24
DRMP-5-17	0	3.4	625	3.00/2.24
DRMP-5-19	0	3.4	625	3.00/2.24
DRMP-5-20	0	3.5	644	3.00/2.24
DRMP-5-21	0	3.6	662	3.00/2.24
DRMP-5-22	0	3.6	662	3.00/2.24
DRMP-5-24	0	3.6	662	3.00/2.24
DRMP-5-25	0	3.6	662	3.00/2.24
DRMP-5-26	0	3.5	644	3.00/2.24
DRMP-5-28	0	3.5	644	3.00/2.24
DRMP-5-29	0	3.6	662	3.00/2.24
DRMP-5-30	0	3.45	635	3.00/2.24
DRMP-5-31	0	3.2	621	3.00/2.24
DRMP-5-32	0	3.5	644	3.00/2.24
DRMP-5-33	0	3.6	662	3.00/2.24
DRMP-5-35	0	3.6	662	3.00/2.24
DRMP-5-36	0	3.5	644	3.00/2.24
DRMP-5-37	0	3.7	653	3.00/2.24
DRMP-5-38	0	3.6	661	3.00/2.24
DRMP-5-44	0	3.5	644	3.00/2.24
DRMP-5-45	0	3.5	644	3.00/2.24
DRMP-5-46	0	3.5	644	3.00/2.24
DRMP-5-47	0	3.6	662	3.00/2.24

(continued)

Parameters	Vernalization sensitivity	Photothermal sensitivity	Thermal time for grain filling	Radiation use efficiency at floral initiation/flowering
DRMP-5-48	0	3.6	662	3.00/2.24
DRMP-5-49	0	3.5	644	3.00/2.24
DRMP-5-50	0	3.6	662	3.00/2.24
DRMP-5-51	0	3.5	644	3.00/2.24
DRMP-5-53	0	3.5	644	3.00/2.24
DRMP-5-54	0	3.5	644	3.00/2.24
DRMP-5-56	0	3.6	645	3.00/2.24
DRMP-5-57	0	3.5	644	3.00/2.24
DRMP-5-58	0	3.5	644	3.00/2.24
DRMP-5-61	0	3.5	644	3.00/2.24
DRMP-5-62	0	3.4	626	3.00/2.24
DRMP-5-64	0	3.6	663	3.00/2.24
DRMP-5-65	0	3.4	626	3.00/2.24
DRMP-5-66	0	3.6	663	3.00/2.24
DRMP-5-71	0	3.4	626	3.00/2.24
DRMP-5-72	0	3.5	644	3.00/2.24
DRMP-5-73	0	3.6	663	3.00/2.24
DRMP-5-74	0	3.5	644	3.00/2.24
DRMP-5-76	0	3.5	644	3.00/2.24
DRMP-5-78	0	3.6	663	3.00/2.24
DRMP-5-80	0	3.6	663	3.00/2.24
DRMP-5-81	0	3.5	644	3.00/2.24
DRMP-5-83	0	3.5	644	3.00/2.24
DRMP-5-84	0	3.6	663	3.00/2.24
DRMP-5-85	0	3.4	626	3.00/2.24
DRMP-5-86	0	3.5	644	3.00/2.24
DRMP-5-87	0	3.6	663	3.00/2.24
DRMP-5-88	0	3.6	663	3.00/2.24
DRMP-5-89	0	3.5	644	3.00/2.24
DRMP-5-90	0	3.4	626	3.00/2.24
DRMP-5-91	0	3.6	663	3.00/2.24
DRMP-5-92	0	3.5	644	3.00/2.24
DRMP-5-94	0	3.6	663	3.00/2.24
DRMP-5-95	0	3.4	626	3.00/2.24
DRMP-5-96	0	3.5	644	3.00/2.24
DRMP-5-97	0	3.5	644	3.00/2.24
DRMP-5-101	0	3.6	663	3.00/2.24
DRMP-5-102	0	3.6	663	3.00/2.24
DRMP-5-103	0	3.4	626	3.00/2.24
DRMP-5-104	0	3.5	644	3.00/2.24
DRMP-5-105	0	3.5	644	3.00/2.24
DRMP-5-107	0	3.6	663	3.00/2.24
DRMP-5-108	0	3.4	626	3.00/2.24

References

Ahmed, M. 2011. Climatic resilience of wheat using simulation approaches. In Pothowar. PhD, PMAS Arid Agricultural University Rawalpindi, Pakistan.

Ahmed, M., M.N. Akram, M. Asim, M. Aslam, F.-U. Hassan, S. Higgins, C.O. Stöckle, and G. Hoogenboom. 2016. Calibration and validation of APSIM-Wheat and CERES-Wheat for spring wheat under rainfed conditions: Models evaluation and application. *Computers and Electronics in Agriculture* 123: 384–401.

Araus, J.L., G.A. Slafer, C. Royo, and M.D. Serret. 2008. Breeding for yield potential and stress adaptation in cereals. *Critical Reviews in Plant Sciences* 27: 377–412.

Asif, M., M.Y. Mujahid, I. Ahmad, N.S. Kisana, M. Asim, and S.Z. Mustafa. 2003. Determining direct selection criteria for identification of high yielding lines in bread wheat (Triticum aestivum). *Pakistan Journal of Biological Sciences* 6: 48–50.

Asim, M. 2008. Determination of spring wheat response to developmental windows using simulation approach. PhD, Quaid-e-Azam university, Islamabad, Pakistan.

Birsin, M.A. 2005. Effects of removal of some photosynthetic structures on some yield components in wheat. *Tarim Bilimleri Dergisi* 11: 364–367.

Chapman, S.C. 2008. Use of crop models to understand genotype by environment interactions for drought in real-world and simulated plant breeding trials. *Euphytica* 161: 195–208.

Chapman, S.C., M.C. Cooper, G.L. Hammer, and D. Butler. 2000. Genotype by environment interactions affecting grain sorghum. II. Frequencies of different seasonal patterns of drought stress are related to location effects on hybrid yields. *Australian Journal of Agricultural Research* 51: 209–222.

Chapman, S.C., M. Cooper, D. Podlich, and G.L. Hammer. 2003. Evaluating plant breeding strategies by simulating gene action and dryland environment effects. *Agronomy Journal* 95: 99–113.

Charles-Edwards, D.A. 1982. *Physiological determinants of crop growth*. Sydney: Academic.

Chenu, K., S.C. Chapman, G.L. Hammer, G. Mclean, and F. Tardieu. 2008. Short-term responses of leaf growth rate to water deficit scale up to whole plant and crop levels. An integrated modelling approach in maize. *Plant, Cell & Environment* 31: 378–391.

Chenu, K., S.C. Chapman, F. Tardieu, G. Mclean, C. Welcker, and G.L. Hammer. 2009. Simulating the yield impacts of organ-level quantitative trait loci associated with drought response in maize: 'Gene-to-phenotype' modeling approach. *Genetics* 183: 1507–1523.

Collard, B.C.Y., M.Z.Z. Jahufer, J.B. Brouwer, and E.C.K. Pang. 2005. An introduction to markers, quantitative trait loci (QTL) mapping and marker-assisted selection for crop improvement: The basic concepts. *Euphytica* 142: 169–196.

Cooper, M., S.C. Chapman, D.W. Podlich, and G.L. Hammer. 2002. The G to P problem: Quantifying gene-to-phenotype relationships. *In Silico Biology* 2: 151–164.

Cooper, M., D.W. Podlich, and O.S. Smith. 2005. Gene-to-phenotype models and complex trait genetics. *Australian Journal of Agricultural Research* 56: 895–918.

Earl, H.J. 2003. A precise gravimetric method for simulating drought stress in pot experiments. *Crop Science* 43: 1868–1873.

Feng, N., Z. He, Y. Zhang, X. Xia, and Y. Zhang. 2013. QTL mapping of starch granule size in common wheat using recombinant inbred lines derived from a PH82-2/Neixiang 188 cross. *The Crop Journal* 1: 166–171.

Hammer, G.L., and D.R. Jordan. 2007. An integrated systems approach to crop improvement. In *Scale and complexity in plant systems research: Gene-plant-crop relations*, Wageningen UR, F. S. N. D, ed. J.H.J. Spiertz, P.C. Struik, and H.H. van Laar, 45–61. Dordrecht: Springer.

Hammer, G. L., E.L. Van-Oosterom, S.C. Chapman, and G. Mclean. 2001. The economic theory of water and nitrogen dynamics and management in field crops. In *Proceedings of the fourth Australian sorghum conference*, eds. Borrell A. K, and R.G Henzell. Kooralbyn, Queensland. 5–8 February 2001. CD Rom Format. Range Media Pvt.Ltd., 2001.

Hammer, G.L., M.J. Kropff, T.R. Sinclair, and J.R. Porter. 2002. Future contributions of crop modelling: From heuristics and supporting decision-making to understanding genetic regulation and aiding crop improvement. *European Journal of Agronomy* 18: 15–31.

Hammer, G.L., S. Chapman, E. Van Oosterom, and D.W. Podlich. 2005. Trait physiology and crop modelling as a framework to link phenotypic complexity to underlying genetic systems. *Australian Journal of Agricultural Research* 56: 947–960.

Hammer, G.L., M. Cooper, F. Tardieu, S. Welch, B. Walsh, F.V. Eeuwijk, S. Chapman, and C. Podlich. 2006. Models for navigating biological complexity in breeding improved crop plants. *Trends in Plant Science* 11: 587–593.

Karamat, F. 2012. *Genetic linkage map construction in a hexaploid wheat population developed for mapping drought tolerance*. http://agris.fao.org/agris-search/search.do?recordID=PK2014000261

Keating, B.A., P.S. Carberry, G.L. Hammer, M.E. Probert, M.J. Robertson, D. Holzworth, N.I. Huth, J.N.G. Hargreaves, H. Meinke, Z. Hochman, G. Mclean, K. Verburg, V. Snow, J.P. Dimes, M. Silburn, E. Wang, S. Brown, K.L. Bristow, S. Asseng, S. Chapman, R.L. Mccown, D.M. Freebairn, and C.J. SMITH. 2003. An overview of APSIM, a model designed for farming systems simulation. *European Journal of Agronomy* 18: 267–288.

Kilic, H., and T. Yagbasanlar. 2010. The effect of drought stress on grain yield, yield components and some quality traits of durum wheat (Triticum turgidum ssp durum) cultivars. *Notulae Botanicae Horti Agrobotanici Cluj-Napoca* 38: 164–170.

Kleijnen, J.P.C. 1995. Verification and validation of simulation models. *European Journal of Operational Research* 82: 145–161.

Letort, V., P. Mahe, P.-H. Cournède, P. De Reffye, and B. Courtois. 2008. Quantitative genetics and functional–structural plant growth models: Simulation of quantitative trait loci detection for model parameters and application to potential yield optimization. *Annals of Botany* 101: 1243–1254.

Li, L., G.S. Mcmaster, Q. Yu, and J. Du. 2008. Simulating winter wheat development response to temperature: Modifying Malo's exponential sine equation. *Computers and Electronics in Agriculture* 63: 274–281.

Ludwig, F., and S. Asseng. 2010. Potential benefits of early vigor and changes in phenology in wheat to adapt to warmer and drier climates. *Agricultural Systems* 103: 127–136.

Mason, R.E., S. Mondal, F.W. Beecher, A. Pacheco, B. Jampala, A.M.H. Ibrahim, and D.B. Hays. 2010. QTL associated with heat susceptibility index in wheat (Triticum aestivum L.) under short-term reproductive stage heatstress. *Euphytica* 174: 423–436.

Meinke, K. 1996. Topological methods for algebraic specification. *Theoretical Computer Science* 166: 263–290.

Messina, C., G. Hammer, Z. Dong, D. Podlich, and M. Cooper. 2009. Modelling crop improvement in a G*E*M framework via gene-trait phenotype relationships. In *Crop physiology: Applications for genetic improvement and agronomy*, ed. V.O. Sadras and D. Calderini, 235–265. Amsterdam/Boston: Academic Press/Elsevier.

Moose, S.P., and R.H. Mumm. 2008. Molecular plant breeding as the foundation for 21st century crop improvement. *Plant Physiology* 147: 969–977.

Nagarajan, S. 2005. Can India produce enough wheat even by 2020. *Current Science* 89: 1467–1471.

Narjesi, V., M. Mardi, E.M. Hervan, A. Azadi, M.R. Naghavi, M. Ebrahimi, and A.A. Zali. 2015. Analysis of Quantitative Trait Loci (QTL) for grain yield and agronomic traits in wheat (Triticum aestivum L.) under normal and salt-stress conditions. *Plant Molecular Biology Reporter* 33: 2030–2040.

Nezhadahmadi, A., Z.H. Prodhan, and G. Faruq. 2013. Drought tolerance in wheat. *The Scientific World Journal* 2013: 610721.

Podlich, D.W., M. Cooper, K.E. Basford, and H.H. Geiger. 1999. Computer simulation of a selection strategy to accommodate genotype environment interactions in a wheat recurrent selection programme. *Plant Breeding* 118: 17–28.

Sinclair, T.R., and N.G. Seligman. 1996. Crop modeling: From infancy to maturity. *Agronomy Journal* 88: 698.

Wang, E., M.J. Robertson, G.L. Hammer, P.S. Carberry, D. Holzworth, H. Meinke, S.C. Chapman, J.N.G. Hargreaves, N.I. Huth, and G. Mclean. 2002. Development of a generic crop model template in the cropping system model APSIM. *European Journal of Agronomy* 18: 121–140.

Yin, X., P. Stam, C.J. Dourleijn, and M.J. Kropff. 1999. AFLP mapping of quantitative trait loci for yield-determining physiological characters in spring barley. *Theoretical and Applied Genetics* 99: 244–253.

Yin, X., S. Chasalow, C.J. Dourleijn, P. Stam, and M.J. Kropff. 2000a. Coupling estimated effects of QTLs for physiological traits to a crop growth model: Predicting yield variation among recombinant inbred lines in barley. *Heredity* 85: 539–549.

Yin, X., M.J. Kropff, J. Goudriaan, and P. Stam. 2000b. A model analysis of yield differences among recombinant inbred lines in barley. *Agronomy Journal* 92: 114–120.

Yu, M., G.-Y. Chen, L.-Q. Zhang, Y.-X. Liu, D.-C. Liu, J.-R. Wang, Z.-E. Pu, L. Zhang, X.-J. Lan, Y.-M. Wei, C.-J. Liu, and Y.-L. Zheng. 2014. QTL mapping for important agronomic traits in synthetic hexaploid wheat derived from Aegiliops tauschii ssp. tauschii. *Journal of Integrative Agriculture* 13: 1835–1844.

Zadoks, J.C., T.T. Chang, and C.F. Konzak. 1974. A decimal code for the growth stages of cereals. *Weed Research* 14: 415–421.

Zhang, L.Y., C. Ravel, M. Bernard, F. Balfourier, P. Leroy, C. Feuillet, and P. Sourdille. 2006. Transferable bread wheat EST-SSRs can be useful for phylogenetic studies among the Triticeae species. *Theoretical and Applied Genetics* 113: 407–418.

Zhang, G., Y. Wang, Y. Guo, Y. Zhao, F. Kong, and S. Li. 2015. Characterization and mapping of QTLs on chromosome 2D for grain size and yield traits using a mutant line induced by EMS in wheat. *The Crop Journal* 3: 135–144.

Chapter 7
Soil and Water Assessment Tool (SWAT) for Rainfed Wheat Water Productivity

Atif Mehmood, Mukhtar Ahmed, Fayyaz-ul-Hassan, Muhammad Akmal, and Obaid ur Rehman

Contents

A. Mehmood (✉) • Fayyaz-ul-Hassan
Department of Agronomy, Pir Mehr Ali Shah Arid Agriculture University, Rawalpindi, Pakistan
e-mail: agronomist31@gmail.com; drsahi63@gmail.com

M. Ahmed
Department of Agronomy, Pir Mehr Ali Shah Arid Agriculture University, Rawalpindi, Pakistan

Department of Biological Systems Engineering, Washington State University, Pullman, WA, USA
e-mail: ahmadmukhtar@uaar.edu.pk; mukhtar.ahmed@wsu.edu

M. Akmal
Department of Soil Science and SWC, Pir Mehr Ali Shah Arid Agriculture University, Rawalpindi, Pakistan
e-mail: akmal@uaar.edu.pk

O. ur Rehman
Space and Upper Atmosphere Research Commission (SUPARCO), Islamabad, Pakistan

M. Ahmed, C.O. Stockle (eds.), *Quantification of Climate Variability, Adaptation and Mitigation for Agricultural Sustainability*,
DOI 10.1007/978-3-319-32059-5_7

Abstract Wheat is sown over a wide range of sowing date in various cropping systems of rainfed and irrigated areas of Pakistan. This variation in sowing time is caused by various factors such as erratic rainfall in rainfed area, late planting or harvesting of preceding crop, lack or unavailability of farm machinery and inputs. At present the greatest challenge which is being faced by the agriculture sector is production of more food from less available water. Increasing water productivity of the crops can help in facing this challenge. For understanding the relationship between water and food, a sound knowledge of crop water productivity (CWP) is important. Dynamics of crop environment proved most important in rainfed areas for crop production. Therefore, to achieve higher CWP (crop water productivity) under changing climatic conditions, increasing WUE (water use efficiency) could be an option by adopting mitigation strategies. These measures might be, adopting good management practices like optimizing the sowing date on long-term basis using simulation modeling as decision support tool. In the present study, Soil and Water Assessment Tool (SWAT) model was parameterized and validated to study the relationship between yield and crop water productivity of rainfed wheat. For this study, field experiment was conducted at three locations using three wheat cultivars and two sowing dates. Crop water productivity was calculated by dividing the grain yield by evapotranspiration during crop growth cycle. Satellite based parameterization of study area for GIS mapping, topographic analysis, vegetation dynamics, land use, and soil mapping was done by using different software packages like ArcGIS 10.1, Erdas Imagine, QGIS, and Swat-CUP. The SWAT model was used to simulate the processes related to the soil-crop-atmosphere interaction in the present study and it showed significant potential to simulate CWP.

Keywords Wheat • SWAT • Crop water productivity • ArcGIS • Erdas imagine • QGIS • Swat-CUP

7.1 Introduction

Pakistan holds an area of 79.61 million hectares out of which around 20.92 million hectares is under cultivation. The agricultural sector upholds the living of the large rural community and guarantees that ample food is accessible for domestic needs. This sector has the principal role in the economy of Pakistan and donates 21.4 % to GDP, provides employment to 45 % of the country's labor force and also backs the growth of other contributing sectors of the economy (Government of Pakistan, 2013).

The area of Pothwar is situated about 32–34° N latitudes and about 70–74° E longitudes. Therefore, it includes the main part of Attock, Chakwal, Jhelum and Rawalpindi (Afshan et al. 2013). The Pothwar region is considered to be the largest block of rainfed agriculture in Pakistan, includes 1.82 million ha rainfed area of the Punjab province. Soil moisture is one of the most limiting factors in this area because of erratic rainfall distribution over space and time. The winter rains are moderate, summer rains could be heavy, accelerating soil erosion. The annual rainfall in the barani areas ranges from 375 mm (of its South West) to 1750 mm (of its North East). About 70 % of the total rainfall in barani areas is received in summer season while remaining 30 % is received in winter season. Additionally the barani tract experience both daily and seasonal temperatures extremes (Ahmed 2011).

In Pakistan, wheat being the staple food is the most important crop and cultivated on the largest acreages among cereals (8.69 million hectares during the growing season 2012–2013) in almost every part of the country. Its contribution to the value addition in agriculture is 10.1 % and 2.2 % is contributed to the GDP (Government of Pakistan, 2013). Under any environment, cultivar used, management practices adopted and the prevailing conditions effect the wheat growth, development and yield. In Pakistan, wheat is sown over a wide range of sowing date in various cropping systems of rainfed and irrigated areas. This variation in sowing time is caused by various factors such as erratic rainfall in rainfed area, late planting or harvesting of preceding crop, lack or unavailability of farm machinery and inputs. The planting window of wheat in Pakistan generally starts from mid of October and extends until the end of December (Ahmed 2011). Delayed planting (generally after 10th November) reduces wheat yield at a rate of 42 kg ha^{-1} day^{-1} after optimum planting time (Khan 2003).

At present the greatest challenge which is being faced by the agriculture sector is production of more food from less available water. Increasing water productivity of the crops could help in facing this challenge (Kijne et al. 2003). For understanding the relationship between water and food, a sound knowledge of crop water productivity (CWP) is important ((Liu et al. 2007; Yang and Zehnder 2007). For crop water productivity estimation three methods are usually used. These three methods are; "rule of thumb", field experiments and the use of crop models. In the "rule of thumb" method crop water productivity is assumed to be an approximately constant value. A common approximation is that a cereal has 1 kg m^{-1} of crop water productivity, i.e., for the production of 1 kg of cereal roughly 1 m^3 of water is consumed (Yang and Zehnder 2007). While the experiments conducted in fields for the determination of crop water productivity, seasonal crop evapotranspiration and crop yield is measured but such type of experiments require more time and cost and are very difficult to be generalized to other climatic conditions and localities (Ines et al. 2002). Using crop growth models, evapotranspiration and crop yield can be simulated instantaneously and then finally the crop water productivity can be estimated (Liu et al. 2007).

Crop growth simulation models are the deliberate representation of a real system and complex biophysical systems can be learned efficiently through the models (Akram 2011). For the evaluation of managing practices to increase the yield and crop water productivity, crop growth models can be effectively used while considering the seasonal variability and weather related risks (Timsina et al. 2008). Therefore, the scientists have lumped together the multi-disciplinary knowledge about a crop in the form of crop simulation models. Models are computer software with mathematical representations of major biological processes and consider systems approach (soil-plant- atmosphere continuum). Models allow us to study at the same time the impact of several combinations of variables related to crop, soil, weather, and management on the growth and yield of a crop while in the real world such studies will need several years, myriad man-hours, and a lot of time and money (Ahmed and Hassan 2011). The rainfed wheat was found to have lesser additional water requirement for a one unit increase in yield as compared to the irrigated wheat. Due to this, there was a larger improvement in crop water productivity when modeled irrigated and rainfed wheat yield and consumptive water use with uncertainty analysis at a sub-basin level in Iran was analyzed by Faramarzi et al. (2010). For the calculation of crop water productivity, computer-generated yield and ET were used. After analyzing the yield and CWP relationship they observed that Model calibration and validation was done against crop yield and ET. For the analysis of uncertainty they used the SUFI-2 program in the SWAT-CUP package. ET values of less than 200 mm were used for the simulation of crop yields. Lack of information like planting and harvest dates, dates of irrigation and pesticide applications, and also consideration of seed variety, which is usually missing from the analysis were the main limitations in this model.

SWAT model was calibrated by using remotely sensed evapotranspiration based on the SEBAL (Surface Energy Balance Algorithm for Land) for the assessment of CWP from the hydrological processes in the Upper Bhima River Basin. To advance the economic water productivity, four scenario analyses was done and diagrammed for different years i.e., wet, normal and dry years. When crop productivity was compared with potential and global values, it was found in lower range. In upstream localities, greater nutrient losses due to heavy runoff and incompetent water usage in water rich command areas are other important reasons for poor water productivity (Kaushal et al. 2012).

It has been revealed by the validation of SWAT model that there was a less productivity gap between modeled productivity of rice and the data observed under normal conditions. The coefficient of variation in rice productivity was higher during La-Nina years compared to El-Nino and normal years. The mean rice productivity was shifted up in both El-Nino and Normal years indicating the possibility of getting higher yields compared to La -Nina years. Analysis of hydrological data and rice productivity indicated that the risk of failure was much lesser during the El-Nino years compared to normal or La-Nina years (Geethalakshmi et al. 2011b).

Advancement in research and technological improvement has the key role in diminishing water stress due to additional crop production. Water requirement for production of crops will definitely increase in the coming years in order to increase the crop water productivity through advancements in research and technological improvement at river basin scale. SWAT model was used to calculate the actual evapotranspiration. For the grouping of regional statistics to basin level a spatial aggregation and disintegration method based on GIS was developed. In order to compare the results, crop water productivities were also validated and observed that the CWPs were within a reasonable range (Huang and Li 2010).

The model was calibrated and used to derive a monthly basin water balance and then CWP and crop water use was assessed by Immerzeel et al. (2008). Innovative integration of remotely sensed ET and a process-based hydrological model was used to assess the water use and CWP in the Upper Bhima catchment. It was observed that the frequency distributions of CWP had low coefficient of variation and limited yield for the current cropping systems. Shifting the crop base might be accomplished by improving the water productivity. In order to evaluate the effect of changing climate on hydrology and yield of rice SWAT model was used by (Geethalakshmi et al. 2011a) in the Bhavani Basin of India. Climate change scenarios were developed using the RegCM3 model with EH5OM GCM (General Circulation Model) output for an A1B scenario, and generated daily climate data were used in the SWAT model. To validate the SWAT model, predicted rice yields for the Bhavani Basin over a period of 11 years were compared with the observed rice yields of Erode district in which the Bhavani Basin is located and the results indicated the satisfactory performance of the model. The SWAT model can be employed under different climate change and management scenarios for developing adaptation strategies to sustain rice production. The SRI (System Rice Intensification) system of rice cultivation was found to be a better adaptive technology for changing climatic conditions than the conventional flooding system of cultivation.

An experimental study was conducted by Luo et al. (2008) to assess the performance of the plant–soil–groundwater modules and the variability and transferability of SWAT2000 at the Yucheng Comprehensive Experimental Station (YCES) of the Chinese Academy of Sciences at Yucheng City. The simulated results by SWAT to the observations showed that SWAT performed quite unsatisfactorily in LAI (Leaf Area Index) predictions during the senescence stage, in yield predictions, and in soil-water estimation under dry soil-profile conditions. Similarly, data collected from the Wild Rice River watershed (North-western Minnesota) was used in study that was carried out by Wang et al. (2006) for the assessment of three PET (Potential Evapotranspiration) methods within SWAT's framework. Three statistics were used to measure the performance of the SWAT model: the Nash-Sutcliffe coefficient, coefficient of determination and performance virtue. The use of the three PET methods resulted in different values for two calibration parameters, namely the soil evaporation compensation factor and SCS curve number. The results indicated that after calibration, using the three PET methods within SWAT produces very similar hydrologic predictions for the study watershed. Furthermore, study was carried out by Gou et al. (2009) to compute annual average ET of different land use types and different regions in Tianjin by two methods, remote sensing inversion and SWAT model simulation. The SWAT model was also used to simulate the ET for dominating crops in Tianjin. While comparing the results of the two methods, the deviation was over 22 %. The results of ET from SWAT model simulation were larger than the ET results of remote sensing inversion. The result of SWAT model should also be validated by local experiment results. The evaporation from soil and transpiration of wheat and corn during their growing period was calculated, and it was found that soil evaporation makes up 43 % of the total ET during wheat's growing period.

In order to study water and crop yield relations several models like APSIM ((Verburg and Bond 2003), APSIM SWIM (APSWIM) (Connolly et al. 2002), GRASP (Rickert and McKeon 1982; McKeon et al. 1982), SWAGMAN Farm

(Edraki et al. 2003; Ranatunga et al. 2008), SWIM (Krysanova et al. 1998; Krysanova et al. 2005), WATBAL (Ranatunga et al. 2008) and SWAGMAN Destiny (Edraki et al. 2003) have been used. These models can be categorized as empirical models and process-based models. Empirical models are mostly based on regression in which a relation is developed among statistical yield of crop and the factors related to native weather, geostatistical condition and management of the crop. So they have the capability to forecast only yield but estimation of water uptake by crop and evaporation through soil is missing. The process-based models are robust in estimation of crop growth or in hydrology (Faramarzi et al. 2010).

Validation and parameterization of crop simulation models, under local conditions is necessary as crop and plant specific parameters have spatial variation with changing environment. Soil and water assessment tool (SWAT model) (Douglas-Mankin et al. 2010) is a basin scale, time continuous and process-based model that functions on a day-to-day basis (Gassman et al. 2007). Processes related to growth of plants, quantity of water and quality of water are simulated through this model (Faramarzi et al. 2010). The present study was conducted by considering following objectives;

(i) Parameterization and validation of the soil and water assessment tool (SWAT model) for rainfed wheat yield and its water productivity.
(ii) Study the relationship between available water, yield and water productivity of rainfed wheat.

7.2 Materials and Methods

Three wheat varieties (Dharabi, NARC-2013, and Chakwal-50) were sown at three different locations, namely, farmer's field (32°92′ N–72°43′ E) Talagang, University Research Farm (URF) (33°11′ N–73°02′E), Chakwal Road and at NARC (National Agriculture Research Centre) (33°67′ N–73°13′ E) Islamabad, on two different sowing dates (SD) (20–30 October (SD1) and 10–20 November (SD2)) during the wheat growing season of 2013–2014. All the operations and inputs were kept uniform for all varieties. The experiment was laid out in three factor factorial Randomized Complete Block Design (RCBD) with three replications. Individual plot size was 6 m×4 m. Seed bed was prepared by following recommended cultural practices. The fertilizers were applied before sowing according to the field requirement and recommended doses. The sowing was done with seed drill. Recommended seed rate @ 125 kg/ha was used with row-to-row distance maintained at 25 cm. Data related to phenological development of wheat crop using the Zadok's scale (Zadoks et al. 1974) was noted by selecting ten plants from each plot randomly and tagging them. These observations were taken from three leaf stage to crop maturity. Three samples of thousand grains were taken at random from seed lot of each plot and weight were taken in grams on digital balance. Grain yield was recorded by harvesting 1 m^2 area per plot and it was converted to get final yield in kg ha^{-1}. Biological yield was determined by harvesting 1 m^2 area per plot and then converting it to get final yield in kg ha^{-1}.

Crop water productivity is a quantifiable term which is used to describe the relationship between crop produced and the amount of water consumed in crop production (Igbadun et al. 2006). In the present study CWP was calculated by using the following formula (Faramarzi et al. 2010) $CWP = Y / ET$, Where Y = Grain Yield and ET = Evapotranspiration. Evapotranspiration was calculated by using Penman-Monteith equation through CROPWAT.

SWAT is a basin-scale, continuous-time model that operates on a daily time step and is designed to predict the impact of management on water, sediment, and agricultural chemical yields in ungauged watersheds. Important components of the model are weather, hydrology, soil temperature and properties, plant growth, nutrients, pesticides, bacteria and pathogens, and land management. In SWAT, a watershed is divided into multiple sub-watersheds, which are then further subdivided into hydrologic response units (HRUs) that consist of homogeneous land use, management, and soil characteristics. The HRUs represent percentages of the sub-watershed area and are not identified spatially within a SWAT simulation. Alternatively, a watershed can be subdivided into only sub-watersheds that are characterized by dominant land use, soil type, and management. In the present study spatial parameterization was performed by distributing the watershed into sub-basins on the basis of topography, dominant soil, landuse, and slope. Delineation and extraction of sub-basins, watershed and topographic analysis was done by using digital elevation model (SRTM DEM 90 m). Vegetation dynamics, landuse mapping and analysis were done with multispectral satellite images of Landsat (15 m) and SPOT-5 (5 m). These tasks were performed using standard remote sensing and GIS software i.e., ArcGIS 10.0, Erdas Imagine. Model validation is the process of demonstrating that a given site-specific model is capable of making sufficiently accurate simulations, although "sufficiently accurate" can vary based on project goals. In the present study the SWAT model was validated according to the validation skill scores like root mean square estimation (RMSE). Data collected from the field experiment was used for the model evaluation. The statistical package used for the computation of mean values was STATISTIX 8.1. The mean values were analyzed through least significant difference (LSD) at $p < 0.05\,\%$. Open source software GNUPlot was used for output graphs.

7.3 Results and Discussion

The experiment was conducted at three locations to get data for the parameterization and validation of SWAT (Soil and Water Assessment Tool) model for rainfed wheat yield, its water productivity and to study the relationship between available soil moisture. Data on a number of parameters for two different sowing dates from all the three locations were recorded and discussed in this chapter.

Table 7.1 Mean number of days at different phenological stages (Zadok Scale) for three wheat cultivars among three varying climatic locations under two sowing dates

	Z-13	Z-47	Z-92
TREATMENTS	Three leaf	Flag leaf	Maturity
SOWING DATES (SD)			
SD1	27.70^{A}	117^{A}	189.37^{A}
SD2	23.44^{B}	101.74^{B}	166.96^{B}
LSD for sowing dates	0.6592	0.9937	1.1246
Cultivars (C)			
Pak – 13	25.222^{A}	107.5^{B}	178.78^{A}
Dharabi	25.611^{A}	108.28^{B}	178.72^{A}
Chakwal – 50	25.889^{A}	112.33^{A}	177^{B}
LSD for genotype	0.8074	1.217	1.3773
LOCATIONS (L)			
NARC, Islamabad	27.94^{A}	112.94^{A}	183.89^{A}
URF, Chakwal Rd.	21.66^{C}	104.94^{C}	170.56^{C}
TALAGANG	27.11^{B}	110.22^{B}	180.06^{B}
LSD for locations	0.8074	1.217	1.3773
INTERACTIONS			
C*L	***	***	NS
C*SD	**	**	NS
L*SD	***	***	***
C*L*SD	**	***	NS

Any two means not sharing a common letter differ significantly at P<5 % level. (*** Significant at P<1 % level, ** Significant at P<5 % Level, * Significant at P<10 % Level, *NS* Non-Significant)

7.3.1 Phenological Development

Phenological development is the most important attribute involved in the crop adaptation to the farming environments. In present study Zadok's scale (Z) was used to determine the phenological development of the wheat crop. Statistical analysis of the data regarding phenological development of three wheat cultivars at three different climatic locations under two sowing dates revealed potential differences for different Zadok developmental stages. Sowing dates and all three locations showed significant variation for days to three leaf stage (Z13) while no significant variation was shown by cultivars. Under SD1, maximum days to Z-13 (32.35) was observed whereas, minimum (20.88) days observed under SD2. Both sowing dates varied 35.34 % for maximum and minimum days to three leaf stage. Similarly, maximum time (27.94 days) taken at Islamabad whereas minimum (21.66) taken at URF, Chakwal road. There was 22.46 % difference among the locations (Table 7.1). Interactive effect of all interactions (C × SD, C × L, L × SD and C × L × SD) was significant on days to three leaf stage (Z-13). Three way interactive effects of all the interactions are presented in Table 7.2.

Table 7.2 Mean number of days at three leaf (Z-13) under Cultivars × Location × Sowing Date Interaction

	NARC Islamabad			URF, Chakwal			Talagang		
Cultivars	SD1	SD2	Mean	SD1	SD2	Mean	SD1	SD2	Mean
Pak-13	28.66^{c}	27.66^{cd}	28.16^{A}	24.33^{e}	19.33^{i}	21.83C	31^{b}	20.33^{hi}	25.66^{B}
Dharabi	28.00^{cd}	26.66^{d}	27.33^{A}	$22fg^{h}$	22.33^{fg}	22.16C	33^{a}	21.66^{fgh}	27.33^{A}
Chakwal-50	27.66^{cd}	29^{c}	28.33^{A}	$21gh^{i}$	21^{ghi}	21C	33.66^{a}	23ef	28.33^{A}
Mean	28.11^{B}	27.77^{B}		22.44^{C}	20.88^{D}		32.55^{A}	21.66^{CD}	
	27.94^{A}			21.66^{C}			27.11^{B}		

Considerable variation among all cultivars, at all climatic locations under both sowing dates for days at flag leaf stage (Z-47) of wheat crop was observed. Maximum number of days to Z-47 (112.33) was taken by wheat cultivar Chakwal-50 and minimum (107.5 days) taken by Pak-13 followed by wheat cultivar Dharabi (108.28 days). There was 4.29 % difference for maximum and minimum days to flag leaf among cultivars. Similarly both sowing dates and locations varied significantly for days to flag leaf stage. SD1 took more days (117) to flag leaf while SD2 took minimum days (101.74) to flag leaf stage. The difference was 13 %. Maximum days to Z-47 were recorded at Islamabad (112.94 days) whereas minimum were recorded at URF, Chakwal road (104.94 days) with a difference of 7.08 % (Table 7.1). Interactive effect of all interactions (C × SD, C × L, L × SD and C × L × SD) was significant on days to Z-47. Three way interactive effects of all the interactions are presented in Table 7.3.

Significant variation was observed among all wheat cultivars, under both sowing dates' and at all three locations for days to maturity (Z-92) of wheat crop. Maximum days to Z-95 were taken by wheat cultivar Pak-13 and Dharabi (178.78 and 178.72) respectively whereas minimum (177) taken by cultivar Chakwal-50. There was a difference of 0.99 % among all the cultivars. Similarly, both sowing dates and all three different climatic locations varied significantly for days to Z-92. Maximum days to Z-92 were recorded at Islamabad (183.89) whereas minimum (170.56) were observed at URF, Chakwal road with a difference of 7.42 % from maximum value. SD1 took more days (189.37) to Z-92 while SD2 took lesser days (166.96) to Z-92 recording a difference of 11.85 % (Table 7.1). Interactive effect of L × SD was significant whereas C × L, C × SD and C × L × SD showed no significant effect on days to Z-92. Three way interactive effects of all the interactions are presented in Table 7.4.

Considerable differences were found among wheat cultivars for phenological development. These variations among cultivars for days at different Zadok developmental stages is regarded to their inherent differences or genetic characteristics (Shahzad et al. 2007) but several environmental factors like temperature (high and low) also effect the growth and development of the wheat crop (Din et al. 2010). The difference among sowing dates might be due to the effect of variable environmental conditions, most probably due to variations in temperature. The results of current study were also supported by (Nahar et al. 2010) who documented that wheat sown under different sowing times is affected by the variations in the temperature. Difference in phenological development due to sowing time was also documented by Rahman et al. (2009); Laghari et al. (2012) and Hussain and Mudasser (2007). As phenological development is affected by the variable climatic conditions, hence the difference in days at different Zadok developmental stages might be attributed to the different climatic variables prevailing at different locations. Inamullah et al. (2011) also reported the effect of prevailing environmental conditions on wheat growth and development.

Table 7.3 Mean number of days at Flag Leaf (Z-47) under Cultivars × Location × Sowing Date Interaction

	Islamabad			URF, Chakwal			Talagang		
Cultivars	SD1	SD2	Mean	SD1	SD2	Mean	SD1	SD2	Mean
Pak-13	195^{NS}	175.67	185.33^{A}	181.33	159.67	170.5^{C}	193	168	180.5^{B}
Dharabi	193.67	176.33	185^{A}	181	160	170.5^{C}	194	167.33	180.67^{B}
Chakwal-50	192	170.67	181.33^{B}	181	160.33	170.67^{C}	193.33	164.67	179^{B}
Mean	193.56^{A}	174.22^{C}		181.11^{B}	160^{E}		193.4^{A}	166.67^{D}	
	183.89^{A}			170.56 ^{C}			180.06 ^{B}		

Table 7.4 Mean number of days at maturity (Z-92) under cultivars × Location × Sowing Date Interaction

	Islamabad			URF, Chakwal			Talagang		
Cultivars	SD1	SD2	Mean	SD1	SD2	Mean	SD1	SD2	Mean
Pak-13	115.33^{cd}	113.67^{de}	114.5A	114^{de}	85.67^{k}	99.83^{F}	117.33^{bc}	99^{i}	108.17^{D}
Dharabi	112.33^{ef}	110.67^{f}	111.5BC	118^{bc}	89^{j}	103.5^{E}	119.33^{b}	100.33^{hi}	109.83^{CD}
Chakwal-50	116^{cd}	109.67^{f}	112.83AB	118^{bc}	105^{g}	111.5B^{C}	122.67a	102.67^{gh}	112.67^{AB}
Mean	114.56^{C}	111.33^{D}		116.67^{B}	93.22^{F}		119.78^{A}	100.67^{E}	
	112.94^{A}			104.94^{C}			110.22^{B}		

Any two means not sharing a common letter differ significantly at $P<5\,\%$ level

7.3.2 Thousand Grain Weight (g)

All the wheat cultivars varied potentially for TGW. Maximum thousand grain weight recorded for PAK-13 (54.83 g) while lowest thousand grain weight recorded for Chakwal-50 (41.33 g) followed by Dharabi (42 g). There was 24.62 % difference among PAK-13 and Chakwal-50 for maximum and lowest thousand grain weight. Similarly, all the locations varied considerably for thousand grain weight of different wheat cultivars. Maximum thousand grain weight recorded for Islamabad (49.5 g) whereas, lowest thousand grain weight recorded for Talagang (42.22 g). Both locations varied 14.70 % for maximum and lowest thousand grain weight of wheat cultivars. Maximum value of thousand grain weight recorded for SD2 (50.55 g) and lowest value of thousand grain weight recorded for SD1 (37.333 g). The percentage variation for maximum and lowest thousand grain weight among SD2 and SD1 was 26.15 % (Table 7.5).

Wheat yield is markedly affected by thousand grain weight. Significant difference among the cultivars for TGW might be related to their genetic make-up. Tahir et al. (2009) and Shahzad et al. (2002), documented that the variation in TGW among cultivars can be attributed to their genetic variability. There was significant variation among all the three locations for TGW which may be due to the variable weather conditions (variability in rainfall distribution, intensity and rainfall pattern and variability in temperature) at the time of grain development at different locations. These results were in line with those of Modarresi et al. (2010), Aggarwal (2008) and Nagarajan (2005) were of the view that variable weather conditions affect different yield components. Sowing dates were also found significantly different for TGW which might be related to the source sink activity of the crop. Rahimian and Banayan (1996) reported significant effects of source sink activity on TGW. The variations among sowing dates can be attributed to the soil moisture regime at specific time. Significant decrease in TGW of wheat crop was observed by Khan et al. (2005) and Qadir et al. (1999) mainly due to the variations in availability of soil moisture. Variation in TGW due to different sowing dates was also reported by Shahzad et al. (2007) and Refay (2011).

7.3.3 Grain Yield (kg ha^{-1})

Statistical analysis of the data regarding grain yield depicted highly significant results for wheat grain yield of different wheat cultivars under both sowing dates at all three locations. Maximum grain yield recorded for PAK-13 (6441.4 kg/ha) followed by the same Cultivar PAK-13 (6295.7 kg/ha), while minimum wheat grain yield exhibited by Chakwal-50 (4174.6 kg/ha). The variation was 35.19 % and 33.69 %. Similarly, all the locations varied considerably for grain yield of different wheat cultivars. Islamabad (5602.2 kg/ha) recorded the maximum wheat grain yield followed by Talagang (5576 kg/ha), while minimum wheat grain yield recorded at

Table 7.5 Yield and yield parameters of three wheat cultivars among three varying climatic locations under two sowing dates

SOWING DATES (SD)	TGW	GY	BY	HI
SD1	43.519B	5144.2 B	14,756 B	0.35 A
SD2	48.778A	5746 A	16,591 A	0.34 A
LSD for Sowing Dates	0.9917	105.84	418.01	NS
CULTIVARS (C)				
Pak-13	52A	6269.1 A	17,203 A	0.36 A
Dharabi	46.056B	5458.7 B	15,985 B	0.34 B
Chakwal-50	40.389C	4607.4 C	13,832 C	0.33 B
LSD for Cultivar	1.2146	129.63	511.95	0.0133
LOCATIONS (L)				
NARC, Islamabad	49.5A	5602.2 A	16,543 A	0.33 B
URF, Chakwal Road	46.722B	5157.1 B	14,694 C	0.35 A
Talagang	42.222C	5576.0 A	15,783 B	0.35 A
LSD for Locations	1.2146	129.63	511.95	0.0133
INTERACTIONS				
C*L	NS	**	NS	*
C*SD	**	*	NS	**
L*SD	***	***	***	***
C*L*SD	***	NS	***	**

Where *TGW* Thousand Grain Weight, *GY* Grain Yield, *BY* Biological Yield, *HI* Harvest Index

URF, Chakwal Road (5157.1 kg/ha). There was 7.94 % reduction in grain yield at URF, Chakwal Road from Islamabad. Both the sowing dates varied considerably for grain yield. Maximum grain yield recorded for SD2 (5872.9 kg/ha) while minimum for SD1 (4674.8 kg/ha). Both sowing dates varied 20.40 % for maximum and minimum grain yield (Table 7.5).

Significant variations were observed in the grain yield of wheat cultivars which might be related to the genetic makeup and their ability to use the available resources properly by transferring the large amount of dry matters to the plant sink (Khabiri et al. 2012). Number of grains per spike is an important contributor to the wheat grain yield and was found more in the wheat Cultivar (PAK-13) which gave higher grain yield, so it might be a reason for the variation in wheat grain yield among cultivars. Same was reported by Slafer and Calderini (2005), who documented that increase in the number of grains per spike, resulted in the enhancement of the grain yield during recent years. The results are also supported by Nayyar et al. (1992). Sowing date is an important factor which affects the grain yield significantly (McLeod et al. 1992). These findings were in line with those of Jackson et al. (2000) who were of the view that rainfall occurrence and its distribution affects the sowing dates of crop, hence the final yield. Ansari (2002) also reported significant effect of sowing dates on the grain yield of wheat cultivars. Variation among the locations might be related to the prevailing climatic conditions. At Islamabad, favorable climatic conditions were available those were not available at other locations in the same period of time. Wheeler et al. (2000) documented that different climatic factors like rainfall and temperature has effect on the grain yield. Araus et al. (2008) also documented significant effect of climatic variables on the grain yield.

7.3.4 Biological Yield

There was a potential difference among all the cultivars under both sowing dates at all the locations for biological yield. Highest biological yield was recorded at Islamabad (16,543 Kg/ha), whereas, lowest biological yield observed at University Research Farm, Chakwal Road (14,694 Kg/ha). The difference in biological yield at URF, Chakwal Road from Islamabad was 11.17 %. Both the sowing dates varied considerably for biomass production. Maximum biological yield observed for SD2 (17,092 kg/ha) while minimum biological yield recorded for SD1 (12,919 kg/ha). The variation among both the sowing dates for biological yield was 24.41 %. Similarly, cultivars showed considerable variation for biological yield. Maximum biological yield recorded for Pak-13 (17,203 kg/ha) and minimum exhibited by Chakwal-50 (13,832 kg/ha). There was 19.5 % difference in maximum and minimum biological yield among the wheat cultivars PAK-13 and Chakwal-50 (Table 7.5).

Dry matter (biological yield) is the final outcome of the crop and is directly influenced by photosynthesis. Photosynthesis is directly related to the available resources, which means if all the resources are available then there will be more photosynthesis. The proper utilization of the prevailing weather variables would affect plant growth positively by increased photosynthesis resulting in increment of plant dry matter in different plant parts and ultimately enhanced the biological yield (Ali et al. 2010). The analysis revealed that the second sowing date resulted in higher biological yield due to optimum temperature and growth period which resulted in good vegetative growth, more dry matter accumulation and higher biological yield. Significant effects of sowing dates on biological yield of the crop were illustrated by Jalota et al. (2010) and Azadbakht et al. (2012). Our results were also in line with Ali et al. (2010) who stated that favorable climatic conditions effects the biological yield. More numbers of tillers m^{-2} and more plant height could also be a reason for more biological yield (Donaldson et al. 2001; Matuz and Aziz 1990). Similarly, at Islamabad there was relatively higher rainfall and no temperature stress on wheat so at Islamabad higher biological yield was recorded. Different wheat cultivars also showed significant variation for biological yield. This difference among the cultivars might be related to their genetic potential and their capability of shifting more amounts of matter to the plant sink (Khabiri et al. 2012).

7.3.5 Harvest Index

Highest harvest index recorded for wheat cultivar PAK-13 (37.17 %) and lowest harvest index recorded for wheat cultivar Chakwal-50 (32.5 %). The difference among both the cultivars was 12.56 %. Similarly, there was considerable difference among locations for highest and lowest harvest index. Highest harvest index was recorded at Talagang (35.39 %) which was statistically in line with that of URF, Chakwal Road (35.22 %), while, lowest harvest index was observed at Islamabad (33.78 %). The difference in harvest index at Islamabad from Talagang and University Research Farm, Chakwal Road was 4.5 % and 4.08 % respectively. There was no significant difference among both the sowing dates for harvest index (Table 7.5).

Harvest index is obtained by dividing grain yield of the crop by above ground biological yield and then multiplied by 100. It is the most important characteristic used to select high yielding cultivars under varying planting/sowing conditions. Raes et al. (2009) reported variations in potential harvest index from one location to another. Climatic conditions were favorable for SD2, due to which there was efficient translocation of photosynthate which resulted in higher harvest index for SD2. Blum et al. (1994) illustrated that whenever there is efficient utilization of photosynthate the harvest index will be at higher side. Our results were also in line with Jalota et al. (2010) who documented significant effects of sowing dates on harvest index.

The interactive effect of sowing dates and cultivars (SD × C) interaction had significant effect on harvest index (Hossain et al. 2011). The increase of harvest index represents the ability of plant to transfer and allocation of material to aerial parts. Harvest index is one of the indices used to assess the proficiency of division of dry matter by plant.

7.3.6 Crop Water Productivity (Kg m^{-3})

Wheat cultivars at all climatic locations under both sowing dates for crop water productivity (CWP) were potentially variable. CWP was highest for cultivar Pak-13 (1.11 Kg m^{-3}) while lowest value recorded for Chakwal-50 (0.75 Kg m^{-3}) with a difference of 32.37 %. Similarly, both sowing dates and all locations showed significant variation. Maximum CWP was observed under SD2 (1.03 Kg m^{-3}) whereas minimum value recorded under SD1 (0.8612 Kg m^{-3}). Both sowing dates varied 17.10 % for CWP. University Research Farm, Chakwal Road (0.95 Kg m^{-3}) gave maximum CWP value whereas Islamabad (0.89 Kg m^{-3}) gave minimum value among all locations. The difference among locations was 6.21 % for maximum and minimum value of CWP. Interactive effects of L × C, L × SD and C × SD interactions were significant whereas L × C × SD interaction had no significant effect on CWP at 5 % level of significance. Three way interactive effects of all the interactions are presented in Table 7.6.

Table. 7.6 Crop water productivity of three wheat cultivars under three location and two sowing date

	Islamabad		Mean	URF, Chakwal Rd.		Mean	Talagang		Mean
	SD1	SD2		SD1	SD2		SD1	SD2	
Pak-13	0.99 NS	1.04	1.02^{B}	1.02	1.21	1.11^{A}	0.99	1.07	1.03^{B}
Dharabi	0.89	0.88	0.89^{D}	0.89	1.03	0.96^{C}	0.85	0.96	0.91^{D}
Chakwal-50	0.71	0.79	0.75^{F}	0.66	0.87	0.76^{EF}	0.75	0.85	0.80^{E}
Mean	0.87^{D}	0.91^{C}		0.86^{D}	1.03^{A}		0.86^{D}	0.96^{B}	
	0.89^{C}			0.95^{A}			0.91^{B}		

7.4 Parameterization and Evaluation of Swat Model

7.4.1 The SWAT Model

SWAT (Soil and Water Assessment Tool) model is a hydro-dynamic, basin-scale, physically based and continuous-time model that has its applications in complex and large basins (Fig. 7.1). It is designed for the prediction of management impacts on water, sediment and agricultural chemical yields in an ungauged watershed and is operated on a daily time step. Basin specific data regarding weather components (temperature, precipitation, evapotranspiration, humidity etc.,) soil properties (physical and chemical), topography, vegetation and land management practices are required to run the model. GIS data can be obtained from different agencies which have already compiled data sets. Some of its components are weather, evapotranspiration, crop growth and irrigation, reach routing, soil properties, land management and hydrology. In SWAT, a watershed is divided into multiple sub-watersheds, which are then further subdivided into hydrological response units (HRUs) that consist of homogeneous land use, management, and soil characteristics. Alternatively, a watershed can be subdivided into only sub-watersheds that are characterized by dominant land use, soil type, and management. It works on the following principle equation:

$$SW_t = SW_o + \sum_{i=1}^{t}\left(R - Q - ET - P - QR\right)$$

Where SW_t is the soil water content at a specific time, SW_0 is for initial water content of the soil, R is for precipitation, Q representing runoffs, ET representing evapotranspiration, P for percolation and QR is for the return flow.

One of the basic objectives behind the development of the SWAT model was the forecasting of the impacts of management decisions regarding climatic and vegetative change on large and ungauged basins.

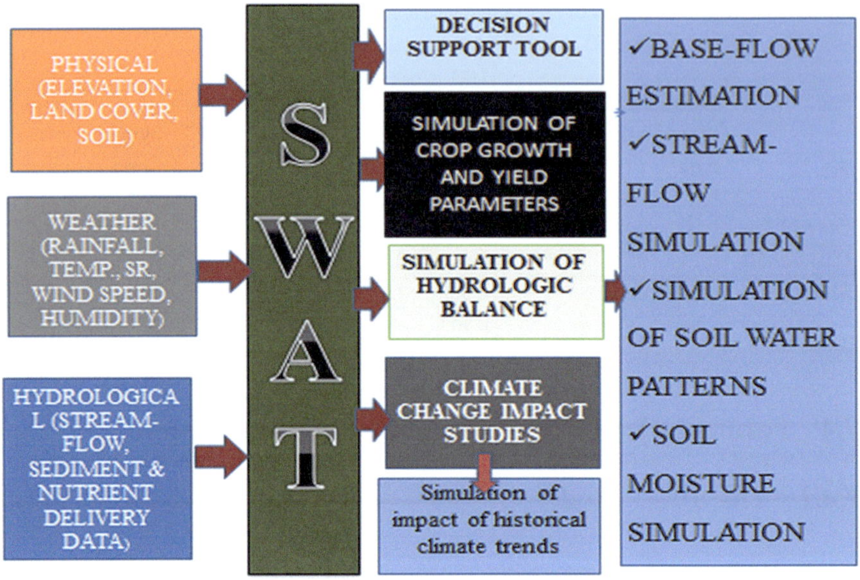

Fig. 7.1 Workflow of the SWAT model

7.4.2 Model Parameterization

In present study spatial parameterization was performed by distributing the watershed into sub-basins on the basis of topography, dominant soil type, landuse, and slope. Delineation and extraction of sub-basins, watershed and topographic analysis was done by using digital elevation model (SRTM DEM 90 m). Vegetation dynamics, landuse mapping and analysis were done with multispectral satellite images of Landsat (15 m) and SPOT-5 (5 m). These tasks were performed using standard remote sensing and GIS software i.e., ArcGIS 10.0, ERDAS IMAGINE 9.1 and QGIS Desktop 2.2.0. The satellite images required for the present study were obtained from the United States department of Geological Survey's website http://earthexplorer.usgs.gov.

For the spatial parameterization, first of all, layer stacking was done using the "Layer stack" function of the standard remote sensing and GIS software ERDAS IMAGINE 9.1. After the layers were stacked and a single image of different bands was obtained, classification of the map was done using "unspecified classification" function of the same remote sensing and GIS software. Then the land use and soil layer were classified using standard remote sensing and GIS software ArcMap 10.1 a component of ArcSWAT.

The required data for this study was obtained from the following sources:

(I) Data regarding soil moisture, crop growth and yield was gathered from the field experiment conducted at three different climatic locations, i.e., Islamabad, University Research Farm, Chakwal Road and Talagang (Fig. 7.2, 7.3, 7.4, 7.5, and 7.6).

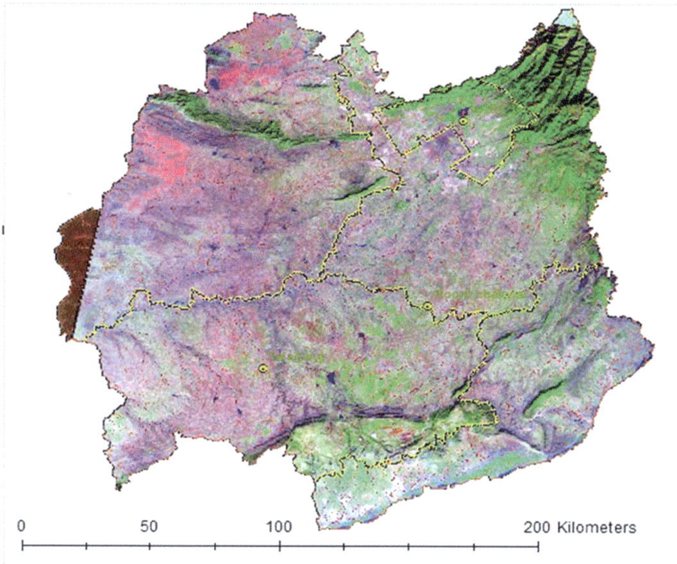

Fig. 7.2 Map of the study Basin

Fig. 7.3 Stacked image of the Basin under study

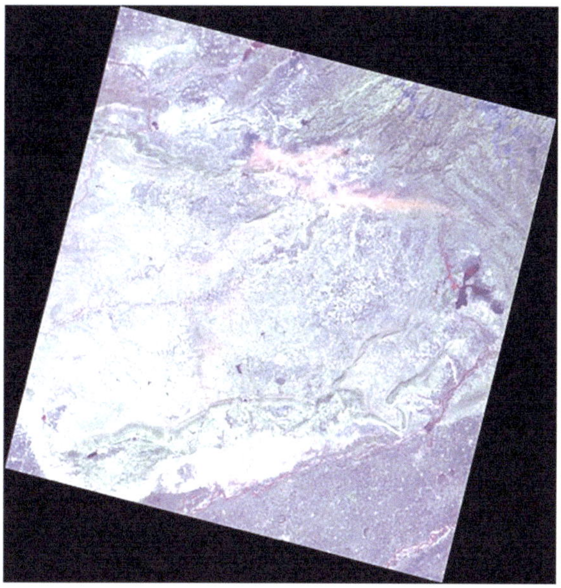

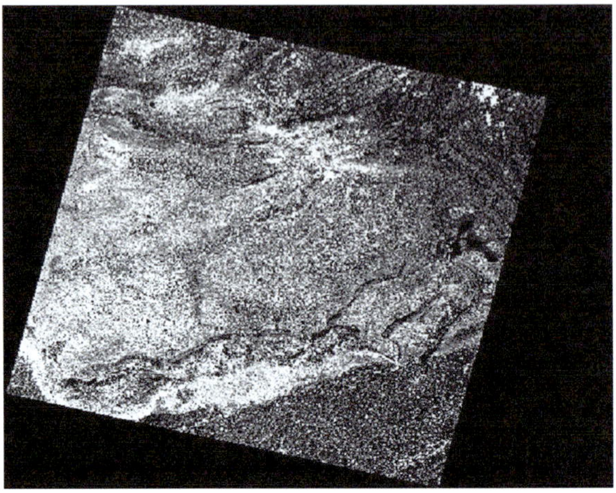

Fig. 7.4 Satellite image of study area after classification into different objects

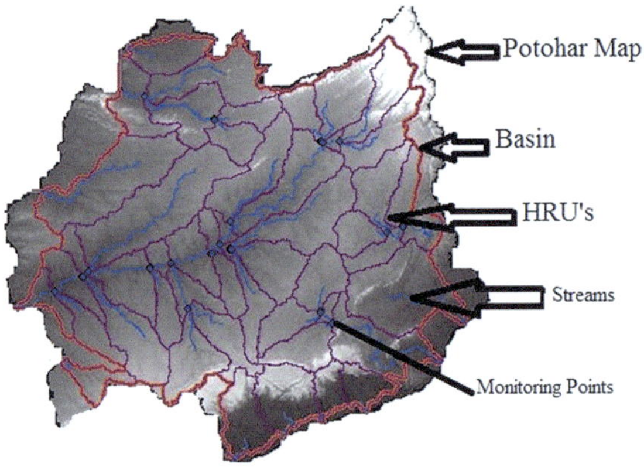

Fig. 7.5 Map of study area classified into Basin, HRU's, streams and monitoring points

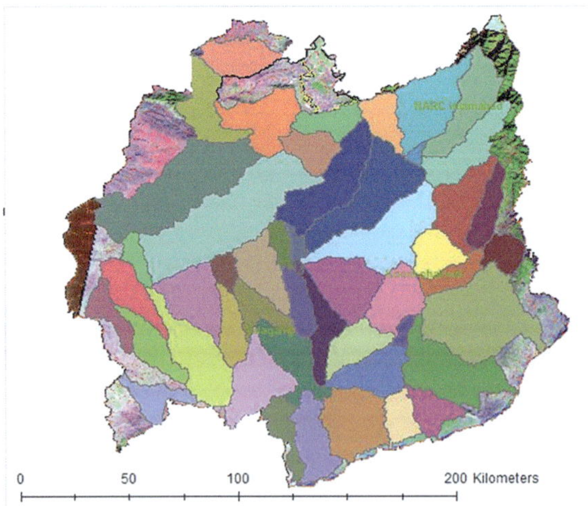

Fig. 7.6 Map of the study area spatially analyzed and delineation into hydrological response units

(II) Weather data of the two locations (Islamabad and URF, Chakwal Road) was obtained from the respective weather stations while of Talagang was interpolated.
(III) Satellite images of the study area were downloaded from the United States Geological Survey www.EarthExplorer.usgs.gov.

7.4.3 Calibration and Uncertainty Analysis

In the present study, for the calibration, validation and uncertainty analysis of the model, land use, crop yield and climatic data was used. There are two groups of parameters which affects the simulated crop yield mostly (Wang et al. 2005). One of these two groups is of the parameters which affect both of the hydrology and crop growth parameters and the other group is of those factors which are sensitive only to the processes of crop growth like harvest index (HI) and sowing and harvesting dates of the crop. In order to model the crop yield, firstly hydrology was calibrated and then the calibrations of yield parameters were performed. For the comparison of the observed and the simulated results the R^2 was used. The crop yield was simulated at the sub-basin level.

7.5 Model Results

7.5.1 Days After Sowing

SWAT model was parameterized to simulate days after sowing for Zadok's scale (Germination, Three leaf, Anthesis and Maturity) under two sowing dates for three wheat cultivars and at three different climatic locations. A close association was witnessed between observed and simulated days after sowing for Zadok's scale measured with validation skill score like R^2 (Table 7.7). The simulated values for days after sowing at germination, three leaf, anthesis and maturity were close to observed with validation skill score of R^2 0.95, 0.97, 0.83 and 0.72 respectively. Simulated days after sowing for germination under two sowing dates *viz.* SD1 and SD2, for three wheat genotypes *viz.* Pak – 13, Dharabi and Chakwal-50 and at three different climatic locations *viz.* NARC Islamabad, Farmer's field Talagang, and URF, Chakwal road were found closely related to observed values with validation skill score of R^2 0.96 and 0.94 for sowing dates, 0.97, 0.96 and 0.95 for genotypes and 0.98, 0.89 and 0.80 for locations, respectively. For days after sowing to three leaf, R^2 values of sowing dates, genotypes and locations showed close relation with the observed values. Similarly at anthesis and maturity stage similar results were observed. Fig. 7.7 represents observed and simulated days after sowing of three wheat cultivars at three different climatic locations under two sowing dates.

Table 7.7 Validation Skill Score (R^2) values for phenology, yield and crop water productivity (cwp) of three wheat genotypes under two sowing dates and at three climatic locations

Factors	R^2					
	Phenology					
	Germination	Tillering	Flag leaf	Maturity	Yield	CWP
Sowing Dates						
SD1	0.9634	0.9863	0.9206	0.9309	0.9131	0.8719
SD2	0.9458	0.9703	0.9506	0.9217	0.9073	0.899
Genotypes						
Pak – 13	0.97	0.9894	0.9561	0.8052	0.7792	0.8525
Dharabi	0.9685	0.9796	0.9887	0.8441	0.7267	0.8238
Chakwal-50	0.9596	0.9893	0.9645	0.935	0.9084	0.9209
Locations						
Islamabad	0.9831	0.9057	0.9394	0.859	0.8967	0.8421
Talagang	0.8913	0.8378	0.9818	0.9209	0.9468	0.9163
URF, Chakwal Rd.	0.8082	0.9981	0.896	0.9468	0.8834	0.8141

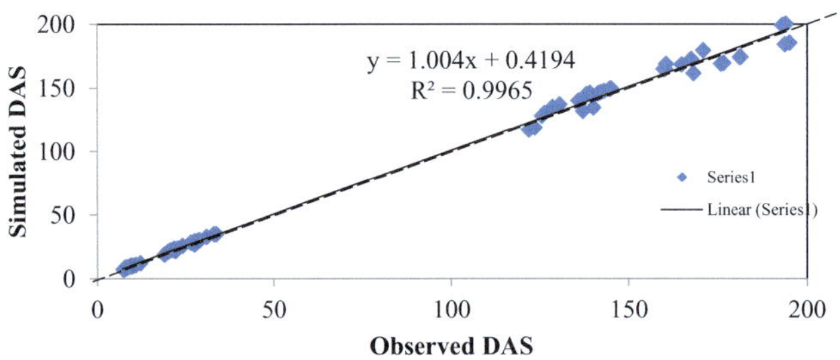

Fig. 7.7 Comparison of observed days after sowing with simulated days after sowing among three wheat cultivars, under two sowing dates and at three locations

7.5.2 *Grain Yield (GY) (kg ha^{-1})*

Grain yield for three wheat cultivars, two sowing dates and three different climatic locations was simulated by parameterization of SWAT model. Results showed close association between observed and modeled grain yield. Fig. 7.8 represents observed and simulated GY of three wheat cultivars at three different climatic locations and under two sowing dates. The simulated results for grain yield with validation skill score of R^2 0.91 and 0.90 for sowing dates, 0.77, 0.72 and 0.90 for cultivars and 0.89, 0.94 and 0.88 for locations, were close to observed results. Maximum simulated GY was for wheat cultivar Pak-13 (6061.6 Kg ha^{-1}) while minimum for Chakwal-50 (4590.7 Kg ha^{-1}). A direct relation with sowing dates calculated in simulated GY by SWAT model. Under SD2 maximum grain yield (5731 Kg ha^{-1}) simulated whereas, minimum GY (5035.1 Kg ha^{-1}) simulated under SD1. Likewise, variation in simulated grain yield was observed at all climatic locations and for three

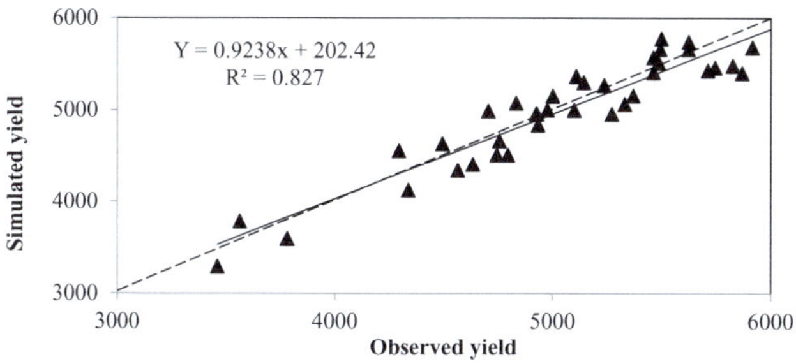

Fig. 7.8 Comparison of observed yield with simulated yield among three wheat cultivars, under two sowing dates and at three locations

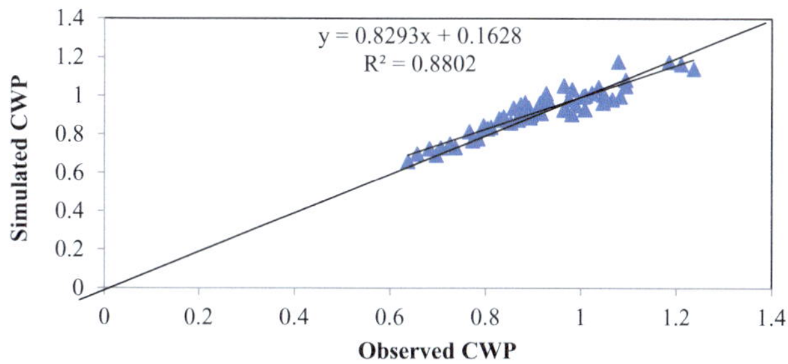

Fig. 7.9 Comparison of observed crop water productivity with simulated crop water productivity among three wheat cultivars, under two sowing dates and at three locations

wheat cultivars. At Islamabad, simulated wheat GY (5519.9 Kg ha^{-1}) was at higher side while at URF, Chakwal its value (5085.5 Kg ha^{-1}) was minimum. These results were close to the observed values with R^2 values of 0.86 and 0.88 respectively.

7.5.3 *Crop Water Productivity (Kg m^{-3})*

Simulated crop water productivity (CWP) under two different sowing dates, for three wheat cultivars and at three different climatic locations was found closely related to the observed CWP (Fig. 7.9). There was a close relationship among observed and simulated CWP with R^2 value of 0.88 for all the wheat cultivars under three different climatic locations among two sowing dates. Wheat cultivars and sowing dates varied potentially for simulating CWP of wheat crop. Among cultivars highest value of CWP was simulated for Pak-13 (1.01 Kg m^{-3}) whereas lowest one simulated for Chakwal-50 (0.79 Kg m^{-3}) while under sowing dates maximum CWP simulated for SD2 (0.97 Kg m^{-3}) and minimum for SD1 (0.87 Kg m^{-3}). Potentially different results were simulated for locations. Islamabad (0.59 Kg m^{-3}) gave maximum value whereas minimum value of CWP simulated for University Research Farm, Chakwal road (0.89 Kg m^{-3}).

References

Afshan, K., M.A. Beg, I. Ahmad, M.M. Ahmad, and M. Qayyum. 2013. Freshwater snail fauna of Pothwar region, Pakistan. *Pakistan Journal of Zoology* 45: 227–233.

Aggarwal, P.K. 2008. Climate change and Indian agriculture: Impacts, adaptation and mitigation. *Indian Journal of Agricultural Sciences* 78.

Ahmed, M. 2011. *Climatic resilience of wheat using simulation approaches in Pothowar.* PhD, PMAS Arid Agricultural University Rawalpindi, Pakistan.

Ahmed, M., and F.U. Hassan. 2011. *APSIM and DSSAT models as decision support tools*. In: 19th international congress on modelling and simulation. Perth, Australia.

Akram, M.N. 2011. *Comparison of different modeling approaches for simulating wheat growth kinetics*. M.Sc. (Hons.), Pir Mehr Ali Shah Arid Agriculture University Rawalpindi, Pakistan.

Ali, M.A., M. Hussain, M.I. Khan, Z. Ali, M. Zulkiffal, J. Anwar, W. Sabir, and M. Zeeshan. 2010. Source-sink relationship between photosynthetic organs and grain yield attributes during grain filling stage in spring sheat (Triticum aestivum). *International Journal of Agriculture and Biology* 12: 509–515.

Ansari, A.H. 2002. Influence of seeding time on grain yield, its components and their interrelation in bread wheat (Triticum aestirum L.) varieties. *Pakistan Journal of Agricultural Research* 17: 7–12.

Araus, J.L., G.A. Slafer, C. Royo, and M.D. Serret. 2008. Breeding for yield potential and stress adaptation in cereals. *Critical Reviews in Plant Sciences* 27: 377–412.

Azadbakht, A., G. Azadbakht, H. Nasrollahi, and Z. Bitarafan. 2012. Evaluation of different planting dates effect on three maize hybrids in Koohdasht region of Iran. *International Journal of Science and Advanced Technology* 2: 34–38.

Blum, A., B. Sinmena, J. Mayer, G. Golan, and L. Shpiler. 1994. Stem reserve mobilization supports wheat-grain filling under heat stress. *Australian Journal of Plant Physiology* 21: 771–781.

Connolly, R.D., M. Bell, N. Huth, D.M. Freebairn, and G. Thomas. 2002. Simulating infiltration and the water balance in cropping systems with APSIM-SWIM. *Soil Research* 40: 221–242.

Din, R., G.M. Subhani, N. Ahmad, M. Hussain, and A. Rehman. 2010. Effect of temperature on development and grain formation in spring wheat. *Pakistan Journal of Botany* 42: 899–906.

Donaldson, E., W.F. Schillinger, and S.M. Dofing. 2001. Straw production and grain yield relationships in winter wheat. *Crop Science* 41: 100–106.

Douglas-Mankin, K., R. Srinivasan, and J. Arnold. 2010. Soil and Water Assessment tool (SWAT) model: Current developments and applications. *Transactions of the ASABE* 53: 1423–1431.

Edraki, M., D. Smith, S. Humphreys, S. Khan, N. O'connell and E. Xevi. 2003. *Validation of SWAGMAN farm and SWAGMAN destiny models*. Commonwealth Scientific and Industrial Research Organisation (CSIRO) land and water technical report 44/03, Griffith, Australia

Faramarzi, M., H. Yang, R. Schulin, and K.C. Abbaspour. 2010. Modeling wheat yield and crop water productivity in Iran: Implications of agricultural water management for wheat production. *Agricultural Water Management* 97: 1861–1875.

Gassman, P.W., M.R. Reyes, C.H. Green, and J.G. Arnold. 2007. The soil and water assessment tool: Historical development, applications, and future research directions. *Transactions of the ASABE* 50: 1211–1250.

Geethalakshmi, V., A. Lakshmanan, R. Jagannathan, Gummidi Sridhar, A.P. Ramaraj, K. Bhuvaneswari, L. Gurusamy and R. Anbhazhagan. 2011a. Climate change impact assessment and adaptation strategies to sustain rice production in Cauvery basin of Tamil Nadu. *International Journal of Current Science* 101: 7

Geethalakshmi, V., A. Lakshmanan, R. Srinivasan, H. Annamalai and N.U. Sekhar. 2011b. *Study of ENSO effects on hydrology and rice productivity in Cauvery basin using SWAT*. Online at http://www.tnau.ac.in/climarice/ASABE/geetha final.pdf.

Gou, S., C. Qiang and Q. Dayong. 2009. *The comparison of distributed hydrological model and remote sensing method for ET calculation in Tianjin*. Online at http://www.seiofbluemountain.com/upload/product/201005/2009shzyhy05a5.pdf.

Government of Pakistan. 2013. *Economic survey of Pakistan 2012–13*. Chapter 2: Agriculture. Ministry of Finance.

Hossain, A., M.A.Z. Sarker, M.A. Hakim, M.V. Lozovskaya, and V.P. Zvolinsky. 2011. Effect of temperature on yield and some agronomic characters of spring wheat (Triticum aestivum l.) genotypes. *Innovation and Technology* 1: 44–54.

Huang, F., and B. Li. 2010. Assessing grain crop water productivity of China using a hydro-model-coupled-statistics approach. Part II: Application in breadbasket basins of China. *Agricultural Water Management* 97: 1259–1268.

Hussain, S.S., and M. Mudasser. 2007. Prospects for wheat production under changing climate in mountain areas of Pakistan—An econometric analysis. *Agricultural Systems* 94: 494–501.

Igbadun, H.E., H.F. Mahoo, A.K.P.R. Tarimo, and B.A. Salim. 2006. Crop water productivity of an irrigated maize crop in Mkoji sub-catchment of the Great Ruaha River Basin, Tanzania. *Agricultural Water Management* 85: 141–150.

Immerzeel, W.W., A. Gaur, and S.J. Zwart. 2008. Integrating remote sensing and a process-based hydrological model to evaluate water use and productivity in a south Indian catchment. *Agricultural Water Management* 95: 11–24.

Inamullah, F.U. Khan, and I.H. Khalil. 2011. Environmental effect on wheat phenology and yields. *Sarhad Journal of Agriculture* 27: 395–402.

Ines, A.V.M., A.D. Gupta, and R. Loof. 2002. Application of GIS and crop growth models in estimating water productivity. *Agricultural Water Management* 54: 205–225.

Jackson, L. F., J. Dubcovsky, L.W. Gallagher, R.L. Wennig, J. Heaton, H. Vogt, L.K. Gibbs, D. Kirby, M. Canevari, H. Carlson, T. Kearney, B. Marsh, D. Munier, C. Mutters, S. Orloff, J. Schmierer, R. Vargas, J. Williams and S. Wright. 2000. *Regional barley and common and durum wheat performance tests in California*, University of California, Agronomy progress report No. 272.

Jalota, S.K., S. Singh, G.B.S. Chahal, S.S. Ray, S. Panigraghy, S. Bhupinder, and K.B. Singh. 2010. Soil texture, climate and management effects on plant growth, grain yield and water use by rainfed maize-wheat cropping system: Field and simulation study. *Agricultural Water Management* 97: 83–90.

Kaushal, K.G., L. Bharati, A. Gaur, B. George, S. Acharya, K. Jella, and B. Narasimhan. 2012. Spatial mapping of agricultural water productivity using swat model in upper Bhima catchment. *Indian Journal of Irrigation and Drainage* 61: 60–79.

Khabiri, E., A.A. Imani, and H. Shahbazi. 2012. Studying the grain yield and yield components in advanced rain fed wheat genotypes. *Annals of Biological Research* 31: 5647–5650.

Khan, A.J., F. Azam, A. Ali, M. Tariq, and M. Amin. 2005. Inter-relationship and path co-efficient analysis for biometric traits in drought tolerant wheat (Triticum aestivumL.). *Asian Journal of Plant Sciences* 4: 540–543.

Khan, M.A. 2003. *Wheat crop Mgt. for yield maximization*, 1–94. Faisalabad: Wheat research Institute.

Kijne, J.W., R. Barker, and D. Molden. 2003. *Water productivity in agriculture: Limits and opportunities for improvement*, Comprehensive assessment of water Mgt. in agriculture, Series 1. Wallingford: CAB7IWMI.

Krysanova, V., F. Hattermann, and F. Wechsung. 2005. Development of the Eco-hydrological model, SWIM for regional impact studies and vulnerability assessment. *Hydrological Processes* 19: 763–783.

Krysanova, V., D. Muller-Wohlfeil, and A. Becker. 1998. Development and test of a spatially distributed hydrological/water quality model for meso-scale watersheds. *Ecological Modelling* 106: 261–289.

Laghari, K.A., M.A. Sial', and M.A. Arain. 2012. Effect of high temperature stress on grain yield and yield components of wheat (Triticum aestivum L.). *Science, Technology and Development* 31: 83–90.

Liu, J., J.R. Williams, A.J.B. Zehnder, and H. Yang. 2007. GEPIC – modelling wheat yield and crop water productivity with high resolution on a global scale. *Agricultural Systems* 94: 478–493.

Luo, Y., C. He, M. Sophocleous, Z. Yin, R. Hongrui, and Z. Ouyang. 2008. Assessment of crop growth and soil water modules in SWAT2000 using extensive field experiment data in an irrigation district of the Yellow River Basin. *Journal of Hydrology* 352: 139–156.

Matuz, J., and J.S. Aziz. 1990. The effect of sowing season on Iraqi and Hungarian wheat varieties. *Cereal Research Communication* 18(1–2): 41–43. (Field Crop Absts., 44 (7): 4424, 1991).

Mckeon, G.M., K.G. Rickert, A.J. Ash, D.G. Cooksley, and W.J. Scattini. 1982. Pasture production model. *Proceedings of the Australian Society for Animal Production* 14: 202–204.

Mcleod, J.G., C.A. Campbell, F.B. Dyck, and C.L. Vera. 1992. Optimum seeding dates of winter wheat in southwestern Saskatchewan. *Agronomy Journal* 84: 86–90.

Modarresi, M., V. Mohammdi, A. Zali, and M. Mardi. 2010. Response of wheat yield and yield related traits of high temperature. *Cereal Research communications* 38: 23–31.

Nagarajan, S. 2005. Can India produce enough wheat even by 2020? *Current science* 89: 1467–1471.

Nahar, K., K.U. Ahamed, and M. Fujita. 2010. Phenological variation and its relation with yield in several wheat (Triticum aestivum L.) cultivars under normal and late sowing mediated heat stress condition. *Notulae Scientia Biologicae* 2: 51–56.

Nayyar, M.M., M. Asghar, and M.J. Tahir. 1992. Effect of sowing dates and seeding rate on wheat grain yield under irrigated conditions. *Journal of Agricultural Research* 30: 329–333.

Qadir, G., M. Saeed, and M.A. Cheema. 1999. Effect of water stress on growth and yield performance of four wheat cultivars. *Pakistan Journal of Biological Sciences* 2: 236–239.

Raes, D., P. Steduto, T.C. Hsiao, and E. Fereres. 2009. Aqua crop-the FAO crop model to predict yield response to water. II. Main algorithms and software description. Special issue on "Yield Response to Water: Examination of the Role of Crop Models in Predicting Water Use Efficiency". *Agronomy Journal* 101: 438–449.

Rahimian, H. and M. Banayan. 1996. *Physical principles of biological breeding*. Jahad Mashhad University Press (translated).

Rahman, M.A., J. Chikushi, S. Yoshida, and A.J.M.S. Karim. 2009. Growth and yield components of wheat genotypes exposed to high temperature stress under control environment. *Bangladesh Journal of Agricultural Research* 34: 361–372.

Ranatunga, K., R.N. Eloise, and G.B. David. 2008. Review of soil water models and their applications in Australia. *Environmental Modeling and Software* 23: 1182–1206.

Refay, Y.A. 2011. Yield and yield component parameters of bread wheat genotypes as affected by sowing dates. *Middle-East Journal of Scientific Research* 7: 484–489.

Rickert, K.G., and G.M. Mckeon. 1982. Soil water balance model: WATSUP. *Proceedings of the Australian Society for Animal Production* 14: 198–200.

Shahzad, K., J. Bakht, W.A. Shah, M. Shafi, and N. Jabeen. 2002. Yield and yield components of various wheat cultivars as affected by different sowing dates. *Asian Journal of Plant Sciences* 1: 522–525.

Shahzad, M.A., W.U. Din, S.T. Sahi, M.M. Khan, Ehsanullah, and M. Ahmad. 2007. Effect of sowing dates and seed treatment on grain yield and quality of wheat. *Pakistan Journal of Agricultural Sciences* 44: 581–583.

Slafer, G.A., and D.F. Calderini. 2005. Importance of breeding for further improving durum wheat yield. In *Durum wheat breeding: Current approaches and future strategies*, ed. C. Royo, M.M. Nachit, N. Di Fonzo, et al., 87–95. New York: The Haworth Press.

Tahir, M., A. Ali, M.A. Nadeem, A. Hussain, and F. Khalid. 2009. Effect of different sowing dates on growth and yield of wheat (Triticum aestivum L.) varieties in district Jhang, Pakistan. *Pakistan Journal of Life and Social Sciences* 7: 66–69.

Timsina, J., D. Godwin, E. Humphreys, Y. Singh, B. Singh, S.S. Kukal, and D. Smith. 2008. Evaluation of options for increasing yield and water productivity of wheat in Punjab, India using the DSSAT-CSM-CERES-Wheat model. *Agricultural Water Management* 95: 1099–1110.

Verburg, K. and W.J. BOND. 2003. *Use of APSIM to simulate water balances of dryland farming system in south eastern Australia*. Technical report 50/03. CSIRO land and water Canberra, Australia. http://www.clw.csiro.au/publications/technical2003/tr50-03.pdf.

Wang, Q., M. Watanabe, and Z. Ouyang. 2005. Simulation of water and carbon fluxes using BIOME-BGC model overcrops in China. *Agricultural and Forest Meteorology* 131: 209–224.

Wang, X., A.M. Melesse, and W. Xang. 2006. Influences of potential evapotranspiration estimation methods on SWAT's hydrologic simulation in a Northwestern Minnesota watershed. *American Society of Agricultural and Biological Engineers* 49: 1755–1771.

Wheeler, T.R., P.Q. Craufurd, R.H. Ellis, J.R. Porter, and P.V.V. Prasad. 2000. Temperature variability and the yield of annual crops. *Agriculture Ecosystem and Environment* 82: 9.

Yang, H., and A. Zehnder. 2007. "Virtual water": an unfolding concept in integrated water resources management. *Water Resources Research* 43: 1–10.

Zadoks, J.C., T.T. Chang, and C.F. Konzak. 1974. A decimal code for the growth stages of cereals. *Weed Research* 14: 415–421.

Chapter 8
Effects of Abiotic Stress in Crop Production

Portrait Pierluigi Calanca

Contents

Abstract Crops respond to stress includes, from molecular to the morphological level. Responses at the whole crop level integrate processes taking place at all the underlying levels. For this reason, their quantitative assessment is not always straight forward. Abiotic stresses already represent one of the key factors limiting worldwide crop production. In poor countries, where agriculture is still practiced at a subsistence level, the livelihood of a large share of the population is constantly challenged by abiotic stress factors and their interactions with biotic stress factors. Climate change is likely to aggravate this situation. Taking into account the expected growth in world population and food demand, finding ways to improve crop tolerance with respect to abiotic stress factors will be essential to further improve agricultural production and enhance food security.

Keywords Biotic and abiotic stress • Crop production • Food security • Sensitivity and stress resistance

P.P. Calanca (✉)
Institute for Sustainability Sciences, Agroscope,
Reckenholzstrasse 191, CH-8046 Zurich, Switzerland
e-mail: pierluigi.calanca@agroscope.admin.ch

M. Ahmed, C.O. Stockle (eds.), *Quantification of Climate Variability, Adaptation and Mitigation for Agricultural Sustainability*,
DOI 10.1007/978-3-319-32059-5_8

8.1 Introduction

Owing to advances in breeding, the introduction of improved farming technologies and, at least in developed countries, relatively cheap access to water, fertilizers and crop protection products, crop yields have risen considerably since the 1950s (Edgerton 2009). While this increase extends to worldwide crop production (World Bank 2015), in many areas progress has not been sufficient to close the gap between actual yields and their climatic potentials (Licker et al. 2010). Various reasons contribute to this state of affairs. Pests and diseases play a role (Oerke 2006), but probably more important has been the impact of abiotic stress factors (Boyer 1982; Bonhert 2007; Devine 2009). Crops experience abiotic stress when environmental conditions depart too strongly from the optimum range for growth and reproduction (Larcher 2003). According to Levitt (1980a) biological stress can be defined as "any environmental factor capable of inducing a potentially injurious strain in living organisms".

In turn, biological strain can defined as either a physical or a chemical change induced by stress on a living organism. As opposed to physical strain, biological strain is therefore "not necessarily [only] a change in dimension" (Levitt 1980a). Various factors can lead to stress in crops (Fig. 8.1). Not all of them are directly linked to climate. In practice, however, the emergence of abiotic stresses is often triggered by anomalous climatic conditions, such critical low and high temperatures, persistent absence of rain, extreme precipitation intensities, or high radiation

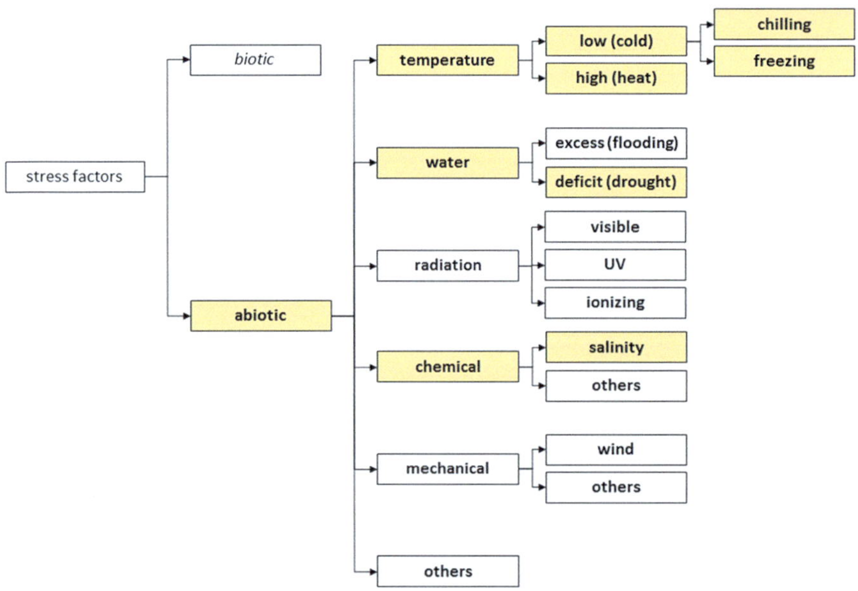

Fig. 8.1 Abiotic stress factors. *Coloured fields* denote those factors often addressed in impact assessments (Modified after Levitt (1980a) and Beck and Lüttge (1990))

Fig. 8.2 Generalized sensitivity of plant processes to water stress and sequence of processes triggered by decreasing water potential in plant tissues (Modified after Hsiao (1973))

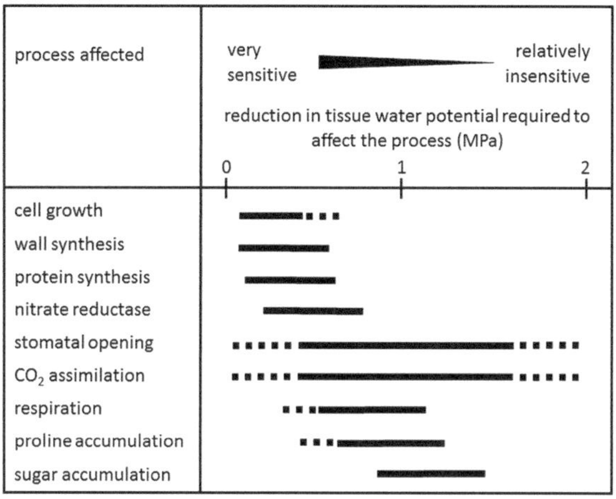

intensities. Problems caused by high salinity are common in arid or semiarid environments (Abrol et al. 1998), where rainfall is too low to prevent accumulations of ions in the soil (Qadir et al. 2014) and where irrigation is the cause of secondary salinization (Ghassemi et al. 1995).

Crops respond to stress at various levels, from the molecular to the morphological (Bonhert 2007). Depending on the process involved, responses to a given stress factor display different sensitivities with respect to the imposed stress signal (Fig. 8.2). Responses at the whole crop level integrate processes taking place at all the underlying levels. For this reason, their quantitative assessment is not always straightforward (Blum 1996).[1]

What happens during stress is essentially determined by the intensity and duration of the factor causing strain. Yet equally important for crops is the timing of stress in relation to development, as crop sensitivities to various stress factors vary according to phenology (Feller and Vaseva 2014). With sorghum exposed to drought, for instance, the largest reduction in grain yield is to be expected when water stress occurs during booting and flowering (Craufurd and Peacock 1993). It is also well known that wheat is particularly sensitive to high temperatures during flowering (Porter and Gawith 1999; Barlow et al. 2015) and that heat stress occurring during the reproductive phase is more harmful than during the vegetative phase (Stone and Nicolas 1995; Farooq et al. 2011).

[1] More information concerning specific responses to various types of abiotic stress can be found elsewhere in the literature and are no further treated here. As a starting point for extending the present discussion one can recommend the textbooks by Levitt (1980a, b), Larcher (2003), various chapters in the book edited by Boote et al. (1994), and several review articles (e.g. Beck and Lüttge 1990; Lichtenthaler 1996; Bonhert 2007; Mittler 2006; Feller and Vaseva 2014; and, Suzuki et al. 2014).

8.2 Resistance to Stress

As with wild plants, crops can, to some extent, resist stress. Stress resistance consists of two components: stress avoidance, i.e. the ability to prevent stress from causing a strain, and stress tolerance, i.e. the ability to cope with a reversible or even irreversible response already triggered by stress (Levitt 1980a, b). The terms "hardiness" and "acclimation" are sometimes used as synonyms to "stress resistance", in particular when discussing the ability of some crops to better survive extreme cold (Snyder and De Melo-Abreu 2005), heat (Paulsen 1994) or drought (Levitt 1980b). For the same reason, the term "hardening" is employed to denote the development of improved tolerance. Acclimation can take place very rapidly. On a hot afternoon, for example, plants are able to shift to higher limiting temperatures within hours (Larcher 2003). In other circumstances, acclimation may require an entire season, as is the case for the development of freezing tolerance in winter cereals (e.g. Pomeroy et al. 1975) and forage grasses (e.g. Larsen 1994). Moreover, the ability to resist adverse environmental conditions is not an enduring feature and can be lost when favourable conditions return. In winter cereals and forage grasses that already underwent acclimation to freezing temperatures, de-hardening can be prompted by a few days of relatively mild temperatures. The consequence is a much higher risk of crop failure from late frosts.

Sensitivity and resistance to stress vary considerably across crops and cultivars (Bray et al. 2000). In cereal crops, resistance to freezing is highest in rye and lowest in oats and durum wheat (Snyder and De Melo-Abreu 2005). When hardening is completed, rye can survive temperatures as low as –40 to –45 °C, whereas the limit is at about −10 °C for durum wheat (Lecomte 1993). This is equivalent to a 30 °C difference in cold tolerance. Likewise, critical temperatures that can impair grain formation during reproductive development barely exceed 30 °C in bean but can reach almost 40 °C in soybean, with intermediate values of about 35 °C in wheat, maize, sorghum, cotton and rice (Hatfield et al. 2011). Different sensitivities also exist with respect to water stress. According to data compiled by Soltani and Sinclair (2012), growth development in sorghum, soybean and maize continues until the fraction of transpirable water in the root zone has dropped to about 0.25, but the development of rice ceases as soon as the fraction of transpirable water in the root zone falls below about 0.6.

8.3 Multiple Stresses

A single abiotic stress seldom befalls a crop. More frequent are situations in which crop development is compromised by the simultaneous occurrence of more than one stress factor (Mittler 2006; Suzuki et al. 2014). In open fields, for example, strong radiation, exceedingly high temperatures, low air humidity and water deficit tend to occur in combination. Common co- occurrences are high salinity in combination with drought, or of high ozone levels in combination with extreme heat. As abiotic stresses have the potential to weaken the defence mechanisms of crops against pathogen and herbivore pests, abiotic stresses are often also precursors of biotic stresses (Suzuki et al. 2014). In many circumstances, crop responses to multiple stresses are unique and cannot be simply inferred by extrapolating responses to individual stress factors. This has clearly been shown concerning molecular responses to heat and drought in tobacco and *Arabidopsis* (Rizhsky et al. 2002, 2004), but similar conclusions hold true also regarding other combinations of stresses (see literature review in Suzuki et al. 2014). When the combined effects of two stress factors are additive, multiple stresses have a higher damaging potential than one would estimate from the sum of the strains induced by the individual factors. This is the case with drought and heat, drought and exceedingly high UV intensities, drought and salinity, heat and ozone, or heat and salinity (Mittler 2006; Suzuki et al. 2014).

Stress enhancement can result even when two (or more) factors act on the same physiological mechanism, if they prompt responses in opposite directions (Feller and Vaseva 2014). Under drought and heat, for instance, a crop initially subjected to high temperatures will open its stomates to increase transpiration and promote cooling. This results in a faster depletion of soil water reserves and onset of water stress. Conversely, a crop subjected to water stress will initially react by closing its stomates, a process that reduces cooling through transpiration and leads to higher foliage temperatures. When compensatory mechanisms exists, the effects of multiple stresses are not cumulative and the overall impact is usually less harmful than the sum of the individual strains (Suzuki et al. 2014). Reduced stomatal conductance in crops suffering from water stress, for example, can enhance the tolerance to ozone stress, and therefore reduce the impact of high ozone doses, which tend to occur with high temperatures during the summer season (Pääkkönen et al. 1998).

8.4 Crop Production and Drought

Drought represents without doubt one of the major threats to worldwide crop production, even in countries where agriculture is highly industrialized (Fig. 8.3). Failure to meet expected production levels can have severe repercussions on prices of agricultural commodities and hence have implications for global food security (IPCC 2014). Also, in poor countries drought has tremendous impacts on livelihood and household economy (Dilley et al. 2005; Sivakumar 2005; Miyan 2015). Especially in Africa, drought has been the reason for food crises and famines.

Often, crops suffering from drought also suffer from heat stress (see discussion in the previous section) which was the case during the drought that affected U.S. agriculture in 2012. Indeed, climatic data reveal that this event was not only exceptional because of the persistence of drought over a large fraction of the cropland (Fig. 8.4) but also because temperatures were higher than normal during most of the summer season, particularly during July (Fig. 8.5c) (GISTEMP Team 2015; EIA 2015).

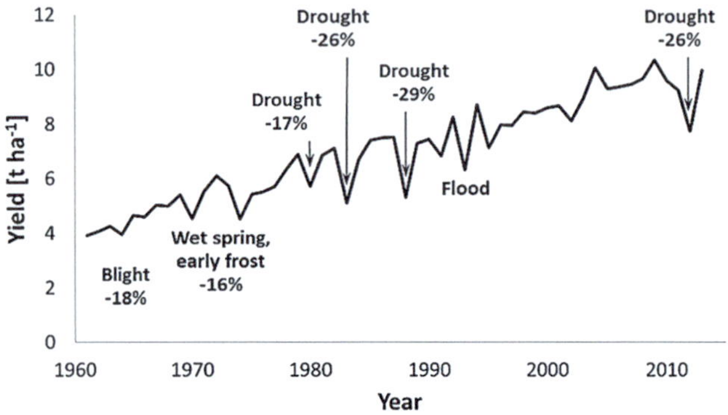

Fig. 8.3 Impact of extreme weather events on maize yields in the US (Adapted from Karl et al. (2009) based on the newest compilation of yields available from FAOSTAT (FAO 2015). The relative loss for 2012 was computed by comparing the actual yield to an estimated potential of ~11 t ha^{-1})

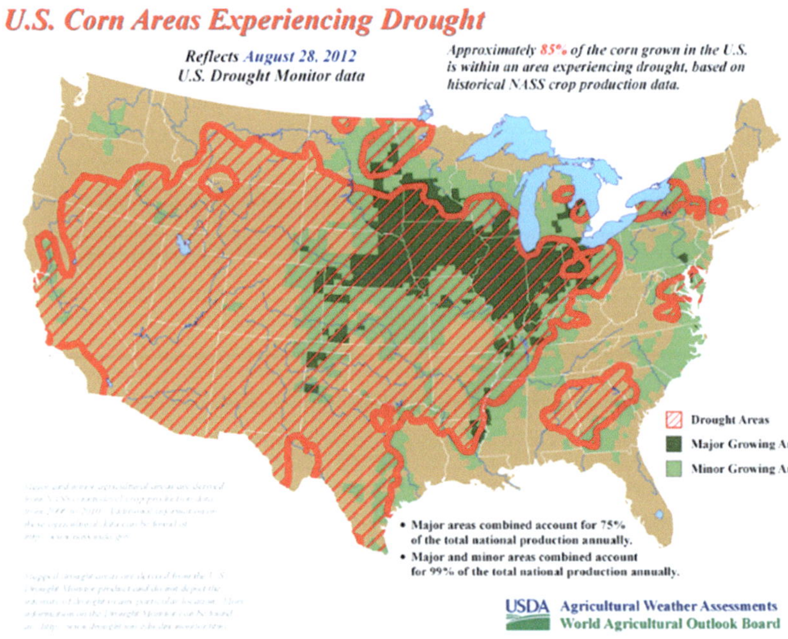

Fig. 8.4 U.S. Corn area in drought at the end of August 2012 (Analysis courtesy of the U.S. Department of Agriculture (EIA 2015))

Thus, the 2012 drought is remembered as "the most extensive drought to affect the U.S. since the 1930s resulting in widespread harvest failure for corn, sorghum and soybean crops, among others, Initial expectations at planting time had suggested [corn] yields averaging a record 166 bushels per acre, but deteriorating growing conditions throughout the summer led USDA to reduce yield expectations. The

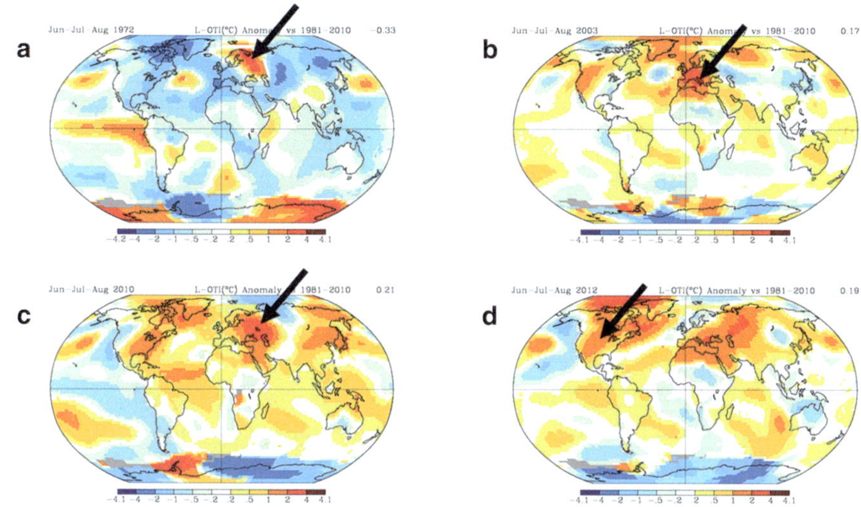

Fig. 8.5 Global temperature anomaly maps for (**a**) 1972, (**b**) 2003, (**c**) 2010 and (**d**) 2012. Courtesy of the National Aeronautics and Space Administration (NASA), Goddard Institute for Space Studies (Hansen et al. 2010; GISTEMP Team 2015). Shown here are the mean anomalies relative to a 1981–2010 baseline for the Northern-Hemisphere summer (June, July and August). Key areas discussed in the text (in the order Ukraine, Western Europe, Russia and the U.S) are indicated with an *arrow*

final 2012 yield estimate was set at 123.4 bushels per acre, the lowest since 1995" (USDA 2015).

As seen in Fig. 8.5, the occurrence of anomalously high temperatures has also been a characteristic of many drought events of relevance for global crop production, e.g. the 1972 event in the Ukraine and, more recently, the two heat waves that struck Western Europe in 2003 and Russia in 2010[2] (Battisti and Naylor 2009; Wegren 2011; Anyamba et al. 2014).

8.4.1 *Crop Exposure to Heat Stress: Recent Trends*

Global temperatures have risen by about 0.8 °C since 1975 (Hartmann et al. 2013). According to IPCC (2014) "negative impacts of climate change on crop yields have been more common than positive impacts (high confidence). The smaller number

[2]The large-scale circulation patterns responsible for the 2010 Russian heatwave eventually led to catastrophic floods in Pakistan. This event affected more than 20 million people (Kirsch et al. 2012) and negatively affected agriculture to an unprecedented scale (FAO 2010; WFP 2010). Undoubtedly, there is an abiotic stress contribution to the damages caused by these floods to crops. Overall, however, the effects of these floods and similar events extend beyond what can be considered as abiotic stress component.

of studies showing positive impacts relate mainly to high-latitude regions, though it is not yet clear whether the balance of impacts has been negative or positive in these regions (high confidence). Climate change has negatively affected wheat and maize yields for many regions and in the global aggregate (medium confidence). Effects on rice and soybean yield have been smaller in major production regions and globally, with a median change of zero across all available data, which are fewer for soy compared to the other crops. Observed impacts relate mainly to production aspects of food security rather than access or other components of food security. Since AR4 [IPCC Fourth Assessment Report], several periods of rapid food and cereal price increases following climate extremes in key producing regions indicate a sensitivity of current markets to climate extremes among other factors (medium confidence)".

The increase in mean growing season temperatures alone has been shown to have had a negative impact on the recent upward trend in crop yields, effectively reducing maize and wheat production by roughly 4 and 6 %, respectively, below what could have potentially been achieved without global warming (Lobell et al. 2011).[3]

In many areas of the world, notably Europe, Asia, Africa and South America, the rise in global mean temperature has been accompanied by an increase in both nighttime minimum as well as daytime maximum temperatures, and by an increase in the frequency of extremely warm conditions (Vose et al. 2005; Donat et al. 2013). The result has been a decrease in exposure to low temperature but an increase in exposure to critically high temperatures and heat stress, in recent decades. Past increase of crop exposure to heat stress during reproductive growth has been confirmed by Gourdji et al. (2013), although the correspondence to trends in growing season mean temperatures has, so far, been weak.[4] According to their analysis, about 10 (soybean and rice) to 30 % (wheat and maize) of the crop area has been exposed to more than 0.1 °C/decade increase in critical high temperatures.

The geographic distribution of crop areas currently at risk of heat stress during reproductive development are easily identified in the maps presented by Gourdji et al. (2013, their Figs. 8.1 and 8.2) and similar maps presented by Teixeira et al. (2013, their Figs. 8.2 and 8.4). For wheat, hot spots are concentrated in southern Russia, Kazakhstan, Pakistan and India; for maize, hot spots are spread across the globe, including Europe (Iberian Peninsula and the Southeast), Africa, and North, Central and South America. These are the regions where the risk of incurring heat stress is expected to further increase in the near future.

---

[3] According to the analysis of Lobell et al. (2011), for maize and wheat, trends in precipitation have worsened the situation, with an additional relative impact of about −0.5 to −1 %.

[4] This is because temperatures have been for the most part below crop critical thresholds and therefore the increase in temperature has yet to be reflected in a significant increase in exceedance probabilities

8.4.2 Global Warming, Heat Stress and Drought

There is little doubt that global change will further alter the conditions for crop production (Lobell and Gourdji 2012). Global climate model simulations suggest that global temperatures will continue to rise during the coming decades (Collins et al. 2013). Depending on which emission scenarios and experiments are being evaluated, the increase in global surface temperature relative to 1986–2005 is expected to reach between +0.3 °C and +4.8 °C by the end of the century. Changes in the shape of the temperature distribution would come on top of the trends in annual or seasonal averages. As a result, by the end of the century growing season temperatures in the tropics and subtropics are expected to exceed current extreme temperatures, and present exceptional temperatures in the temperate zones, such as those recorded during the 2003 heat wave in Western Europe, are expected to become the norm (Battisti and Naylor 2009).

In more detail, daily maximum temperatures are projected to increase by +1.5 to +5.5 °C until the end of the century (Collins et al. 2013; Sillmann et al. 2013). Exposure to critically high temperatures during the reproductive period is expected, therefore, to be more common in the future. Without adaptation, there could be an increase in the fraction of the total harvested area exposed to heat stress (Gourdji et al. 2013). For maize, for instance, this fraction could triple by 2050 as compared to today, with serious implications for global production. Changes in land utilization and management could reduce the global exposure to heat stress. Critical high temperatures in wheat production could e.g. be avoided by shifting sowing dates (Teixeira et al. 2013).

Less certain is the future exposure of cropland to agricultural droughts. In fact, projected changes in total precipitation amounts, seasonality of precipitation, and duration of wet and dry spells vary considerably depending on model and emission scenario (Collins et al. 2013). The question of whether changes in the atmospheric branch of the hydrological cycle will be dominated by thermodynamics (intensification reflecting a higher energy content of the lower atmosphere) or shifts in the circulation patterns, including possible shifts in global teleconnection patterns such as the El Niño-Southern Oscillation, is also not settled.

According to Collins et al. (2013), there is nevertheless some confidence that some of the current agricultural areas will experience a decrease in soil moisture. In the words of Trenberth et al. (2014), "the contrast in precipitation between wet and dry regions and between wet and dry seasons will probably increase, although there may be regional exceptions. Climate change is adding heat to the climate system and on land much of that heat goes into drying. A natural drought should therefore set in quicker, become more intense, and may last longer. Droughts may be more extensive as a result. Climate change may not manufacture droughts, but it could exacerbate them and it will probably expand their domain in the subtropical dry zone."

8.4.3 *Effects of Elevated CO_2 Concentrations*

For the discussion of abiotic stresses under future climatic conditions, it is important to bear in mind that the positive effects of elevated atmospheric CO2 concentrations (Körner 2006; Lobell and Gourdji 2012) could partially offset the negative effects of higher temperatures and decreased water availability. Results of so-called Free-Air CO_2 Enrichment (FACE) experiments have shown that higher CO_2 levels stimulate photosynthesis and net primary production (along with dark respiration, though), improve nitrogen use efficiency and decrease water use at both the leaf and canopy scale (Leakey et al. 2009).[5]

Increased water use efficiency under high CO_2 levels would result from a reduction in stomatal conductance (Bunce 2004) and transpiration (Vanuytrecht et al. 2012),[6] which should potentially lead to decreased incidence of water stress under future climatic conditions. Reduced evapotranspiration would also help control the salinity problem since reduced transpiration would improve the water status of the soil and limit the necessity for irrigation.

However, as indicated earlier, changes in stomatal conductance also affect the thermal balance of crops, and reduced stomatal conductance could therefore lead to higher heat stress if water is insufficient to maintain transpiration for a longer time at an adequate level. Clearly, the consequences of elevated CO_2 for crop exposure to multiple stresses need to be more systematically examined (cf. Lobell 2014).

An additional pathway by which elevated CO_2 concentrations could alter the sensitivity of crops to water shortage is by increasing the root: shoot ratio (Vanuytrecht et al. 2012). The processes by which assimilates would be preferentially allocated to the roots are not fully understood (Passioura 1994), but undoubtedly a relative increase in root biomass would improve the ability of crops to exploit soil water and nutrients alike, which could help reduce the susceptibility of crops to nutrient stress.

8.5 Adaptation

Given that the probability of extreme climatic conditions is likely to increase under climate change, options to cope with a higher incidence of some abiotic stress factors are necessary to maintain or even increase crop productivity (IPCC 2014). There are various options by which the impact of abiotic stress can be reduced. With regard to heat stress, changes in field calendars (e.g. earlier sowing dates), the use of early-ripening cultivars, or the replacement of sensitive with less sensitive crops

[5] Because of the different photosynthetic pathways, overall responses to high levels of CO_2 in C3 and C4 crops are expected to differ, though perhaps not as distinctly as the direct effect of CO_2 on assimilation (Vanuytrecht et al. 2012).

[6] Note that in grasslands water savings are almost fully responsible for the observed biomass responses to elevated CO_2 (Körner 2006).

are among those most often addressed in impact assessments when considering the farm scale (e.g. Trnka et al. 2014). Some of these options are not without side effects, though. An example is the cultivation of early-ripening varieties. On the one hand, this would help reduce exposure to critical temperatures during summer. On the other hand, it would entail an overall shortening of the growing season and could eventually lead to lower yields.

Improved soil management can also help cope with abiotic stresses, as shown by the outcomes of an experiment conducted in Switzerland during the record-breaking heatwave of 2003 (Feller and Vaseva 2014). In this experiment, leaf temperature and stomatal conductance in sugar beet were monitored during sunny days on till and no-till plots. Under conventional tillage, midday temperatures in leaves were 2 to 3 °C higher than under conservation soil management, whereas stomatal conductance was reduced by roughly a factor of two.

The impact of abiotic stresses can also be reduced by improving stress tolerance. This is a primary goal of ongoing breeding programs. The reader is referred to e.g. Vinocur and Altman (2005); Witcombe et al. (2008) or Devine (2009) for good overviews, and to e.g. Tardieu (2003); Tardieu and Tuberosa (2010) and Semenov et al. (2014) for an appreciation of how breeding efforts can be supported by mathematical modelling. So far, experiences indicate that there is potential for breeding to improve heat and low temperaturetolerance, as well as tolerance to multiple stresses (Devine 2009). Breeding for drought and salinity tolerance appears to be more difficult, but not without possibilities (Witcombe et al. 2008). It has been shown that breeding could help adapt crops to low nutrient levels while retaining the ability to respond to fertilization (Witcombe et al. 2008).

Concerning drought, changes in the hydrological cycle and a reduction in global water availability for the agricultural sector (Milly et al. 2005; Strzepek and Boehlert 2010) leave little doubt that in many areas of the world the need for irrigation is going to increase in the future (Vörösmarty et al. 2000). Even though in some areas sustained irrigation could be possible without unintended consequences, consideration of the environmental impacts of irrigation is necessary. Salinization of agricultural soils is a problem that already has reached critical levels (Ghassemi et al. 1995) and that needs to be solved to make crop production sustainable. Depletion of groundwater is a problem in major crop production areas in the U.S., Europe, China and India and the Middle East (Wada et al. 2010). Again, options to limit groundwater extractions are required to limit the impacts of agriculture on the global environment.

8.6 Concluding Remarks

Abiotic stresses already represent one of the key factors limiting worldwide crop production. In poor countries, where agriculture is still practiced at a subsistence level, the livelihood of a large share of the population is constantly challenged by abiotic stress factors and their interactions with biotic stress factors. Climate change is likely to aggravate this situation. Taking into account the expected growth in world population and food demand, finding ways to improve crop tolerance with respect to abiotic stress factors will be essential to further improve agricultural production and enhance food security. Various options are currently being explored, some of them showing promising results. A proper assessment of the net effects of such measures can deliver the basis for an objective discussion (Lobell 2014).

Acknowledgements The preparation of this contribution has been facilitated by the author's participation in the project "Modelling vegetation response to EXTREM Events" (MODEXTREME, http://modextreme.org), which has received funding from the European Community's Seventh Framework Programme – FP7 (KBBE.2013.1.4-09) under Grant Agreement No. 613817, 2013–2016.

References

Abrol, I.P., J.S.P. Yadav, and F.I. Massoud. 1998. Salt-affected soils and their management. *FAO Soil Bulletin* 39. *Food and Agriculture Organization (FAO) of the United Nations, Rome*. Available online at http://www.fao.org/docrep/x5871e/x5871e00.htm. Accessed 20 July 2015.

Anyamba, A., J.L. Small, S.C. Britch, C.J. Tucker, and E.W. Pak. 2014. Recent weather extremes and impacts on agricultural production and vector-borne disease outbreak patterns. *PLoS One* 9(3): e92538. doi:10.1371/journal.pone.0092538.

Barlow, K.M., B.P. Christy, G.J. O'leary, P.A. Riffkin, and J.G. Nuttall. 2015. Simulating the impact of extreme heat and frost events on wheat crop production: A review. *Field Crops Research* 171: 109–119.

Battisti, D.S., and R.L. Naylor. 2009. Historical warnings of future food insecurity with unprecedented seasonal heat. *Science* 323: 240–244.

Beck, E., and U. Lüttge. 1990. Stress bei Pfalnzen. *Biol Unserer Zeit* 20: 237–244.

Blum, A. 1996. Crop responses to drought and the interpretation of adaptation. *Plant Growth Regulation* 20: 135–148.

Bonhert, H.J. 2007. Abiotic stress. In *Encyclopedia of life sciences (eLS)*. Chichester: John Wiley & Sons Ltd. http://www.els.net/[doi: 10.1002/9780470015902.a0020087]. Accessed 20 July 2015

Boote, K.J., J.M. Bennett, T.R. Sinclair, and G.M. Paulsen (eds.). 1994. *Physiology and determination of crop yield*. Madison: American Society of Agronomy, Crop Science Society of America, and Soil Science Society of America.

Boyer, J.S. 1982. Plant productivity and environment. *Science* 218: 443–448.

Bray, E.A., J. Bailey-Serres, and E. Weretilnyk. 2000. Responses to abiotic stress. In *Biochemistry & molecular biology of plants*, ed. B. Buchanan, W. Gruissem, and R. Jones. Rockville: The American Society of Plant Physiologists.

Bunce, J.A. 2004. Carbon dioxide effects on stomatal responses to the environment and water use by crops under field conditions. *Oecologia* 140: 1–10.

Collins, M., R. Knutti, J. Arblaster, J.L. Dufresne, T. Fichefet, P. Friedlingstein, X. Gao, W.J. Gutowski, T. Johns, G. Krinner, M. Shongwe, C. Tebaldi, A.J. Weaver, and M. Wehner. 2013. Long-term climate change: Projections, commitments and irreversibility. In *Climate change 2013: The physical science basis. Contribution of working group I to the fifth assessment report of the intergovernmental panel on climate change*, ed. T.F. Stocker, D. Qin, G.-K. Plattner, M. Tignor, S.K. Allen, J. Boschung, A. Nauels, Y. Xia, V. Bex, and P.M. Midgley. Cambridge/New York: Cambridge University Press.

Craufurd, P.Q., and J.M. Peacock. 1993. Effect of heat and drought stress on sorghum (*Sorghum Bicolor*). II: Grain yield. *Experimental Agriculture* 29: 77–86.

Devine, M.D. 2009. Enhancing crop productivity through increased abiotic-stress tolerance and biomass broduction. In *Adapting Agriculture to Climate Change (Proceedings of the twenty-first annual conference of the National Agricultural Biotechnology Council, Saskatoon, Saskatchewan, June 24–26, 2009)*, ed. Eaglesham A, and R.W.F. Hardy, 65–72. NABC Report 21, National Agricultural Biotechnology Council, Ithaca, New York.

Dilley, M., R.S. Chen, U. Deichmann, A.L. Lerner-Lam, and M. Arnold. 2005. *Natural Disaster Hotspots: A Global Risk Analysis*. Washington, DC: World Bank. https://openknowledge.worldbank.org/handle/10986/7376. Accessed 20 July 2015.

Donat, M.G., L.V. Alexander, H. Yang, I. Durre, R. Vose, R.J.H. Dunn, K.M. Willett, E. Aguilar, M. Brunet, J. Caesar, B. Hewitson, C. Jack, A.M.G. Klein Tank, A.C. Kruger, J. Marengo, T.C. Peterson, M. Renom, C. Oria Rojas, M. Rusticucci, J. Salinger, A.S. Elrayah, S.S. Sekele, A.K. Srivastava, B. Trewin, C. Villarroel, L.A. Vincent, P. Zhai, X. Zhang, and S. Kitching. 2013. Updated analyses of temperature and precipitation extreme indices since the beginning of the twentieth century: The HadEX2 dataset. *Journal of Geophysics Research and Atmosphere* 118: 2098. doi:10.1002/jgrd.50150.

Edgerton, M.D. 2009. Increasing crop productivity to meet global needs for feed, food, and fuel. *Plant Physiology* 149: 7–13.

EIA (U.S. Energy Information Administration). 2015. *Drought has significant effect on corn crop condition, projected ethanol production*. http://www.eia.gov/todayinenergy/detail.cfm?id=7770. Accessed 20 July 2015.

Farooq, M., H. Bramley, J.A. Palta, and K.H.M. Siddique. 2011. Heat stress in wheat during reproductive and grain-filling phases. *Critical Reviews in Plant Sciences* 30: 1–17.

Feller, U., and I.I. Vaseva. 2014. Extreme climatic events: Impacts of drought and high temperature on physiological processes in agronomically important plants. *Front Environment Science*. 06 October 2014. http://dx.doi.org/10.3389/fenvs.2014.00039. Accessed 20 July 2015.

Food and Agriculture Organization of the United Nations. 2010. FAO emergency response to Pakistan floods. Food and Agriculture Organization of the United Nations, Update on Agriculture (Pakistan Floods), Issue No 3, 19 August 2010. http://www.fao.org/fileadmin/user_upload/emergencies/docs/Pakistan_Update_Agriculture_3.pdf. Accessed 20 July 2015.

Food and Agriculture Organization of the United Nations. 2015. Food and Agriculture Organization of the United Nations, Statistics Division. http://faostat.fao.org/. Accessed 20 July 2015.

Ghassemi, F., A.J. Jakeman, and H.A. Nix. 1995. *Salinization of land and water resources: Human causes, extent, management and case studies. Centre for resources and environmental studies*. Wallingford: Australian National University. CAB International.

GISTEMP Team. 2015. GISS Surface Temperature Analysis (GISTEMP). NASA Goddard Institute for Space Studies. http://data.giss.nasa.gov/gistemp/. Accessed 20 July 2015.

Gourdji, S.M., A.M. Sibley, and D.B. Lobell. 2013. Global crop exposure to critical high temperatures in the reproductive period: Historical trends and future projections. *Environment Research Letters* 8(024041): 10. doi:10.1088/1748-9326/8/2/024041.

Hansen, J., R. Ruedy, M. Sato, and K. Lo. 2010. Global surface temperature change. *Reviews of Geophysics* 48: RG4004. doi:10.1029/2010RG000345.

Hartmann, D.L., A.M.G. Klein Tank, M. Rusticucci, L.V. Alexander, S. Brönnimann, Y. Charabi, F.J. Dentener, E.J. Dlugokencky, D.R. Easterling, A. Kaplan, B.J. Soden, P.W. Thorne, M. Wild, and P.M. Zhai. 2013. Observations: Atmosphere and surface. In *Climate change 2013: The physical science basis. Contribution of working group I to the fifth assessment report of the intergovernmental panel on climate change*, ed. T.F. Stocker, D. Qin, G.-K. Plattner, M. Tignor, S.K. Allen, J. Boschung, A. Nauels, Y. Xia, V. Bex, and P.M. Midgley. Cambridge/New York: Cambridge University Press.

Hatfield, J.L., K.J. Boote, B.A. Kimball, L.H. Ziska, R.C. Izaurralde, D. Ort, A.M. Thomson, and D. Wolfe. 2011. Climate impacts on agriculture implications for crop production. *Agronomy Journal* 103: 351–370.

Hsiao, T.C. 1973. Plant responses to water stress. *Annual Review Plant Physiology* 24: 519–570.
IPCC. 2014. Summary for policymakers. In *Climate change 2014: Impacts, adaptation, and vulnerability. Part A: Global and sectoral aspects. Contribution of working group II to the fifth assessment report of the intergovernmental panel on climate change*, ed. C.B. Field, V.R. Barros, D.J. Dokken, K.J. Mach, M.D. Mastrandrea, T.E. Bilir, M. Chatterjee, K.L. Ebi, Y.O. Estrada, R.C. Genova, B. Girma, E.S. Kissel, A.N. Levy, S. MacCracken, P.R. Mastrandrea, and L.L. White. Cambridge/New York: Cambridge University Press.
Karl, T.R., J.M. Melillo, and T.C. Peterson (eds.). 2009. *Global climate change impacts in the United States*. Cambridge/New York: Cambridge University Press.
Kirsch, T.D., C. Wadhwani, L. Sauer, S. Doocy, and C. Catlett. 2012. Impact of the 2010 Pakistan floods on rural and urban populations at six months. *PLOS Currents Disasters*, Aug 22, 2012, Ed 1. doi:10.1371/4fdfb212d2432.
Körner, C. 2006. Plant CO_2 responses: An issue of definition, time and resource supply. *New Phytologist* 172: 393–411.
Larcher, W. 2003. *Physiological plant ecology*. Berlin: Springer.
Larsen, A. 1994. Breeding winter hardy grasses. *Euphytica* 77: 231–237.
Leakey, A.D.B., E.A. Ainsworth, C.J. Bernacchi, A. Rogers, S.P. Long, and D.R. Ort. 2009. Elevated CO_2 effects on plant carbon, nitrogen, and water relations: Six important lessons from FACE. *Journal of Experimental Botany* 60: 2859–2876.
Lecomte, C. 1993. Seuils de sensibilite au gel hivernal en grandes cultures. In *Le gel en agriculture*, ed. C. Rious, 83–100. Paris. http://prodinra.inra.fr/record/100990. Accessed 20 July 2015.
Levitt, J. 1980a. *Responses of plants to environmental stresses. Volume I: Chilling, freezing, and high temperature stresses*. New York: Academic.
Levitt, J. 1980b. *Responses of plants to environmental stresses. Volume II: Water, radiation, salt, and other stresses*. New York: Academic.
Lichtenthaler, H.K. 1996. Vegetation stress: An introduction to the stress concept in plants. *Journal of Plant Physiology* 148: 4–14.
Licker, R., M. Johnston, and J.A. Foley. 2010. Mind the gap: How do climate and agricultural management explain the 'yield gap' of croplands around the world? *Global Ecological Biogeography* 19: 769–782.
Lobell, D.B. 2014. Climate change adaptation in crop production: Beware of illusions. *Global Food Security* 3: 72–76.
Lobell, D.B., and S.M. Gourdji. 2012. The influence of climate change on global crop productivity. *Plant Physiology* 160: 1686–1697.
Lobell, D.B., W. Schlenker, and J. Costa-Roberts. 2011. Climate trends and global crop production since 1980. *Science* 333: 616–620.
Milly, P.C.D., K.A. Dunne, and A.V. Vecchia. 2005. Global pattern of trends in streamflow and water availability in a changing climate. *Nature* 438: 347–350.
Mittler, R. 2006. Abiotic stress, the field environment and stress combination. *Trends in Plant Science* 11: 15–19.
Miyan, M.A. 2015. Droughts in Asian least developed countries: Vulnerability and sustainability. *Weather and Climate Extremes* 7: 8–23.
Oerke, E.C. 2006. Crop losses to pests. *Journal of Agricultural Science* 144: 31–43.
Pääkkönen, E., J. Vahala, M. Pohjola, T. Holopainen, and L. Kärenlampi. 1998. Physiological, stomatal and ultrastructural ozone responses in birch (Betula pendula Roth.) are modified by water stress. *Plant, Cell and Environment* 21: 671–684.
Passioura, J.B. 1994. The yield of crops in relation to drought. In *Physiology and determination of crop yield*, ed. K.J. Boote, J.M. Bennett, T.R. Sinclair, and G.M. Paulsen, 343–359. Madison: American Society of Agronomy, Crop Science Society of America, and Soil Science Society of America.
Paulsen, G.M. 1994. High temperature responses of crop plants. In *Physiology and determination of crop yield*, ed. K.J. Boote, J.M. Bennett, T.R. Sinclair, and G.M. Paulsen, 365–389. Madison: American Society of Agronomy, Crop Science Society of America, and Soil Science Society of America.

Pomeroy, M.K., C.J. Andrews, and G. Fedak. 1975. Cold hardening and de-hardening responses in winter wheat and winter barley. *Canadian Journal of Plant Science* 55: 529–535.

Porter, J.R., and M. Gawith. 1999. Temperatures and the growth and development of wheat: A review. *European Journal of Agronomy* 10: 23–36.

Qadir, M., E. Quillérou, V. Nangia, G. Murtaza, M. Singh, R.J. Thomas, P. Drechsel, and A.D. Noble. 2014. Economics of salt-induced land degradation and restoration. *Natural Resources Forum* 38: 282–295.

Rizhsky, L., H. Liang, and R. Mittler. 2002. The combined effect of drought stress and heat shock on gene expression in tobacco. *Plant Physiology* 130: 1143–1151.

Rizhsky, L., H. Liang, J. Shuman, V. Shulaev, S. Davletova, and R. Mittler. 2004. When defense pathways collide: The response of Arabidopsis to a combination of drought and heat stress. *Plant Physiology* 134: 1683–1696.

Semenov, M.A., P. Stratonovitch, F. Alghabari, and M.J. Gooding. 2014. Adapting wheat in Europe for climate change. *Journal of Cereal Science* 59: 245–256.

Sillmann, J., V.V. Kharin, F.W. Zwiers, X. Zhang, and D. Bronaugh. 2013. Climate extremes indices in the CMIP5 multimodel ensemble: Part 2. Future climate projections. *Journal of Geophysics Research* 118: 2473–2493.

Sivakumar, M.V.K. 2005. Impacts of natural disasters in agriculture, rangeland and forestry: An overview. In *Natural disasters and extreme events in agriculture*, ed. M.V.K. Sivakumar, R.P. Motha, and H.P. Das, 1–22. Berlin: Springer.

Snyder, R.L., and J.P. de Melo-Abreu. 2005. *Frost protection: Fundamentals, practice and economics*. Rome: Food and Agriculture Organization of the United Nations.

Soltani, A., and T.R. Sinclair. 2012. *Modeling physiology of crop development, growth and yield*. Wallingford: CAB International.

Stone, P.J., and M.E. Nicolas. 1995. Effect of timing of heat stress during grain-filling on two wheat varieties differing in heat tolerance. I. Grain growth. *Australian Journal of Plant Physiology* 22: 927–934.

Strzepek, K., and B. Boehlert. 2010. Competition for water for the food system. *Philosophical Transactions of the Royal Society, B: Biological Sciences* 365: 2927–2940.

Suzuki, N., R.M. Rivero, V. Shulaev, E. Blumwald, and R. Mittler. 2014. Abiotic and biotic stress combinations. *New Phytologist* 203: 32–43.

Tardieu, F. 2003. Virtual plants: Modelling as a tool for genomics of tolerance to water deficit. *Trends in Plant Science* 8: 9–14.

Tardieu, F., and R. Tuberosa. 2010. Dissection and modelling of abiotic stress tolerance in plants. *Current Opinion in Plant Biology* 13: 206–212.

Teixeira, E.I., G. Fischer, H. van Velthuizen, C. Walter, and F. Ewert. 2013. Global hot-spots of heat stress on agricultural crops due to climate change. *Agricultural and Forest Meteorology* 170: 206–215.

Trenberth, K.E., A. Dai, G. van der Schrier, P.D. Jones, J. Barichivich, K.R. Briffa, and J. Sheffield. 2014. Global warming and changes in drought. *Nature Climate Change* 4: 14–22.

Trnka, M., R.P. Rötter, M. Ruiz-Ramos, K.C. Kersebaum, J.E. Olesen, Z. Žalud, and M.A. Semenov. 2014. Adverse weather conditions for European wheat production will become more frequent with climate change. *Nature Climate Change* 4: 637–643.

U.S. Department of Agriculture. 2015. http://www.ers.usda.gov/topics/in-the-news/us-drought-2012-farm-and-food-impacts.aspx. Accessed 20 July 2015.

Vanuytrecht, E., D. Raes, P. Willems, and S. Geerts. 2012. Quantifying field-scale effects of elevated carbon dioxide concentration on crops. *Climate Research* 54: 35–47.

Vinocur, B., and A. Altman. 2005. Recent advances in engineering plant tolerance to abiotic stress: Achievements and limitations. *Current Opinion in Biotechnology* 16: 123–132.

Vörösmarty, C.J., P. Green, J. Salisbury, and R.B. Lammers. 2000. Global water resources: Vulnerability from climate change and population growth. *Science* 289: 284–288.

Vose, R.S., D.R. Easterling, and B. Gleason. 2005. Maximum and minimum temperature trends for the globe: An update through 2004. *Geophysical Research Letters* 32: L23822. doi:10.1029/2005GL024379.

Wada, Y., L.P.H. van Beek, C.M. van Kempen, J.W.T.M. Reckman, S. Vasak, and M.F.P. Bierkens. 2010. Global depletion of groundwater resources. *Geophysical Research Letters* 37: L20402. doi:10.1029/2010GL044571.

Wegren, S.K. 2011. Food security and Russia's 2010 drought. *Eurasian Geography and Economics* 52: 140–156. http://dx.doi.org/10.2747/1539-7216.52.1.140. Accessed 20 July 2015.

Witcombe, J.R., P.A. Hollington, C.J. Howarth, S. Reader, and K.A. Steele. 2008. Breeding for abiotic stresses for sustainable agriculture. *Philosophical Transactions of the Royal Society B* 363: 703–716.

World Bank. 2015. Statistical data. http://data.worldbank.org/indicator. Accessed 20 July 2015.

World Food Programme. 2010. Pakistan – Flood impact assessment, September 2010. http://documents.wfp.org/stellent/groups/public/documents/ena/wfp225987.pdf. Accessed 20 July 2015.

Chapter 9
Drought Tolerance in Cereal Grain Crops Under Changing Climate

Zohra Aslam, Jabar Zaman Khan Khattak, and Mukhtar Ahmed

Contents

Z. Aslam (✉) • J.Z.K. Khattak
Department of Bioinformatics, International Islamic University Islamabad, Islamabad, Pakistan
e-mail: zohra.msbi37@iiu.edu.pk; zzohra.aslam@gmail.com; jabar.khattak@iiu.edu.pk

M. Ahmed
Department of Agronomy, Pir Mehr Ali Shah Arid Agriculture University, Rawalpindi, Pakistan

Department of Biological Systems Engineering, Washington State University, Pullman, WA, USA
e-mail: ahmadmukhtar@uaar.edu.pk; mukhtar.ahmed@wsu.edu

M. Ahmed, C.O. Stockle (eds.), *Quantification of Climate Variability, Adaptation and Mitigation for Agricultural Sustainability*,
DOI 10.1007/978-3-319-32059-5_9

Abstract Earth life is greatly dependent on function and properties of water. A major threat to agricultural production is drought. Drought is a multidimensional stress and world spread problem that cause substantial losses by influencing the yield and production seriously. Tolerance to drought is a principal target for molecular strategies to crop enhancement. The plants ability to resist drought conditions is important for agricultural production globally. Current progress in responses to drought has been made in our comprehending of signal transduction, gene expression and transcriptional regulation in plants. Plants have developed a diverse variety of drought resistance mechanisms in front of water limiting conditions at physiological, metabolic and molecular level. Water uptake and development of healthier root, WUE, osmotic adjustment, and mineral nutrients also have important consequences on adaptation to drought. This chapter is organized around the concept of "drought tolerance in rice and maize crops". Some innovative tactics are discussed. This chapter summarizes different aspects of crop breeding for drought tolerance and analyses how conventional breeding, genetics, biotechnology tools, micro arrays, MAS, QTL, bioinformatics and transgenic crops as well as mineral nutrients, plant growth regulations can participate to advancing the emancipation of drought-resistant rice and maize cultivars. We foresee the functional and genetic of drought resistance based on such premises. Novel opportunities for tailoring new genotypes will be generated 'by design'. Harnessing the genomics-assisted breeding's potential will need an integrated knowledge of physiological and molecular processes and a multidisciplinary approach influencing drought tolerance.

Keywords Drought • Tolerance • Abiotic stress • Phenotyping • Transgenic plants • Functional genomics • Bioinformatics • QTL • Crop improvement • Mineral nutrients and root growth

Abbreviations

WHO	World health organization
ROS	Reactive oxygen specie
IAA	Indole-3-Acetic Acid
GA_3	Gibberellic Acid
BL	Brassinolide
IRRI	The International Rice Research Institute
CIMMYT	The International Maize and Wheat Improvement Center
IITA	International Institute for Tropical Agriculture
SSA	Sub-Saharan Africa
PGRs	Plant growth regulators
CIAT	The International Center for Tropical Agriculture
ABA	Abscisic acid
CO_2	Carbon dioxide
GB	Glycine betaine

DI	Deficit irrigation
WUE	Water-use efficiency
EUW	Effective use of water
RNA	Ribonucleic Acid
DNA	Deoxyribonucleic Acid
OA	Osmotic adjustment
ATP	Adenosine triphosphate
QTL	Quantitative trait loci
eQTL	Expression quantitative trait loci
SNP	Single nucleotide polymorphism
MAS	Marker-assisted selection
DPE	Drought-prone environments
TE	Transpiration efficiency
HI	Harvest index
WU	Water uptake
IRFGC	International Rice Functional Genomics Consortium
ICIS	International Crop Information System
GCP	Generation Challenge Program
TF	Transcription factors
GST	Glutathione-S-transferases

9.1 Introduction

The Earth land area covers a 140 million km^2 less than one third of the Earth's surface. Land sources are non-renewable, fragile and finite; which include landscapes which are important for human welfare and habitat; land cover, important for environment; and soil, important for agriculture. The main dynamic force on land resources since 1972 has been increasing and growing food production. Cropping is the largest world's source of livelihood and employment in developing countries. For agriculture, water is the major element. In many regions of the world, water originates agricultural production. Indeed, in reducing hunger, the 'green revolution' was effective because the increased irrigation use behind the successful increase in production of crop was one of the reasons. Nevertheless, the twenty-first century demands will reduce the availability of water for irrigated agriculture (Hong-Bo et al. 2006).

The Worldwide human population is projected to increase by 2050 and will demand more water for environmental, municipal, domestic, and industrial needs. In 2002 for 2220 million people, food is needed than in 1972. It means that pressure on land will remain to be severe predominantly in Asia and Africa. Undoubtedly, in the developing world, environmental stresses are main cause of food security where sufficient food production is a major challenge. A large proportion of the agriculture world depends on precipitation for irrigation. The world major food crops, wheat, maize and rice have been negatively influenced by drought conditions. In many regions (China, Central

Asia, South American and African countries, the Middle East, United States regions and Australia), crops are affected negatively by drought stress conditions (http://www.globalresearch.ca/index.php?context¼va&aid¼412252), ultimately decline in significant food production. The expected climate changes in the upcoming ages may exacerbate the negative effects of drought stress in economically important crops as well as in food crops. Presently in globe, the study of drought has been one of the core directions in biological breeding and plant biology (Hong-Bo et al. 2006).

Currently, drought episodes have resulted in blanket fire in various regions of the world including Central and North America, Northern China, Russia, India, Africa, Central Australia, Canadian prairies and England producing contagious diseases, millions of human death and famines. According to WHO, water deficit is the death cause for about half the people who exterminated by natural catastrophe. Drought season can be forecasted; however irregular precipitation is modulated by changes in climate such as by the rise in global temperature as well as the EI Nino Southern Oscillation and imbalance in the heat cycle. All these variations are directly related to human interventions (Xoconostle-Cázares et al. 2010).

For sustainable agriculture, the need for new technologies or alternatives (accelerating the natural varieties selection and genes insertion from other plant species or varieties) will provide a real-world solution to alleviate the drought problems such as drought tolerant plants (Xoconostle-Cázares et al. 2010). The goals of this chapter are to describe the current progress and research in crop improvement on drought tolerance in maize and rice and to review the recent knowledge of physiological processes and key traits involved in reproductive stage, and growth regulation processes under drought stress, regarding integrated mechanisms for drought tolerance improvement in maize and rice.

9.1.1 Drought

Drought can be deliberated as climate's pressures set. It is a physical-chemical complex process, almost connected with all biological aspects which include: DNA, microRNA, RNA, proteins, lipids, carbohydrates, and mineral elements also related to signal transduction, growth, sell-circle, development and so on (Hong-Bo et al. 2006). In Agriculture, "Drought is due to a water shortage in the root zone; ultimately declined in crop yield". Tolerance comprises of dehydration tolerance or drought avoidance that is measured ultimately by reproductive success of crops. Drought tolerance is "The potential of crop to display, flower and grow economic yield under limited supply of water" (Farooq et al. 2009).

Drought involves several changes namely transient increases in levels of ABA, reduced growth, accumulation of protective enzymes and compatible solutes, increased antioxidant levels, and transcriptional inactivation/activation of particular genes. Drought reduces the productivity of plants by hindering photosynthesis and growth. Water supply is triggered by metabolic and stomatal effects. Water deficiency produces stomatal cessation and thus decreases intercellular concentration of CO_2, whereas mesophyll cells dehydration damages/impairs the photosynthetic equipment's. Cell growth and photosynthesis are amongst the crucial processes to

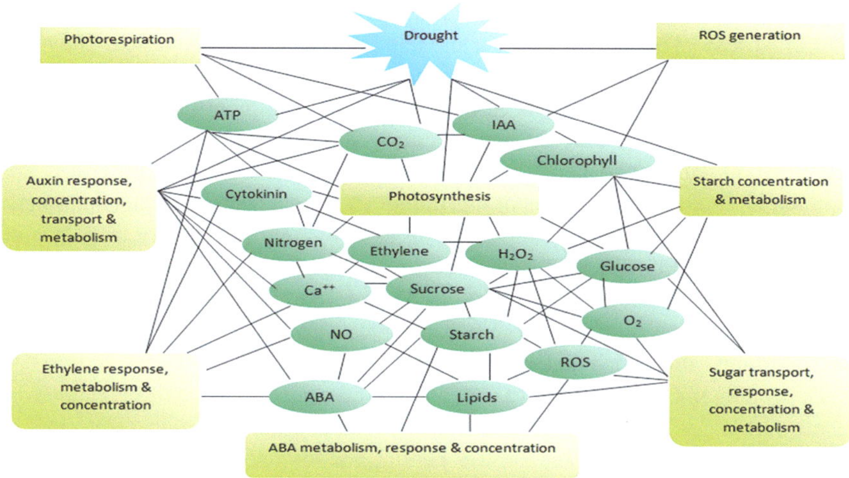

Fig. 9.1 Biotic linkage generated for photosynthesis and drought interactions

be influenced by drought (Chaves and Oliveira 2004; Chaves et al. 2009). Under drought conditions water deficiency in tissues of plant may lead to stomatal closing resulting in lower intake of CO_2 and finally photosynthesis will be affected negatively (Lawlor and Cornic 2002; Lawlor and Tezara 2009; Chaves et al. 2009). Under mild drought stress conditions, stomatal closure is the key factor, limiting activity of photosynthesis and under severe drought conditions; metabolic damages takes place (Waseem et al. 2011). On photosynthesis, drought deleterious effects will be mediated or facilitated by the responsiveness of:

(i) Protein synthesis and gene expression
(ii) Stress metabolites accumulation (Waseem et al. 2011)
(iii) ATP synthesis, respiration system and electron transport in mitochondria (Atkin and Macherel 2009).

The biological linkage generated for photosynthesis and drought interactions is depicted in (Fig. 9.1).

In recent years the extensive progress has occurred in revealing the nature of several factors affecting photosynthesis subjected to drought stress in plants. However, when use publically accessible data to found which events are controlled by photosynthesis, the deficiency of stress characterization is revealed instantaneously, impairing the probability to integrate and compare data (Waseem et al. 2011).

9.1.2 Importance of Rice and Maize as Cereal Food Crops

As a cereal grain, rice (*Oryza sativa*) is the most extensively consumed staple food especially in West Indies and Asia for a large human population in global world. It is the second highest grain in production worldwide after maize. It is the predominant source of energy of the world, provides 20 % energy supply for 8 countries in Africa, 17 in Pacific and Asia and 9 in South and North America. The nutrient contents of rice food are water, energy, protein, fat, carbohydrates, fiber, sugar, calcium, iron, folate, vitamin B6, A, E, K, beta-carotene, lutein + zeazanthin, pantothenic acid, riboflavin, thiamin, magnesium, selenium, copper, manganese, zinc, sodium, potassium, phosphorus and fatty acids (saturated, polyunsaturated and monounsaturated). With regards of nutritional value, rice is not a complete protein. It does not contain sufficient amount of all essential amino acids for good health. Fish and animal products are useful to complete the amino acid profile as they provide essential amino acids in large amount and micronutrients pulses such as lentils, groundnuts and beans. Leafy green vegetables and fruits also supply essential micronutrients and enrich dietary diversity.

Depending upon availability of water, rice crop can be grown in diverse environments. Generally, rice crop can survive flooding; it does not bloom in a waterlogged area, however it can grow and survive herein. For food security and rural population, rice is a major mainstay. For the nutrition, rice is vital of the population in Latin America as well as in Asia and the Africa and in Caribbean. Developing countries account for 95 % of the total rice production, with Asian countries account for 92 %; India and China only accountable for nearly half of the world rice production output. Today, the vast bulk of rice emanates from countries including China, India, Indonesia, Bangladesh, Vietnam, Thailand, Myanmar, Pakistan, Philippines and Japan as illustrated in (Fig. 9.2).

Rice is produced in geographic regions in a diverse range of climate. The world top 20 rice producers in 2010 as depicted in (Table 9.1) serves a broad range of demands including basic food for gross proportion of poor consumers and farmers of the world. The three largest rice producers were China (197 Mt), India (131 Mt) and Indonesia (64 Mt) in 2009. Focusing to the future, the rice demand and production is projected to boost from 200 Mt in1960 to 678 Mt in 2009 as population of the world is envisaged to boost steadily to around nine billion in 2050 compared to

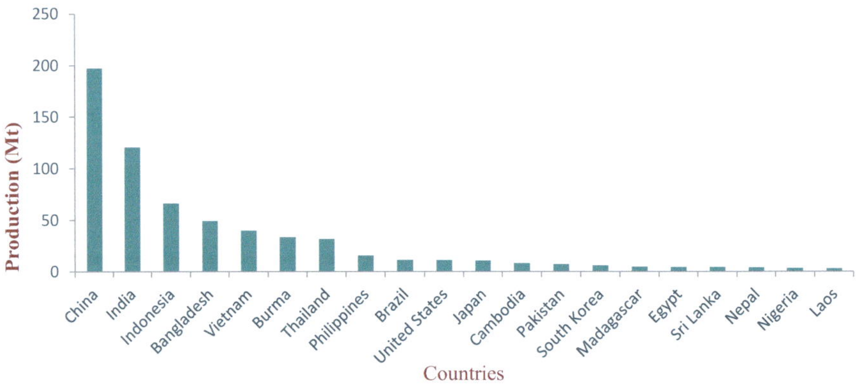

Fig. 9.2 Production of rice worldwide

Table 9.1 Rice production by top 20 producers in 2010

Countries	Production (Mt)
China	197.2
India	120.6
Indonesia	66.4
Bangladesh	49.3
Vietnam	39.9
Burma	33.2
Thailand	31.5
Philippines	15.7
Brazil	11.3
United States	11
Japan	10.6
Cambodia	8.2
Pakistan	7.2
South Korea	6.1
Madagascar	4.7
Egypt	4.3
Sri Lanka	4.3
Nepal	4
Nigeria	3.2
Laos	3

Source: FAOSTAT 2013

the past (FAO 2006, 2007). Maize (*Zea mays*) is one of the top three cereal crop that originated in Central America, grown in the world along with wheat and rice. More than 7000 years ago from wild maize, it was domesticated in Central America and Mexico. Maize is the most important staple food and cereal crop in Latin America and Sub-Saharan Africa (SSA) for almost 1.2 billion people. In the early sixteenth centuries, explorers and traders introduced corn in other countries because of its ability to cultivate in miscellaneous agroecological environments. It is used for human consumption, important in dishes of Central America, ground into meal or flour and eaten as a snack and popped. Corn oil is used in industries as well as in cooking, obtained from grain. Corn both in the form of ethanol and corn oil, has becoming an important biofuel. The corn demand and production as biofuel is projected to boost by 42 % over the past decades worldwide. The 40 % of the world maize produces by the United States; others topmost ten countries producing maize includes China, Brazil, Mexico, Indonesia, India, France, Argentina, South Africa and Ukraine as depicted in (Fig. 9.3). Worldwide production of maize as depicted in (Table 9.2) in 2009 was 817 Mt, more than rice (678 Mt). Maize constitutes a staple food in several regions of the ecosphere. It is major source of corn oil (cooking oil) and maize flour (cornstarch). It is consumed as vegetable, rich in essential amino acids, carbohydrates, vitamins A, E, C, minerals and proteins (9 %), also rich in calories and dietary fiber. The nutrient contents of maize food are water, energy, protein, fat, carbohydrates, fiber, sugar, calcium, iron, folate, vitamin B6, A, E, K,

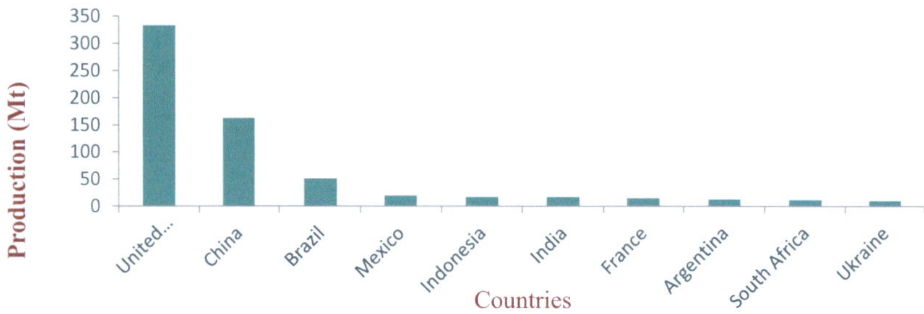

Fig. 9.3 Production of maize worldwide

Table 9.2 Maize production by top 10 producers in 2009

Countries	Maize production (Mt)
United States	333
China	163
Brazil	51
Mexico	20
Indonesia	17
India	17
France	15
Argentina	13
South Africa	12
Ukraine	10
World	817

(Source: FAO 2006)

beta-carotene, lutein + zeazanthin, pantothenic acid, riboflavin, thiamin, magnesium, selenium, copper, manganese, zinc, sodium, potassium, phosphorus and fatty acids (saturated, polyunsaturated and monounsaturated). Maize starch can be enzymatically treated and hydrolyzed to produce syrups. It is sometimes used for beer as the starch source. In the Canada and the United States, it is mostly grown for livestock to feed, as silage, grain or forage.

9.2 Institutes Working on Drought Tolerance

9.2.1 Institutes Working on Maize Drought Tolerance

In the developing world, The International Maize and Wheat Improvement Center (CIMMYT) located in Mexico acts as leader and catalyst in a wheat and maize worldwide innovative network that deals and helps the poor (Ortiz et al. 2008). Since 1970, the researchers have developed hybrids show both higher stable productivity and drought resistance depending on the seasonal and site conditions

(Banziger et al. 2006). These are employed in 13 countries of Africa which includes: Benin, Angola, Kenya, Nigeria, Ghana, Zimbabwe, Zambia, Uganda, Tanzania, Mali, Malawi, Ethiopia and Mozambique in the project frame "Drought Tolerant Maize for Africa" led by the International Institute for Tropical Agriculture (IITA) and CIMMYT (Rovere et al. 2009). A number of QTL's for drought response has been identified in maize (Ribaut and Ragot 2007; Sonev et al. 2009; Chen et al. 2007); associated to yield, flowering and plant height. Based on the selective markers expression, the generation of transgenic maize will impact positively the maize production market. The maize trancriptomic analysis to obtain cisgenic lines has identified candidate genes under drought stress (Zheng et al. 2010). The bacterial gene's overexpression encoding choline dehydrogenase provided higher resistance to renovated plants when linked with maize of wild type (Quan et al. 2004).

Beyond CIMMYT's focus on value added germplasm and higher grain yield, it plays an "integrative" role in crops management research, lower production costs, promoting the proficient water usage and other inputs, well management of biotic stresses and improved resilience and diversity of system (Ortiz et al. 2008). Furthermore, the CIMMYT needs to confirm that its products reach end-users and their livelihoods improve. CIMMYT in this esteem is the main public, transnational source of wheat seed embedded technology to alleviate poverty, reduce vulnerability and serving breeder/agriculturalist move from subsistence to income-generating production systems (Ortiz et al. 2008).

9.2.2 *Institutes Working on Rice Drought Tolerance*

As millinery crop, rice importance in feeding a great proportion of global population; this represents a landmark for breeding. The International Rice Research Institute (IRRI) using different approaches is focused on this task, developed rice hybrids under drought stress with high yield by conventional crosses. Markers for grain productivity as well as QTLs associating root length and drought resistance have been identified. New varieties obtained with stacked properties: higher productivity, earlier flowering and drought resistance (Bernier et al. 2009). High tolerant variety of rice in upland India generated based on QTLs, showing grain quality superior and also higher productivity. The genetic engineering has permitted the genes overexpression obsessed by drought-induced promoters for trehalose accumulation, also providing tolerance to salinity and cold (Wu and Garg 2003).

9.3 CIAT Strategies Towards Improvement of Crops for Integrating Genomics Approach

The International Center for Tropical Agriculture (CIAT) combines physiology, plant breeding and genomic approaches for crop improvement in upland rice, common bean, tropical grasses and cassava to exploit and understand underlying genetic mechanism of adaptation of abiotic stress. The completion of rice genome sequence in 2002 (Goff et al. 2002) has caused much exhilaration amongst plant scientists. The collecting information explores new standards, allow scientists to address practical and fundamental questions/problems in a multidisciplinary manner. CIAT aimed at improving varieties of upland rice, cassava and beans, has developed a biotechnology team (http://www.ciat.cgiar.org/). The team presently comprises specialists in genetics, cellular biology, breeding, entomology, molecular biology, statistics, pathology, and plant physiology. The technology in center is equipped including facilities of cDNA microarray for large-scale analyses of gene expression, facilities of molecular marker for genotyping and for genetic conservation; tissue culture-cryo conservation (Ishitani et al. 2004).

The CIAT program still includes conventional breeding approaches for crops genetic improvement, including germplasm screening for new traits, producing new crosses in new genotypes to recombine variation sources. In these activities, the comparative advantages include:

- Operational cost is relatively low
- Linkages to broad collaborators network in advanced institutes as well as developing countries
- Diverse germplasm holdings in large (Ishitani et al. 2004).

For applying genomics, these comparative advantages of CIAT program are invaluable to crop improvement. The CIAT strategy by using gene pools as breeding tool resources combines both (i) phenotype to gene (top-down) and (ii) gene to phenotype (bottom-up) approaches (Ishitani et al. 2004).

(i) The top-down strategy uses in multiple environments by beginning with characteristic analysis for studying abiotic stress tolerance (Ishitani et al. 2004) as depicted in (Fig. 9.4).

This involves agronomic traits and for stress physiology, analyzing crop phenotypes in different locations (Ishitani et al. 2004).

(ii) The bottom-up (gene to phenotype) strategy is also an important component of CIAT's (Ishitani et al. 2004).

To recognize candidate genes which are accountable for specific traits, a technique or tool is required for selection of candidate genes from gene's large pool available from resources of genetics. For example, the genome of rice plant contains 28,000 genes (Kikuchi et al. 2003). Starting with this large pool of gene, genetic data accessible for crops will be filtered by physiological analysis, biochemical screening, and phenotypic screening (Ishitani et al. 2004) as depicted in (Fig. 9.5). Two major gaps remain to crop improvement in successfully applying genomic approaches. The first gap concerns in the field, understanding the crops phenotypic traits and through genomics enhancing that knowledge. The second lies in mechanism to attain improved crop phenotypes by applying genomic approaches. Furthermore, challenge is to combine effectively different genomics information,

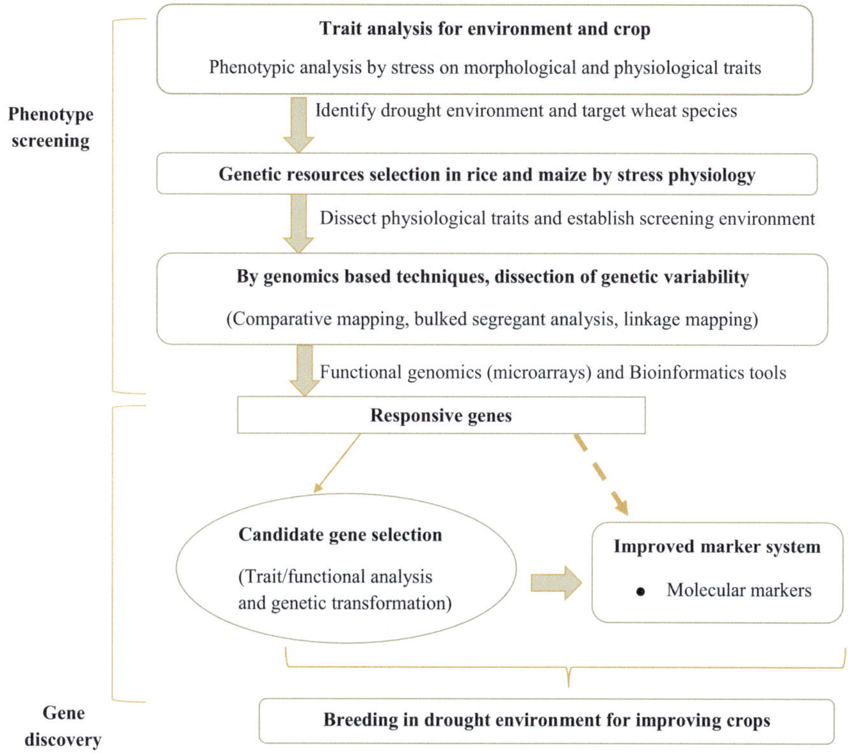

Fig. 9.4 Crop improvement strategy program of CIAT

integrating that information to maximize the efforts of crop improvement (Ishitani et al. 2004).

9.4 Drought Effects on Rice and Maize Crop Plants Yield

Scarcity of water resources was the catalyst of great food shortage of the historical time. It is the critical single threat to food security of the world. The drought severity is unpredictable because it depends on many factors: evaporative demands, distribution and occurrence of rainfall and dampness storing ability or aptitude of soils. Even though crop responses to drought are well-known relatively, crop performance is fragmentary where various stresses co-occur under multifaceted environment. That's why the crops have to respond instantaneously to numerous stresses (excessive heat, drought and light) which in the field may coincide (Zhou et al. 2007).

In plants many physiological processes of yield-determining respond to water deficit. Yield integrates in a complex way of these physiological processes. Thus, it is difficult over the whole life cycle of crops to elucidate how plants combine,

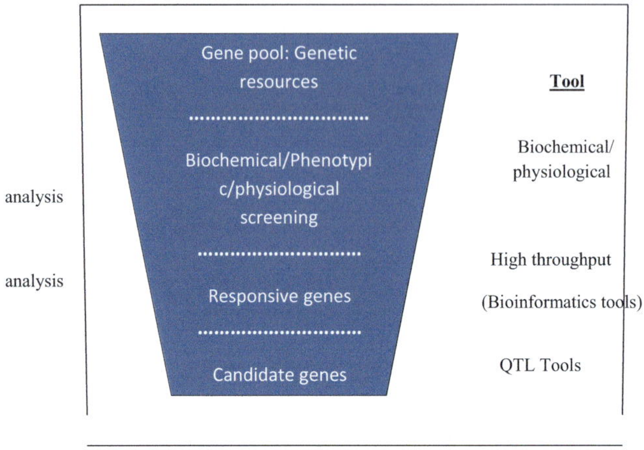

Fig. 9.5 Gene's filtration for traits

display and accumulate the ever-changing and indefinite physiological processes. For water severity, stress, timing and duration of stress and plant responses after removal of stress and interaction among other factors are highly important (Plaut 2003). The drought stress effects are obvious at all phenological phases of plant growth range from morphological to molecular levels at whatever phase the water deficiency takes place. The various drought stress effects: crop growth and yield, nutrient relations, water relations, photosynthesis, assimilate partitioning, respiration and oxidative damage. In crops species such as maize and rice, drought-induced reduction in yield has been reported as depicted in (Table 9.3). It depends on the duration and severity of stress period (Farooq et al. 2009). Reduction in the grain growth rate in the water-stressed wheat resulted from decline activity of sucrose synthase, although growth cessation caused from adenosine diphosphate-glucose-pyrophosphorylase inactivation (Ahmadi and Baker 2001).

Under the altering environments, it is imperative to improve, advance and progress the drought resistance of crops. However, development of drought tolerant plants in meeting the future food demands; might be a promising approach. Among other things to enhance the development of drought tolerance crops, requires the physiological mechanisms knowledge and genetic control of traits that contributing in different developmental stages of plants (Farooq et al. 2009).

9.5 How Rice and Maize Can Be Adapted to Drought Stress?

Plants adapt to, respond and survive under drought stress by initiation of several physiological, morphological and biochemical responses. Drought stress disturbs the plants water reactions at organ, cellular and tissue level, causing adaptation reactions, damage, specific and unspecific reactions. To survive with drought conditions, resistant plants induce defense mechanisms (Farooq et al. 2009).

Table 9.3 Yield reduction by drought in rice and maize crops

Species	Growth stages	Reduction in yield	References
Maize	Reproductive	32–92 %	Atteya (2003)
Maize	Vegetative	25–60 %	Atteya (2003)
Rice	Grain filling (in case of mild stress)	30–55 %	Basnayake et al. (2006)
Rice	Grain filling (in case of severe stress)	60 %	Basnayake et al. (2006)
Maize	Reproductive	70–47 %	Farooq et al. (2009)
Maize	Reproductive	63–87 %	Kamara et al. (2003)
Rice	Reproductive (in case of mild stress)	53–92 %	Lafitte et al. (2007)
Rice	Reproductive (in case of severe stress)	48–94 %	Lafitte et al. (2007)
Maize	Grain filling	79–81 %	Monneveux et al. (2006)
Rice	Reproductive	24–84 %	Venuprasad et al. (2007)

A slow step in enlightening the drought resistance mechanism in the improvement of drought resistance of rice and maize plants has vulnerable both use of conventional breeding & selection approaches and use of modern biotechnological, genomics and genetic strategies. Mechanism of drought resistance through approaches of conventional breeding, molecular, genetics, biotechnological, genomics, transcription factors, bioinformatics, role of mineral nutrients, and root studies are presented.

9.5.1 Conventional Breeding

The major goal of plant breeding is to increase yield. On a universal scale, crop breeding has been immensely successful in increasing yield; such as after the green revolution, the introduction of dwarf rice varieties and hybrid maize development. However for irrigated agriculture, the green revolution was mainly driven. This has generally resulted in insignificant breeding resources for enhanced productivity in saline or water deficit ecosystems (Peleg et al. 2012).

For drought-prone environments (DPE), conventional breeding has been complemented by adopting exotic germplasm (to amplify crop gene pool) and physiological mechanisms that include harvest index (HI), water uptake (WU) and water-use efficiency (WUE) as yield drivers (Reynolds and Tuberosa 2008). WUE under stress is considered an important component of drought tolerance of crop and yield determinant. It has been indicate that rainfed plant production can be enhanced per unit used of water, out coming in "more crop per drop". As long as photosynthesis biochemistry cannot be better genetically, WUE and Transpiration efficiency (TE) are driven by traits of plant that minimize crop water-use and transpiration which are extremely significant for plant production. As production of biomass is linked to transpiration tightly, breeding for transpiration for capture maximal soil moisture under drought stress is the most important target for yield enhancement/improvement. Effective use of water (EUW) for transpiration implies capture maximized soil moisture which involves minimal loss of water by soil evaporation and reduced non-stomatal transpiration. High harvest index (HI) in terms of assimilates partitioning and reproductive functions towards reproduction expresses successful yield and plant production. By improving status of plant water, EUW helps sustain reproductive success and assimilate partitions, therefore, EUW is a major target in water-limited conditions for yield improvement (Blum 2009).

Although to rice production, drought has been identified as major constraint in the rainfed ecosystem. Considering the extremely adaptable nature of rainfed ecosystem, breeding requires the genotype's development that meets the farmer's preference; possess resistance of widespread biotic stresses and combine ability of high-yielding with better levels of drought tolerance. This aim can be achieved with product-oriented, long-term, and large scale breeding program intended for rainfed environment (Todaka et al. 2015).

9.5.2 The Era of '-OMICS'/Introducing New Technologies

Traditional breeding has major restraints, including the need for various backcrosses to eliminate detrimental traits. Therefore, presently the focus is on quantitative trait loci (QTL), marker-assisted breeding (MAS), genetics and genomics approaches, biotechnology, and omics era which permit 'pyramiding' of necessary traits for fast improvement in crop with little input/response of resources (Kantar et al. 2011). For plant adaptations to abiotic stress, newly developed approaches will help or facilitate the cloning, use of QTL and mapping related to stress conditions:

- Single feature polymorphism, new molecular platform, for example diversity array (DArT), array based technology, and single nucleotide polymorphism are becoming progressively accessible
- Development of mapping tools, such as advance mapping software, association mapping and consensus maps
- To test environmental effects, development of statistical models and high throughput advance phenotyping
- Tools, such as microRNAs, RNA interference and TILLING for candidate gene's functional analysis and growing availability of sequence information, such as "deep" mRNA and DNA sequencing
- Proficient transformation techniques will permit stress related major QTLs to be deliberated for positional cloning, objective to more directly manipulate the target trait by genetic engineering (Kantar et al. 2011).

9.5.3 Drought Tolerance Through Genomics Approach

The new '-omics' (genomics, proteomics, sequencing and bioinformatics) platform have added new dimensions as illustrated in (Table 9.4) for deciphering and manipulating the genetics basis of tolerance to drought.

Table 9.4 Genomic approaches related to drought

Crop	Approach	Main characteristics	References
Maize	Transcriptome analysis	Under drought, transcriptional profiles of tissues of placenta-pedicel and endosperm in developing kernels	Hajheidari et al. (2005)
Rice	Transcriptome analysis	Microarray expression based study of almost 21,000 genes for osmotic adjustment in phenotypically differing accession	Hazen et al. (2005)
Rice	Transcriptome analysis	Putative 589 drought responsive genes were confined on the physical map and discussed their correspondence with Quantitative trait loci	Gorantala et al. (2005)
Maize	Proteome analysis	Basal portion analysis of growing leaves	Riccardi et al. (2004)

The genomics-based approaches provide route to desirable agronomic traits that effects such responses at quantitative trait loci (QTLs), thus enabling us to improve crops yield and drought tolerance under water deficit conditions more effectively as compared to conventional approaches. Marker-assisted selection (MAS) already improves drought related traits and is helping breeders. Analysis of gene products and sequence data should facilitate the cloning and identification of genes at target QTLs. The genomics-based approaches contribute novel information under water-limited conditions to identify and analyze candidate genes and elucidate their regulation and function. Further information can be obtained on the candidate genes role at target loci and ascertaining/determining their effect on the phenotype through EcoTILLING, a platform for classifying single nucleotide polymorphism (SNP) haplotypes (Tuberosa and Salvi 2006). To enhance drought tolerance, the successful exploitation of genomics will only be possible within an interdisciplinary, coherent context able to provide a complete understanding of limiting factors of crop yield in drought-prone environment (Tuberosa and Salvi 2006).

'Genetical Genomics' approach has been developed currently, based on expression profiling of gene in a segregating population and fingerprinting of related individuals based on marker to analyze trans- and cis-acting factors, and to delineate a genetic network that related to a trait (Jansen and Nap 2001; Jansen 2003). Scientists have modified and adopted the Jansen and Nap concept of genetical genomics to identify trait related sets of pathways and genes controlling/monitoring storage events in evolving or developing seeds. Now extending this (genetical and genomic approach) to decipher or interpret molecular regulatory linkages underlying both yield and tolerance, using introgression lines as a substitute of a segregating population (Sreenivasulu et al. 2007).

Introgression lines offer definite advantages for crop improvement by track or path characteristics for both tolerance and yield. To detect expression QTLs (eQTLs), extensive expression profiling is ongoing from introgression lines. Such information can be used to advance and develop direct transgenic approaches and

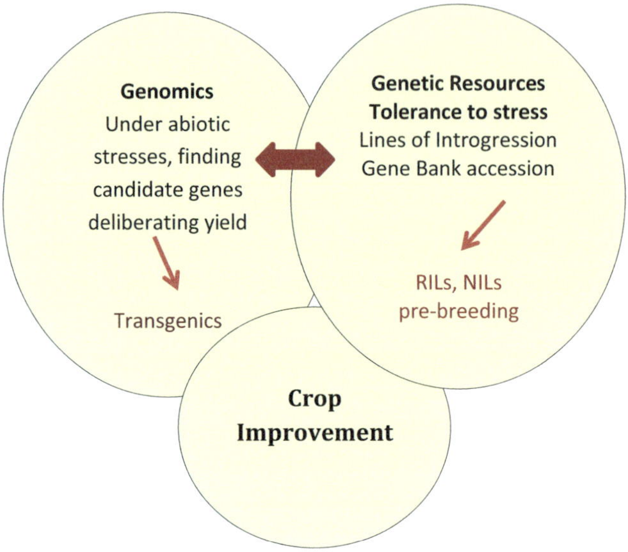

Fig. 9.6 Gap bridging for abiotic stress resistance between genomics and breeding approaches

molecular markers related to trait. These both are main component of molecular breeding approaches (Sreenivasulu et al. 2007) as depicted in (Fig. 9.6).

Although the approach ('genetical genomics') is still in its early stages, struggles are underway in plant species in this direction. In addition, the integrating information obtained from genomic genetic-based breeding approach to divulge developmental programs that will enhance the grain quality, accelerate yield stability under abiotic stresses conditions (Sreenivasulu et al. 2007).

9.5.4 Drought Tolerance Through Applied Biotechnology

Recent efforts by genetic transformation to increase stress tolerance in plants have resulted in various significant achievements. Nevertheless, the genetically complicated systems/mechanism of drought, salinity, cold, heat stress tolerance extremely makes the task challenging. Therefore, applied biotechnology should be integrated fully with breeding and classical physiology (Vinocur 2005).

Because of the multigenic nature, the enhancement by classical breeding of plant abiotic stress tolerance is fraught with complications. Further difficulties ascend from the large inconsistency in sensitivity of stress during life the cycle at different periods of a plant. Of the several types of crop response to drought stress, avoidance mechanisms result at whole-plant level from physiological and morphological changes while resistance mechanism are triggered or activated by molecular biochemical and cellular modifications that advance themselves to manipulation at

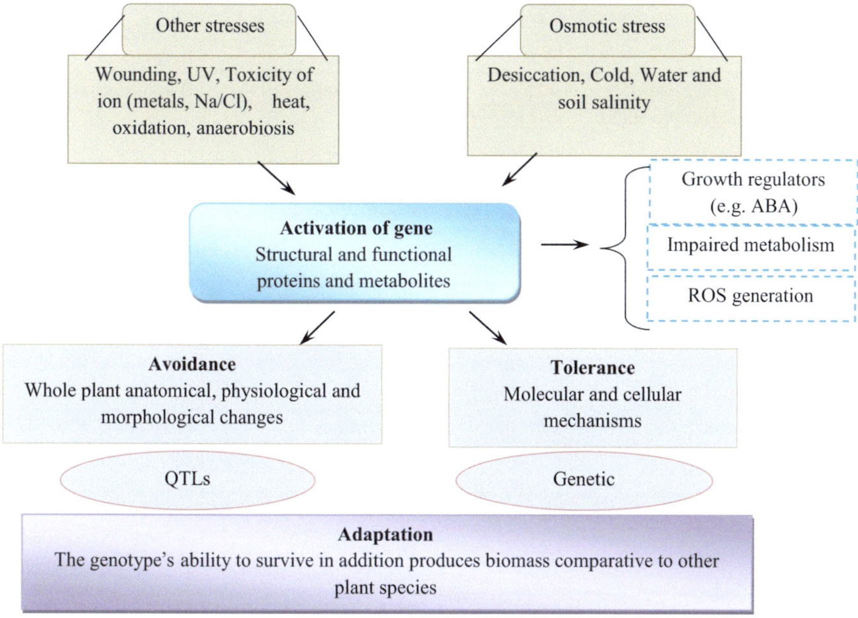

Fig. 9.7 Applied biotechnology: the interacting factors for crop tolerance in conventional and molecular breeding

biotechnological level (Vinocur 2005). The interacting factors as illustrated in (Fig. 9.7) are presented.

The QTL mapping application is one way to dissecting the complex question of crop stress resistance. QTL approach when fully developed will be of great importance to breeding for plant stress resistance. QTLs linked with abiotic stress resistance in many plant species have been identified for example drought stress in cotton and salt stress in rice (Vinocur 2005).

9.5.5 *Drought Tolerance Through Transgenic Technology*

Genetic improvement or enrichment through transgenic approaches in rice plant complements traditional breeding when the preferred or chosen gene is unavailable in gene pool, thus demanding regulatory element's modification. Transgenic approach allows gene's functional validation studies accountable for molecular mechanisms. Two major genes groups have been employed generally to improve stress tolerance as depicted in (Table 9.5) by transfer of gene (DNA segment):

(i) DNA segment encoding for regulatory proteins of signal transduction and transcription includes: detoxification enzymes, osmoprotectants, proteases, chaperones and water channel proteins (Bhatnagar-Mathur et al. 2008).

Table 9.5 Mechanism, gene and compound for drought resistance in rice crop

Putative mechanism	Gene	Gene product	References
Chaperones (protective proteins of macromolecules)	*HVA1*	LEA protein	Battaglia et al. (2008)
	OsLEA3-1		
	PMA		
Water channel transporters	*RWC3*	Aquaporin	Afzal et al. (2016)
Regulatory proteins (signaling factors or mechanism)	*MAPK*	Protein kinases	Todaka et al. (2015)
	CDPK		
Regulatory proteins	*NAC*	Transcription factors	Todaka et al. (2015)
	WRKY		
	HD-zip		
	DREB/CBF		
Osmoprotectants	*adc*	Polyamines	Slama et al. (2015)
	P5CS	Proline	
	TPSP; TPP and TPS	Trehalose	
	BADH	Glycinebetaine	
	cox		
Detoxification enzymes	*MnSOD*	ROS (Scavenging protein)	Martinez et al. (2016)
Regulatory proteins (signaling factors or mechanism)	*LOS5*	ABA (biosynthesis key enzyme)	Verslues (2016)
	NCED		

(ii) DNA segment encoding for functional proteins with structural or enzymatic function includes: protein kinases, ABA biosynthesis, transcription factors and phospholipid metabolism enzymes (Bhatnagar-Mathur et al. 2008).

LEA proteins are major proteins group act as chaperone (protective protein of macromolecules). Typically, accumulate during the low temperature, in dehydration response, salinity and dawn phases of embryogenesis. Transgenic lines of rice with improved stress resistance expressing this (LEA) protein. Plants accumulate ROS as an outcome of dehydration stress, which impair cellular structures. Leaves are equipped with metabolites and antioxidant enzymes under ideal/optimal conditions to cope with ROS. The enzymes accumulation such as catalase, superoxide dismutases, glutathione-S-transferases (GST) and ascorbate peroxidases has been reported in abiotic stress conditions. Aquaporin gene (water channel transporter) has been observed also to improve drought tolerance in rice plant (Todaka et al. 2015).

9.5.6 Drought Tolerance Through Transcription Factors

To enhance rice stress tolerance by gene transfer, one of the promising approach through the use of DNA segments encoding protein factors that are intricate in signal transduction and expression of gene). Since, when combine with a suitable promoter, they can regulate many downstream genes that are involved in stress tolerance (Todaka et al. 2015).

Transcription factors CBF/DRE have been reported in rice to be beneficial in enhancing stress tolerance in transgenic plants through inducing or manipulating the stress-relation expression of target genes. In response to stress, CRT/DRE and ABRE are cis-acting elements where CBF/DRE transcription factors (TF) bind. Furthermore, they are involved in gene expression ABA-dependent and gene expression ABA-independent respectively. Under field conditions, SNAC1, stress-responsive NAC protein or TF was characterized in rice plant. SNAC1 showed 22–34 % higher seed at reproductive stages under severe drought conditions in the field. This gene was induced in rice guard cells specifically under drought conditions. In rice, gene's overexpression resulted in drought tolerance significantly and better stomatal closure in the drought-stressed conditions though the transgenic plant's yield and photosynthesis rate under normal conditions of growth were not affected (Serraj et al. 2009).

Members of various different TFs classes have been associated in stress responses, including MYB, bZIP, MYC, AP2, zinc-finger proteins and homeodomain-leucine zipper (HD-ZIP). HD-ZIP encodes proteins that have been reported in plants as well as assumed/said to regulate responses and development to environmental clues. Genes involved in signaling pathways of ABA have been shown valuable for enhancing drought tolerance in rice. LOS5 and NCED both regulate genes related to stress in transgenic plants; their overexpression led to improved drought tolerance in transgenic plants. In response to high salinity and primarily to drought stress, ABA is de novo synthesized. More extra regulatory factors, such as MAPK and CDPK also involved in biosynthesis of ABA and were identified to improved drought tolerance in rice (Serraj et al. 2009). Generally, transgenic approach allows the gene function's validation and recent progress illustrates that transgenic technology can be complementary for conventional or to other breeding strategies if the phenotypic assessment is conducted properly (Serraj et al. 2009).

9.5.7 Drought Tolerance Through Signaling Pathways

Adaptations of plants to environmental abiotic stresses are controlled by molecular network cascades as depicted in (Fig. 9.8), trigger stress responsive mechanism to protect membrane and damaged proteins, repair, and re-establish homeostasis. For abiotic stress tolerance, engineering strategies of plant rely on genes that translate proteins conferring tolerance to stress or enzymes in pathways prominent to the synthesis of structural and functional metabolites or genes expression that is involved in regulatory and signaling pathways.

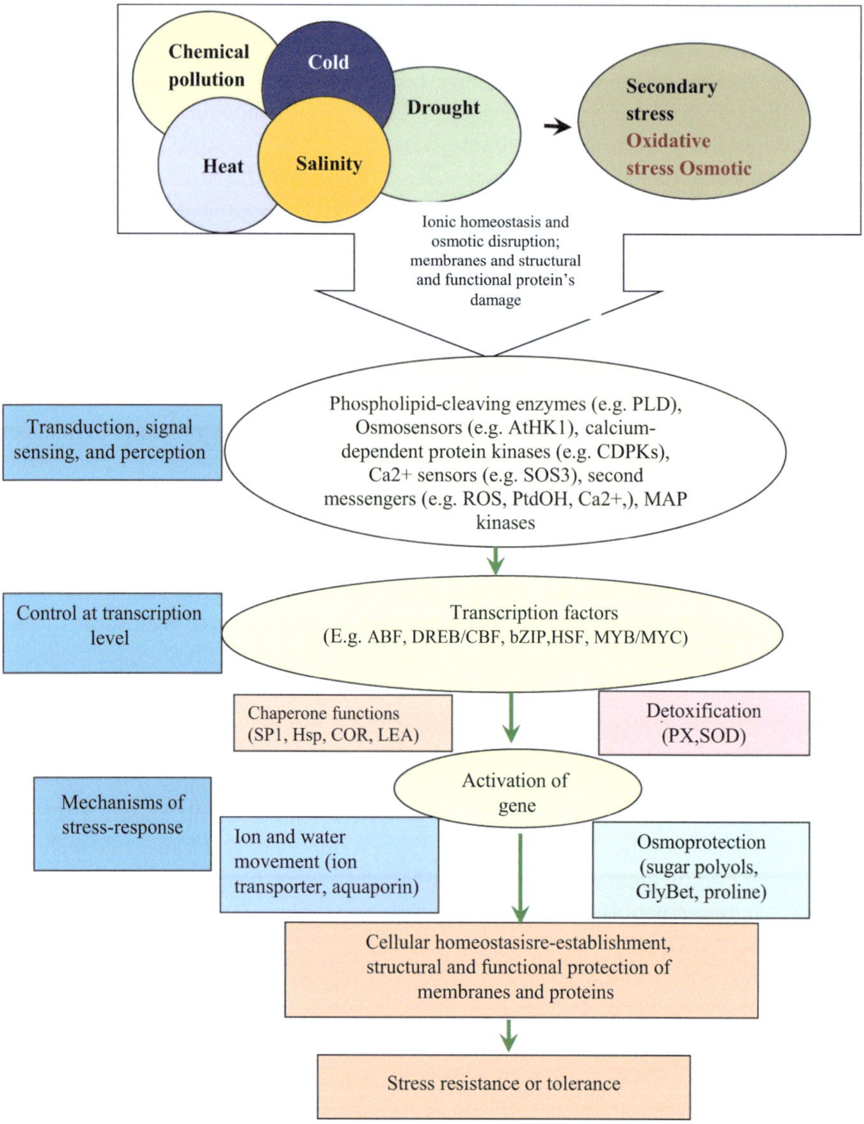

Fig. 9.8 The intricacy of plant response to environmental abiotic stress

9.5.8 Drought Tolerance Through Bioinformatics and Functional Analysis of Gene

Bioinformatics development and exploration offer a rich combination of tools, protocols, computing infrastructure and databases that can be used to answer and help biological research queries, and often a significant savings in laboratory resources and time. Bioinformatics integrate and incorporate information of crop data across diverse collection about phenotype, genotype, germplasm, growth characteristics, cellular expression, environmental conditions, and applied treatments (Serraj et al. 2009).

The Generation Challenge Program (GCP) http://www.generationcp.org, are directed toward improvement in crop through comparative biology and genomics across species. The research theme of GCP is also the phenotypic as well as molecular genetic resource's characterization to discover relevant or valued alleles for crop enhancement. One major development and research subprogram of the GCP concentrates on crop informatics. This program is endeavoring to advance public standards globally for crop management information as well as inclusive plant scientific domain model http://pantheon.generationcp.org and a tool platform for analyzing and accessing information obtainable from internationally networked GCP partner's databases including other data sources (Serraj et al. 2009).

The Comparative Plant Stress-Responsive Gene Catalog http://dayhoff.generationcp.org is about drought research, developed to facilitate the knowledge's integration across crops about drought-responsive genes. This catalog is a compendium of multiple sequence alignments, associated experimental evidence, phylogenetic trees and protein families. The principal objective is to elucidate paralogous and orthologous relationship between genes of plant that involved in response to abiotic stress mainly drought. The International Crop Information System (ICIS) is computerized database suite and system of tools http://www.icis.cgiar.org for characterization data for crops, evaluation of nomenclature, genealogy use, and integrated management generally. In addition to ICIS, many excellent publically/online accessible crop and plant databases are available as illustrated in (Table 9.6). The founding of the International Rice Functional Genomics Consortium (IRFGC) http://www.iris.irri.org/IRFGC goals to organize research in 'functional genomics' era by building common strategies and explore ways to merge resources of international rice functional genomics. The consortium is struggling for rice gene characterization in areas of expression arrays, bioinformatics, genomic stock, and proteomics (Serraj et al. 2009).

9.6 Role of Inorganic Nutrients, Organic Osmolytes and Plant Growth Regulators in Drought Tolerance

Cellular osmotic adjustment (OA) is most common response in water stress conditions which facilitate plants to thrive under drought conditions (Blum 2009). By aggregation of several inorganic and organic solutes in cells, osmotic adjustment takes place. Further, osmotic adjustment enabling plants to absorb water in adequate amount from its exterior medium to tolerate working of normal metabolic processes and therefore growth (Chimenti et al. 2006). Instantaneously, plants produce antioxidants variety that counteract the ROS generation in response to water deficit conditions (Munne-Bosch and Penuelas 2003). These consist of nonenzymatic antioxidants namely: carotenoids, phenolics, glutathione, ascorbic acid, and tocopherols as well

Table 9.6 Publicly accessible web resources related to plants

URL	Database	Crop
http://rapdb.lab.nig.ac.jp	RAPDB	Oryza sativa
http://www.tigr.org/plantProjects.shtml	TIGR plant genomes	Oryza sativa
http://www.maizegdb.org	MaizeGDB	Maize
http://www.barleybase.org	Barleybase	Barley
http://www.grin.usda.gov	GRIN	Plant genetic resources
http://www.singer.cgiar.org	SINGER	Plant genetic resources
http://www.mips.gsf.de	MATDB	Arabidopsis thaliana
http://www.nasc.org	NASC	Arabidopsis thaliana
http://www.arabidopsis.org	TAIR	Arabidopsis thaliana

as enzymatic antioxidants namely: glutathione cycle enzymes/ascorbate, and superoxide dismutase catalase (Alscher et al. 2002; Jaleel et al. 2008). Another important adaptation of plant under drought stress is water-use efficiency (WUE), to develop plant with enhanced drought stress tolerance (Sequera-Mutiozabal et al. 2016; Ray et al. 2004). Moreover, the evapotranspiration control to counter excessive water loss, deficit irrigation (DI) exploitation strategy to increase utilization of water, use of synthetic and natural conditioners to retain content of soil moisture, effective use of water (EUW) for yield enhancement/improvement, genetic improvement of water stress resistance in established crops, and growing drought-resistant crop species (Levi et al. 2009). Minerals nutrients, organic osmolytes and plant growth regulators (PGRs) as depicted in (Table 9.7) also play key roles in modulating growth of plant and development under on-stress and stress conditions (Ashraf et al. 2011).

The organic solutes referred to as compatible solutes or compatible osmolytes contribute to osmoregulation as well as protect the membrane structures and biomolecules also protect DNA from ROS damaging effects. It is considered that glycine betaine (GB) is an important osmoprotectant against drought. The structure of GB inside plant is highly stable. GB can penetrate easily through leaf epidermis and progress/modify to other organs to enhanced water stress tolerance effectively. Proline amino acid is another strong osmoprotectant. Under stress conditions, it can stabilize proteins and membrane structure; regulate/control cytoplasmic PH, and ROS scavenger. In summary, the role of organic osmolytes (glycerol, trehalose, sorbitol, mannitol, proline, and GB) in drought stress tolerance is osmoregulation maintenance in plants.

Additionally, organic osmolytes play key roles in cellular functions such as proteins stabilization, ROS scavenging, and protection of structure of membrane. Plant growth regulators (PGRs) have been considered to play active roles in metabolic processes, plant adaptation, plant development and growth to stressful and nonstress environments including water stress conditions (Huang et al. 2008). PGRs including gibberellins, auxins, abscisic acid (ABA), ethylene, and cytokinins are involved in water regulation movement at shoot and root levels by altering the cell membranes permeability and finally cell turgor.

Table 9.7 Role of inorganic nutrients, organic osmolytes and plant growth regulators to enhance droughttolerance

Inorganic nutrients	Organic osmolytes	Plant growth regulators
Nitrogen	Glycerol	Brassinolide (BL), Ascorbic acid, Salicylic acid, Jasmonic acid, Ethrel, Benzylaminopurine, Polyamines, Ethylene, Abscisic acid, Gibberellic acid (GA_3), Auxin, Indole-3-Acetic acid (IAA)
Manganese	Sorbitol	
Magnesium	Mannitol	
Zinc	Trehalose	
Potassium	Proline	
Phosphorus	Glycinebetaine	
Calcium		

The availability of important inorganic mineral nutrients including N, P, K, Ca, Mg, Zn, and Mn is perturbed by water stress in the soil, leading to imbalances or nutritional deficiencies in plants; may occur by poor nutrient mobility or poor root growth in the soil. However, in plants, availability of impaired nutrition under water stress may be triggered by several factors: reduced transpirational stream and interference in unloading mechanisms and uptake of nutrients. Under drought stress or in drought suffering areas, inclusive knowledge of the organic nutrients and their roles will help improve or advance management of fertilization in plant growth. Further, as significant is a good understanding of drought-stress effects on nutrient absorption, availability, accumulation, partitioning, and transportation in plants. Additional, potential interactions to drought stress between plant response and nutrient application (Ashraf et al. 2011).

9.7 The Root

In Asia, drought is affecting 20 % of the entire rice-growing regions (Pandey and Bhandari 2008). Roots are the major organ of plant for uptake of water and nutrients. To cope with scarcity stress, plants use different mechanisms: drought recovery, drought avoidance, drought escape, and drought tolerance. Among these four, the roots are associated with mechanism of drought avoidance. Genotypes having higher root to shoot ratio, deep roots with a high capacity of penetration and branching, cuticular resistance high, early stomatal closure and elasticity in leaf rolling are described as component qualities of drought avoidance (Samson et al. 2002; Wang and Yamauchi 2006). Achieving drought tolerance in crops for meeting the growing challenges in water shortage of the world will be necessary.

A set of root traits/parameters are of considerable functional significance that includes: hard penetrability, maximum root depth, root to shoot ratio, root anatomy, hydraulic conductance of root, maximum depth of root, root volume, root branching, root diameter and elongation rate of root (Wu and Cheng 2014) as depicted in (Table 9.8). Direct role of coarse and deep roots in drought tolerance has been hypothesized because larger roots diameter are related to branching and penetration ability and they have lower axial resistance and greater radii of xylem vessel to water flux.

Table 9.8 Root qualities and their functional features

Root parameters	Functional features
Hardpan penetrability	Penetration ability to subsurface hardpans
Maximum root depth	Potential for assimilation of soil moisture
	Nutrients absorption in deep layer of soil
Root branching	Soil exploration power
Root number	Potential to architecture of root system
	Physical strength
Root volume	The strength to filter a large soil volume
Root diameter	The Potential for
	Branching
	Hydraulic conductivity
	Penetration ability
Root to shoot ratio	Assimilate/incorporate allocation
Deep root to shoot ratio	Root growth vertically
	Potential for soil moisture absorption & nutrients/minerals in deeper layers of soil
Total surface area/root length	Size of total system of root
Specific root length	Branching degree
	Root materials density
	Porosity due to development of aerenchyma
Weight density/root length	Nutrients and water uptake rate

The "Composite Transport Model" for transport in roots and uptake of water in which transcellular, symplastic and apoplastic pathways contribute to transport and uptake of water. Exchange between paths helps possibly the roots to adjust their ability of water uptake according to leaves transpiration demand. The adaptation of rice plant to grow in flooded conditions, its roots exhibit unique apoplastic barriers information related with other plants but effect of apoplastic barriers in rice on transport and uptake of water are unclear since rice genotypes can be adapted to non-flooded or flooded conditions. The scientists reported that under drought conditions, maize reduced the water potential of soil to lower leaves than rice. Rice having a smaller length of root than maize had a lower ability to uptake per unit root length water (Wu and Cheng 2014).

For root architecture, rice plant includes great genetic diversity. Rice has shallow growth of root and for drought improvement should emphasize on coarse, deep root growth. For deep root growth, many genotypes of rice have the potential but it is controlled strongly by the environment (i.e. drought stress severity and hardpans presence) (Wu and Cheng 2014).

Furthermore, in all conditions the fine roots presents a large percentage of entire length of root. Thus, it is strongly expected to greatly contribute to take up of water by the total root system. Finally, the inconsistencies between rice roots function and spatial distribution are poorly understood under drought stress; need to be talked with more emphasis on the function of root biology under water deficit conditions (Wu and Cheng 2014).

Simulation models, historical climate series and system analysis can also play key roles in enhancing and characterizing the integration and precision of phenotyping either by approaches of heuristically guiding assimilated phenotypes or directly linking/associating model coefficients as well as in the target environment, access plant response to the drought patterns and to evaluate the significance of candidate physiological traits (Serraj et al. 2009).

From an evolutionary sight, all plant species comeback or responses to stress and all resistance mechanism are genotype-specific and programmed. Still, adaptation to abiotic stress has environmental and ecological advantages. Therefore, efficient and well-organized plant breeding can be achieved by linking conventional and molecular breeding for abiotic stress resistance (Vinocur 2005).

9.8 Conclusions and Future Perspective

There are some of the questions and issues that plant scientists and research managers must now confront to choose an ideal portfolio of strategic rice and maize genetic improvement research for the upcoming years. Till the dramatic demand's expansion for maize biofuels and rice as well as climate-induced glitches in the past ages, the prediction/vision for a reversal of the sturdy/steady decrease of the cereals real prices including rice and maize seemed poor. There are various predicted factors and trends on which decision base. The continuously increasing global population demand more energy, and food to supply an ever growing world need for animal products; reducing supplies of water for agriculture and the climate change effects are mounting the levels of abiotic environmental stress mainly drought across major rice and maize-growing areas.

The biotechnology and bioinformatics applications for enhance use of rice and maize genetic resources in the crop improvement is likely to offer new prospects to enhance yield. As presented in this chapter, ICIS, CIMMYT, IRRI, CIAT, IRFGC and global research partners are conserving the rice and maize genetic endowment and enhancing stability and plants yield across the areas/cropping system where rice and maize thrives. Generally, it is accepted that the drought syndrome complexity can be tackled with holistic strategy integrating crop breeding with tolerance trait's physiological dissection and tools of molecular genetics together with agronomical practices that lead to increase conservation and matching crop genotypes and soil moisture use with the environment.

There is a look of optimism that the pledge ushered by the current technologies by genomics-assisted breeding and functional genomics of stress resistance could generate relevant or valued information for use in maintainable and sustainable agriculture for engineering stress-resistant crops.

References

Afzal, Z., T. Howton, Y. Sun, and M.S. Mukhtar. 2016. The roles of aquaporins in plant stress responses. *Journal of Developmental Biology* 4: 9.

Ahmadi, A., and D.A. Baker. 2001. The effect of water stress on the activities of key regulatory enzymes of the sucrose to starch pathway in wheat. *Plant Growth Regulation* 35: 81–91.

Alscher, R.G., N. Erturk, and L.S. HEATH. 2002. Role of superoxide dismutase (SODs) in controlling oxidative stress in plants. *Journal of Experimental Botany* 53: 133–141.

Ashraf, M., N.A. Akram, F. Al-Qurainy, and M.R. Foolad. 2011. Drought tolerance: Roles of organic osmolytes growth regulators, and mineral nutrients. *Advances in Agronomy* 111: 249–296.

Atkin, O.K., and D. Macherel. 2009. The crucial role of plant mitochondria in orchestrating drought tolerance. *Annals of Botany* 103: 581–597.

Atteya, A.M. 2003. Alteration of water relations and yield of corn genotypes in response to drought stress, Bulg. *Journal of Plant Physiology* 29: 63–76.

Banziger, M., P. Setimela, D. Hodson, and V. Biniganavile. 2006. Breeding for improved abiotic stress tolerance in maize adapted to Southern Africa. *Agriculture Water Management* 80: 212–224.

Basnayake, J., S. Fukai, and M. Ouk. 2006. Contribution of potential yield, drought tolerance and escape to adaptation of 15 rice varieties in rainfed lowlands in Cambodia. In *Proceedings of the Australian Agronomy conference*, Australian Society of Agronomy, Birsbane, Australia.

Battaglia, M., Y. Olvera-Carrillo, A. Garciarrubio, F. Campos, and A.A. Covarrubias. 2008. The enigmatic LEA proteins and other hydrophilins. *Plant Physiology* 148: 6–24.

Bernier, J., A. Kumar, R. Venuprased, S. Impa, and R.P. Gowdaa. 2009. The large effect drought-resistance QTL qtl12.1 increases water uptake in upland rice. *Field Crops Research* 110: 39–46.

Bhatnagar-Mathur, P., V. Vadez, and K.K. Sharma. 2008. Transgenic approaches for abiotic stress tolerance in plants: Retrospect and prospects. *Plant Cell Reports* 27: 411–424.

Blum, A. 2009. Effective use of water (EUW) and water-use efficiency (WUE) is the target of crop yield improvement under drought stress. *Field Crops Research* 112: 119–123.

Chaves, M.M., and M.M. Oliveira. 2004. Mechanisms underlying plant resilience to water deficits: Prospects for water-saving agriculture. *Journal of Experimental Botany* 55: 2365–2384.

Chaves, M.M., J. Flexas, and C. Pinheiro. 2009. Photosynthesis under drought and salt stress: Regulation mechanisms from whole plant to cell. *Annals of Botany* 103: 551–560.

Chen, M., Q.Y. Wang, X.G. Cheng, Z.S. XU, and L.C. Li. 2007. GmDREB2, a soybean DRE-binding transcription factor, conferred drought and high-salt tolerance in transgenic plants. *Biochemistry Biophysics Research Communication* 353: 299–305.

Chimenti, C.A., M. Marcantonio, and A.J. Hall. 2006. Divergent selection for osmotic adjustment results in improved drought tolerance in maize (Zea mays L.) in both early growth and flowering phases. *Field Crops Research* 95: 305–315.

FAO. 2006. World agriculture: Towards 2030/2050, Interim report. Global Perspective Studies Unit, Food and Agriculture Organization, Rome

FAO. 2007. Commission on genetic resources for food and agriculture. Food and Agriculture Organization, Rome. http://www.fao.org/ag/cgrfa/itpgr.htm. (January 2007)

FAOSTAT, Data. 2013. *Food and agriculture organization of the United Nations. Statistical database*.

Farooq, M., A. Wahid, N. Kobayashi, D. Fujita, and S.M.A. Basra. 2009. Plant drought stress: Effects, mechanisms and management. *Agronomy Sustainable Development* 29: 185–212.

Goff, S.A., D. Ricke, T.H. Lan, G. Presting, R. Wang, M. Dunn, J. Glazebrook, A. Sessions, P. Oeller, H. Varma, D. Hadley, D. Hutchison, C. Martin, F. Katagiri, B.M. Lange, T. Moughamer, Y. Xia, P. Budworth, J. Zhong, T. Miguel, U. Paszkowski, S. Zhang, M. Colbert, W.L. Sun, L. Chen, B. Cooper, S. Park, T.C. Wood, L. Mao, P. Quail, R. Wing, R. Dean, Y. Yu, A. Zharkikh, R. Shen, S. Sahasrabudhe, A. Thomas, R. Cannings, A. Gutin, D. Pruss, J. Reid, S. Tavtigian, J. Mitchell, G. Eldredge, T. Scholl, R.M. Miller, S. Bhatnagar, N. Adey, T. Rubano, N. Tusneem,

R. Robinson, J. Feldhaus, T. Macalma, A. Oliphant, and S. Briggs. 2002. A draft sequence of the rice genome (Oryza sativa L. ssp. japonica). *Science* 296: 92–100.
Gorantla, M., P.R. Babu, V.B.R. Lachagari, F.A. Feltus, A.H. Paterson, and A.R. Reddy. 2005. Functional genomics of drought stress response in rice: Transcript mapping of annotated unigenes of an indica rice (Oryza sativa L. cv. Nagina 22). *Current Science* 89: 496–514.
Hajheidari, M., M. Abdollahian-Noghabi, H. Askari, M. Heidari, S.Y. Sadeghian, E.S. Ober, and G.H. Salekdeh. 2005. Proteome analysis of sugar beet leaves under drought stress. *Proteomics* 5: 950–960.
Hazen, S., M.S. Pathan, A. Sanchez, I. Baxter, M. Dunn, B. Estes, H.-S. Chang, T. Zhu, J. Kreps, and H. Nguyen. 2005. Expression profiling of rice segregating for drought tolerance QTLs using a rice genome array. *Functional & Integrative Genomics* 5: 104–116.
Hong-Bo, S., C. Xiao-Yan, C. Li-Ye, Z. Xi-Ning, W. Gangh, Y. Yong-Bing, Z. Chang-Xing, and H. Zan-Min. 2006. Investigation on the relationship of proline with wheat anti-drought under soil water deficits. *Colloids and Surfaces. B, Biointerfaces* 53: 113–119.
Huang, D., W. Wu, S.R. Abrams, and A.J. Cutler. 2008. The relationship of droughtrelated gene expression in Arabidopsis thaliana to hormonal and environmental factors. *Journal of Experimental Botany* 59: 2991–2997.
Ishitani, M., I. Rao, P. Wenzl, S. Beebe, and J. Tohme. 2004. Integration of genomics approach with traditional breeding towards improving abiotic stress adaptation: Drought and aluminum toxicity as case studies. *Field Crops Research* 90: 35–45.
Jaleel, A.C., B. Sankar, P.V. Murali, M. Gomathinayagam, G.M.A. Lakshmanan, and R. Panneerselvam. 2008. Water deficit stress effects on reactive oxygen metabolism in Catharanthus roseus; impacts on ajmalicine accumulation. *Colloids Surface* 62: 105–111.
Jansen, R.C. 2003. Studying complex biological systems using multifactorial perturbation. *Nature Revolution in Genetics* 4: 145–151.
Jansen, R.C., and J.P. Nap. 2001. Genetical genomics: The added value from segregation. *Trends in Genetics* 17: 88–391.
Kamara, A.Y., A. Menkir, B. Badu–Apraku, and O. Ibikunle. 2003. The influence of drought stress on growth, yield and yield components of selected maize genotypes. *Journal of Agricultural Science* 141: 43–50.
Kantar, M., Stuart J. Lucas, and H. Budak. 2011. Drought stress: Molecular genetics and genomics. *Approaches Advances in Botanical Research* 57: 445–493.
Kikuchi, S., K. Satoh, T. Nagata, N. Kawagashira, K. Doi, N. Kishimoto, J. Yazaki, M. Ishikawa, H. Yamada, H. Ooka, I. Hotta, K. Kojima, T. Namiki, E. Ohneda, W. Yahagi, K. Suzuki, C.J. Li, K. Ohtsuki, and T. Shishiki. 2003. Collection, mapping, and annotation of over 28,000 cDNA clones from japonica rice. *Science* 301: 376–379.
Lafitte, H.R., G. Yongsheng, S. Yan, and Z.K. Li. 2007. Whole plant responses, key processes, and adaptation to drought stress: The case of rice. *Journal of Experimental Botany* 58: 169–175.
Lawlor, D.W., and G. Cornic. 2002. Photosynthetic carbon assimilation and associated metabolism in relation to water deficits in higher plants. *Plant, Cell and Environment* 25: 275–294.
Lawlor, D.W., and W. Tezara. 2009. Causes of decreased photosynthetic rate and metabolic capacity in water-deficient leaf cells: A critical evaluation of mechanisms and integration of processes. *Annals of Botany* 103: 561–579.
Levi, A., L. Ovnat, A.H. Paterson, and Y. Saranga. 2009. Photosynthesis of cotton nearisogenic lines introgressed with QTLs for productivity and drought related traits. *Plant Science* 177: 88–96.
Martinez, V., T.C. Mestre, F. Rubio, A. Girones-Vilaplana, D.A. Moreno, R. Mittler, and R.M. Rivero. 2016. Accumulation of flavonols over hydroxycinnamic acids favors oxidative damage protection under abiotic stress. *Frontiers in Plant Science* 7: 838.
Monneveux, P., C. Sanchez, D. Beck, and G.O. Edmeades. 2006. Drought tolerance improvement in tropical maize source populations: Evidence of progress. *Crop Science* 46: 180–191.
Munne-Bosch, S., and J. Penuelas. 2003. Photo and antioxidative protection, and a role for salicylic acid during drought and recovery in field-grown Phillyrea angustifolia plants. *Planta* 217: 758–766.
Ortiz, R., H. Braun, J. Crossa, J.H. Crouch, G. Davenport, J. Dixon, S. Dreisigacker, E. Duveiller, H.E. Zhonghu, H. Julio, K.J. Arun, K. Masahiro, P. Kosina, Y. Manes, M. Mezzalama, A. Morgounov, J. Murakami, J. Nicol, G. Ortiz Ferrara, I. Ortiz-Monasterio, T.S. Payne,

R. Javier Pen, M.P. Reynolds, K.D. Sayre, R.C. Sharma, R.P. Singh, J. Wang, M. Warburton, H. Wu, and I. Masa. 2008. Wheat genetic resources enhancement by the international maize and wheat improvement center (CIMMYT). *Genetic Resources in Crop Evolution* 55: 1095–1140.

Pandey, S., and H. Bhandari. 2008. Drought: Economic costs and research implications. In *Drought frontiers in rice: Crop improvement for increased rainfed production,* ed. Serraj, R., Bennett, J., Hardy, B. World Scientific Publishing and Los Ba˜nos (Philippines): International Rice Research Institute, Singapore.

Peleg, Z.V.I., H. Walia, and E. Blumwald. 2012. Integrating genomics and genetics to accelerate development of drought and salinity tolerant crops. *Theoretical Applied Genetics* 125(4): 625–645.

Plaut, Z. 2003. *Plant exposure to water stress during specific growth stages, Encyclopedia of water science*, 673–675. Encyclopedia of Water Science: Taylor & Francis.

Quan, R., M. Shang, H. Zhang, Y. Zhao, and J. Zhang. 2004. Engineering of enhanced glycinebetaine synthesis improves drought tolerance in maize. *Plant Biotechnology Journal* 2: 477–486.

Ray, I.M., A. Segovia-Lerma, and L.W. Murray. 2004. Diallel analysis of carbon isotope discrimination and its association with forage yield among nine historically recognized alfalfa germplasm. *Crop Science* 44: 1970–1975.

Reynolds, M., and R. Tuberosa. 2008. Translational research impacting on crop productivity in drought-prone environments. *Current Opinion in Plant Biology* 11: 171–179.

Ribaut, J.M., and M. Ragot. 2007. Marker-assisted selection to improve drought adaptation in maize: The backcross approach, perspectives, limitations and alternatives. *Journal of Experimental Botany* 58: 351–360.

Riccardi, F., P. Gazeau, M.P. Jacquemot, D. Vincent, and M. Zivy. 2004. Deciphering genetic variations of proteome responses to water deficit in maize leaves. *Plant Physiology Biochemistry* 42: 1003–1011.

Rovere, L., M. Sudarshan, D. John, A.M. Pedro, G. Kamala, P.H. Dave, and F.V. Dagoberto. 2009. *Economic and livelihood impacts of maize research in hill regions in Mexico and Nepal: Including a method for collecting and analyzing spatial data using Google Earth.* http://ideas.repec.org/p/ags/cimmis/56090.html.

Samson, B.K., H. Hasan, and L.J. Wade. 2002. Penetration of hardpans by rice lines in the rainfed lowlands. *Field Crops Research* 76: 175–188.

Serraj, R., A. Kumar, K.L. McNally, I. Slamet-Loedin, R. Bruskiewich, R. Mauleon, J. Cairns, and R.J. Hijmans. 2009. Chapter 2 - Improvement of drought resistance in rice. In *Advances in agronomy*, ed. L.S. Donald, 41–99. Burlington: Academic Press.

Sequera-Mutiozabal, M., A.F. Tiburcio, and R. Alcázar. 2016. Drought stress tolerance in relation to polyamine metabolism in plants. In *Drought stress tolerance in plants*, Physiology and Biochemistry, vol. 1, ed. A.M. Hossain, H.S. Wani, S. Bhattacharjee, J.D. Burritt, and P.L.-S. Tran, 267–286. Cham: Springer International Publishing.

Slama, I., C. Abdelly, A. Bouchereau, T. Flowers, and A. Savouré. 2015. Diversity, distribution and roles of osmoprotective compounds accumulated in halophytes under abiotic stress. *Annals of Botany* 115(3): 433–447. mcu239.

Sonev, S., E. Todorovska, V. Avramova, S. Kolev, N. Abu-Mhadi, and N.K. Christov. 2009. Genomics assisted improvement of drought tolerance in maize: QTL approaches. *Biotechnology* 23: 1410–1413.

Sreenivasulu, N., S.K. Sopory, and P.B. Kavi Kishor. 2007. Deciphering the regulatory mechanisms of abiotic stress tolerance in plants by genomic approaches. *Gene* 388: 1–13c.

Todaka, D., K. Shinozaki, and K. Yamaguchi-Shinozaki. 2015. Recent advances in the dissection of drought-stress regulatory networks and strategies for development of drought-tolerant transgenic rice plants. *Frontiers in Plant Science* 6: 84.

Tuberosa, R., and S. Salvi. 2006. Genomics-based approaches to improve drought tolerance of crops. *Trends in Plant Science* 8: 405–412.

Venuprasad, R., H.R. Lafitte, and G.N. Atlin. 2007. Response to direct selection for grain yield under drought stress in rice. *Crop Science* 47: 285–293.

Verslues, P.E. 2016. ABA and cytokinins: challenge and opportunity for plant stress research. *Plant Molecular Biology:* 1–12.
Vinocur, B.A. 2005. Altman Recent advances in engineering plant tolerance to abiotic stress: Achievements and limitations. *Current Opinion in Biotechnology* 16: 123–132.
Wang, H., and A. Yamauchi. 2006. Growth and function of roots under abiotic stress in soil. In *Plant-environment interactions*, 3rd ed, ed. B. Huang. New York: CRC Press.
Waseem, M., A. Ali, M. Tahir, M.A. Nadeem, M. Ayub, A. Tanveer, R. Ahmad, and M. Hussain. 2011. Mechanism of drought tolerance in plant and its management through different methods continental. *Journal of Agricultural Science* 5(1): 10–25.
Wu, W., and S. Cheng. 2014. Root genetic research, an opportunity and challenge to rice improvement. *Field Crops Research* 165: 111–124.
Wu, R., and A. Garg. 2003. Engineering rice plants with trehalose-producing genes improves drought resistance under field conditions. *Theoretical Applied Genetics* 115: 35–46.
Xoconostle-Cázares, B., F.A. Ramirez-Ortega, L. Flores-Elenes, and R. Ruiz-Medrano. 2010. Drought tolerance in crop plants. *American Journal of Plant Physiology* 5: 1–16.
Zheng, J., J. Fu, M. Gou, J. Huai, and Y. Liu. 2010. Genome-wide trancriptome analysis of two maize inbred lines under drought stress. *Plant Molecular Biology* 72: 407–421.
Zhou, Y., H.M. Lam, and J. Zhang. 2007. Inhibition of photosynthesis and energy dissipation induced by water and high light stresses in rice. *Journal of Experimental Botany* 58: 1207–1217.

Chapter 10
Wheat Physiological Response Under Drought

Raseela Ashraf, Fayyaz-ul-Hassan, Mukhtar Ahmed, and Ghulam Shabbir

Contents

R. Ashraf (✉) • Fayyaz-ul-Hassan
Department of Agronomy, Pir Mehr Ali Shah Arid Agriculture University, Rawalpindi, Pakistan
e-mail: raseelaashraf@gmail.com; drsahi63@gmail.com

M. Ahmed
Department of Agronomy, Pir Mehr Ali Shah Arid Agriculture University, Rawalpindi, Pakistan

Department of Biological Systems Engineering, Washington State University, Pullman, WA, USA
e-mail: ahmadmukhtar@uaar.edu.pk; mukhtar.ahmed@wsu.edu

G. Shabbir
Department of Plant Breeding and Genitics, Pir Mehr Ali Shah Arid Agriculture University, Rawalpindi, Pakistan
e-mail: drgshabbir@uaar.edu.pk

M. Ahmed, C.O. Stockle (eds.), *Quantification of Climate Variability, Adaptation and Mitigation for Agricultural Sustainability*,
DOI 10.1007/978-3-319-32059-5_10

Abstract Drought affects wheat crop adversely resulting in reduced growth and productivity. The physio-biochemical adaptations of wheat were studied using ten wheat genotypes namely V-4178, PDW-34, Chakwal-97, Shorawaki, 4098805, Baluchistan, Yecora-70, S-24, SARC-1 and Pasban-90 were sown in pots by using completely randomized design two factorial with three replications. Wheat genotypes were raised at two soil moisture levels i.e. soil moisture level maintained at 100 % field capacity and soil moisture level maintained at 40 % of field capacity. The effects of the two soil moisture levels were studied on wheat genotypes by considering the parameters photosynthetic rate (An), transpiration rate (E), stomatal conductance (g_s), stomatal resistance (r_s), leaf membrane stability index (LMSI), leaf succulence, relative water content (RWC), epicuticular wax, proline contents, chlorophyll content (SPAD), grain yield, biological yield and harvest index. Drought was considered responsible for the enhanced production of proline and epicuticular wax, reduced stomatal conductance, high stomatal resistance, low photosynthetic and transpiration rate in genotypes as a mechanism to bear the harsh conditions. Low harvest index, biological yield and grain yield were also recorded as a result of drought. Hence, it is concluded that the genotypes resistant to drought performed better under drought conditions due to better physio-biochemical adaptations.

Keywords Wheat • Leaf membrane stability index • Relative water content • Epicuticular wax • Proline • Photosynthetic rate • Transpiration rate • Stomatal conductance • Stomatal resistance and drought

10.1 Introduction

Agriculture the climate sensitive sector has a dominant role in the world and Pakistan's economy. Almost 45 % people of the country are directly or indirectly related to agriculture. It contributes 21.4 % share to the gross domestic product (GDP) (GOP 2013). Out of 19.8 million hectares of area under agriculture, rainfed area contributes about 5 million hectares mainly located in the Pothwar uplands, northern mountains and northeast plains of Punjab and accounts 24 % of the total area under cultivation. Out of which approximately 15 % land is disposed to drought mainly comprising rainfed areas (Mujtaba and Alam 2002). Rainfed agriculture contributes 10 % of the national wheat production and it constitutes 14 % of the total wheat production of Punjab province.

The genetic yield potential of the wheat cultivars prevailing in country is 6–8 t ha^{-1} while the national average yields are approximately 2.7 t ha^{-1} pointing towards a yield gap of almost 60 %. There are different factors contributing to this yield gap. The major causes of low output are late harvesting of Kharif crops causing delayed sowing of wheat, unavailability of developed inputs like seed, inefficient use of fertilizer, shortage of irrigation water, drought and terminal heat stress.

Zhu (2002) has defined drought as the lack of sufficient moisture obligatory for the normal growth and completion of life cycle of a plant. Drought is said to be a condition where a dry soil (due to absence of rain or late irrigation) or a hot dry wind (high *ETo*) causes a sizeable decline in crop performance with respect to plant persistence or economic yield or crop quality). Hassan (2006) reported that drought stress has severe effects on growth and development of crops. It not only limits the plant growth and development but also reduces productivity and results in yield losses when occurring at reproductive stages (Hays et al. 2007). In cereals like wheat, drought is an important abiotic stress that affects grain filling by reducing grain number and size. A plant suffers from water stress, when root to shoot water supply turns to be problematic and when there is a very high transpiration rate. These two situations often correspond where the climate is arid and semiarid.

The rate of photosynthesis is decreased by water deficit (Lawlor and Cornic 2002) and triggered leaf senescence (Martinez et al. 2003). Under water stress conditions, the photosynthetic rate of the ear parts was less affected than the flag leaf. This relative droughttolerance seemed to be related to the maintenance of higher relative moisture contents in lemmas, glumes and awns (Tambussi et al. 2005). Relative water content (RWC) of leaves has an important role and direct relation with soil water content, being important indicator of leaf water stress (Merah 2001). Relative water contents of leaves were lowered immediately in plants exposed to stress. Similarly, osmotic potential and leaf water potential were decreased (Grover et al. 2004). In general, leaf water potential decreased with stress intensity (Yurekli et al. 2001).

Photosynthesis is the fundamental process that influences crop productivity and it may decrease due to water stress (Chaves and Oliveira 2004). The stomatal check is considered to be the major cause of decreased photosynthesis under water deficit (Cornic 2000). Reduction in RWC has been identified to stimulate stomatal closure and caused a parallel reduction in photosynthetic rate (Cornic 2000). The production of various organic solutes is one of the general responses of plants towards abiotic stress (Serraj and Sinclair 2002) including small molecules like proline (Szabados and Savoure 2010). Those professed osmotic regulators or harmonious osmolytes provide protection to the plants from stresses by adjustments at cellular level through maintenance of membranes integrity and enzymes stability (Farooq et al. 2009). Various studies on transgenic plants established that proline has a composite influence on stress responses, proposing important role of proline in stress tolerance (Mattioli et al. 2008). Proline accumulation helps plant in maintenance of low water potential. This lowering water potential along with the accumulation of osmolytes permits surplus water uptake and thus buffer the instant effects of water deficit in the plant body (Kumar et al. 2003). In wheat, higher proline aggregates were found in stress enduring cultivars than in stress sensitive cultivars (Nayyar and Walia 2003). Proline can develop stress forbearance by different ways and it was observed by Sharma and Dubey (2005) that proline is capable to act as molecular chaperone keeping protein integrity and inhibiting protein accumulation and equilibrium and also by protecting the nitrate reductase in osmotic stress situations. It helps in stabilizing membranes and proteins (Ashraf and Foolad 2007). It also plays a role as antioxidant (Sharma and Dietz 2006) and controls the cytosolic acidity (Sivakumar et al. 2000). Osmoregulation is the fundamental mechanism through which plants confront the devastating effects of water stress by manufacturing compatible solutes, usually certain sugars, polyols, amino acids, betaines and associated compounds (Ramanjulu and Bartels 2002). Scientists reported amino acid proline is eminently present in plants and frequently accumulates in large amounts as a consequence of different environmental stresses (Kishore et al. 2005). Large amounts of proline facilitate a plant to retain low water potentials. Due to low water potentials, the increased amounts of suitable osmolytes which were involved in osmoregulation permit the uptake of additional water and buffer the instant effect of water deficit within the plant body. Besides, role of proline has been recognized to induce the manifestation of stress responsive genes, having proline receptive features in their promoters (Chinnusamy et al. 2005). Climate change resulting in rise in temperature and rainfall variability in different parts of world caused cereal grain yield reduction and increase in yield variability (Olesen et al. 2011). Global warming is likely to increase evapotranspiration (ET) and decrease crop physiological functions (Hassan 2006) and also results in water stress conditions for crop plants and these conditions have a direct effect on plant growth and productivity. The global water crisis, seriously influences crop productivity particularly in most of the Asian countries where irrigated agriculture accounts for 90 % of total diverted fresh water (Huaqi et al. 2002). Drought impacts include growth, yield, membrane integrity, pigment content, osmotic adjustment water relations, and photosynthetic activity (Praba et al. 2009). Drought stress is affected by climatic, edaphic and agronomic factors. The susceptibility of plants to drought stress varies in dependence of stress degree, different accompanying stress factors, plant species, and their developmental stages (Demirevska et al. 2009). Acclimation of plants to water deficit is the result of different events, which lead to adaptive changes in plant growth and physio-biochemical processes, such as changes in plant structure, growth rate, tissue osmotic potential and antioxidant defenses (Duan et al. 2007). Photosynthesis, which is the basic process influencing crop productivity, is hindered by water stress. The physiological processes including alleviation of photosynthetic efficiency, oxidative damage, uptake of water and nutrients by crop are severely affected under continuously changing temperature and moisture deficit (Wang et al. 2011). Chlorophyll is the most important pigments active in the photosynthetic process. In photosynthesis, antenna pigments in leaf chloroplasts absorb solar radiations, through resonance transfer the resulting excitation is channeled to the pigments of reaction center, which release electrons as a result the photochemical process set in motion. The chlorophyll is the most essential of these pigments, thus virtually necessary for the oxygenic conversion of light energy to the stored chemical energy. From a physiological perspective, leaf chlorophyll content is therefore a parameter of significant interest in its own right. The accessory pigment, carotenoids also have a very important role in photosynthesis. Biosynthesis of carotenoids in plants is a genetic characteristic; however, environmental conditions also play a significant role.

Gupta et al. (2001) studied physiological and yield attributes of two wheat genotypes with stress at boot and anthesis. They reported that number of grains, grain yield, biological yield and harvest index decreased to a greater extent when water stress was imposed at anthesis stage. The deficiency of water led to severe decline in yield probably by disrupting leaf gas exchange properties which not only limited the size of the source and sink tissues but the phloem loading, assimilate translocation and dry matter partioning were also impaired (Farooq et al. 2009).

Accumulation of proline under stress in many plant species has been correlated with stress tolerance, while concentration has been shown to be generally higher in stress-tolerant than in stress-sensitive plants. It influenced protein solvation and preserved the quarternary structure of complex proteins, maintained membrane integrity under dehydration stress and reduced oxidation of lipid membranes or photoinhibition (Demiral and Turkan 2004). Furthermore, it also contributed to stabilize sub-cellular structures, scavenging free radicals and buffering cellular redox potential under stress conditions (Ashraf and Foolad 2007).

Many abiotic stresses like water stress, salinity and cold direct to major transformations in carbohydrate metabolism (Kaur et al. 2000). Sugars have a major role in growth and development of plants facing abiotic stresses by amending the carbohydrate metabolic rate. A variety of stress responsive genes are stimulated by glucose (Price et al. 2004). Increased sugar levels in various parts of plants are a result of different ecological stresses (Prado et al. 2000; Gill et al. 2001). Metabolic profiling revealed that plants exposed to drought and heat stresses have accumulation of sucrose and several other sugars like maltose and glucose (Rizhsky et al. 2004). In view of the above scenario the research was conducted with the following objectives:

- To study the Physio-biochemical response of wheat genotypes to water stress (drought).
- To evaluate different wheat genotypes exposed to induced water stress

10.2 Methodology

The physio-biochemical adaptations of wheat in response to drought was studied through pot experiment conducted at PMAS Arid Agriculture University, Rawalpindi, Pakistan (latitude 33° 42/N and longitude 73° 10/E) from 5th November, 2011 to 3rd May, 2012. Seeds of ten wheat genotypes namely V-4178, PDW-34, Chakwal-97, Shorawaki, 4098805, Baluchistan, Yecora-70, S-24, SARC-1 and Pasban-90 were obtained from National Agricultural Research Center (NARC). The experiment was conducted in earthen pots (15 cm height and 9 cm diameter). These pots were filled with 8 Kg of soil taken from the field of Agronomy department. The inert matter was removed from soil by using 2 mm sieve.

Experiment was arranged in completely randomized design two factorial with three replications of each treatment. Ten seeds per genotype per pot were sown in total sixty pots. NPK was applied at the time of sowing at the rate of 100:50:0 kg ha^{-1}. Twenty treatments were applied by maintaining two moisture levels for each genotype. The two moisture levels were: M_1 = Soil maintained at field capacity during complete crop life cycle and M_2 = Soil maintained at 40 % of field capacity during complete crop life cycle. Pots were irrigated after adding soil and covered with polythene sheet to prevent increase in soil moisture by precipitation. Field capacity was measured by the standard procedure given in laboratory manual of ICARDA. Field capacity was maintained by using tensiometer.

10.2.1 Physiological Parameters

Infrared Gas Analyzer (IRGA) was used to measure photosynthetic rate, transpiration rate, stomatal conductance, stomatal resistance, at Anthesis stage (Long and Bernacchi 2003). Leaf membrane stability index (LMSI) was taken according to the method given by Chandrasekar et al. (2000). Leaf strips (0.2 g) of uniform size were taken in test tubes containing 10 ml of double distilled water in two sets. Test tube in one set were kept at 400 °C in water bath for 30 min and test tubes in second set were incubated at 100 °C in boiling water bath for 15 min. LMSI was calculated by using the formula;

$$LMSI = (1 - C1 / C2)$$

where, C1 = electrical conductivity of water containing the sample in test tube one and C2 = electrical conductivity was measured in test tube two. Leaves were taken randomly from each plant at anthesis stage. Fresh leaf weight was taken and leaf area was measured. Further leaves were oven dried at 60 °C for 2 days and dry weight was taken. Succulence was calculated by using following formula; Succulence = fresh weight – dry weight/leaf area. Fully expended flag leaves were taken and fresh weight (FW) was instantly noted, then leaves were saturated for 4 h in distilled water at room temperature, and turgid weight (TW) was recorded. After drying for 24 h at 80 °C total dry weight (DW) was recorded. Relative water content were measured according to the formula of Barrs and Weatherley (1962).

$$RWC(\%) = ((FW - DW) / (TW - DW)) \times 100$$

Leaves (0.5 g) were randomly taken from the plant and their area was measured. Three leaf samples were washed three times in 10 ml carbon tetrachloride for 30 s per wash. The extract was filtered, evaporated to dryness and the remaining wax was weighed. Wax contents were expressed on the basis of leaf area only, i.e. wax content mg cm^{-2} (Silva-Fernandes et al. 1964).

10.2.2 Biochemical Parameters

Fresh leaf tissue (0.5 g) was taken from plants at flag leaf stage from each pot, were homogenized in 10 ml of 3 % sulfosalicyclic acid and then filtered. Proline was estimated spectrophotometericaly following the ninhydrin method (Bates et al. 1973) using pure proline. Chlorophyll contents were measured by SPAD-chlorophyll meter by taking three readings from the flag leaf.

10.2.3 *Yield Parameters*

Biological and grain yield was recorded per pot and it was converted in kg ha^{-1}. Harvest index was calculated using the formula given by Donald (1962).

$$HI = (\text{Grain Yield} / \text{Biological Yield}) \times 100$$

10.3 Results and Discussion

10.3.1 *Physiological Parameters*

10.3.1.1 Photosynthetic Rate (μ mol m^{-2} s^{-1})

The two moisture levels (M_1 soil maintained at field capacity level and M_2 soil maintained at 40 % field capacity level) had a significant effect on photosynthetic rate with the higher value 23.64 μ mol m^{-2} s^{-1} for moisture level kept at field capacity against 18.35 μ mol m^{-2} s^{-1} at 40 % field capacity (Table 10.1). The comparison of genotypes for photosynthetic rate revealed that all the genotypes vary from one another in a significant manner. The maximum photosynthetic rate (24.63 μ mol m^{-2} s^{-1}) was exhibited by Pasban-90, which was at par with SARC-1 (24.04 μ mol m^{-2} s^{-1}) and Yecora-70 (22.84 μ mol m^{-2} s^{-1}). Contrarily the minimum photosynthetic rate (17.61 μ mol m^{-2} s^{-1}) was observed in V-4178 that did not differ significantly from PDW-34 (17.74) and Chakwal-97 (18.76).

The difference between the maximum and minimum value was 40 %. The interaction between genotypes and moisture levels was found to be non-significant

Table 10.1 Effect of two soil moisture levels on photosynthetic rate (μmol$m^{-2}s^{-1}$) of different wheat genotypes (Different letters indicate a statistical difference among the treatments at $P<0.05$)

Genotypes	M_1	M_2	Means
V-4178	19.56	15.65	17.61^{e}
PDW-34	19.70	15.77	17.74^{e}
Chakwal-97	20.83	16.69	18.76$^{d, e}$
Shorawaki	22.17	17.74	19.95^{d}
4098805	22.30	17.85	20.07$^{c, d}$
Baluchistan	24.93	19.43	22.18^{b}
Yecora-70	25.80	19.87	22.84$^{a, b}$
S-24	25.27	19.04	22.15$^{b, c}$
SARC-1	27.30	20.77	24.04$^{a, b}$
Pasban-90	28.57	20.69	24.63^{a}
Means	23.64a	18.35^{b}	

Table 10.2 Effect of two soil moisture levels on transpiration rate (mol m^{-2} s^{-1}) of different wheat genotypes (Different letters indicate a statistical difference among the treatments at P<0.05)

Genotypes	M_1	M_2	Means
V-4178	10.3	9.27	9.79[a]
PDW-34	9.85	8.86	9.35[a, b]
Chakwal-97	9.67	8.70	9.18[a, b]
Shorawaki	9.46	8.50	8.97[a, b, c]
4098805	9.19	8.27	8.73[b, c, d]
Baluchistan	8.67	7.80	8.23[c, d, e]
Yecora-70	8.28	7.45	7.86[d, e]
S-24	7.80	7.02	7.41[e, f]
SARC-1	7.06	6.54	6.71[f, g]
Pasban-90	6.72	6.05	6.38[g]
Means	8.70[a]	7.83[b]	

(Table 10.1). Water deficit primarily affects photosynthesis through reduced stomatal conductance, carbon dioxide diffusion to the chloroplast and metabolic restraints. Drought stress straightly alters the leaf carbohydrate contents by minimizing the rate of photosynthesis. Results showed that photosynthetic rate was highly dependent on moisture availability. All genotypes had better performance under proper moisture conditions showing higher rate of photosynthesis while they performed poorly at low moisture levels. The photosynthetic rate decreased with the extension of the drought phase. Drought stress decreased the rate of photosynthesis significantly (Kawamitsu et al. 2000; Azimi et al. 2010).

10.3.1.2 Transpiration Rate (mol m^{-2} s^{-1})

Soil moisture content had a significant effect on transpiration rate of wheat crop having higher value (8.70) at field capacity level; however, lower value (7.83) was observed in genotypes at 40 % field capacity. The transpiration rate was significantly affected by the wheat genotypes. The genotypes V-4178 exhibited the maximum transpiration rate (9.79) and did not differ significantly from PDW-34 (9.35), Chakwal-97 (9.18) and Shorawaki (8.97). On the other hand the minimum transpiration rate (6.38) was recorded in Pasban-90 which was at par with SARC-1 (6.71). The interaction between soil moisture levels and genotypes was found to be insignificant (Table 10.2).

Transpiration is the loss of water from cell walls inside the leaves and diffusion from leaf interior to outer climate. The factors which significantly affect transpiration rate are temperature, stomatal conductance, humidity (Tullus et al. 2012), stomatal density, leaf area, leaf orientation and soil moisture. Transpiration is also an important energy dissipation mechanism of crops. The genotype which can reduce transpiration under moisture deficit environment could be a better option for drought conditions to get more productivity. Similar, to current study Mafakheri et al. (2010) reported reduced transpiration under moisture deficit conditions.

Table 10.3 Effect of two soil moisture levels on stomatal conductance (mol m^{-2} s^{-1}) of different wheat genotypes (Different letters indicate a statistical difference among the treatments at $P<0.05$)

Genotypes	M_1	M_2	Means
V-4178	0.84	0.76	0.80^{a}
PDW-34	0.77	0.69	0.73^{b}
Chakwal-97	0.75	0.68	0.71^{b}
Shorawaki	0.73	0.65	$0.69^{b,c}$
4098805	0.71	0.64	$0.67^{b,c}$
Baluchistan	0.67	0.60	$0.63^{c,d}$
Yecora-70	0.64	0.57	0.61^{d}
S-24	0.60	0.54	$0.57^{d,e}$
SARC-1	0.54	0.49	$0.52^{e,f}$
Pasban-90	0.52	0.47	0.49^{f}
Means	0.68^{a}	0.61^{b}	

10.3.1.3 Stomatal Conductance (mol m^{-2} s^{-1})

Soil moisture content had a significant impact on stomatal conductance of wheat crop. The plants maintained at field capacity had higher stomatal conductance (0.68) as compared to those maintained at 40 % field capacity (0.61) (Table 10.3). The genotypes also showed high degree of variation in terms of stomatal conductance. The maximum stomatal conductance was observed in V-4178 (0.79) while minimum was measured in Pasban-90 (0.49).

The interaction between genotypes and treatments was non-significant (Table 10.3). Stomatal conductance in wheat decreased with decrease in days of drying and leaf water potential. Stomatal conductance reduced under drought conditions to save water losses. The mean CO_2 assimilation (Stomatal conductance) for wheat genotypes over different moisture levels revealed the maximum stomatal conductance value for genotype V-4178 (0.8) followed by PDW-34 (0.73). The maximum stomatal conductance for V-4178 was due to early vigor which resulted in good CO_2 influx and stomatal efficiency. Similar conclusion was reported by Liao et al. (2004) who observed good growth rate due to efficient assimilation of CO_2 in early vigor genotypes. In the current study stomatal conductance was reduced in some genotypes when exposed to drought while genotype V-4178 and PDW-34 behaved better and stomatal conductance was higher which led to better crop development. Stomatal conductance was directly related to the availability of moisture contents in the soil. Similar, to current study Liang et al. (2002) reported decreased water level of the rhizosphere caused reduction of stomatal conductance. However, Rebetzke et al. (2001) reported that stomatal conductance was the genetic feature of crop genotypes.

Table 10.4 Effect of two soil moisture levels on stomatal resistance (m^2 s mol^{-1}) of different wheat genotypes (Different letters indicate a statistical difference among the treatments at P<0.05)

Genotypes	M_1	M_2	Means
V-4178	0.37	0.59	0.48^{d}
PDW-34	07	0.57	0.47^{d}
Chakwal-97	0.37	0.60	0.48^{d}
Shorawaki	0.43	0.66	0.55^{d}
4098805	0.47	0.71	0.59^{d}
Baluchistan	0.57	0.97	$0.77^{c, d}$
Yecora-70	0.60	0.95	0.79^{d}
S-24	0.73	1.18	$0.96^{b, c}$
SARC-1	0.90	1.43	$0.52^{e, f}$
Pasban-90	1.00	1.55	0.49^{f}
Means	0.58^{a}	0.92^{b}	

10.3.1.4 Stomatal Resistance (m^2 s mol^{-1})

A significant distinction was observed between the two treatments for stomatal resistance. Plants in M_1 (soil maintained at field capacity level) had more stomatal resistance with the average value 0.92 than M_2 (soil maintained at 40 % field capacity level) with the average 0.58. The genotypes varied significantly from each other for stomatal resistance. The maximum stomatal resistance (1.28) was measured in Pasban-90 and was at par with SARC-1 (1.16) while the minimum stomatal resistance (0.48) in V-4178 which did not differ significantly from PDW-34 (0.47), Chakwal-97 (0.71), Shorawaki (0.69), 4098805 (0.67), Baluchistan (0.63) and Yecora-70 (0.61). The genotypes and soil moisture levels interacted non-significantly (Table 10.4).

Stomatal resistance is opposite to the stomatal conductance. Stomatal resistance is the ability of a plant to resist the gaseous exchange. It is a critical physiological parameter for drought assessment, dependent on genetic and microclimatic factors as well as moisture availability. Stomatal resistance of the tolerant genotypes was high under the stress conditions. Under stress conditions plant close their stomata to conserve water and control water losses, resulting in water saving for the maintenance of plant metabolic processes. The current study results indicated that the genotype with more stomatal resistance has better stress tolerance. The results of the study were similar to Iqbal and Bano (2009) who observed detrimental effect of stresses on crop productivity due to change in moisture contents.

10.3.1.5 Leaf Membrane Stability Index (%)

The two treatments varied significantly for leaf membrane stability index. Leaf membrane stability index was found higher (77.82) in plants maintained at field capacity than those raised at 40 % field capacity (66.63) (Table 10.5). Leaf

Table 10.5 Effect of two soil moisture levels on leaf membrane stability index (%) of different wheat genotypes (Different letters indicate a statistical difference among the treatments at P<0.05)

Genotypes	M_1	M_2	Means
V-4178	69.28	61.60	65.44[g]
PDW-34	71.08	62.73	66.91[f, g]
Chakwal-97	73.38	63.86	68.62[e, f, g]
Shorawaki	75.19	64.98	70.09[d, e, f]
4098805	77.89	66.12	72.00[c, d, e]
Baluchistan	79.69	67.13	73.41[b, c, d]
Yecora-70	81.19	68.26	74.73[a, b, c]
S-24	82.89	69.38	76.14[a, b, c]
SARC-1	83.29	70.51	76.90[a, b]
Pasban-90	84.29	71.74	78.02[a]
Means	77.82[a]	66.63[b]	

Table 10.6 Effect of two soil moisture levels on leaf succulence (mg m^{-2}) of different wheat genotypes (Different letters indicate a statistical difference among the treatments at P<0.05)

Genotypes	M_1	M_2	Means
V-4178	12.12[f, g]	6.63[j]	9.37[e]
PDW-34	12.41[f, g]	6.87[j]	9.64[e]
Chakwal-97	12.71[e, f]	7.14[j]	9.93[e]
Shorawaki	13.12[d, e, f]	8.60[i]	10.83[d]
4098805	13.71[c, d, e]	9.31[h, i]	11.51[c, d]
Baluchistan	14.12[c, d]	9.82[h]	11.97[b, c]
Yecora-70	14.41[b, c]	10.13[h]	12.27[c, d]
S-24	15.32[a, b]	1.54[k]	8.43[f]
SARC-1	15.42[a, b]	11.56[g]	13.49[a]
Pasban-90	15.62[a]	12.79[e, f]	14.21[a]
Means	13.90[a]	8.44[b]	

membrane stability index in case of genotypes varied considerably. The genotype Pasban-90 had maximum leaf membrane stability index (78.02) and the minimum (65.44) was recorded in V-4178. The interaction table illustrated a non-significant interaction between genotypes and soil moisture levels (Table 10.5).

Leaf membrane stability index is the ability of plant to thrive under drought conditions. Leaf membrane stability index is a key parameter, for determining the ability of a plant to survive under water deficit situation. It was deduced from the trial results that leaf membrane stability has dependence on the soil moisture level.

It is high under normal conditions and reduced when plants were under stressed environment. The results of the study were found to be concurrent to the results given by Iqbal and Bano (2009). LMSI reduced linearly as the stress period prolonged (Chakraborty et al. 1993).

Table 10.7 Effect of two soil moisture levels on relative water content (%) of different wheat genotypes (Different letters indicate a statistical difference among the treatments at P<0.05)

Genotypes	M_1	M_2	Means
V-4178	77.29	61.71	69.50[f]
PDW-34	77.79	62.63	70.21[f]
Chakwal-97	78.99	62.73	70.86[f]
Shorawaki	80.79	63.34	72.07[f]
4098805	83.70	63.75	73.73[e, f]
Baluchistan	85.49	69.38	77.44[d, e]
Yecora-70	87.40	71.43	79.42[c, d]
S-24	89.80	74.50	82.15[b, c]
SARC-1	90.90	79.72	85.31[a, b]
Pasban-90	91.90	82.17	87.04[a]
Means	84.40[a]	69.14[b]	

10.3.1.6 Leaf Succulence (mg m^{-2})

Individual as well as interactive effects of soil moisture levels and whet genotypes on leaf succulence were significant. Pasban-90 grown at soil field capacity exhibited the maximum leaf succulence (15.62) and was at par with SARC-1 (13.49) and S-24 (8.43) raised at soil field capacity (Table 10.6). On the other hand the minimum leaf succulence (6.63) was recorded in V-4178 which did not differ significantly from PDW-34 (9.64) and Chakwal-97 (9.93). Results point towards a significant relation between leaf succulence and moisture availability. Leaf succulence reduced with declined moisture availability (Qi et al. 2009). Less succulent leaves were found under stress situations. Whereas small sized succulent leaves helped the plant to avoid transpirational losses to ensure its survival under moisture deficit environment. The drought resistant genotypes had more succulent leaves than those susceptible to drought. That is why they performed better under moisture deficit environment. In an experiment conducted by Razzaq et al. (2013) it was found that the cultivars relatively suitable for drought conditions have higher leaf succulence and those with lower leaf succulence were vulnerable to drought.

10.3.1.7 Relative Water Content (%)

There was a highly significant variation found between the two soil moisture levels. However, their interactive effect was found non-significant. Main effects of genotypes and soil moisture levels on relative water content (RWC) of wheat crop were significant. The wheat plants grown at soil field capacity exhibited 22% higher RWC (84.40) than the wheat genotypes raised at 40% of the soil field capacity (69.14). Maximum RWC (87.04) were recorded in Pasban-90 which was at par with SARC-1 (85.31) (Table 10.7). Contrarily the minimum RWC (69.50) were observed in V-4178 which did not differ significantly from PDW-34 (70.21), Chakwal-97 (70.86), Shorawaki (72.07) and 4098805 (73.73).

Table 10.8 Effect of two soil moisture levels on epicuticular wax (g cm^{-2}) of different wheat genotypes (Different letters indicate a statistical difference among the treatments at $P<0.05$)

Genotypes	M_1	M_2	Means
V-4178	5.72	7.37	6.54^{f}
PDW-34	5.79	7.47	6.63^{f}
Chakwal-97	5.84	7.57	6.71^{f}
Shorawaki	5.90	7.57	6.79^{f}
4098805	7.11	7.57	7.50^{e}
Baluchistan	7.21	7.57	7.95^{d}
Yecora-70	7.31	7.57	8.11$^{c, d}$
S-24	7.91	7.57	8.46$^{b, c}$
SARC-1	8.01	7.57	8.61^{b}
Pasban-90	8.11	7.57	9.33^{a}
Means	6.89^{b}	8.43^{a}	

Relative water content is a better measure of plant water status. It is an indicator of the plant physical and physiological activities. RWC can be used instead of plant water potential as it is related to cell volume. Genotypes with high RWC have more resistance to drought stress, as the maintenance of higher RWC helps plants to maintain their normal metabolic activities. Hence the genotypes with higher RWC had an edge over others under stressed condition. Results of the current study illustrated drop in RWC with increasing drought stress. RWC contributes towards drought resistance and the genotype with higher RWC under stress under water stress condition has more resistance towards drought. Keyvan (2010) found reduction in RWC of various genotypes under stressed conditions. Similarly, drop in RWC of various wheat genotypes were reported by Nouri et al. (2011).

10.3.1.8 Epicuticular Wax (g cm^{-2})

The current study exhibited significant difference between two soil moisture levels with respect to epicuticular wax accumulation. The maximum epicuticular wax contents (8.43) were found for M_2 (soil maintained at 40 % field capacity) whereas M_1 (soil maintained at field capacity) produced minimum epicuticular wax contents (6.88). The two soil moisture levels had a variation of 22 % for the accumulation of epicuticular wax. In case of wheat genotypes there was a significant variation regarding epicuticular wax. The genotype Pasban-90 had maximum epicuticular wax (9.32) while minimum epicuticular wax (6.54) was observed in V-4178 which did not differ significantly from PDW-34 (6.63), Chakwal-97 (6.71) and Shorawaki (6.79). A significant interaction was found between genotypes and soil moisture levels for epicuticular wax at 5 % probability level. Pasabn-90 had the maximum epicuticular wax (10.54) at 40 % field capacity whereas minimum epicuticular wax was recorded in V-4178 (Table 10.8). Epicuticular wax is produced under drought conditions to cope water losses from the leaf surface. It is an important factor for droughttolerance. Epicuticular wax enhances stomatal resistance to avoid the water

Table 10.9 Effect of two soil moisture levels on proline content (μg g^{-1}) of different wheat genotypes (Different letters indicate a statistical difference among the treatments at P<0.05)

Genotypes	M_1	M_2	Means
V-4178	41.31	47.82	44.56^{g}
PDW-34	42.81	48.12	$45.46^{f, g}$
Chakwal-97	44.11	48.32	$46.21^{e, f, g}$
Shorawaki	44.31	48.92	$46.61^{d, e, f}$
4098805	44.91	49.62	$47.27^{c, d, e}$
Baluchistan	45.31	49.92	$47.62^{c, d, e}$
Yecora-70	45.91	50.42	$48.17^{c, d}$
S-24	46.31	51.22	$48.77^{b, c}$
SARC-1	47.22	52.72	$49.97^{a, b}$
Pasban-90	47.32	53.92	50.62^{a}
Means	44.95^{b}	50.10^{a}	

losses under water deficit conditions. Hence, the water is available for the plants to maintain its proper physiology. Epicuticular wax load is higher in the drought tolerant genotypes (Elham et al. 2012). The results of the study were found to be in accordance with Elham et al. (2012).

10.3.2 Biochemical Parameters

10.3.2.1 Proline Content (μg g^{-1})

Proline content was measured at flag leaf stage. The moisture levels have a significant effect on proline content. Moisture level at 40 % field capacity produced more proline content than the moisture level at field capacity. Maximum value (50.09) was obtained for M_2 and the minimum proline contents (44.95) were produced at M_1. There was 11 % difference between the two moisture levels.

The results of the study have illustrated that the genotypes varied in significant manner for proline contents. Maximum proline accumulation (50.61) was recorded in Pasban-90 which did not differ significantly from SARC-1 (49.96) and V-4178 had the minimum proline accumulation of 44.56 which was at par with PDW-34 (45.46) and Chakwal-97 (46.21). There was 14 % difference between the maximum and the minimum values for proline content (Table 10.9).

Results established a non-significant interaction between genotypes and moisture levels. The most generalized response of plants to cope the drought stress is proline accumulation in the plant leaves (Sayed et al. 2012). All the genotypes behaved differently under moisture regime to accumulate proline content. Drought resistant genotypes accumulated more proline content than drought susceptible genotypes. Similar to our findings Chandra et al. (2004) reported that resistant cultivars accumulated more proline than drought susceptible genotypes.

Table 10.10 Effect of two soil moisture levels on chlorophyll content (SPAD value) of different wheat genotypes (Different letters indicate a statistical difference among the treatments at $P<0.05$)

Genotypes	M_1	M_2	Means
V-4178	35.22	28.18	31.70[e]
PDW-34	35.49	28.39	31.94[e]
Chakwal-97	37.54	30.03	33.79[d, e]
Shorawaki	39.92	31.93	35.92[d]
4098805	40.15	32.12	36.14[c, d]
Baluchistan	44.89	34.27	39.93[b]
Yecora-70	45.50	34.98	41.11[a, b]
S-24	46.44	35.77	39.88[b, c]
SARC-1	49.12	37.25	43.25[a, b]
Pasban-90	51.38	37.39	44.32[a]
Means	42.57[a]	33.03[b]	

10.3.2.2 Chlorophyll Content (SPAD Value)

The two moisture levels had a significant variation for the chlorophyll content. Plants maintained at field capacity had the higher chlorophyll content (42.57) than those grown at 40 % field capacity (33.03) (Table 10.10). The two moisture levels had a variation of 29 % approximately. The genotypes also varied significantly for the chlorophyll contents. Pasban-90 had the maximum chlorophyll content (44.32) and was at par with SARC-1(43.25) and Yecora-70 (41.11), however, chlorophyll content were found to be lower most (31.70) in V-4178, preceded by PDW-34 (31.94) and Chakwal-97 (33.79). Both genotypes had a variation of 40 % for chlorophyll contents. Non-significant interaction for chlorophyll content was found between soil moisture levels and genotypes (Table 10.10). Chlorophyll is the fundamental photosynthetic pigment found in the plant leaves. Any reduction in chlorophyll content causes poor growth and development resulting in low yield. Chlorophyll content tends to fall under moisture deficit condition due to photo- oxidation and degradation of chlorophyll pigment. Plants able to maintain high chlorophyll content when subjected to stress are drought resistant (Homayoun et al. 2011).

In the current study the genotypes maintaining higher chlorophyll contents produced better results and were able to maintain the near to normal photosynthetic activity even under drought conditions. The findings of Anosheh (2012) were similar to our results. Decline in chlorophyll content was observed under drought stress whereas in some genotypes chlorophyll content increased and thus termed as drought resistant (Homayoun et al. 2011; Anosheh 2012).

Table 10.11 Effect of two soil moisture levels on biological yield (kg ha^{-1}) of different wheat genotypes (Different letters indicate a statistical difference among the treatments at P<0.05)

Genotypes	M_1	M_2	Means
V-4178	4300.00	3248.57	3864.29^{f}
PDW-34	4485.71	3471.43	3978.57$^{e, f}$
Chakwal-97	4671.43	3585.71	4128.57$^{e, f}$
Shorawaki	4864.29	4042.86	4454.14$^{c, d, e}$
4098805	4992.86	3842.86	4421.43$^{d, e, f}$
Baluchistan	5350.00	4550.00	4942.86$^{b, c, d}$
Yecora-70	5650.00	4371.43	5007.14$^{a, b, c}$
S-24	5557.14	4300.00	4928.57$^{b, c, d}$
SARC-1	5857.14	4650.00	5257.14$^{a, b}$
Pasban-90	6550.00	4542.86	5550.00^{a}
Means	5253.71^{a}	4078.57^{b}	

10.3.3 *Yield Parameters*

10.3.3.1 Biological Yield (kg ha^{-1})

The two soil moisture levels vary extensively with respect to biological yield. Higher (5235.71 kg ha^{-1}) biological yield was recorded in M_1 (at field capacity) as compared to M_2 (at 40 % field capacity) where it was 4078.57 kg ha^{-1}. The results revealed great variation among the genotypes for biological yield. The higher biological yield (5550.00 kg ha^{-1}) was recorded in Pasban-90 which was at par with SARC-1(5257.14 kg ha^{-1}) and Yecora-70 (5007.14 kg ha^{-1}) while minimum biological yield (3864.29 kg ha^{-1}) was observed in V-4178 which did not differ significantly from PDW-34 (3978.57 kg ha^{-1}) and Chakwal-97 (4128.57 kg ha^{-1}). Non-significant interaction was established between the soil moisture levels and genotypes (Table 10.11). Drought stress affected growth and development of plants resulting in reduced leaf area that indicates low photosynthetic activity. Photoassimilates are not partitioned properly and hence both grain yield and biological yield move to decline. According to Khalili et al. (2013), increasing water deficit decreases number of seeds as well as 100 grain weight causing a decline in grain yield and biological yield. Khakwani et al. (2011) reported drop in biological yield of wheat genotypes grown under stress conditions. Results of our study were concomitant to Chandler and Singh (2008).

10.3.3.2 Grain Yield (kg ha^{-1})

The results in table expressed significant difference between the two soil moisture levels for grain yield, M_1 (at field capacity) produced comparatively higher yields (1842.86 kg ha^{-1}) than M_2 (at 40 % field capacity) (1428.57 kg ha^{-1}). The percentage change between the two soil moisture levels was calculated to be 29 %. Significant

Table 10.12 Effect of two soil moisture levels on grain yield (kg ha^{-1}) of different wheat genotypes (Different letters indicate a statistical difference among the treatments at P<0.05)

Genotypes	M_1	M_2	Means
V-4178	1512.43	1221.43	1378.57[e]
PDW-34	1535.71	1228.57	1385.71[e]
Chakwal-97	1621.43	1300.00	1464.29[d, e]
Shorawaki	1728.57	1378.57	1557.14[d]
4098805	1671.43	1392.86	1564.29[c, d]
Baluchistan	1942.86	1514.29	1728.57[b]
Yecora-70	2007.14	1550.00	1785.71[a, b]
S-24	1971.43	1485.71	1728.57[b, c]
SARC-1	2121.43	1614.29	1871.43[a, b]
Pasban-90	2221.43	1614.29	1914.29[a]
Means	1842.86[a]	1428.57[b]	

variation among the genotypes was witnessed at 5 % probability level. Pasban-90 gave maximum grain yield (1914.29 kg ha^{-1}) followed by SARC-1 (1871.43 kg ha^{-1}) whereas minimum grain yield was taken from V-4178 (1378.57 kg ha^{-1}) preceded by PDW-34 (1385.71 kg ha^{-1}). The difference calculated for grain yield between Pasban-90 and V-4178 was 40 %. The interaction between soil moisture levels and genotypes was non-significant (Table 10.12). Grain yield is directly related to the number of spikes per plant and hundred grain weight. Grain yield was affected badly due to drought stress. Less number of spikes and lesser hundred grain weight both resulted in lower grain yield. The results exhibited that the genotypes have low grain yield at 40 % field capacity due to moisture deficiency. The genotypes producing higher grain yield even under drought are considered to be drought resistant. The results were found in harmony with Keyvan (2010) and Akram et al. (2012).

10.3.3.3 Harvest Index

The relationship between genotypes and two soil moisture levels was described in terms of harvest index. There was no significant difference between the two water regimes in terms of harvest index. Similarly, the difference between the genotypes in terms of harvest index was not significant. All the genotypes produced similar results. Hence the interaction between genotypes and soil moisture levels was also found non-significant (Table 10.13). Photosynthetic efficiency of a plant can be determined by the harvest index. It indicates the strength of source to sink relation. Many scientists have reported an obvious decline in HI under drought stress due to weaker source to sink relationship (Ghaderi et al. 2009). However, the results of our study were in accordance with Ahmadizadeh et al. (2011) who proposed significant effect of drought on all traits except harvest index.

Table 10.13 Effect of two soil moisture levels on harvest index of different wheat genotypes (Different letters indicate a statistical difference among the treatments at P<0.05)

Genotypes	M_1	M_2	Means
V-4178	0.36	0.36	0.36
PDW-34	0.35	0.35	0.35
Chakwal-97	0.35	0.36	0.36
Shorawaki	0.36	0.34	0.35
4098805	0.35	0.36	0.36
Baluchistan	0.36	0.33	0.35
Yecora-70	0.36	0.36	0.36
S-24	0.35	0.34	0.35
SARC-1	0.36	0.35	0.36
Pasban-90	0.34	0.36	0.35
Means	0.35	0.35	

References

Ahmadizadeh, M., A. Nori, H. Shahbazi, and M. Habibpour. 2011. Effects of drought stress on some agronomic and morphological traits of durum wheat (*Triticum durum* Desf.) landraces under greenhouse condition. *African Journal of Biotechnology* 10: 14097–14107.

Akram, H.M., A. Sattar, A. Ali, and M. Nadeem. 2012. Agro-physiolohical performance of wheat genotypes under normal moisture and drought conditions. *Iranian Journal of Plant Physiology* 2: 361–369.

Anosheh, H.P. 2012. Exogenous application of salicylic acid and chlormequat chloride alleviates negative effects of drought stress in wheat. *Advanced Studies in Biology* 4: 501–520.

Ashraf, M., and M.R. Foolad. 2007. Roles of glycine betaine and proline in improving plant abiotic stress resistance. *Environmental and Experimental Botany* 59: 206–216.

Azimi, M.R., H. Alizadeh, G.H. Salekdeh, and M.R. Naghavi. 2010. Association of yield and flag leaf photosynthesis among wheat RILs under drought condition. *Journal of Food, Agriculture and Environment* 8: 409–411.

Barrs, H.D., and P.E. Weatherley. 1962. Are-examination of relative turgidity technique for estimating water deficit in leaves. *Australian Journal of Biological Sciences* 15: 413–428.

Bates, L.S., R.P. Waldren, and I.D. Teare. 1973. Rapid determination of free proline for water stress studies. *Plant and Soil* 39: 205–207.

Chandler, S.S., and T.K. Singh. 2008. Selection criteria for drought tolerance in spring wheat (*Triticum aestivum* L.). Series: Coping with wheat in a changing environment abiotic stresses. In *Proceedings of the 11th international wheat genetics symposium*, ed. R. Appels, R. Eastwood, E. Lagudah, P. Langridge and M. Mackay Lynne, 1–3. Australia: Sydney University Press.

Chakraborty, U., B.N. Chakraborty, and M. Kapoor. 1993. Changes in the levels of peroxidase and phenylalanine ammonia-lyase in Brassica napus cultivars showing variable resistance to Leptosphaeria maculans. *Folia Microbiologica* 38: 491–496.

Chandrasekar, V., R.K. Sairam, and G.C. Srivastava. 2000. Physiological and biochemical responses of hexaploid and tetraploid wheat to drought stress. *Journal of Agronomy and Crop Science* 185: 219–227.

Chandra, A., P.S. Pathak, R.K. Bhatt, and A. Dubey. 2004. Variation in drought tolerance of different *Stylosanthes* accessions. *Biologia Plantarum* 48: 457–460.

Chaves, M.M., and M.M. Oliveira. 2004. Mechanisms underlying plant resilience to water deficits: Prospects for water-saving agriculture. *Journal of Experimental Botany* 55: 2365–2384.

Chinnusamy, V., A. Jagendorf, and J.K. Zhu. 2005. Understanding and improving salt tolerance in plants. *Crop Science* 45: 437–448.

Cornic, G. 2000. Drought stress inhibits photosynthesis by decreasing stomatal aperture not by affecting ATP synthesis. *Trends in Plant Science* 5: 187–188.
Demiral, T., and I. Turkan. 2004. Does exogenous glycine betaine affect antioxidative system of rice seedlings under NaCl treatment? *Journal of Plant Physiology* 161: 1089–1110.
Demirevska, K., D. Zasheva, R. Dimitrov, L. Simova-Stoilova, M. Stamenova, and U. Feller. 2009. Drought stress effects on Rubisco in wheat: Changes in the Rubisco large subunit. *Acta Physiologiae Plantarum* 31: 1129–1138.
Donald, C.M. 1962. In search of yield. *Journal of the Australian Institute of Agricultural Science* 28: 171–178.
Duan, B., Y. Yang, Y. Lu, H. Korpelainen, F. Berninger, and C. Li. 2007. Interactions between drought stress, ABA and genotypes in *Picea asperata. Journal of Experimental Botany* 58: 3025–3036.
Elham, F., R.A.K. Nejad, G.H. Salekdeh, and F. Najafi. 2012. Evaluation of cuticular wax deposition, stomata and carbohydrate of wheat leaves for screening drought tolerance. *Advances in Environmental Biology* 6: 4035–4040.
Farooq, M., A. Wahid, N. Kobayashi, D. Fujita, and S.M.A. Basra. 2009. Plant drought stress: Effects, mechanisms and management. *Agronomy for Sustainable Development* 29: 185–212.
Ghaderi, M.G., H. Zeinalikhanghah, A.H. Hosseinzadeh, A.R. Taleei, and M.R. Naghavi. 2009. Evaluation of relationships between grain yield, yield components and the other characteristics associated with grain yield in bread wheat using multivariate statistical analysis. *Iranian Journal of crop Research* 7: 573–582.
Gill, P.K., A.D. Sharma, P. Singh, and S.S. Bhullar. 2001. Effect of various abiotic stresses on the growth, soluble sugars and water relations of sorghum seedlings grown in light and darkness. *Bulgarian Journal of Plant Physiology* 27: 72–84.
Government of Pakistan. 2013. Economic survey, 2012–2013. Govt. of Pakistan. Ministry of Finance, Economic Advisor's Wing, Finance Div. Islamabad. HALL, A. E. 2001. Crop responses to environment. CRC Press: New York.
Grover, A., A. Kapoor, D. Kumar, H.E. Shashidhar, and S. Hittalmani. 2004. Genetic improvement for abiotic stress responses. In *Plant Breeding-Mendelian to molecular approaches*, ed. H.K. Jain and M.C. Kharkwal, 167–193. New Delhi: Narosa Publishing House.
Gupta, N.K., S. Gupta, and A. Kumar. 2001. Effect of water stress on physiological attributes and their relationship with growth and yield of wheat cultivars at different stages. *Journal of Agronomy and Crop Science* 186: 55–62.
Hassan, J.A. 2006. Effects of water stress and high temperature on gas exchange and chlorophyll fluorescence in *Triticum aestivum* L. Photosynthetica. *Journal of Plant Physiology* 44: 312–315.
Hays, D.B., J.H. Do, R.E. Mason, G. Morgan, and S.A. Finalyson. 2007. Heat stress induced ethylene production in developing wheat grains induces kernel abortion and increased maturation in a susceptible cultivar. *Plant Science* 172: 1113–1123.
Homayoun, H., M.S. Daliri, and P. Mehrabi. 2011. Effect of drought stress on leaf chlorophyll in corn cultivars (*Zea mays*). *Middle-East Journal of Scientific Research* 9: 418–420.
Huaqi, W., B.A.M. Bouman, D. Zhao, W. Changgui, and P.F. Moya. 2002. Aerobic rice in northern China: Opportunities and challenges. Workshop on water wise rice production, 8–11 April IRRI, Los Banos, Philippines.
Iqbal, S., and A. Bano. 2009. Water stress induced changes in antioxidant enzymes, membrane stability and seed protein profile of different wheat accessions. *African Journal of Biotechnology* 8: 6576–6587.
Kaur, S., A.K. Gupta, and N. Kaur. 2000. Effect of GA_3, kinetin and indole acetic acid on carbohydrate metabolism in chickpea seedlings germinating under water stress. *Plant Growth Regulation* 30: 61–70.
Kawamitsu, Y., T. Driscoll, and J.S. Boyer. 2000. Photosynthesis during desiccation in an intertidal alga and a land plant. *Plant Cell Physiology* 41: 344–353.

Keyvan, S. 2010. The effects of drought stress on yield, relative water content, proline, soluble carbohydrates and chlorophyll of bread wheat cultivars. *Journal of Animal and Plant Sciences* 8: 1051–1060.

Khakwani, A.A., M.D. Dennett, and M. Munir. 2011. Drought tolerance screening of wheat varieties by inducing water stress conditions. *Songklanakarin Journal of Science and Technology* 33: 135–142.

Khalili, M., M.R. Naghavi, A.P. Aboughadareh, and H.N. Rad. 2013. Effects of drought stress on yield and yield components in maize cultivars (*Zea mays* L.). *International Journal of Agronomy and Plant Production* 4: 809–812.

Kishore, P.B.K., S. Sangam, R.N. Amrutha, P.S. Laxmi, K.R. Naidu, K.R.S.S. Rao, S. Rao, K.J. Reddy, P. Theriappan, and N. Sreenivasulu. 2005. Regulation of proline biosynthesis, degradation, uptake and transport in higher plants: its implications in plant growth and abiotic stress tolerance. *Current Science* 88: 424–438.

Kumar, S.G., A. Mattareddy, and C. Sudhakar. 2003. NaCl effects on proline metabolism in two high yielding genotypes of mulberry (*Morus alba* L.) with contrasting salt tolerance. *Plant Science* 165: 1245–1251.

Lawlor, D.W., and G. Cornic. 2002. Photosynthetic carbon assimilation and associated metabolism in relation to water deficits in higher plants. *Plant, Cell and Environment* 25: 275–294.

Liang, Z., F. Zhang, M. Shao, and J. Zhang. 2002. The relations of stomatal conductance, water consumption, growth rate to leaf water potential during soil drying and rewatering cycle of wheat (*Triticum aestivum*). *Botanical Bulletin of Academia Sinica* 4: 187–192.

Liao, M., I. Fillery, and J. Palta. 2004. Early vigorous growth is major factor influencing nitrogen uptake in wheat. *Functional Plant Biology* 31: 121–129.

Long, S.P., and C.J. Bernacchi. 2003. Gas exchange measurements, what can they tell us about the underlying limitation to photosynthesis? Procedure and sources of error. *Journal of Experimental Botany* 54: 2393–2401.

Mafakheri, A., A. Siosemardeh, B. Bahramnejad, P.C. Struik, and E. Sohrabi. 2010. Effect of drought stress on yield, proline and chlorophyll contents in three chickpea cultivars. *Australian Journal of Crop Science* 4: 580–585.

Martinez, D.E., V.M. Luquez, C.G. Bartoli, and J. Guiamet. 2003. Persistence of photosynthetic components and photochemical efficiency in ears of water-stressed wheat (*Triticum aestivum*). *Physiologia Plantarum* 119: 519–525.

Mattioli, R., D. Marchese, S.D. Angeli, M.M. Altamura, P. Costantino, and M. Trovato. 2008. Modulation of intracellular proline levels affects flowering time and inflorescence architecture in Arabidopsis. *Plant Molecular Biology* 66: 277–288.

Merah, O. 2001. Potential importance of water status traits for durum wheat improvement under Mediterranean conditions. *Journal of Agricultural Science* 137: 139–145.

Mujtaba, S.M., and S.M. Alam. 2002. Drought phenomenon and crop growth. http://www.Pakistaneconomist.com/issue13/i&e4.htm.

Nayyar, H., and D.P. Walia. 2003. Water stress induced proline accumulation in contrasting wheat genotypes as affected by calcium and abscisic acid. *Biologia Plantarum* 46: 275–279.

Nouri, A., A. Etminan, J.A.T. Silva, and R. Mohammadi. 2011. Assessment of yield, yield-related traits and drought tolerance of durum wheat genotypes (*Triticum turjidum* var. durum Desf.). *Australian Journal of Crop Science* 5: 8–16.

Olesen, J.E., M. Trnka, K.C. Kersebaum, A.O. Skjelvag, B. Seguin, P.P. Sainio, F. Rossi, J. Kozyra, and F. Micale. 2011. Impacts and adaptation of European crop production systems to climate change. *European Journal of Agronomy* 34: 96–112.

Praba, M.L., J.E. Cairns, R.C. Babu, and H.R. Lafitte. 2009. Identification of physiological traits underlying cultivar differences in drought tolerance in rice and wheat. *Journal of Agronomy and Crop Science* 195: 30–46.

Prado, F.E., C. Boero, M. Gallardo, and J.A. Gonzalez. 2000. Effect of NaCl on germination, growth and soluble sugar content in *Chenopodium quinoa* wild seeds. *Botanical Bulletin of Academia Sinica* 41: 27–34.

Price, J., A. Laxmi, S.K. Martin, and J.C. Jang. 2004. Global transcription profiling reveals multiple sugar signal transduction mechanisms in Arabidopsis. *Plant Cell* 16: 2128–2150.
Qi, C.H., M. Chen, J. Song, and B.S. Wang. 2009. Increase in aquaporin activity is involved in leaf succulence of the euhalophyte *Suaeda salsa*, under salinity. *Science* 176: 200–205.
Ramanjulu, S., and D. Bartels. 2002. Drought- and desiccation-induced modulation of gene expression in plants. *Plant, Cell and Environment* 25: 141–151.
Razzaq, A., Q. Ali, A. Qayyum, I. Mahmood, M. Ahmad, and M. Rasheed. 2013. Physiological responses and drought resistance index of nine wheat (*Triticum aestivum* L.) cultivars under different moisture conditions. *Pakistan Journal of Botany* 45: 151–155.
Rebetzke, G.H., A.G. Condon, R.A. Richards, and J.J. Read. 2001. Phenotypic variation and sampling for leaf conductance in wheat (*Triticum aestivum* L.) breeding populations. *Euphytica* 121: 335–341.
Rizhsky, L., H. Liang, J. Shuman, V. Shulaev, S. Davletova, and R. Mittler. 2004. When defense pathways collide. The response of Arabidopsis to a combination of drought and heat stress. *Journal of Plant Physiology* 134: 1683–1696.
Sayed, M.A., H. Schumann, K. Pillen, A.A. Naz, and J. Leon. 2012. AB-QTL analysis reveals new alleles associated to proline accumulation and leaf wilting under drought stress conditions in barley (*Hordeum vulgare* L.). *BMC Genetics* 13: 61–75.
Serraj, R., and T.R. Sinclair. 2002. Osmolyte accumulation: Can it really help increase crop under drought conditions? *Plant, Cell and Environment* 25: 333–341.
Sharma, S.S., and K.J. Dietz. 2006. The significance of amino acids and amino acid-derived molecules in plant responses and adaptation to heavy metal stress. *Journal of Experimental Botany* 57: 711–726.
Sharma, P., and R.S. Dubey. 2005. Modulation of nitrate reductase activity in rice seedlings under aluminium toxicity and water stress: role of osmolytes as enzyme protectant. *Journal of Plant Physiology* 162: 854–864.
Silva-Fernandes, A.M., E.A. Baker, and J.T. Martin. 1964. Studies of plant cuticle iv. The isolation and fractionation of cuticular waxes. *The Annals of Applied Biology* 53: 43–58.
Sivakumar, P., P. Sharmila, and P.P. Saradhi. 2000. Proline alleviat es salt-stress induced enhancement in Rubisco-1,5-Bisphosphate oxygenase activity. *Biochemical and Biophysical Research Communications* 279: 512–515.
Szabados, L., and A. Savoure. 2010. Proline: A multifunctional amino acid. *Trends in Plant Science* 15: 89–97.
Tambussi, E.A., S. Nogues, and J.L. Araus. 2005. Ear of durum wheat under water stress: Water relations and photosynthetic metabo-lism. *Planta* 221: 446–458.
Tullus, A., P. Kupper, A. Sellin, L. Parts, J. Sober, T. Tullus, K. Lohmus, A. Sober, and H. Tullus. 2012. Climate change at northern latitudes: Rising atmospheric humidity decreases transpiration, N-uptake and growth rate of hybrid aspen. *PLoS One* 7: e42648.
Wang, X., J. Cai, D. Jiang, F. Liu, T. Dai, and W. Cao. 2011. Pre-anthesis high-temperature acclim ation alleviates damage to the flag leaf caused by post-anthesis heat stress in wheat. *Journal of Plant Physiology* 168: 585–593.
Yurekli, F., I. Turkan, B. Porgali, and S.F. Topcuoglu. 2001. Indole acetic acid gibbrellic acid, abscisic acid levels in NaCl-treated tomato species differing in salt tolerance. Israel. *Journal of Plant Sciences* 49: 269–277.
Zhu, J.K. 2002. Salt and drought stress signal transduction in plants. *Annual Review of Plant Physiology* 53: 243–273.

Chapter 11
Silvopastoral Systems: Best Agroecological Practice for Resilient Production Systems Under Dryland and Drought Conditions

S.F.J. Solorio, J. Wright, M.J.A. Franco, S.K. Basu, S.L. Sarabia, L. Ramírez, B.A. Ayala, P.C. Aguilar, and V.J.C. Ku

Contents

S.F.J. Solorio (✉) • S.L. Sarabia • L. Ramírez • B.A. Ayala • P.C. Aguilar • V.J.C. Ku
Campus de Ciencias Biológicas y Agropecuarias, Universidad Autónoma de Yucatán, Carretera Mérida-Xmatkuil Km. 15.5. C.P. 97100, Mérida, Yucatán, Mexico
e-mail: ssolorio@correo.uady.mx

J. Wright
Centre for Agroecology and Food Security, Coventry University, Coventry, UK

M.J.A. Franco
Departamento de economía, Universidad de Extremadura, Carretera de Cáceres s/n, Badajoz 06071, Spain

S.K. Basu
Department of Biological Sciences, University of Lethbridge, Lethbridge, AB T1K 3M4, Canada

M. Ahmed, C.O. Stockle (eds.), *Quantification of Climate Variability, Adaptation and Mitigation for Agricultural Sustainability*,
DOI 10.1007/978-3-319-32059-5_11

Abstract Intensified agriculture systems have had enormous negative consequences on ecosystems, particularly contributing towards unrestricted drought and desertification. In fact, the expansion of agriculture is the main cause of ecosystem degradation. The regions most vulnerable to such degradation are drylands, comprising 40 % of total land area and where 42 % of the global population resides. It is well known that climate change impact rainfed crops and water storage; which in turn impact the water availability for irrigation in dryland regions. Soils are also greatly affected by climate change: changes in rainfall and temperature affect crop growth, nutrient cycles, plant biodiversity and soil organic matter. Also, livestock production in tropical regions faces serious limitations, including inadequate management, the low quality and irregular availability of forage resources and, ultimately, the consequences of climate change. Among other reasons, low soil fertility and the irregularity of rain distribution have caused the majority of pastures to deteriorate. In general, tropical pastures are large contributors of greenhouse gas emissions, especially methane, which is associated with their high fiber content. To counter climate change requires: linking adaptation with mitigation. Silvopastoral systems are presented here as a set of strategies to enhance productivity whist reducing input costs and increasing environmental sustainability that also enhance carbon sequestration and build the resilience of the system to cope with the impacts of climate change.

Keywords Drought • Desertification • Nutrient cycling • Silvopastoral systems • Mitigation

Abbreviations

C	Carbon
CH_4	Methane
CO_2	Carbon dioxide
CP	Crude Protein
CT	Condensed Tanning
DM	Dry Matter
FAO	Food and Agriculture Organization
GHG	Greenhouse Gasses
H_2	Hydrogen
IFAD	International Fund for Agricultural Development
IPCC	International Panel for Climatic Change
IFPRI	International Food Policy Research Institute
Kcal	Kilocalories
N	Nitrogen
N_2O	Nitrous oxide
NDF	Neutral Detergent Fiber
SPS	Silvopastoral Systems
SRES	Special Report on Emissions Scenarios

UNEP United Nation Environment Program
WFP World Food Programme
WMO World Meteorological Organization

11.1 Introduction

Climate trend over the last years have been rapid in agricultural regions around the world (Lobell and Gourji 2012). Direct impacts will cause changes in the environment, less and uneven distributed rainfall (drought or inundation) affect agriculture productivity. Furthermore, frequent warm will increase more frequent drought period in most of the major cereal cropping regions around the world (Lobell and Gourji 2012). Agricultural production systems are currently under pressure to make important changes in order to increase food production and reduce the impact on natural resources and the environment. During periods of drought, traditional livestock production systems (extensive and intensive) have serious forage limitations in terms of both the availability and quality, affecting direct animal consumption and digestibility of forage and, consequently, the production of milk and meat. In extensive systems in the tropics (without trees or shrubs), high temperatures directly affect animal performance, and lead to decreased production of milk and meat and a higher incidence of disease (Herrero et al. 2009; Nardone et al. 2010). Moreover, in those situations, ruminants depend on low quality forage diets that are high in cellulose and hemicellulose content and favor high ratio of acetate: propionate production in the rumen; consequently those diets tend to produce more methane per unit of feed consumed (Canul-Solis et al. 2014).

Agriculture systems are the major sources of greenhouse gases (methane, nitrous oxide and carbon dioxide). Methane from enteric fermentation and from manure left on pasture makes up for 55 % of all emissions (FAOSTAT 2014). Table 11.1, show data (GHG emissions are measured in CO_2 equivalent) by continent of GHG emission for a period 2001–2010. Continent emissions and production profiles vary widely; Asia and the Americas have the highest level of emissions (almost 3.7 billion tones CO_2 eq).

Table 11.1 Agriculture GHG and methane emissions from enteric fermentation

	Agriculture		Methane (enteric fermentation)	
Continent	(Billion t CO_2 eq)	(%)	(Billion t CO_2 eq)	(%)
Africa	0.81	14	0.01	14
Americas	1.35	25	0.44	33
Asia	2.38	43	0.86	37
Europe	0.65	14	0.10	12
Oceania	0.22	4	0.01	4

Source: FAOSTAT Division (2014)

Recently, several studies have demonstrated that silvopastoral systems (SPS) can play a key role in the mitigation of climate change (Murgueitio et al. 2011; Cuartas et al. 2014). Adding legumes into grazing land promote atmospheric nitrogen (N) fixation and increase animal feed quality while decreasing methane emission. They can also provide important ecosystem services and reduce the pressure for tropical deforestation. Silvopastoral systems can also mitigate greenhouse gasses emissions. These gases can be reduced in terms of fluxes, by managing carbon and Nitrogen flows in the farming system. In the SPS, the standing stock of carbon above ground is usually higher than the equivalent land use without trees, and planting trees also may increase soil carbon sequestration (Murgueitio et al. 2011).

Silvopastoral systems involve the integration of different tree and shrub species with grass species and ruminants. The combination of shrub legumes associated with forage grasses makes these systems less susceptible to sudden changes in production, maintaining production and increases forage quality compared to grasses grown in monocultures. In the SPS, legume species are successfully combined, particularly *Leucaena leucocephala* (Leucaena) with grasses, such as *Panicum maximum* (guinea grass). The combination of both species, with other species of trees with multiple uses, can more efficiently tap into the available soil resources, which results in the increased production and profitability of such systems (Murgueitio et al. 2011; Nair 2012). The SPS improve environmental conditions for animal production, increased nutrient recycling and improved water efficiency, greater input of organic matter, N fixation and transference, carbon sequestration and reduction of methane emissions from soil (Cuartas et al. 2014). The major root biomass of various species in symbiosis and mutualism in SPS encourage better and healthier plants, more resistant to unpredictable changes of climate, pests and diseases in plants.

The main potential of SPS relates to the increased yield and quality of animal fodder and subsequent milk production and as well as daily body weight gain and maintenance. Additionally, SSP is one of the most important approaches to offsetting agricultural emission (Peters et al. 2013).

In the SPS, stocking rates are higher in Leucaena pastures than in Guinea grass pasture monocrops, resulting in up to four times higher production of milk (Table 11.2). Due to positive interactions between leguminous shrubs and grasses, annual liveweight gains from Leucaena pastures can be very high in different tropical countries (Murgueitio et al. 2011; Shelton and Dalzell 2007). It has been demonstrated that Leucaena forage can reduce partially the use of expensive concentrates in lactating cows (Peniche-González et al. 2014).

Plant biodiversity, typical of SPS, generates the conditions for greater resilience of livestock production systems in the tropics. The different strata formed by shrubs and trees produce a microclimate that reduces temperatures and water stress of grasses growing under the canopy of trees. When it rains the different plant structures (aerial and root biomass) enable more efficient use of water. Forage production of different species (grasses and legumes) is more evenly distributed throughout the year with less variation in the availability and quality under conditions of prolonged drought than conventional grass monocrop (Cuartas et al. 2014).

Table 11.2 Two productive indicators from conventional grassland and silvopastoral systems

Indicator	Conventional grassland systems (monocrops)	Silvopastoral systems
Milk production (kg/day)	3–4	7–9
Daily weight gains (kg/day)	0.3–0.4	0.8–1.0
Stocking rate (AU/ha)	1	2–4
Forage yield (t DM/ha/year)	6–10	15–25
Protein production (kg/ha)	360 a 600	2100–3500
N fixation (kg/ha/year)	0	300–500
C-sequestration (ton ha/year)	3 a 4	5 a 10
Methane emission (% year/animal)		15–20 (less compared to pastures)

Source: Compiled by author

The SPS meets the most important requirements in terms of providing an animal production system that is resilient to climate change. They require fewer external inputs which is more conducive to the management and improvement of animal health conditions. Greenhouse gases emissions are reduced mainly methane (CH_4) Nitrous oxide (N_2O) and carbon dioxide (CO_2), and biodiversity is promoted and preserved. In the SPS, the use of local species widely adapted to extreme drought conditions and that have a lower photoperiod sensitivity is encouraged. In addition, a diverse range products (milk, meat, wood and fruits) and a diversity of forage species is also promoted. All the above features are essential to the design requirements of a SPS and create more resilient animal production systems to climate change (Dumont et al. 2014).

11.2 Quantitative Impact of Heat and Drought Stress on Agricultural Systems

There is general agreement that, climate change have a negative impact on agriculture, (Nelson et al. 2009; FAO 2013). Climate change will affect direct crop production through induced higher temperatures and therefore less soil water availability. Table 11.3 shows the effect of two climate changes scenarios on rain fed and irrigated crop yield in developing and developed countries (Nelson et al. 2009). The two models, National Center for Atmospheric Research (NCAR) model and the Commonwealth Scientific and Industrial Research Organization, Australia (CSIRO) model have been used to simulate future climate change (Nelson et al. 2009). According with the data shown in Table 11.3, the impact on the crop´s yield reduction will be greater in developing than in developed countries. Wheat and rice, which are staple foods, are expected to be the most affected crops. Agriculture in developing countries is carried usually under rainfed conditions; therefore, it is mostly affected by climatic variability.

Table 11.3 Effect of climate change on crop yield (%) using two model scenarios

Region/crop	CSIRO	NCAR
Developing countries		
Maize	−2.0	−2.8
Rice	−14.4	−18.5
Wheat	−28.3	−34.3
Developed countries		
Maize	−1.2	−8.7
Rice	−3.5	−5.5
Wheat	−5.7	−4.9

Source: Modified from Nelson et al. (2009)
NCAR Atmospheric Research model and *CSIRO* Commonwealth Scientific and Industrial Research Organization

The Fifth Assessment Report of IPCC projected potential future impacts climate change on agriculture due to enhanced GHG emissions to the atmosphere (IPCC 2014). The report suggested that for an average increase of ~1–3 °C of the global mean temperature (by 2100, relative to the 1990–2000 average level) there could be decrease in the productivity of certain cereals at the lower latitudes and subsequent higher productivity at higher latitudes.

11.3 Livestock and Climate Change

Livestock production impact considerably the emission of greens house gases (Gerber et al. 2013). The increase in domestic animal inventories, because of the increasing demand for animal products (the livestock revolution) has resulted in greater emissions of greenhouse gases over the last few years. The environmental impact of ruminant production systems has been a matter of concern recently due to the increase in the emissions of greenhouse gases (GHG) such as methane (Havlick et al. 2014) and nitrous oxide (Lessa et al. 2014). Global emissions from agriculture continued to increase by almost 10 % in the last 10 year (FAOSTAT 2014), the largest emitters in the agriculture are the ruminants (thought enteric fermentation and manure left on pasture with 40 % and 16 % respectively. Table 11.4, shows the FAOSTAT database (average) of GHG emissions of five countries who contributed with the highest in the last 10 year.

Carbon dioxide (CO_2), methane (CH_4) and nitrous oxide (N_2O), are the main GHG emitted by ruminant production systems and the global contribution to anthropogenic GHG emissions represents between 7 and 18 % (Gerber et al. 2013). CH_4 has a global warming potential 23 times greater than that of CO_2 (IPCC 2006). In ruminants, CH_4 represents an energy loss between 5 and 18 % of the gross energy consumed (Moss et al. 2000). Energy losses vary depending on the type of feed consumed, fibrous and low-quality rations give rise to greater energy losses in cattle (Kurihara et al. 1999). In tropical regions, ruminant production systems are charac-

Table 11.4 Greenhouse gasses emissions (thousand gigagrams CO_2eq) from agriculture highest five countries producer in 2012

	Year		Average
Country	2000	2011	1990–2012
China	600	800	660
India	580	680	568
Brazil	360	450	370
USA	360	370	354
Australia	195	200	156

Source: FAOSTAT (2014)

terized by the grazing of native and introduced grasses, which present variations in the quantity and quality (chemical composition, digestibility) available throughout the year. Anaerobic fermentation of structural carbohydrates inside the rumen produces volatile fatty acids (such as acetic, propionic and butyric), carbon dioxide, heat and methane (Briceño-Poot et al. 2012).

The intake of grasses with a high concentration of fibrous components (cellulose, hemicellulose) gives rise to a pattern of rumen fermentation with a high proportion of acetic acid (Kurihara et al. 1999); while in ruminants fed starch-based rations the pattern of rumen fermentation results in an increased molar proportion of propionic acid (Johnson and Johnson 1995) in rumen liquor. Methane is formed by the reduction of CO_2 to CH_4 by means of the metabolic hydrogen (H_2) present in the rumen. Archimede et al. (2013)) recently proposed that tropical grasses tend to result in higher emissions of methane from the rumen of sheep than temperate grasses, and that this may be related to the rumen microbiome under different feeding conditions.

11.4 Meeting Climate Change and Food Security Goals

The world faces the challenge of feeding nine billion people by 2050. Consequently, food production will need to be increased by about 70 % (FAO 2014). This task of meeting the burgeoning demand for vegetable and animal products in a sustainable way to 2050 is made further complicated by the associated effects of climate change, energy constraints and vegetation and soil degradation. According to FAO, IFAD and WFP (2013), this challenge may be overcome by: sustainable soil management practices, increase the availability of nutrient for crops, cultivating in association a wider diversity of species and adapted cultivars, implementing rotational and sequential plantations, employing good quality seeds of adapted high producing genotypes; agro ecological pest, weed and disease control, including the reduction of the water footprint.

According to Bruinsma (2009), cereal production would need to rise by 33 % (3 billion tonnes) and meat production by 43 % (470 million tonnes). This means that at least the 90 % of the crop production has to be intensified by using more sustainable practices (Bennett and Carpenter 2005). Organic agriculture is a sustainable practice, which has been proposed as an approach to produce healthy food. According to the FAO, conversion to organic farming would mitigate up to 20 % of GHG emissions by abandoning synthetic fertilizers. However, even though, it has been suggested that organic agriculture has the potential to feed the world population by 2050 (Erb et al. 2009), it will require of more agricultural land.

However, organic agriculture it is not necessarily compatible with lower emissions of GHG nor always employ sustainable energy management systems particularly in industrialized regions. In developing countries, it is important to increase efficiency of production by means of improving management agricultural practices and incorporating technologies locally developed, such as the use of agro industrial byproducts for animal feeding. The adoption of agro silvopastoral systems can improve the quality of livestock diet and increase animal production and welfare, promote environmental services and biodiversity (Murgueitio et al. 2011; Broom et al. 2013).

Regarding animal production systems, the challenges of meeting the increasing demand for meat and milk will be best achieved, according to Dumont et al. (2014), by following five ecological principles: Improvement of animal health through appropriate management practices, reduce inputs required for production, increased metabolic and feed efficiency to reduce animal wastes; maintain and promote through agroecological practices the biological diversity and adapting management practices orientated to preserve biological diversity.

11.5 Livestock Production and Grazing Systems

Livestock production in many areas of the world is highly dependent on grazing systems, which vary widely among the distinct geographical zones. It is estimated that about 45 % of the earth's usable surface is covered by pastoral systems (Reid et al. 2008), which supports about 360 million cattle and over 600 million sheep and goats, distributed in the arid, semi-arid, sub humid, humid, temperate and tropical zones (De Haan et al. 1997). Globally, livestock products contribute 17 % to kilocalorie consumption and 33 % to protein consumption, but there are large differences between rich and poor countries (Rosegrant et al. 2009). According to projections, in the next three decades 30 % more grass will be required to meet the global demand for meat, and milk and improved management and use of fertilizers will be necessary to meet these increases (Bouwman et al. 2005).

Despite the vast extension of grazing systems, only 9 % of global beef production and 30 % of global sheep and goat meat comes from them. Whereas 54 % of total meat and 90 % of milk comes from mixed crop-livestock farming systems (Herrero et al. 2010). Further, developing countries produce 50 % of the beef, 41 % of the milk and 72 % of the lamb globally (Steinfeld et al. 2006; Herrero et al. 2009). Significant increases are expected to occur in 2050 as rates of growth of livestock production in the developing world rise higher than those in developed countries

(>2 %/year and <1 %/year, respectively) (Rosegrant et al. 2009; Herrero et al. 2009). It has been reported that 1.3 billion poor rural households are at least partly dependent on livestock systems for their livelihoods, and one fifth of the global population is employed in the global livestock industry (LID 1999). Livestock systems are important users of natural resources and have been identified as being partly responsible for environmental decline. Negative effects of grazing systems include land conversion and land degradation. It is estimated that 20 % of the world's rangelands have been degraded to some extent due to livestock impacts (soil erosion and, compaction, overgrazing), and livestock production is the cause of 18 % of greenhouse gas emissions (Steinfeld et al. 2006).

11.6 Climate Change and Pasture Degradation

Studies carried out by Merlo (2008) on forage yield of a tropical grass widely used in the tropical region of Mexico, *Brachiaria brizantha*, indicate that forage production capacity of this grass species with frequency of use every 35 days, is 11.9 tons of MS/ha. Another study (Flores 2009), carried out in the dry tropics region of Mexico (Yucatan), reported that *B. brizantha* pasture quality managed irrigated not allow higher weight gains of 341 g/animal/day even with the low stocking rate (1.5 AU/ha) where animal selectivity is allowed since the efficiency of grazing was only 66 % of the feed material. This low animal productivity was associated with the low rate of forage ruminal digestion of *B. brizantha*. This denotes the limited productive capacity of monoculture tropical grasses and the need to improve animal diet to improve indicators of productivity in the tropical Livestock.

Grassland resources are the main source of food for ruminant animal and play an important role in soil and water conservation. However, fodder production from grasslands (monoculture) is susceptible to unexpected changes in rainfall distribution and distribution pattern, high temperatures, which contribute to increases in plant lignification. Shifts in rainfall amounts and patterns, and changes in the frequency and severity of droughts and floods severely affect forage production and quality. Lack of water available in the soil increases fiber content of cell walls which decreases digestibility and feed intake of pastures while protein content also decreases. Both rainfall pattern and soil fertility influence forage growth. In this sense, when forage production is reduced, animal carrying capacity is limited; animals have to consume more forage and walk further to meet their nutritional requirements, and animals are obliged to overgraze. The decrease on nutritive value of grass with increased maturity is related to the increment on the proportion of stem and the corresponding decreased proportion of leaves.

Tropical forages demonstrate greater vulnerability to even slight changes in temperature or rainfall patterns, which can have devastating effects on quality and production (FAO 2008a, b). For instance, frequent or prolonged drought periods contribute to premature pasture degradation. Prolonged or repeated drought periods throughout the year have degrading effects and in particular on land near watering points where overgrazing destroys the soil plant cover, therefore increasing soil compaction and erosion resulting in low plant nutrients reducing forage quality and availability in the long dry season (FAO 2008a, b). As a result, there is a low animal carrying capacity (Animal unit per ha), low weight gain (<300 g day), and long time to slaughter (30–36 months), and, with high production costs.

In short, prolonged dry periods affects grass physiology, increasing fiber content and limiting its digestion by ruminants. As millions of cattle graze low nutritive quality tropical pastures, contributes to global CH_4 emissions. Similarly, changes in rainfall patterns contribute to increased levels of invasive species, woody species and weeds species that characterize the pasture degradation process. Subsequently, pasture degradation encourages farmer to open new forest areas to convert into pasture/grazing areas.

11.7 Silvopastoral Model for Climate Change Resilience

SPS are ecosystems of high conservation value because of their richness in vegetation types embedded into a dynamic shrubs-grassland mosaic. Furthermore, SPS are among the most promising approaches for sustainable management of tropical regions. In the perspective of ongoing climate and land-use changes, it is crucial to assess the resilience and the adaptive capacity at different spatial and temporal scales.

Thanks to various research projects developed in Mexico related with SPS on leguminous shrubs associated with pastures, the interactions between different type's vegetation patterns (Fig. 11.1), cattle activity, grassland dynamics and regeneration are well known. All this information will be useful for the modeling of SPS, however, critical features of climate change (temperature fluctuations and heat waves, precipitation shift and drought) will be considered into this strategic model. In the proposed model, SPS involve the use of multipurpose legumes, mainly *L. leucocephala* at high densities and trees to supply quality fodder for animals. Leucaena is a shrub legume suitable to a wide range of edafoclimatic conditions. Leucaena foliage contains 20–25 % protein, which produces extraordinary weight gain in cattle. Its roots help in N fixation in the soil, which improves soil rebuilding and surface texture as well as reducing run-off (Fig. 11.1).

Leucaena in combination with grasses can support from 2.5 to 4 cattle per hectare. One hectare can produce up to 15–20 tons of edible dry matter and is highly resistant to pest and diseases. Two major advantages of Leucaena are its longevity and droughttolerance. During dry periods, the deep roots enable it to continue to produce high quality forage, which enables graziers to better handle drought and to reduce their dependent on supplementation of protein/urea.

SPS successfully combine leguminous species, mainly trees and shrubs with grasses which interac in the same space and time. The combination of both species along with other multipurpose tree species more efficiently exploit the available resources of soil and atmosphere, reflecting higher production and profitability of

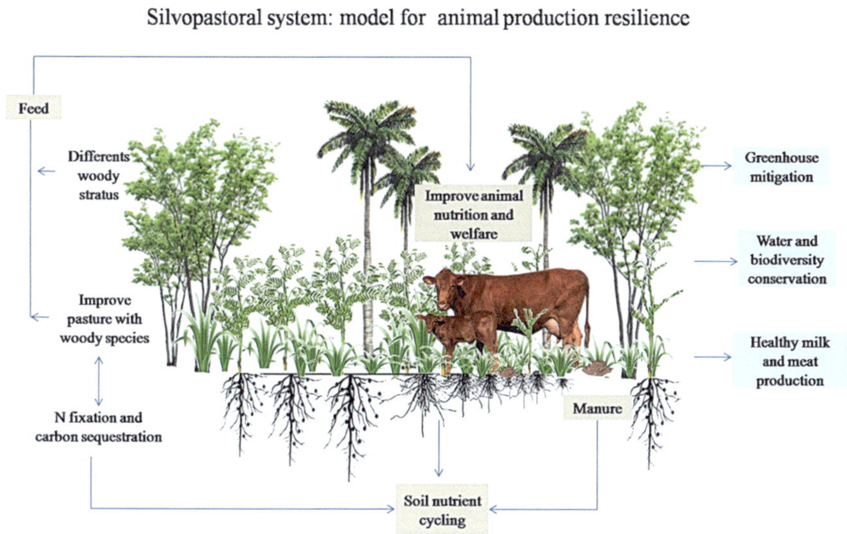

Fig. 11.1 Model of silvopastoral systems for the resilience and the adaptive capacity to climate change (Author own source)

the systems. Additionally, the SPS promote reforestation, increase soil fertility especially the N content and carbon sequestration avoiding the emission of carbon dioxide and methane.

The successful management of SPS model is determined by four major components:

Nutrient cycling, in SPS there is a continuous state of dynamic transfer of plant nutrients, with plants using soil minerals for metabolic purposes and returning them back to the soil litter, or through root senescence.

Nitrogen fixation and Carbon Sequestration, SPS enable recovery of degraded areas and a viable business for farmers -They have high potential for carbon sequestration. Net carbon flux and primary productivity increased significantly due to integration of trees and shrubs with grasses. Compared to 'grass-only' systems, soil organic matter, biological productivity and carbon storage is greater in the SPS. Furthermore, a more cost effective and sustainable solution to pasture degradation is the addition of vigorous forage legumes to the pasture. This boosts soil N levels by biological N fixation. More than 150 kg N/ha/year (equivalent to 320 kg urea/ha/year), can be fixed by Leuceana pastures, and some of this is returned to the pasture through deposition of urine and dung during grazing.

Greenhouse Gasses Mitigation, Methane emission from ruminants is one of the sector´s largest greenhouse gas emissions. SPS is one of the most important approaches to offsetting agricultural emission.

Microclimatic conditions, the shade within silvopastoral systems reduces temperature, ameliorate environments and is beneficial to animal performance. Milk yields of dairy cattle and liveweight gains of cattle in feed lots has been increased through the use of shading in hot climates.

11.7.1 Silvopastoral Systems: Environmental Impact

SPS have a positive impact on the environment since methane emissions from ruminant animals tend to be reduced when the foliage of legumes, which contain condensed tannins, is incorporated in the ration (Soltan et al. 2013) or when ground pods of *E. cyclocarpum* are incorporated in the ration of sheep (Ayala et al. 2014). Furthermore, legumes such as leucaena have the capacity to uptake atmospheric N, thus increasing the efficiency of N utilization at farm level. However, under certain circumstances SPS may contribute to emissions of nitrous oxide from the urine, (Lessa et al. 2014) if the intake of crude protein (CP) exceeds the capacity of the rumen microbial population to capture fermentable N into their protein (i.e. 210 g CP/kg digestible organic matter), for example, when the intake of foliage of leucocephala exceeds 50 % of dry matter of the ration. Under such conditions it is then recommended to use sources of readily fermentable energy (for example cane molasses, *Manihot esculenta roots,* and citrus byproducts) to balance the lack of energy by matching the excess N in the rumen arising from the CP of the legume. Nevertheless, the impact of SPS on the environment will be negligible, with the positive effects counterbalancing by far any possible negative effects.

Soil organic matter (SOM) from plays an important role in tropical forests where soils are highly weathered, through an improvement of soil functioning and forest sustainability. When forests are clear-cut, SOM is almost immediately lost, and this triggers a series of soil degradation processes. The rate at which the system can re-establish itself can be slow and may require the use of land restoration techniques. SPS with fast-growing leguminous nitrogen-fixing trees can, in a short to medium time, recuperate degraded land. The use of leguminous trees can increase efficiency in re-establishing the nutrient cycling processes of the system; based on stocks of litter, soil C and N. Recovering degraded land with SPS is effective in sequestering carbon dioxide from the atmosphere at high rates.

11.7.2 Silvopastoral Systems: Increased Adaptive Capacity for Climate Change

SPS is the production of livestock on land which combines multipurpose trees and fodder shrubs at high densities with grasses to improve yield and quality of fodder as well as add trees for fodder, timber, or firewood (Fig. 11.2). These systems can also provide important ecosystem services and reduce the pressure for tropical deforestation. SPS can mitigate greenhouse gas emissions. In the SPS, the standing stock of carbon above ground is usually higher than the equivalent land use without trees. Further, planting trees in association with grasses improves nutrient cycling and nutrient use efficiency. Adding legumes into grazing land can therefore promote atmospheric N fixation and increase animal feed quality while decreasing methane emissions. Also, legumes association improves soil rebuilding and surface texture as well as reducing run-off.

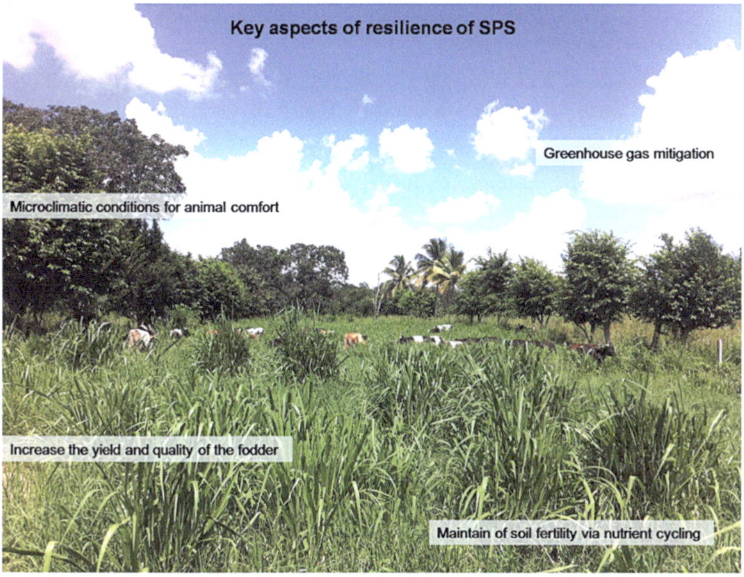

Fig. 11.2 Silvopastoral systems combine multipurpose trees and fodder shrubs with grasses to improve yield and quality of fodder for livestock systems resilience (Author own source)

SPS is a clear alternative to current agricultural practices by reconciling conservation and development needs. In the SPS, fodder species are established at high density (of 30,000 to 50,000 shrubs/ha) for direct browsing by animals. Species included in the SPS: leucaena, *Tithonia diversifolia* and *Guazuma ulmifolia*; in addition to fodder, they can supply timber or fruit for an additional farm income. Pruning of shrubs is usually at a height of 1 m; animals are allowed to move in a rotational grazing system. With the implementation of SPS it is possible to reduce the amount of chemical N fertilizers required by crops and for reducing emission of N_2O. Several countries of Latin America such as Argentina, Brazil, Colombia, Mexico, Nicaragua and Panama, are practicing this innovative approach to cattle management.

SPS play an important role in tropical livestock systems and increase resilience by reducing vulnerabilities and increasing adaptive capacity (Cuartas et al. 2014). There are several ways in which these systems build resilience as follow:

- Increase the yield and quality of fodder, and subsequently increase carrying capacity.
- Increase adaptive capacity; provide better microclimatic conditions for animal comfort.
- Develop and use efficient local feed sources of protein and energy.
- Maintain soil fertility via nutrient cycling and protect soils from erosion.
- Develop environmental livestock systems compatible with sustainable agriculture.

- Improve soil fertility through the increase in organic matter, atmospheric N fixation and better nutrient cycling.
- Reduce inter-annual and seasonal variation in forage availability
- Sensitivity to drought can be reduced by using diverse vegetal species forming different layers.

Environmental Benefits Include Control of soil erosion, improvement in water management, improvement in soil fertility (through N fixation), improved carbon (C) sequestration above and belowground including conservation of biodiversity and greenhouse gasmitigation. In SPS, temperatures can be 4–8 °C lower under the tree canopy compared to temperatures measured outside the tree canopy which improve animal health and productivity, adding resilience for adaptation to climate variability.

11.7.3 Best Practices for Silvopastoral Systems Resilience and Recommendations

There is a need to build adaptive capacity and strengthen the resilience of agriculture systems to climate changes. SSP hold a greater diversity of multipurpose woody species that are mainly local to specific agro-ecological regions, well adapted to drought conditions and they contribute increasing resilience to climate change by livestock systems (Fig. 11.2).

Due to the positive interactions, trees and shrubs in association with grasses contribute to improve productivity and forage quality. Livestock can benefit from steady access to food and adequate diets in a better environment as animals receive less heat stress thus contributing to improve productivity. Additional, the diversity of leguminous multipurpose woody species can help livestock systems that are based on grass monocultures to restore, improving forage value, productivity and body conditions of animals and reducing shortage of feed in prolonged dry conditions as well as land degradation through reforestation and improved nutrient cycling.

Improved livestock management holds the greatest potential for climate proofing farming systems. SPS increase ground cover, improve soil water use and reduce N leaching and volatilization. SPS can enhance the resilience and protect livestock production against extreme weather fluctuations in temperature and rainfall (Fig. 11.3). In SPS, water loss (soil evaporation, infiltration and plant transpiration) is reduced by the tree and shrubs shade, therefore grasses growth beneath trees are more resistant to droughts thus, the water is used more efficiently. SPS are based mainly on the use of well adapted local plant resources and with dependence from external inputs.

In the design of SPS, it is recommended to use plant species already present and adapted to each region (species that better tolerate drought and produce fodder during the dry season). Local species, generally, are well adapted to local environment, thus labor and dependence on external resources are reduced, to achieve maximum

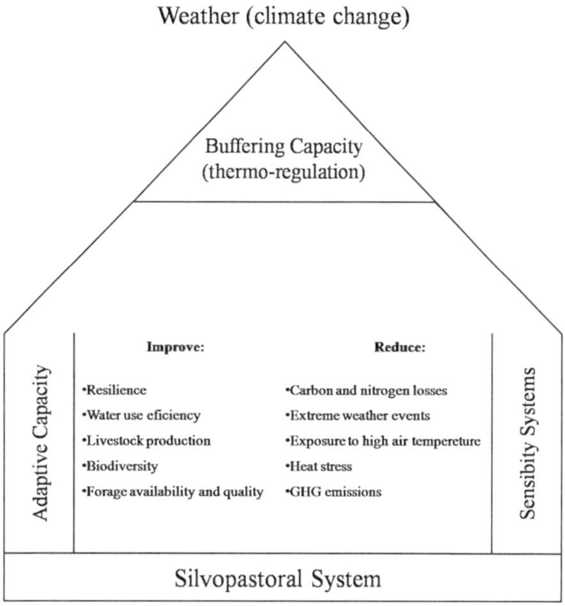

Fig. 11.3 Silvopastoral systems build livestock systems resilience based on three fundaments: buffering capacity, adaptive capacity and sensibility systems (Author own source)

benefit and minimal environmental impact. The use of diverse woody species is important to increase and support production through drought and adverse climatic conditions. Multipurpose leguminous and non-leguminous trees and shrubs help to protect soil and pasture productivity by protecting the soil from the sun, while retaining soil moisture. The higher the SOM content, the higher the water holding capacity and water infiltration, which also increases resilience to drought, heavy precipitation and extreme temperatures for the protection of macro and micro soil fauna and reduced erosion.

11.8 Conclusion

Pastures grown in monocrops suffer more damage and exhibit less recovery, decrease regional ecosystems integrity and biodiversity, and increase vulnerability to climate change. On the other hand, SPS with fast-growing leguminous nitrogen-fixing trees can, in a short to medium time, recuperate degraded land. SPS contribute to animal livestock production and further reduction in emission of greenhouse gases. Stocks of litter, C and N enable leguminous trees to increase the efficiency of nutrient cycling processes. The recovery of degraded land and sequestering of carbon dioxide at higher rates are also both benefits of the SPS. SPS possess very important environment and productive characteristic that makes them as one of the best strategies for livestock production. SPS, increase adaptive capacity and decrease vulnerability to climate change.

Acknowledgments This publication was partially supported by the Fundacion Produce Michoacan and the SAGARPA-CONACYT (Grant Number 2011-14-175169).

References

Archimede, H., M. Rira, M. Eugene, D.P. Morgavi, C. Anais, F. Pericarpin, B. Calif, C. Martin, C. Marie-Magdeleine, and M. Doreau. 2013. Intake, total-tract digestibility and methane emission of Texel and Black-belly sheep fed C4 and C3 grasses tested simultaneously in a temperate and a tropical area. In *Advances in animal biosciences. Proceedings of the 5th Greenhouse Gases and Animal Agriculture Conference (GGAA 2013)*. Dublin, Ireland. pp 285.

Ayala, A., S. Alvores-Moreno, G.J. Alayon, P.C. Aguilar, S.F. Solorio, A.L. Ramirez, M.J. Magaña, and V.J. Ku. 2014. Effect of the ground fruit of *Enterolobium cyclocarpum* Jacq, Griseb on the population of protozoa and methane production in the rumen of air sheep. Primera conferencia de gases de efecto invernadero en sistemas agropecuarios de Latinoamérica (GALA), Chile.

Bennett, E., and S. Carpenter. 2005. Scenarios for ecosystems services, rationale and overview. Chapter 5. Scenarios assessment of the full report. Millennium ecosystem assessment. Island Press, Washington, DC. http://www.millenniumassessment.org/en/Scenarios.aspx

Bouwman, A.F., K.W. van der Hoek, B. Eickhout, and I. Soenario. 2005. Exploring changes in world ruminant production systems. *Agricultural Systems* 84: 121–153.

Briceño-Poot, G.E., A. Ruiz-Gonzalez, A. Chay-Canul, A.J. Ayala-Burgos, C.F. Aguilar-Perez, F.J. Solorio-Sanchez, and J.C. Ku-Vera. 2012. Voluntary intake, apparent digestibility and prediction of methane production by rumen stoichiometry in sheep fed pods of tropical legumes. *Animal Feed Science* 176: 117–122.

Broom, D.M., F.A. Galindo, and E. Murgueitio. 2013. Sustainable, efficient livestock production with high biodiversity and good welfare for animals. *Proceedings of the Royal Society B* 280: 2013–2025.

Bruinsma J. 2009. The resource outlook to 2050, how much do land, water and crop yields need to increase by 2050? Paper presented at the FAO Expert Meeting, 24–26 June 2009, Rome on how to feed the world in 2050. FAO, Rome.

Canul-Solis, J.R., A.T. Piñeiro-Vázquez, E.G. Briceño-Poot, A.J. Chay-Canul, J.A. Alayón-Gamboa, A.J. Ayala-Burgos, C.F. Aguilar-Pérez, F.J. Solorio-Sánchez, O.A. Castelán-Ortega, and J.C. Ku-Vera. 2014. Effect of supplementation with saponins from *Yucca schidigera* on ruminal methane production by Pelibuey sheep fed *Pennisetum purpureum* grass. *Animal Production Science* 54: 1834–1837.

Cuartas, C.C.A., R.J.F. Naranjo, M.A.M. Tarazona, R.E. Murgueitio, O.J.D. Chará, V.J. Ku, S.F.J. Solorio, E.M.X. Flores, S.B. Solorio, and R.R. Barahona. 2014. Contribution of intensive silvopastoral systems to animal performance and to adaptation and mitigation of climate change. *Revista Colombiana de Ciencias Pecuarias* 27(2): 76–94.

De Haan, C., H. Steinfeld, and H. Blackburn. 1997. *Livestock & the environment: Finding a balance*. Rome/Washington, DC: European Commission Directorate-General for Development/ Development Policy Sustainable Development and Natural Resource. Development.

Dumont, B., E. González-García, M. Thomas, L. Fortun-Lamothe, C. Ducrot, J.Y. Dourmad, and M. Tichit. 2014. Forty research issues for the redesign of animal production systems in the 21st century. *Animal* 8(8): 1382–1393.

Erb, K.-H., H. Haberl, F. Krausmann, C. Lauk, C. Plutzar, J.K. Steinberger, C. Müller, A. Bondeau, K. Waha, and G. Pollack. 2009. Eating the planet: Feeding and fuelling the world Sustainably, fairly and humanely – A scoping study. Social Ecology Working Paper 116. Institute of Social Ecology, Vienna.

FAO (Food and Agriculture Organization of the United Nations). 2008a. *Climate change and food security, a framework document*. Rome: FAO.

FAO. 2008b. *The genetic improvement of forage grasses and legumes to enhance adaptation of grasslands to climate change*. Rome: FAO.

FAO, IFAD and WFP. 2013. *The state of food insecurity in the world 2013. The multiple dimensions of food security*. Rome: FAO.

FAO. 2014. *Greenhouse gas emissions from agriculture, forestry and other land use*. Rome: Food and Agriculture Organization of the United Nations.

FAOSTAT. 2014. Food and Agriculture Organization of the United Nation, Statistics Division. http://faostat3.fao.org/browse/G2/*/E.

Flores, C.S.A. 2009. Efecto de la carga animal sobre la disponibilidad, calidad del forraje y cambio de peso de toretes en pastoreo de *Brachiaria brizantha* durante la época de nortes en Yucatán. Tesis M. en C.P.A.T. opción Nutrición. FMVZ-UADY, Mérida, México

Gerber, P.J., H. Steinfeld, B. Henderson, A. Mottet, C. Opio, J. Dijkman, A. Falcucci, and G. Tempio. 2013. *Tackling climate change through livestock: A global assessment of emissions and mitigation opportunities*. Rome: Food and Agriculture Organization of the United Nations (FAO).

Havlick, P., H. Valin, M. Herrero, M. Obersteiner, E. Schmid, M.C. Rufino, A. Mosnier, P.K. Thornton, H. Böttcher, R.T. Conant, S. Frank, S. Fritz, S. Fuss, F. Kraxner, and A. Notenbaert. 2014. Climate change mitigation through livestock system transitions. *Proceedings of the National Academy of Science of the United States of America* 111: 3709–3714.

Herrero, M.P.K., P. Thornton, S. Gerber, and R.S. Reid. 2009. Livestock, livelihoods and the environment: Understanding the trade-offs. *Current Opinion in Environment Sustainability* 1: 111–120.

Herrero, M., P.K. Thornton, A.M. Notenbaert, S. Wood, S. Msangi, H.A. Freeman, D. Bossio, J. Dixon, M. Peters, J. van de Steeg, J. Lynam, P. Parthasarathy Rao, S. Macmillan, B. Gerard, J. McDermott, C. Sere, and M. Rosegrant. 2010. Smart investments in sustainable food production: Revisiting mixed crop-livestock systems. *Science* 327: 822–825.

IPCC. 2006. 2006 IPCC Guidelines for National Greenhouse Gas Inventories. Prepared by the National Greenhouse Gas Inventories Programme, ed. H.S. Eggleston, L. Buendia, K. Miwa, T. Ngara, and K. Tanabe. Toyama: IGES.

IPCC. 2014. Climate change 2014: Impacts, adaptation, and vulnerability. Part A: Global and sectoral aspects. Contribution of Working Group II to the Fifth Assessment Report of the Intergovernmental Panel on Climate Change, ed. C.B. Field, V.R. Barros, D.J. Dokken, K.J. Mach, M.D. Mastrandrea, T.E. Bilir, M. Chatterjee, K.L. Ebi, Y.O.Estrada, R.C. Genova, B. Girma, E.S. Kissel, A.N. Levy, S. MacCracken, P.R. Mastrandrea, and L.L. White. New York: Cambridge University Press.

Johnson, A., and E. Johnson. 1995. Methane emissions from cattle. *Journal of Animal Science* 73: 2483–2492.

Kurihara, M., T. Magner, R.A. Hunter, and G.J. McCrabb. 1999. Methane production and energy partition of cattle in the tropics. *British Journal of Nutrition* 81: 227–234.

Lessa, A.C.R., B.E. Madari, D.S. Paredes, R.M. Boddey, S. Urquiaga, C.P. Jantalia, and B.J.R. Alves. 2014. Bovine urine and dung deposited on Brazilian savannah pastures contribute differently to direct and indirect soil nitrous oxide emissions. *Agriculture, Ecosystems and Environment* 190: 104–111.

Livestock in Development. 1999. *Livestock in poverty-focused development*. Crewkerne: LID, 96 pp.

Lobell, D.B., and S.M. Gourji. 2012. The influence of climate change on global crop productivity. *Plant Physiology* 160: 1686–1697.

Merlo, M.F.E. 2008. Rendimiento y valor nutritivo de *Brachiaria brizantha* a diferentes edades y épocas del año en Yucatán, México. Tesis M. en C.P.A.T. opción Nutrición. FMVZ-UADY, Mérida, México.

Moss, A.R., J.P. Jouany, and J. Newbold. 2000. Methane production by ruminants: Its contribution to global warming. *Annales de Zootechnia* 49: 231–253.

Murgueitio, E., Z. Calle, F. Uribe, A. Calle, and B. Solorio. 2011. Native trees and shrubs for the productive rehabilitation of tropical cattle ranching lands. *Forest Ecology and Management* 261(10): 1654–1663.

Nair, P.K.R. 2012. Climate change mitigation: A low-hanging fruit of agroforestry. In *Agroforestry – The future of global land use*, ed. P.K.R. Nair and D. Garrity, 31–67. Dordrecht: Springer.

Nardone, A., B. Ronchi, N. Lacetera, M.S. Ranieri, and U. Bernabucci. 2010. Effects of climate changes on animal production and sustainability of livestock systems. *Livestock Sciences* 130: 57–69.

Nelson, G.C., M.W. Rosegrant, J. Koo, R. Robertson, T. Sulser, T. Zhu, C. Ringler, S. Msangi, A. Palazzo, M. Barka, M. Magalhaes, R. Valmonte-Santos, M. Ewing, and D. Lee. 2009. *Climate change: Impact on agriculture and costs of adaptation*, Food policy report. Washington, DC: IFPRI.

Peniche-González, I.N., Z.U. González-López, C.F. Aguilar-Pérez, J.C. Ku-Vera, A.J. Ayala-Burgos, and F.J. Solorio-Sánchez. 2014. Milk production and reproduction of dual-purpose cows with a restricted concentrate allowance and access to an association of *Leucaena leucocephala* and *Cynodon nlemfuensis*. *Journal of Applied Animal Research* 42(3): 345–351.

Peters, M., M. Herrero, M. Fisher, K.H. Erb, I. Rao, G.V. Subbarao, A. Castro, J. Arango, J. Chará, E. Murgueitio, R. van der Hoek, P. Läderach, G. Hyman, J. Tapasco, B. Strassburg, B. Paul, A. Rincón, R. Schultze-Kraft, S. Fonte, and T. Searchinger. 2013. Challenges and opportunities for improving eco-efficiency of tropical forage-based systems to mitigate greenhouse gas emissions. In *Proceedings of the 22nd international grassland congress*, Australia.

Reid, R.S., K.A. Galvin, and R.L. Kruska. 2008. Global significance of extensive grazing lands and pastoral societies: An introduction. In *Fragmentation in semi-arid and arid landscapes: Consequences for human and natural systems*, ed. K.A. Galvin, R.S. Reid, R.H. Behnke, and N.T. Hobbs, 1–24. Dordrecht: Springer.

Rosegrant, M.W., M. Fernandez, A. Sinha, J. Alder, H. Ahammad, C. de Fraiture, B. Eickhout, J. Fonseca, J. Huang, O. Koyama, A.M. Omezzine, P. Pingali, R. Ramirez, C. Ringler, S. Robinson, P. Thornton, D. van Vuuren, H. Yana-Shapiro, K. Ebi, R. Kruska, P. Munjal, C. Narrod, S. Ray, T. Sulser, C. Tamagno, M. van Oorschot, and T. Zhu. 2009. Looking into the future for agriculture and AKST. In *International assessment of agricultural knowledge, science and technology for development (IAASTD): Agriculture at a crossroads, global report*, ed. B.D. McIntyre, H.R. Herren, J. Wakhungu, and R.T. Watson, 307–376. Washington, DC: Island Press.

Shelton, M., and S. Dalzell. 2007. Production, economic and environmental benefits of leucaena pastures. *Tropical Grass-lands* 41: 174–190.

Soltan, Y.A., A.S. Morsy, S.M.A. Sallam, R.C. Lucas, H. Louvandini, M. Kreuzer, and A.L. Abdalla. 2013. Contribution of condensed tannins and mimosine to the methane mitigation caused by feeding *Leucaena leucocephala*. *Archives of Animal Nutrition* 67: 169–184.

Steinfeld, H., P. Gerber, T. Wassenaar, V. Castel, M. Rosales, and C. de Haan. 2006. *Livestock's long shadow: Environmental issues and options*. Rome: FAO.

Chapter 12
Climate Variability Impact on Wheat Production in Europe: Adaptation and Mitigation Strategies

Salem Alhajj Ali, Luigi Tedone, and Giuseppe De Mastro

Contents

S.A. Ali (✉) • L. Tedone • G. De Mastro
Faculty of agriculture, Department of agriculture and environmental science,
University of Bari (ALDO MORO), Via Amendola 165/A, 70126 Bari, Italy
e-mail: salem_grb@yahoo.com; luigi.tedone@uniba.it; giuseppe.demastro@uniba.it

M. Ahmed, C.O. Stockle (eds.), *Quantification of Climate Variability, Adaptation and Mitigation for Agricultural Sustainability*,
DOI 10.1007/978-3-319-32059-5_12

Abstract Increased carbon dioxide concentration, rise in temperature and drought stress are important key factors causing frequent occurrence of climate events. Important adaptation strategy such as modification of phenological pattern to avoid stressful period during plant development will be key feature in crop plants. In addition, comprehensive understanding of plants response to elevated CO_2 concentration, temperature and drought stress alone or in combination will be needed to acclimatize crop plant to these changes. Study of climate variability impact on wheat production concerning mitigation strategies is need of time in order to reduce the risk of climate change on crop yield and growth. Similarly, information about the time in which climate variable(s) occurred in the field is important as the severity of its effect/their combined effect can vary largely. Agronomic practices such as cultivar choice, water and nitrogen supply, nutrients availability and growing conditions should be taken into account to design adaptation options. The failure of agriculture to adapt to climatic variability will impact global food, especially wheat production. A holistic approach will be paramount to sustaining agriculture and the vitality of the world in the face of climate change.

Keywords Carbon dioxide • Temperature • Drought • Climate change • Wheat • Adaptation • Mitigation

12.1 Introduction

The necessity to increase cereal yield is needed to meet the projected increased demand for the increasing population worldwide. In fact, in recent years, cereal yield and quality were incredibly improved due to modernized agriculture with intensive inputs, high mechanization and new technology. However, the use of these intensive input has environmental implications that must not forgotten. To keep up with this level, yield safety has gained more attention recently (Barnabás et al. 2008). Special attention was given to the effect of unpredicted climate events and their frequent occurrence on cereal yields in Europe. Therefore, understanding of the major stress factors, which have been found to be unfavorable for crop yield stability, would be the key for sustaining crop productivity.

Frequent occurrences of climate events such as increasing carbon dioxide concentration in the atmosphere, increases in temperature (Rosenzweig and Tubiello 2007) and drought (water deficit) stress (Barnabás et al. 2008) are considered amongst the most important key factors reported in the literature. Their effects (alone or in combination) are expected to influence food production (Abraha and Savage 2006) by causing yield reduction in many parts of the world Benlloch-Gonzalez et al. (2014). On wheat yield, for example, heat has an effect in shortening the growth phases of the crop resulted in decrease in wheat yield estimated to reach 6 percent for each 1 °C increase in temperature (i.e. Asseng et al. 2015). In addition, changing in precipitation patterns, especially in the dry areas, could affect the availability of surface water, which is necessary for crop germination. Whereas, elevated carbon dioxide (eCO_2) affect positively and, therefore, promotes the development of crops with C3 photosynthetic pathway.

Nowadays, the impact of climate change on food safety and quality is becoming the center of any regional, national and international political dialogue. Assessment studies of the potential impacts on crop development and yield have been intensively carried out with the help of crop models. However, these estimates are depending on many factors such as the uncertainties in climate change scenarios, region of study and crop model used (Aggarwal and Mall 2002). In Europe, findings of different studies are in agreement that unexpected climate events could cause a severe damage to the second widely grown cereal crop (wheat) (Trnka et al. 2014). Because of wheat's importance in this area, the effects of environmental changes on wheat yield need to be assessed (Amthor 2001) in the framework of sustainability. Understanding the social, economic and environmental dimensions of wheat production is essential to develop appropriate and targeted adaptation efforts. Such efforts must take into consideration that the impacts, due to the environmental changes, have to be integrated in the frame of multi-criteria context.

This chapter is not a collection review of the literature that has a contribution to the climate change impact on wheat production, but an attempt has been made to update and outline the available information on wheat production, under climate variability, to better understand how climate variables affecting wheat production in order to give a prospect towards potential adaptation and mitigation strategies for sustainable crop production.

12.1.1 Wheat Global and Country Scenario

The cultivation of wheat (Emmer wheat) goes back to thousands of years as a first domesticated crop to provide major dietary components for ancient civilizations. It is well adapted to varying climate conditions. Nowadays, because of its adaptability with the ability to grow in difficult weathers and geographic areas, wheat occupies more land than any other crop (FAOSTAT 2013) in order to keep up with the increasing population worldwide. It represents the most favoured staple food with large share (1/5) of the total humanity's food. Its nutritional value (as a source of calories, carbohydrate and protein (Braun et al. 2010)) makes its trade the highest when compared to other crops together. Wheat is a temperature crop therefore grows best under Mediterranean-like conditions. Nevertheless, relatively low temperature is required with dew formation at early stage for best crop establishment. In general, wheat is adopted to a wide range of temperature (Briggle 1980), moisture, precipitation (Leonard and Martin 1963) and soils. However, high moisture content resulted from excessive rainfall with high temperature can cause yield losses due to the spread of some common diseases (i.e. root rot). Whereas poorly drained soil with high soil acidity can cause wheat growth to fail.

Most economically important crops are belonging to either C_3 or C_4 plant group. The classification came according to whether the first stable organic compound formed during plant photosynthesis is the glyceraldehyde 3-phosphate with three carbon atoms or oxaloacetate with four carbon atoms (Wolfe and Erickson 1993). Wheat is considered one of the C_3 plants. Generally, the term "wheat" means species of wheat, which include the durum and common. The importance of wheat came from its easy storage and conversion to flour or semolina for making different food products especially bread and pasta. The most common and widely known wheat species are belonging to three groups based on their genomes (Smale et al. 1996). Among them, common bread wheat (*Triticum aestivum*) and durum wheat (*T. turgidum*) as international commodities are now playing an important role in feeding the world.

In general, the cultivation of cereals, according to the data collected between 2008 and 2013, covered an area of more than 700 million hectare worldwide with 2,6 billion tons of production (s.1). In wheat production, however, the evolution of the production in the last 50 years is impressive (Fig. 12.2). It shows an increase from about 200 million ton in 1961 to the 714 million tons in 2013. Although the surface used for wheat cultivation has been stable in the last 50 years, yield stability has increased substantially across environments. In the last years (Table 12.1), total wheat was cultivated in more than 219 million hectares (31 % of the total cereal production).

In terms of surface, the average of the last years (Table 12.1), wheat (common and durum) represents the main cereal cultivation (almost 220 million ha) followed by Maize (170 million) and rice (161 million) (FAOSTAT 2013). However, in terms of productivity, Maize (880 million tons) represents the highest total production worldwide followed by rice (714 million tons) and wheat with 684 million tons (Fig. 12.3). Consequently, the productivity resulted the highest in Maize (5.1 t/ha) and rice (4,4 t/ha) while in wheat the average production comes the third with 3,1 t/ha (Table 12.1).

In the last year (2013), worldwide (Table 12.1 & Fig. 12.2) and European wheat production (Fig. 12.5) has reached the highest production ever. Average across the last years (Fig. 12.1) show that wheat growth rate (%) was the highest in Oceania followed by Europe respect to cereal growth rate. This was mainly due to the improvement of wheat cultivars through breeding initiatives, agriculture techniques, and mechanization. Total growth for both cereal and wheat production, however, reached the maximum in Africa and Asia, respectively (Fig. 12.1). In addition to the mentioned previously, the adoption of best agricultural practices has also played an important role in enhancing the sustainable production of cereal. For example, compared to conventional tillage system, the adoption of conservation tillage found to increase soil moisture storage efficiency, which therefore result in crop yield increase (Al-Issa and Samarah 2006). According to the experiences of some countries i.e. Turkey, the adoption of conservation agriculture practices with the ability to hold up water were applied, especially in the dray seasons, resulted in impressive yield increase in some regions (Curtis 1982; Dalrymple 1986). This has led the FAO to urge key producer countries to change their agriculture practices in order to meet sustainability criteria.

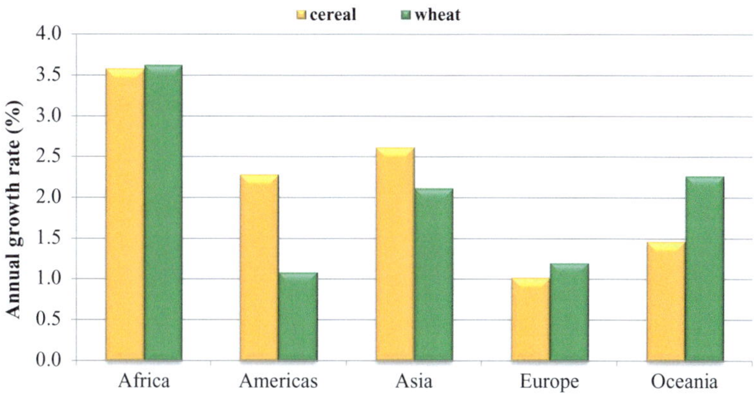

Fig. 12.1 Total cereal to wheat production growth rates by region (2000–2013) (Source: FAOSTAT (2013))

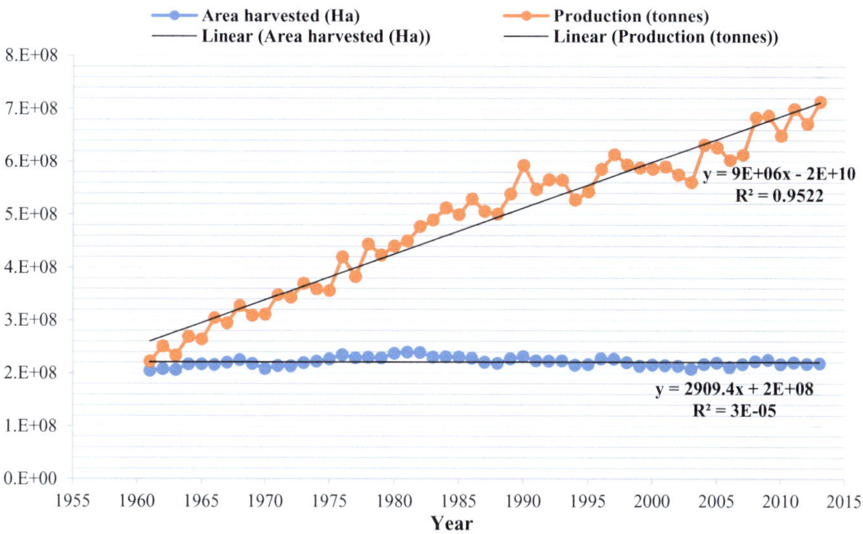

Fig. 12.2 Evolution of cultivated areas and yield of wheat

Although intensive agriculture may ensure high yields, it can cause serious environmental damages. For example, the use of farm inputs such as nitrogen fertilizer and irrigation water has increase dramatically to maintain or improve crop yield, however they represent direct contributions to the environmental impacts. In some key producer countries (i.e. India and China), total cereal yields was doubled and tripled respectively for the two countries, due to the dramatic increase of these inputs (Borlaug and Dowswell 1996 and Smale et al. 1996).

Table 12.1 Production, yield and harvested areas of main cereal commodities worldwide (2008–2013)

Area harvested (Ha)										
year	Wheat	Maize	Rice, paddy	Barley	Sorghum	Millet	Others	Oats	Rye	Total
2008	222.279.147	162.689.152	159.992.624	55.532.588	45.381.971	35.639.572	12.649.149	11.278.727	6.751.458	712.194.388
2009	224.634.012	158.743.228	158.130.441	53.955.565	40.944.505	34.026.599	12.276.773	10.194.476	6.622.357	699.527.956
2010	216.965.233	164.029.760	161.188.783	47.368.145	41.583.161	35.851.148	11.832.675	9.096.364	5.035.287	692.950.556
2011	220.195.869	172.256.930	162.799.640	48.488.187	42.311.669	33.819.227	12.149.371	9.637.787	5.132.617	706.791.297
2012	217.319.740	178.551.622	162.317.207	49.573.245	38.157.854	31.722.594	12.558.698	9.580.377	5.331.985	705.113.322
2013	218.460.701	184.192.053	164.721.663	49.781.046	42.120.446	32.916.261	12.549.745	9.758.714	5.758.584	720.258.913
Total	219.975.784	170.077.124	161.525.060	50.783.129	41.749.934	33.995.900	12.336.069	9.924.408	5.771.998	706.139.405

Yield (Hg/Ha)										
year	Wheat	Maize	Rice, paddy	Barley	Sorghum	Millet	Others	Oats	Rye	Total
2008	30.736	51.055	43.028	27.873	14.659	9.660	15.733	22.931	26.851	26.947
2009	30.571	51.669	43.442	28.131	13.906	7.663	16.478	22.900	27.624	26.932
2010	29.928	51.897	43.551	26.111	14.442	9.095	16.061	21.682	23.754	26.280
2011	31.762	51.542	44.602	27.419	13.765	8.043	16.244	23.155	25.358	26.877
2012	30.899	48.882	45.478	26.931	14.946	9.519	17.055	22.245	27.411	27.041
2013	32.646	55.200	45.271	29.078	14.574	9.075	17.837	24.410	28.994	28.565
Total	31.090	51.708	44.229	27.591	14.382	8.843	16.568	22.887	26.665	27.107

Production (tonnes)										
year	Wheat	Maize	Rice, paddy	Barley	Sorghum	Millet	Others	Oats	Rye	Total
2008	683.207.030	830.611.273	688.414.632	154.788.319	66.524.443	34.429.033	25.724.419	25.862.841	18.128.210	2.527.690.200
2009	686.720.279	820.202.618	686.957.597	151.782.372	56.937.770	26.073.631	27.712.840	23.344.913	18.293.688	2.498.025.708
2010	649.325.445	851.270.850	701.998.667	123.683.001	60.056.057	32.607.751	25.028.105	19.722.987	11.960.960	2.475.653.823
2011	699.389.500	887.854.782	726.121.583	132.950.642	58.242.536	27.200.373	25.671.440	22.316.322	13.015.534	2.592.762.712
2012	671.496.872	872.791.598	738.187.643	133.506.664	57.029.990	30.197.305	27.117.618	21.312.011	14.615.719	2.566.255.420
2013	713.182.914	1.016.736.092	745.709.788	144.755.038	61.384.559	29.870.058	28.510.775	23.821.207	16.695.636	2.780.666.067
Total	683.887.007	879.911.202	714.564.985	140.244.339	60.029.226	30.063.025	26.627.533	22.730.047	15.451.625	2.573.508.988

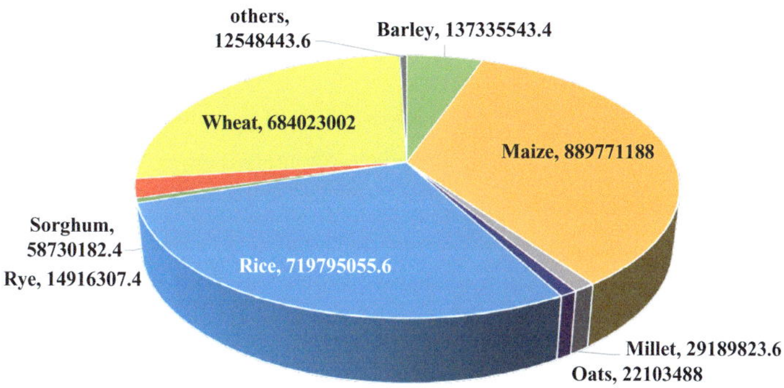

Fig. 12.3 Main cereals cultivation by production (million tonnes) worldwide

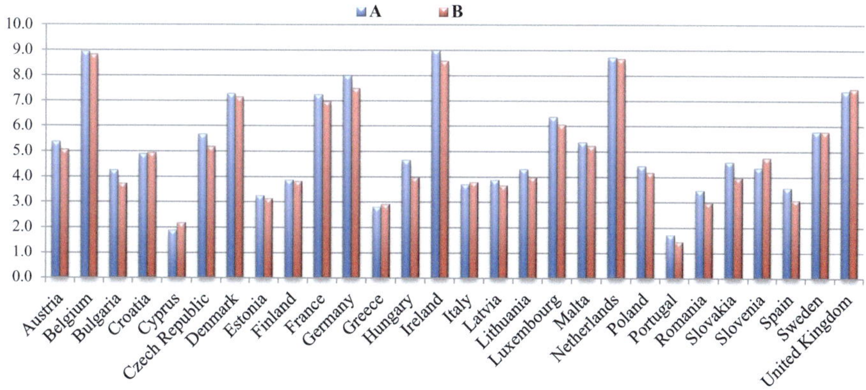

Fig. 12.4 Wheat yield (t/ha) in Europe (**a**) in the last year (2013) versus (**b**) Average of the last five years (2009–2013)

Despite these increasing yields, United States Department of Agriculture (USDA 2015) estimated wheat production in European Union (EU) (2015–2016 growing season) to be reduced by 8.6 million tons, and cultivated area by 0.2 million hectares from previous year. Whereas 4.7 % of yield reduction is estimated compared to the last year (2014–2015), but it remains above the 5- year average by 2.6 %. The combination of dryness and heat has reduced estimated EU wheat yields after unfavourable weather conditions. Wheat production is reduced by 1.3 million tons in Germany, 0.6 million in France, 0.6 million in Spain and 0.3 million in Hungary. Low rainfall levels, minimal soil moisture and high temperatures contributed to deterioration in crop conditions in the wheat.

Average across last years (2008–2013), Table 12.2 shows that wheat production is mainly concentrated in Asia (100 million hectares and 318 million tons) followed by Europe (57 million hectares and 225 million tons) and America (37 million hectares and 117 million tons) (FAOSTAT 2013). Oceania and Africa are the lowest producers with lowest cultivation area, respectively. The average annual world farm yield for wheat was 3.3 tonnes per hectare, in 2013. In the same year, New Zealander wheat farms were the most productive with a nationwide average of 9.1 tonnes per

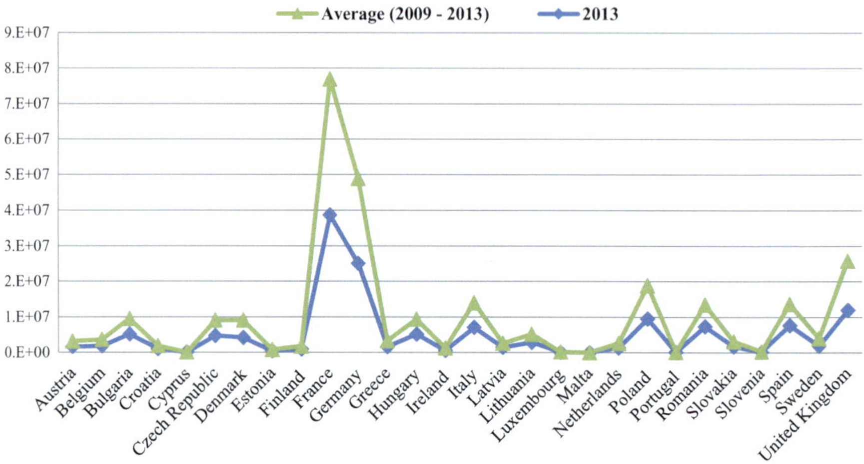

Fig. 12.5 Wheat production (million tonnes) in Europe; last year (2013) versus Average of the last five years (2009–2013)

hectare, whereas Ireland was a close second (FAO 2013). In terms of yield, Europe presents the highest yield with 3,8 t/ha (Table 12.2).

According to the FAO (Table 12.4), China, India, USA, Russian Federation, France and Canada are ranked as the most wheat producer countries in the last ten years. The same ranking order was maintained in 2013 (FAOSTAT 2014). Globally, India occupies more land for wheat production, whereas China's share is almost one-sixth of the world production with highest yield (Table 12.3). However, in absolute terms, the highest value of yield is found in the Northern countries of Europe; Ireland, Netherlands and Belgium where the average yields reach more than 8.7 t/ha (average of last five years (2009–2013)) (Fig. 12.4) (FAOSTAT 2015).

In Europe, during the last five years (2009–2013), wheat production increased from 139 million tonne in 2009 to 143 million tonne in 2013, making the last year more productive than the average of five year. Most of cultivated wheat in Europe produced by five countries (more than 60 % of total wheat in Europe). Among them is, in order, France (38), Germany (24), UK (14), Poland (9) and Italy (7) million tonnes (Fig. 12.5). The figure shows that France and Germany are the most wheat producer in Europe with almost 50 % of the total production.

- **DurumWheatProduction**

In general, durum what is the most common cultivar among cereals in Europe, and in Italy. Globally, wheat known as spring wheat and winter wheat according to their sowing date and temperature requirements. In Italian environment, planting usually takes place in autumn, between the end of October to early December, while in North Europe, for the low winter temperature, sowing is almost done in spring period. Within the total wheat production, durum wheat production represents marginal part of total production. In fact, the total surface of durum wheat cultivation is about 13,5 million hectares with production of about 35,8 million tons (data 2013/2014), which represents about 5,6 % of total world wheat production.

Table 12.2 Cereal production, harvested area and yield by continent

Area harvested (Ha)					
	2008	2009	2010	2011	2012	2013
Africa + (Total)	8.614.597	10.335.513	9.285.580	9.744.623	10.224.952	10.083.907
Americas + (Total)	41.539.474	38.096.909	36.351.181	36.134.979	36.451.175	36.496.507
Asia + (Total)	96.951.474	101.271.033	101.391.278	101.307.735	102.842.312	101.748.086
Europe + (Total)	61.601.068	61.088.661	56.375.422	59.555.921	54.246.541	57.582.991
Oceania + (Total)	13.572.534	13.841.896	13.561.772	13.452.611	13.554.760	12.549.210

Production (tonnes)						
	2008	2009	2010	2011	2012	2013
Africa + (Total)	19.681.726	26.099.406	21.413.455	25.049.032	24.704.201	28.072.592
Americas + (Total)	119.032.855	109.924.267	113.665.816	108.966.549	108.954.670	117.504.487
Asia + (Total)	274.582.549	300.071.097	290.094.084	313.678.202	311.417.956	318.834.326
Europe + (Total)	248.146.360	228.566.037	201.569.189	223.902.371	196.036.413	225.468.124
Oceania + (Total)	21.763.540	22.059.472	22.582.901	27.793.346	30.393.632	23.303.385

Yield (Hg/Ha)						
	2008	2009	2010	2011	2012	2013
Africa + (Total)	22.847	25.252	23.061	25.705	24.161	27.839
Americas + (Total)	28.655	28.854	31.269	30.155	29.891	32.196
Asia + (Total)	28.322	29.631	28.611	30.963	30.280	31.336
Europe + (Total)	40.283	37.415	35.755	37.595	36.138	39.155
Oceania + (Total)	16.035	15.937	16.652	20.660	22.423	18.570

Table 12.3 First wheat producer countries worldwide (2008–2013)

Area harvested (Ha)						
country	2008	2009	2010	2011	2012	2013
India	28.038.600	27.752.400	28.457.400	29.068.600	29.860.000	29.650.000
China, mainland	23.617.200	24.291.000	24.256.000	24.270.380	24.268.300	24.100.000
Russian Federation	26.070.300	26.632.900	21.639.800	24.835.500	21.277.900	23.371.410
United States of America	22.540.828	20.191.200	19.270.930	18.496.360	19.797.644	18.274.206
Kazakhstan	12.906.300	14.329.400	13.138.000	13.686.400	14.410.900	12.953.500
Australia	13.530.196	13.788.000	13.507.000	13.400.000	13.500.000	12.500.000
Canada	10.031.700	9.638.200	8.268.700	8.543.600	9.497.200	10.441.500
Pakistan	8.549.800	9.046.000	9.131.600	8.900.700	8.666.000	8.693.000
Turkey	8.090.000	8.100.000	8.103.400	8.096.000	7.529.600	7.772.600
Yield (Hg/Ha)						
country	2008	2009	2010	2011	2012	2013
India	28.022	29.071	28.395	29.886	31.775	31.538
China, mainland	47.620	47.390	47.486	48.376	49.869	50.506
Russian Federation	24.459	23.182	19.181	22.645	17.727	22.288
United States of America	30.175	29.897	31.167	29.418	31.154	31.720
Kazakhstan	9.715	11.900	7.336	16.609	6.829	10.762
Australia	15.831	15.706	16.390	20.455	22.152	18.284
Canada	28.521	27.855	28.017	29.568	28.646	35.943
Pakistan	24.514	26.568	25.528	28.328	27.086	27.874
Turkey	21.980	25.432	24.279	26.927	26.695	28.369

Production (tonnes)						
country	2008	2009	2010	2011	2012	2013
India	78.570.200	80.679.400	80.803.600	86.874.000	94.880.000	93.510.000
China, mainland	112.464.000	115.115.000	115.181.000	117.410.000	121.023.000	121.720.000
Russian Federation	63.765.140	61.739.750	41.507.580	56.239.990	37.719.640	52.090.797
United States of America	68.016.100	60.365.730	60.062.410	54.413.310	61.677.387	57.966.658
Kazakhstan	12.538.200	17.052.000	9.638.400	22.732.070	9.841.300	13.940.800
Australia	21.420.177	21.656.000	22.138.000	27.410.076	29.905.009	22.855.576
Canada	28.611.100	26.847.600	23.166.800	25.261.400	27.205.200	37.529.600
Pakistan	20.958.800	24.033.000	23.310.800	25.213.800	23.473.000	24.231.000
Turkey	17.782.000	20.600.000	19.674.000	21.800.000	20.100.000	22.050.000

Table 12.4 Ranking of the top wheat producers countries worldwide (10-year average (2004–2013) expressed in million metric tons)

Ranking	Country	Mean
1	China, mainland	111
2	India	80
3	United States of America	58
4	Russian Federation	50
5	France	38
6	Canada	27
7	Germany	24
8	Pakistan	23
9	Australia	22
10	Turkey	20
11	Ukraine	19
12	United Kingdom	15
13	Kazakhstan	14
14	Iran	14
15	Argentina	13
16	Poland	9
17	Italy	8
	WORLD	658

Source: FAOSTAT data, 2015

Durum wheat (*Triticum turgidum* L. var. durum) is a staple crop in the Mediterranean Basin, where Italy and Spain are among the main European producers (EUROSTAT 2015). The cultivation of durum wheat is typical of Mediterranean climate, but it is also widespread in several other countries. Globally, Canada is the first country by surface of durum wheat cultivation followed by Italy (12.1 %), Algeria (8.8 %), Russia (8.8 %), US (7.7 %) and Turkey & Syria (7.6 %). In Europe, durum wheat cultivation area is about 3,5 million ha equal to 25.6 % of world area. Almost 86 % of durum wheat cultivated area in EU is occupied by Italy (48 %) followed by Spain (25 %) and Greece with 13 %.

12.1.2 Climate Variability, Population Growth and Crop Production: An Overview

In fact, climate variables are, somehow, interrelated. Among the most important variables including elevated temperature, water stress and increasing atmospheric CO_2 concentration which represent potential threat to crop growth and development (Bita and Gerats 2013). The unpredictable occurrence and severity of these variables, alone or in combination have an effect on the morphological, physiological, and biochemical parameters of crop plants and, therefore, their productivity. With technological and mechanisation improvements in agriculture to increase yields,

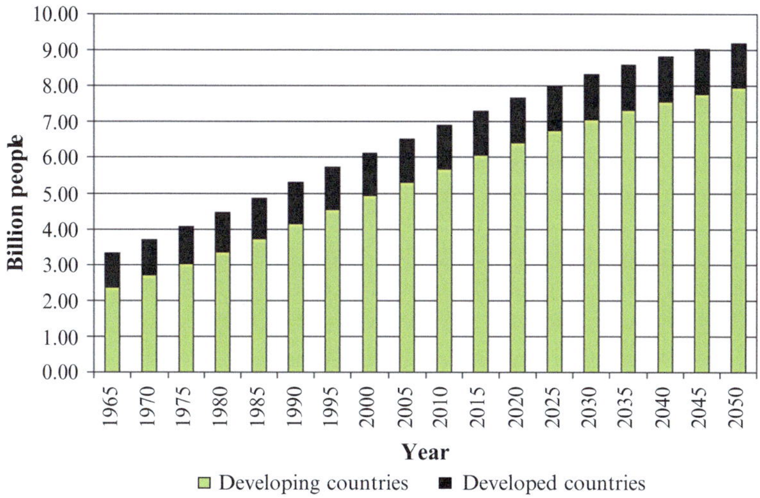

Fig. 12.6 World population (1965–2050) (Source: Population Division of the Department of Economic and Social Affairs of the United Nations Secretariat (2007))

climate variability is still significant factors in agricultural productivity (Kang et al. 2009). Increasing the uncertainty with respect to food production under changing climate has drawn more attention to the risks associated with these changes (Reddy and Pachepsky 2000).

The projected increase in world population from 7.2 billion to 8.4 billion by 2030 (UN 2013) (Fig. 12.6) would lead to increasing demand for food by 35 % (NIC 2012) and for wheat by 70 %, by 2050 (CIMMYT 2014). With the reduction of agriculture output by 6 and 20 % in industrial and least developed countries by 2080 respectively, yields could decrease by 15 % on average (Masters et al. 2010). Hisas (2011) has predicted wheat deficit by 14 % because of global climate change by 2020. This will lead global food production to unlikely satisfy future demand under predicted climate change scenarios. Such a pressure could be further exacerbated by the changing climate, particularly in the longer-term (Falloon et al. 2015). Consequently, the ever-increasing population has alarmed food security, therefore attempts have been initiated to integrate modern technology and biotechnology tools to improve yields of internationally important agricultural commodities (i.e. wheat and rice).

Achieving sustainable agriculture for food security and poverty alleviation, many obstacles such as climate change, natural resource availability, a growing global population, and increasing demand for goods and services need to be overcome, so that a holistic approach that includes stress-tolerant germplasm, sustainable farm and natural resource management, and sound policy interventions can be followed in order to maintain food security.

12.1.3 *Wheat and Food Security Under Climate Variability*

Although intensive agriculture resulted in substantial increase in global food production, future food security remains uncertain due to depletion of natural resources. According to the World Bank (2008), world population is expected to reach 9 billion by 2050 with grow rate of 34 % (Fig. 12.6). In this regard, the necessity to increase global food supply will be a major challenge to satisfy food demand for this increase in world population. Some studies (i.e. Anon 2009 and Semenov et al. 2014) expected an increase in global food supply to reach 70 % by 2050. However, this expectation remains uncertain and will continue to threaten food security unless proper actions are made to satisfy human needs. In the light of global efforts to ensure food security, annual global wheat production rate, including production quality, needs to be increased by 2 % until 2020 (Reynolds et al. 2008). In order to satisfy the nutritional needs for the increasing population, factors affecting global food production need to be considered in the frame of a holistic approach. The unforeseen changes in climate patterns, for example, represent a real threat to agricultural activities (Lepetz et al. 2009). Across the globe, this threat has become of a great concern to many farmers as well as agricultural scientists (Bosello and Zhang 2005) trying, in the frame of international efforts, to better understand, in order to avoid the unfavorable and unpredicted crop yield shortage that affect food security. For example, climate variables have shown to have a crucial role in the production of food including wheat. Given the importance of wheat, and considering the high sensitivity of wheat to climatic and environmental variations (Porter and Semenov 2005), the effect of climatic variability on crop yields has become an important issue within climatic impact community (Semenov and Porter 1995; Wilks and Riha 1996).

In order to sustain food supply, the effects of key biotic and abiotic factors on food production have to be well studied and results should be integrated in any political agenda regarding food security. Elevated temperature, for example, effect plant growth negatively through increase crop respiration, decreases evapotranspiration and accelerates nutrient mineralization in soils. Due to unforeseen increase in temperature, changes in land use pattern may be expected due to drought incidence, which can cause a serious menace to food security (Lobell et al. 2013). In addition, the widespread of pathogens and other biotic variables in the agroecosystem, due to the increase in temperature, can cause large direct impact on crop production. Increase temperature can affect also nutrients availability, mineral mobilization processes and microbial activities leading to disorder in plant growth and therefore affecting crop productivity in an indirect way. Therefore, the integration of any efforts including agronomic and breeding options to avoid or counteract the effect of increasing temperature on plant growth and productivity would count as a global effort to insure food security.

12.1.4 Institutes Working on Wheat: Climate Variability Research and Outcomes

In the framework of international efforts to ensure food supply under the unforeseen climate changes, many institutions are working for the development of options to be adopted in key food production regions, especially cereals. The International Center for Agricultural Research in the Dry Areas (ICARDA), for example, is conducting field researches in the dry regions to ensue good delivery of its research outcomes to the small farms and poor farmers helping them to face future challenges in agriculture.

According to ICARDA research group, the key production regions worldwide are among the most vulnerable areas to climate change (ICARDA 2014).

In the framework of its research activities, ICARDA carried out a project with the aim to increase knowledge about the potential impacts of changing climate on crops productivity in Central Asia. Using crop simulation models, they assessed the environmental impacts of different wheat varieties cultivated in key production zones in central Asia with the aim to understand the potential change in total aboveground biomass, yield and nutrients uptake as effected by various IPCC climate change scenarios. Their published results (Sommer et al. 2013) predict long-term, however small (0.1–0.5 t ha^{-1}), increase in wheat yield across the study areas. This small increase was found to account for 12 % cross central Asia as a whole. However, the response of wheat yield in each study site was found to be highly dependent on crop management, nutrients and water availability.

Within its initiative to insure food security, ICARDA works to help local farmers using their innovations techniques to provide animal feeds for livestock production by improving rainfed barley productivity with little water availability (ICARDA 2014). Being characterized by low annual rainfall, the use of local crop management techniques (i.e. zero tillage, crop rotation and water-harvesting) in the dry areas of Jordan and Iraq found to be promising in terms of crop yield. Compared to conventional tillage, barley yields were increased by 20 and 50 % in Jordan and Iraq, respectively during the growing seasons (2011–2013).

Following the international efforts, the International Maize and Wheat Improvement Centre (CIMMYT) works to overcome the negative impacts of climate change on wheat yield. Through its breeding program, the development of wheat germplasm resulted in high-yielding and stress-resistant new wheat cultivars to suit the needs of diverse wheat growing environments which have been successful to increase yield potential (i.e. Araus et al. 2002 and Trethowan Mvan Ginkel and Rajaram 2002) and to respond to climate change in certain environments.

Breeding program includes also the creation of new varieties with resistant to abiotic stresses such as diseases. Example came from international cooperation efforts when wheat stem rust (Ug99) first found in Uganda in late nineties. It represents one of the most damaging disease of wheat in the history. Due to the susceptibility of most wheat varieties to this disease, its spread could cause a real threat to the world's wheat crops. In this regard, CIMMYT, ICARDA and FAO along with national research organizations have worked together to breed and distribute resistant varieties. This international cooperation resulted in the creation of the Borlaug Global Rust Initiative (BGRI) (FAO 2009) to counteract this threat. Recently, with the wide spread of stripe rust (or yellow rust) in different parts of Africa, the Middle East and Asia, CIMMYT has developed a number of wheat cultivars resistant to this epidemic and, at the same time, resistant to Ug99 with high yield potential (10–15 %) than grown varieties. In addition, CIMMYT also collaborate with key national agricultural research system partners to identify important physiological traits that have value as predictors of yield at high temperatures. This collaboration resulted in the establishment of International Heat Stress Genotype Experiments at a 4-year basis. Experimental locations were selected based on a classification of temperature and humidity during wheat growing cycle. "Hot" and "very hot" locations were defined as having mean temperatures above 17.5 and 22.5 °C, respectively, during the coolest month. Whereas "Dry" and "Humid" locations were defined as having mean vapor pressure deficits above and below 1.0 kPa, respectively. Asseng et al. (2015) used data from seven of the original 12 locations to represent a range of temperatures. Some of their findings can be found later in this chapter.

In Italy, Cereal Research Centre (CRA) is working mainly on the collection and maintenance of cereal varieties suitable for the cultivation in the Mediterranean area with the aim for the preservation and enhancement of biodiversity. In particular, the centre works to improve wheat varieties in order to create wheat resistant cultivars to different abiotic and biotic stresses. The improvement of wheat yield and quality represents also fundamental aspects in its genetic improvement and breeding program with reference to climate change. Under climate change, the centre works also to develop agronomical adoption options to counteract the unforeseen changes in weather events and to ensure cereal yields for human and animal consumption. For example, using breeding approach Annicchiarico and Pecetti (1993) studied the role of some agronomic traits to select durum wheat for a dry Mediterranean region in Northern Syria. Using two different rainfall amount and distribution locations, the authors found that the influence of certain characters on grain yield was affected by environmental variables. Another work lead by CRA investigates the correlation between climatic variables and morpho-physiological variation of genetic resources in durum wheat landraces of four countries of origin (Ethiopia, Morocco, Syria, and Turkey) (Annicchiarico et al. 1995). Their study highlighted that climatic variables (specially drought stress and high temperature) and agronomic traits (in particular earliness of heading) were highly correlated. For example, temperature was found to have positive effect on potential yield as well as earliness within Turkish germplasm, whereas lower drought and heat stress caused varieties from Ethiopian and Syrian gene pools to have longer spike. These findings can lead to the development of agriculture policies through the integration of genetic, breeding as well as agronomic options to fight climate change and ensure food security.

12.1.5 Challenges: Limitation of and All Related Problem to Global Wheat Production

There are certain things that all human beings need to survive and food is one of them. The biggest concern about climate change is being the main influential factor in food security worldwide. This would be the primary challenge in the 21st century in order to satisfy human needs from food within the frame of sustainability (Lal 2005).

Wheat plays an important role in feeding the world as it represents a major stable food for more than 2.5 billion rural people in the world. Nevertheless, several studies demonstrate that climate change threatens its future harvest in the key production areas. This implies the need to improve wheat management systems to ensure productivity under changing climate. However, the adoption of new management strategies has to be in accordance with future change in heat and other climatic stresses.

In Europe, scientists predict annual mean temperature to rise more than at the global mean. The occurrence of such climate extreme event in Europe will be altered by climate change. Heat waves, droughts and heavy rainfalls (Christensen et al. 2007), for instance, are expected to increase in frequency, intensity and duration leading to increase the uncertainty with respect to food production (Reddy and Pachepsky 2000). These risks will have severe impact on food security threatened food supply for people live in susceptible parts of the world.

In fact, global researches outcomes predicted a decline in grain yield for most regions in the world in response to climate change (Tubiello et al. 2000; Luo et al. 2005; Meza and Silva 2009; Kristensen et al. 2011). For example, the Natural Resources Institute of Finland NRIF (2015) reported that future global wheat harvest is expected to decline on average by 6 % for each degree Celsius of temperature increase if improved management to face environmental extremes is not adapted. Worldwide, this would correspond to 42 million tons of yield reduction (Asseng et al. 2015), which equals to a quarter of current global wheat trade. However, the main issue that every country has to face today is the choice of proper crop management which has to ensure better adoption to the future change in climate and, at the same time, sustain crop productivity. To this purpose, adaptive options might include water management (water use efficiency) (Kang et al. 2009), the introduction of new cultivars through breeding and the optimization of crop management practices (Reynolds et al. 2009; Alhajj Ali et al. 2015) all of which are thought to present viable options to reduce environmental impacts from cropping and counteract the effects of climate change (Grasty 1999) on crop productivity. In fact, the documented increase in crop yield was found to be in relation to these options, however, yield increase rate may vary according to the different climatic areas, mainly in relation to the occurrence of drought (Araus et al. 2002) and other climatic variables. Recently, some of these options were found to increase, by three times, European wheat yields compared to the second half of the last century (Ewert et al. 2005). Globally, similar increase in cereal production was obtained as a result of breeding and agronomic improvements.

12.2 Quantification of Climate Variability

12.2.1 Crop Responses to Abiotic Stresses – Overview

12.2.1.1 Plant Response to High Temperature

One of the most limiting factors affecting crop growth is the temperature (Porter and Moot 1998). Elevated air temperature was found to be responsible for the acceleration of plant metabolism (Larcher 2003) resulted in disturbance of some physiological processes including plant development, fruit formation and the functioning of the photosynthetic apparatus, respiration and transpiration (Farrar and Williams 1991). However, the severity of heat stress was found to be higher when stages such as flowering and grain filling occur during sensitive growth. Several studies (Giannakopoulos et al. 2009; Battisti and Naylor 2009; Supit et al. 2010) highlighted that rising temperatures may also shortened the growing season which resulted in time reduction for biomass accumulation (i.e. decreased of leaf area production) during grain-filling period, thus reducing yields (Wheeler et al. 1996b). In this regard, a decline (3–10 %) in grain yield was observed for each ° C increase in mean temperature (Jones et al. 2003 and Berntsen et al. 2003). Using simulation models (see, **section III**), Tubiello et al. (2000) reported similar trend of yield reduction (10–40 %) due to the acceleration in plant phenology due to increase in temperatures. In rainfed areas, Leemans and Solomon (1993) reported a reduction in global crop yields of approximately 15 % temperature variability. Reynolds et al. (2010) indicate that the most significant factors associated with yield reduction under heat stress are increased sterility, shortened life cycle, reduced light interception and the disorder in carbon assimilation processes (photosynthesis, transpiration, and respiration). The effects of heat stress on phonological development and growth of different wheat cultivars is well documented (Nawaz et al. 2013). The authors found that, at reproductive stages, heat stress was more severe and affect agronomic, physiological and yield related traits of all studied wheat cultivars.

In fact, the single effect of either high or low temperatures occurs and affects plant growth Grace (1988) and therefore productivity. In addition, the cumulative effect resulted from high temperature occurrence in some days during growing season appear to have a significant influence on yield than change in temperature by ± 2 °C during the whole growing season (Rawson 1992). The variability in temperature during growing season can reduce the necessary time for the crop to enjoy the optimal temperatures for photosynthesis and development. Consequently, variation in crop responses to temperature can be expected depending on whether the optimum temperature for photosynthesis, crop growth and potential yield is reached (Conroy et al. 1994). In nature, the optimum temperature for growth differs between the growth stages of a crop and between crops. The one suitable for crop emergence will not be suitable for flowering or grain-filling and the vice versa. One of the main effects of heat stress (increases in above-optimum air temperatures) is on photosynthetic sink capacity of plants (Wahid et al. 2007; Asseng et al. 2011). During flowering, Wilks and Riha (1996) found that the exposure to below the optimal temperatures can reduce sink capacity of the plant leading to lower photosynthesis rate. Conroy et al. (1994) detailed the effect of increasing temperature on wheat development and found that the low photosynthesis rate caused higher leaf appearance rates therefore less time for flowering which lead to decrease in leaf area production and grain yield. Therefore, any defect in photosynthesis process resulted from unfavorable increase/decrease in air temperature is expected to affect grain yield. This include the production of oxidative reactive species (Wang et al. 2011), pollen mortality (Saini et al. 2010), increasing grain abortion (Hays et al. 2007), causes metabolic limitations and causes oxidative damage to the chloroplast (Farooq et al. 2011).

In wheat, the impact of temperature variability on yields has been studied intensively (i.e. Asseng et al. 2011). Some studies found that any slight increase in temperature, out of the specific optimum range for crop growth, can affect wheat yield (Ortiz et al. 2008; Lobell et al. 2008) as well as grain quality (Borghi et al. 1997; Porter and Semenov 2005; Fedoroff et al. 2010). However, the severity of the impact depends on the time and duration in which plants are exposed to the high temperature. Nawaz et al. (2013) found elevated temperature mostly affect plant at booting and heading stages with higher severity than anthesis and grain-filling stages. Whereas, optimal protein quality was reached when plant experienced high temperature (≥32 °C) during grain-filling for roughly 60 h (Graybosch et al. 1995).

Going deep in the analysis of increase temperature effects on plant growth at different stages, DuPont and Altenbach (2003) found environmental factors (e.g. high temperature) mainly affect formation and therefore number of spikelets prior to anthesis. At anthesis, high temperature stress (>30 °C) can cause floral abortion by disturbs grain fertilization (Saini and Aspinall 1982) and grain set (Ferris et al. 2000) therefore decreased the grain number per ear at maturity in spring wheat (Ferris et al. 1998) especially when plant expose to this temperature over the mid-anthesis period. This found to have significant effect on crop yields (Semenov and Shewry 2011) however the severity of the impact will depend on the cultivar (Dupont and Altenbach 2003). Results from field experiments conducted in Australia to test the response of different wheat cultivars to high temperature (Tashiro and Wardlaw 1989) show that a temperature of≥27 °C could significantly reduce grain size of several Australian wheat cultivars, resulting in yield reduction. After anthesis, heat stress impact found to primarily influence kernel size and composition especially when a crop expose to a temperature of 35 °C for one day resulted in 18–35 % of yield reduction (Alexander et al. 2010; Talukder et al. 2010). This yield reduction can be explained by the fact that the exposure to ≥35 °C limits photosynthesis rate (Crafts-Brandner and Law 2000; Griffin et al. 2004) due to the temperature sensitivity of Rubisco activase content and therefore Rubisco activation. High temperature (Jenner 1994) also influences the activity of enzymes in the pathway of starch synthesis. The author found that rate of starch deposition increases with the temperature up to 30 ° C and decreases above this temperature. During wheat grain development, however, the exposure to this high temperature range can affect protein storage in the seeds, which modify dough quality (Blumenthal et al.

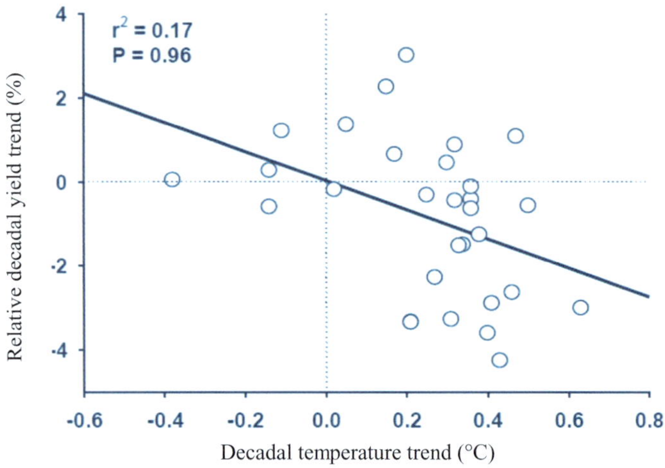

Fig. 12.7 Relative decadal yield trend based on simulated 30-year model as affected by temperature trend between 1981 and 2010 for 30 global scenarios

1993; Irmak et al. 2008) and cause a significant reduction in the starch accumulation (Hurkman et al. 2003). Altenbach et al. (2003) found a reduction in starch content by shortening the duration of starch accumulation when expose to high temperature. However, this effect was more pronounced when high nighttime temperatures increased.

At field level, the distribution of heat waves through the growing season can reduce moisture content in the soil, which disrupts vegetative growth and cause crop to fail. While short time exposure to high temperature in wheat (≥ 35 °C for some hours or one day) affect grain quality due to the decrease in the activity of soluble starch synthase (Keeling et al. 1993; Denyer et al. 1994; Jenner 1994) lead to lower grain weight and number (Wollenweber et al. 2003; Schapendonk et al. 2007) therefore less grain yield (Mullarkey and Jones 2000; Tewolde et al. 2006).

Recently, a meta-analysis was conducted by Asseng et al. (2015) to understand the effect of elevated air temperature on global wheat production. The authors used date from different parts in the world to test the accuracy of 30 different wheatcrop models against field experiments. Their findings indicate that worldwide increase in air temperature will poses a real threat to wheat yields reducing global wheat production by six percent for each °C increase (Fig. 12.7).

12.2.1.2 Plant Response to Drought

It is generally accepted that water scarcity reduces crop productivity in most of vulnerable parts in the world. Drought (water deficit) is considered to be the main environmental factor limiting plant growth processes (Vanaja et al. 2011). Limitation of water supply during vegetative stage can affect plant development (Volaire 2003) and grain formation during reproductive stage (Li et al. 2000).

Hypotheses indicate that changing in climate events will permit the widespread of drought in different parts of the world (Foulkes et al. 2007; Witcombe et al. 2008) due to lower precipitation levels and increase in air temperature. Several studies (e.g. Barnabás et al. 2008; Jäger 2010) reported that the increasingly frequent occurrence of water deficiency due to climate change will have substantial effects on the development of cereal plants. These effects thought to have an influence on the biology, physiology, biochemistry and therefore productivity of the crop (Jones et al. 2003; Sobhkhizi et al. 2014). It is well documented that the primary physiological impacts of drought is on photosynthesis (Lawlor 1995). Nezhadahmadi et al. (2013) detailed the physiological response of plant to drought stress. Some of its effect on plant physiology includes change in size, number and aperture/closure of stomata, which resulted in reduction of plant growth rates. The biochemical responses of plants to water stress was discussed intensively in the literature and include, for example, reduction in chlorophyll content and photochemical reactions efficiency of photosynthesis. Leaf and root, instead, are the main influential morphological characteristics of wheat crop (Denčić et al. 2000) during drought stress. Different studies (Szegletes et al. 2000; Zhu 2002; Lawlor and Cornic 2002; Yordanov et al. 2003; Ji et al. 2010) reported the main biological and chemical processes and/or parameters affected most by water supply limitation. This covers, for example, aspects related to growth inhibition, pollen sterility, accumulation of abscisic acid in spikes, enzymes production and hormone composition. However, many factors can affect plants' response to drought stress (Chaves et al. 2003; Flexas et al. 2004; Denby and Gehring 2005; Ribas-Carbo et al. 2005; Pradhan et al. 2012) including duration and time of the stress, plant genotype and the interaction with other environmental variables (McDonald and Davies 1996; Rizhsky et al. 2002).

In general, several studies highlighted remarkable results concerning the effect of drought stress on plant including early maturity, growth deterioration and leaf area reduction (Rizza et al. 2004; Rucker et al. 1995). Besides, in the nature, water stress can limit leaf extension of the plant in order to balance the water absorbed with the one lost through transpiration (Passioura 1996). In specialized study, however, Lonbani and Arzani (2011) found increase in the length and area of flag leaf in wheat, whereas width of the flag leaf was not affected under drought stress. Crop-growth stages are differing according to their susceptibility to drought. Certain plant stages are more susceptible to water limitation, whereas some others can deal with it more effectively (Akram 2011). For instance, the occurrence of drought during specific plant growth stage found to determine the severity of the impact on single wheat yield component (Day and Intalap 1970). Pradhan et al. (2012) indicate that drought stress impedes crop performance at all phenological stages, however the reproductive (Abayomi and Wright 1999) and grain-filling phases are the most sensitive (Dickin and Wright 2008; Pradhan et al. 2012). Therefore, the occurrence of water stress during these two susceptible stages will cause a substantial yield (Araus et al. 2002; Turner 2004; Farooq et al. 2014) and qualitative losses (Rharrabtia et al. 2003b; Yang and Zhang 2006) through reducing test weight and increasing ash content. Figure 12.8 compare the effect of drought on wheat yield losses (%) at vegetative growth and reproductive stage. Comparison data were taken from different

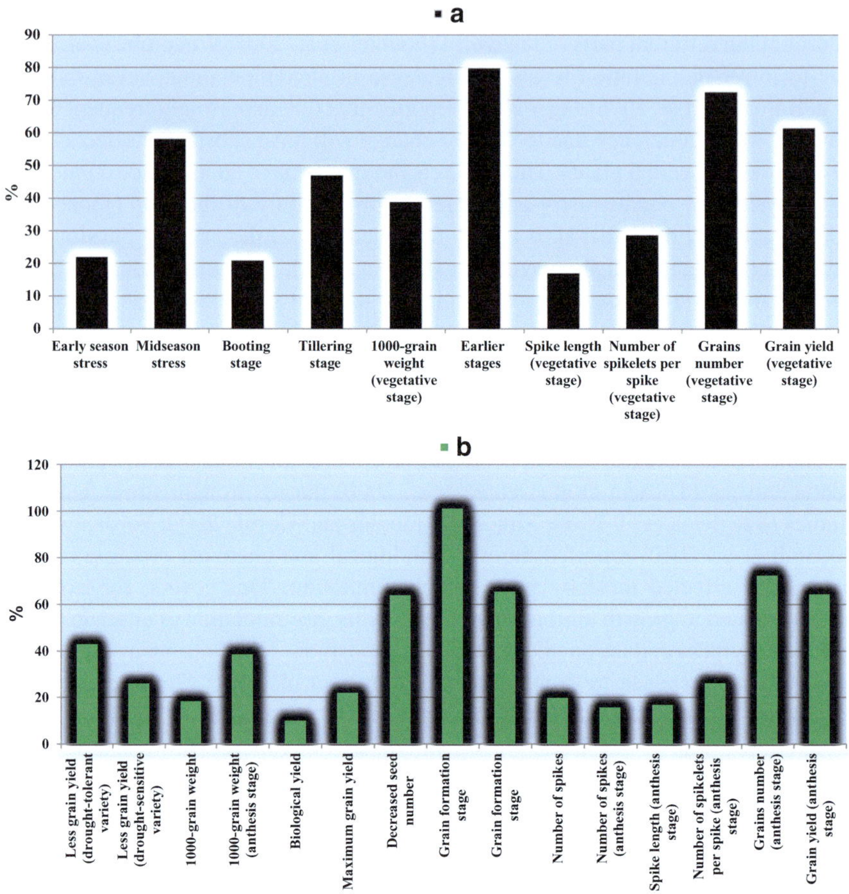

Fig. 12.8 Yield losses (%) at (**a**) vegetative growth and (**b**) reproductive stages under drought in wheat as reported by different studies (Milad et al. 2011; Schneekloth et al. 2012; Rizza et al. 2004; Tuberosa and Salvi 2006; Sivamani et al. 2000; Smirnoff 1993; Ingram and Bartels 1996; Salekdeh et al. 2002). Modified from Nezhadahmadi et al. (2013)

studies (Smirnoff 1993; Ingram and Bartels 1996; Salekdeh et al. 2002; Sivamani et al. 2000; Rizza et al. 2004; Tuberosa and Salvi 2006; Milad et al. 2011; Schneekloth et al. 2012). In the analysis of these data, Nezhadahmadi et al. (2013) found a reduction in grain number and therefore grain yield due to lower pollination resulted from water stress occurrence at anthesis (Ashraf 1998). In addition to increase photosynthesis rate, adequate available water during or after anthesis improves grain size due to increase available time for carbohydrate translocation to grains (Zhang and Oweis 1998) thereby leads to increase grain yield. Drought stress found also to have significant effect on grain weight after anthesis, even after the increase in water availability afterwards (Gooding et al. 2003). Saini and Westgate (1999) found water deficit during grain development to cause premature cessation

Table 12.5 Water stress effects on yield components and leaf area index of wheat at different growth stages

	Stress period			
Parameter	Control	Pre-anthesis	Anthesis	Grain filling
Leaf area index[a]	5.0	3.3	5.0	5.0
Fertile tillers/m^2	513	658	434	485
Grains/spike	32.7	13.0	27.1	31.4
1000 grain weight (g)	56.3	55.2	53.7	49.2
Grain yield (g/m^2)	779	559	498	658
Harvest index	0.52	0.50	0.53	0.53

[a] Leaf area index was measured at booting stage. Source: Hochman 1982

of grain filling. However, in dry environments, with a likely drought at the end of the growing season, higher yields may be achieved with a slower water uptake (Manschadi et al. 2006).

The occurrence of water stress during the critical spike growth period can cause a significant decline in grain number (Hochman 1982). Table 12.5, however, show a maximum reduction in grain per spike when water stress occur at pre-anthesis. At the same stage, leaf area index was the one affected most by water stress. Whereas fertile tillers and grain yield were mostly effected by water stress at anthesis.

When compared to heat stress, Li et al. (2013) indicate that studies on the effects of drought on wheat quality are limited. One of the recent studies found significant effect of water scarcity on protein composition with varied severity based on the time in which the stress occurred (Flagella et al. 2010). When water deficit occurred all the way through the growing season, the authors found a reverse relationship between the increase in protein content and grain yield. However, when water stress occurred in winter but followed by heavy rainfall in spring good protein content can be expected without affecting yield. The positive correlation between water deficit throughout the season and protein content increase was already spotted elsewhere by other authors (e.g. Guttieri et al. 2005 and Garrido-Lestache et al. 2005). In the study of Rharrabtia et al. (2003a) water stress also found to increase protein content but thousand-kernel weight was reduced consequently.

12.2.1.3 Plant Response to Elevated CO_2

The second half of the last century witnessed remarkable increase in carbon dioxide concentration (Alcamo et al. 1996) since the onset of the industrial revolution. This increase is expected to double by the mid of this century (IPCC 1996). However, as atmospheric CO_2 represents the main source of carbon for plant growth, CO_2 enrichment thought to have fertilizer effects (Dhakhwa et al. 1997) which could, in the long run, improve biomass productivity and grain quality (Kimball et al. 2002). The main effect of elevated atmospheric CO_2 on plant growth though to be the increasing assimilation rates of CO_2 in C_3 plants (Conroy et al. 1994). Nevertheless, the positive effect on grain yield is usually limited to other factors. Results from simulation studies reported remarkable improvement of shoot production under elevated CO_2. Although this positive effect influenced by temperature and precipitation, Ewert et al. (2007) reported that the increase in grain yield under increased CO_2 concentration did not exceed 4 %.

In fact, both biomass and grain yield increased with doubling CO_2. This CO_2 effect, and in particular on grain yield, varied largely depending on years and temperature (range of 7 to 168 % for grain yield). Despite the positive effect of elevated CO_2 concentration on crop productivity, climate change, in general, affect crop production in negative way. At farm level, the state in which equilibrium point between these two effects can be reached is dependent on other factors (i.e. nutrients and water limitation), which thought to be affected by changing in climate. In fact, different crops have been tested for their response to elevated CO_2 levels (i.e. Cure and Acock 1986; Kimball et al. 2002). In each case study, the response level was found to be dependent on environmental (Bowes 1993) as well as management factors (Rosenzweig and Tubiello 2007).

It is well known that carbon dioxide is the main chemical compound for photosynthesis process. The increase in atmospheric CO_2 concentration will, consequently, lead to higher photosynthesis rates and therefore, higher productivity. Since the evidence suggests that C3 plants require higher concentrations of atmospheric CO_2 than present levels to complete photosynthesis process for best growth (Stitt 1991), C4 plants are expected to respond to elevated CO_2 with less extent. However, it is worth mentioning that in each plant group and among species the biochemical processes of photosynthesis are different and have been found to effect plant response to elevated CO_2 accordingly. In controlled experiments, for example, the C_3 species including wheat tend to show significant gains in net photosynthesis from increased CO_2 because of photorespiration. This effect on plants grown at higher CO_2 concentration include also decrease the rate of water evaporation from plant through transpiration which enhance water use efficiency (Tuba et al. 1994, 1996). However, lower transpiration rates may reduce nitrogen uptake from the soil (Manderscheid et al. 1995; Kimball and Bernacchi 2006) and affect productivity.

Review study by Fuhrer (2003) reported that average C_3 plant yield simulated to increase by 30 % with double CO_2 concentration whereas lower average yield was reported when field experiment data was used. The benefits which plants can gain from doubling CO_2 (Qaderi and Reid 2009) could be explained by its influence on different aspects related to plant metabolism including slow plant transpiration by promoting stomatal closure to enhance water use efficiency (Morison 1985; Allen 1990; Bowes 1993). In some cases, transpiration rate of wheat was reduced by 70 % under increasing CO_2 concentration (Conroy et al. 1994). These remarkable results on crop growth and yield deserve to be mentioned and their impacts on global food production need to be integrated in future action plan concerning food production strategy. However, as mentioned previously, the effects of other factors on plant response to elevated CO_2 must not be forgotten. To improve the photosynthetic process, in order to increase water use efficiency and grain yield of cereals grown at elevated CO_2, Dahal et al. (2014) found some variables such as temperature and

Table 12.6 Effects of elevated CO_2 on wheat biomass, grain yield and grain protein at maturity compared to ambient CO_2

CO_2 µmol mol^{-1}	Aboveground dry biomass (g m^{-2})	Yield (g m^{-2})	Protein (%)	Reference
Ambient (390)	1511	582	11.2	Buchner et al. (2015)
Elevated (550)	1968	656	10.8	
Ambient (380)	–	515,2	–	Dahal et al. (2014)
Elevated (700)	–	828,8	–	
Ambient (384)	–	–	17	Fernando et al. (2014)
Elevated (550)	–	–	16	
Ambient (350)	1074,6	–	–	Ma et al. (2007)
Elevated (550)	1219,3	–	–	
Ambient (409)	1174	481.1	–	Högy et al. (2010)
Elevated (537)	1295	532.4	–	
Ambient (370)	1348	605	–	Ewert et al. (2002)
Elevated (550)	1583	724	–	
Ambient (380)	1249	529	–	Ewert et al. (2002)
Elevated (690)	1373	647	–	

cultivar to be dependent. Other study by Wall (2001) found that enhanced CO_2 level reduce the damage to wheat even under dry conditions. Reasons might be due to the fact that a 200 µmol mol^{-1} increase in CO_2 level resulted in a 30 % increase in carbon supplies to wheat leaves (Wall 2001; Wall et al. 2000) which acted as carbon fertilizer. However, this very much depends on stomatal density which controls gases movement through plant leaves making CO_2 available for photosynthesis. Under experiment conditions, Woodward and Bazzaz (1988) found that increase in CO_2 concentration up to about 310 µ l/l decreased the stomatal density, but no effect was found above this CO_2 level. Under water limitation, Doheny-Adams et al. (2012) further found less stomatal density usually associated with higher leaf temperature (which imply reduced transpiration rate) at 200 ppm CO_2. This would lead to better plant growth due to increase water use efficiency and CO_2 assimilation rate.

Different studies found that some plant parameters including root (length, diameter and number) (Lee-Ho et al. 2007), leaf area and leaf thickness (Bowes 1993; Bray and Reid 2002) were influenced by elevated CO_2 level to the greatest extent. With higher carbon dioxide gain, Pritchard and Rogers (2000) found higher stimulation rate to lateral root production in plants resulting in positive consequences for plant production. Higher CO_2 concentration was also found to increase seed yield of various wheat cultivars, but negatively affect grain and flower protein (Ziska et al. 2004). Similar findings were reported by different authors and can be seen in the following table (Table 12.6).

Across the globe, Table 12.7 show that wheat grain yield increases in response to elevated CO_2 with substantial variation. The increase rate has been reported to range from a minimum of 8.4 in the USA to a maximum of 35 % across Europe as compared to ambient CO_2. The table shows the same trend of aboveground biomass which ranged from 10.4 to 58 %, whereas grain protein was always decreased at all the study sites with maximum reduction rate of 28 % in UK.

Table 12.7 Effect of elevated CO_2 on wheat grain, protein and above ground biomass at maturity as compared to ambient CO_2 in different parts of the world

Reference study	Grain yield	%	Grain protein	%	Aboveground biomass	%	Country
Han et al. (2015)	increased	11.4	–	–	–	–	China
Buchner et al. (2015)	increased	11	Decrease	4	increase	20–30	Australia
Kimball et al. (1995)	increased	8.4	–	–	–	–	USA
Erbs et al. (2010)	–	–	Decrease	4–13	–	–	–
Thompson and Woodward (1994)	increased	30	Decrease	28	–	–	UK
Ma et al. (2007)	–	–	–	–	increase	20,15	China
Cardoso-Vilhena and Barnes (2001)	–	–	–	–	increase	44	–
Lam et al. (2012)	–	–	–	–	increase	27–58	–
Fernando et al. (2014)	–		Decrease	10	–		–
Högy et al. (2010)	increased	10.7	–	–	increase	10.4	Germany
Högy and Fangmeier (2008)	–	–	Decrease	3.9-14.1	–	–	–
Wang et al. (2013)	increased	24	Decrease	11	increase	28	–
Bender et al. (1999)	increased	35	–	–	–	–	Different sites in Europe

Findings of recent Free Air CO_2 Enrichment (FACE) experiment in Australia reported greater above ground dry biomass (20–30 %) and grain yield (11 %), whereas grain protein was lowered (~ 4 %) in plants grown under elevated CO_2 compared to ambient CO_2 (Buchner et al. 2015). This magnitude of stimulation by CO_2 enrichment of both, biomass and yield, is in agreement with an earlier meta-analysis of FACE data by Ainsworth and Long (2005) while Erbs et al. (2010) reported a similar reduction (4 %) of grain protein in wheat under elevated CO_2.

Grain quality of some plants found to be also influenced under elevated CO_2. The affect includes mainly changes in lipid metabolism Qaderi and Reid (2009), especially decreases in the major chloroplast lipids (MGDG and PG) (Ekman et al. 2007). In wheat, this effect leads to changes in grain lipids and doubled the number of mitochondria in leaves Williams et al. (1994) resulted in reduction in N concentration Conroy (1992) but double alanine concentration in wheat leaves Conroy et al. (1994). In fact, wheat leaves play a fundamental role in protein synthesis process (Barneix and Guitman 1993) therefore any lack in nitrogen concentration in leaves will lead to change in grain quality (Hocking and Meyer 1991).

12.2.1.4 High Temperature and Drought Interaction

The interaction between the most defused abiotic stresses present a real concern in cereals key production areas worldwide. In fact, as discussed previously, the increase in air temperature usually combined with lose of water from the soil through evaporation and from plant tissue via transpiration. The occurrences of these natural processes, in the presence of high temperature, cause water shortage and lessen water availability for the plant threatened its survival (Moffat 2002). Information about the time in which the two stresses occurred in the field is important as the severity of their combined effect can vary largely. In cereals, the effect of their combined effect was found to be more severe at later growth stages (Blum 1998; Yang et al. 2004; Barnabás et al. 2008) resulting in remarkable yield losses (Fig. 12.9). Reasons behind this large yield reduction found to be due to the reduction in starch accumulation rate as a result of lower endosperm cells number (Nicolas et al. 1985).

In wheat, the interaction of these two stresses was widely investigated (Dupont and Altenbach 2003; Yang and Zhang 2006; Barnabás et al. 2008). Their combined effect shown to be more severe on crop development and productivity (Asseng et al. 2011) than each stress occurs alone (Savin and Nicolas 1996; Wang and Huang 2004). For example, Rizhsky et al. (2002) found increase temperature to cause stomata opening for transpiration, which cool down the plant. Whereas, when heat stress occurs and combined with drought the plant is not be able to open its stomata resulted in higher leaf temperature and therefore dysfunction. More evident results suggest further decrease in yields is expected under heat and drought stresses during grain filling (Dickin and Wright 2008). Compared to heat stress alone (with yield losses of 31 %), Balla et al. (2011) indicate that this percentage would reach 76 % after drought and heat stress in combination.

Although, the reproductive and grain-filling stages are the most sensitive to heat and drought (Saini and Westgate 1999), the occurrence of theses stresses during flowering result also in large reduction in grain weight (Fig. 12.9) and grain number (Semenov et al. 2014).

In three-way interaction between climate variables, Chen et al. (1996) investigate the effect of heat, water limitation and CO_2 on net primary production of wheat. Positive response was observed with doubled CO_2 and water availability, while a reduction in net primary production was associated with limited water and CO_2 at ambient level.

12.2.1.5 High Temperature and [eCO_2] Interaction

In fact, plant growth and development are connected to the environment via a combination of linear and non-linear responses (Campbell and Norman 1989). Reach the point of equilibrium between these sophisticated interactions would be an important task to scientists as well as farmers to ensure sustainable production of crop plants. However, any disruption of this equilibrium via climate change would

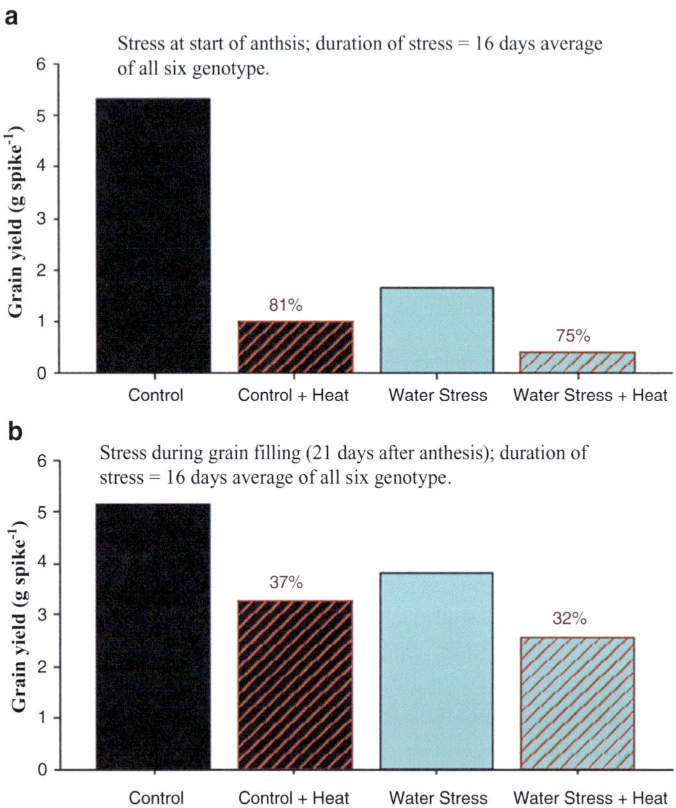

Fig. 12.9 Measured mean (mean of six cultivars) wheat grain yield impact with increased temperature (optimum day/night temperature of 21/15 °C and high temperature stress of 36/30 °C) with and without water stress for (**a**) 16 days of high temperature stress starting from anthesis and (**b**) for 16 days of high temperature stress during grain filling starting 21 days after anthesis (Source: Asseng et al. (2015))

lead to substantial losses in agriculture crops (Watson et al. 1996; Wolf et al. 2002). The main effects of changing the magnitude of climate events will be on photosynthesis process and phenological development of the plant (Albert et al. 2011; Semenov et al. 2012) resulting in crop yield decrease.

It is well known that the increase in atmospheric CO_2 concentration is associated with rise in ambient temperature (IPCC 2007a, b). Several studies (i.e. Carter et al. 2007) suggest that this increase would reach around 4 °C in the future. These changes will, therefore, lead to considerable impact on the productivity of many crops (Watson et al. 1996) including wheat (Farooq et al. 2011).

Although the effects of increasing CO_2 and temperature on plant productivity separately have been widely discussed in the literature, consequences of their interaction have received scant attention. One of the few studies on their interactions (Batts et al. 1997) reported positive as well as negative impact on biomass and grain yield. While others (Vu 2005 and Borjigidai et al. 2006), found only positive interaction on photosynthesis process of C3 plants. In the analysis of their combined effect, Wheeler et al. (1996a) found a decline in wheat yield with increasing temperature, but this declining was lessen in the presence of elevated CO_2. Fuhrer (2003) further confirmed that temperature variation controls the effect of increase CO_2 concentrations. Results by Idso et al. (1987) indicate that CO_2 enrichment will reduce canopy development and affected plant growth as cold air temperatures occur over winter. In this regard, plant grown under highly temperature fluctuation may not benefit from the fertilizer effect due to increased carbon dioxide. Studies found that cereals grown in areas with high to moderate temperatures to benefit more from increasing levels of CO_2 concentrations (Baker and Allen 1993; Rawson 1995) reflected in grain yields increase. The level of response to elevated CO_2 for each 1 °C rise in temperature was predicted to be 1.8 % (Rawson 1995). Nevertheless, this effect may turn to be negative and no further yield can be expected when temperature exceed certain level regardless the increase in CO_2 concentrations. When temperature goes beyond 35 °C the interactive effects between these two variables will be difficult to predict given the fact that their interaction may be effected by other variables such as drought and nutrient availability. Similar conclusion can be expected when temperature goes beyond the average minimum optimal level. The identification of the minimum and the maximum optimal temperature in which fertilizer effect of elevated CO_2 can be expected will be difficult task as this might varied according to the cultivar, region and weather conditions. Wolf and Kempenaar (1998) concluded that when different cultivars were compared, they all showed substantial increases in biomass and grain yield with increased CO_2. However, in the interactions between them they found no different with two cultivars showing no difference in CO_2 response across the 2–3 °C range of mean seasonal temperature and the other two showing reduced CO_2 response at higher temperatures.

Some efforts to investigate the effect of out-range optimal temperature under elevated CO_2 on plant growth found that exposure to temperature higher than 36 °C or lower than 18 °C will experience reduction in carbohydrate export through phloem (Reddy et al. 1998). However, at elevated CO_2, Bunce (1998) found 63 % stimulation in photosynthesis rate of wheat at 10 °C temperature whereas the simulation rate was 115 % at 30 °C. Negative response to elevated CO_2 was reported by other authors (Idso et al. 1987) when temperatures went below 18 °C. Under FACE experimental conditions and elevated CO_2 concentrations, photosynthesis rate was 11 % higher when plant exposed to temperature above 25 °C compared to temperatures below this level (Ainsworth and Long 2005). In other experiments (Ewert et al. 2002), anthesis and maturity dates were reached earlier under elevated CO_2 compared to ambient CO_2 when higher temperature (+ 0.7 °C) was recorded (Table 12.8). The higher biomass and grain yields were reported between temperature of 13.8 and 15.1 °C except for the year of 1999 in Braunschweig, Germany. Whereas, the lowest yields under elevated CO_2 were reported at higher temperature (18.9–19.9) (Table 12.8).

Between the optimal minimum and maximum temperature in which plant can benefit more from fertilization effect of elevated CO_2, even slight increase in temperature during growing season was found to annul grain yield stimulation (Batts et al. 1997; Amthor 2001). This effect was also reported on root growth and multi-

Table 12.8 Effects of location, growing season, elevated CO_2 and temperature on the productivity and phonological stage duration of spring wheat

Experiment location	year	Treatment CO_2 (μmol mol^{-1})		Temperature* °C	Phenological stage Anthesis (day)	Maturity (day)	Biomass (g m^{-2})	Grain (g m^{-2})
Maricopa (Arizona, US)	1992/1993	Low	370	14.5	84	126	1528	648
		High	550		81	124	1721	759
Maricopa (Arizona, US)	1993/1994	Low	370	13.8	93	133	1348	605
		High	550		91	130	1583	724
Braunschweig (Germany)	1998	Low	380	15.1	151	208	1201	512
		High	670		151	208	1763	755
Braunschweig (Germany)	1999	Low	410	14.2	155	205	960	434
		High	680		155	205	1392	654
Giessen (Germany)	1998	Low	370	18.9	173	209	845	401
		High	650		173	209	1337	657
Giessen (Germany)	1999	Low	380	19.9	183	224	1249	529
		High	690		183	224	1373	647

Modified from Ewert et al. (2002)

plications (Benlloch-Gonzalez et al. 2014), and grain quality (Fernando et al. 2014). However, the effect of elevated CO_2 on grain quality was more linked to the time in which heat stress occurs. Benlloch-Gonzalez et al. (2014) evaluate the effect of elevated CO_2 and higher mean temperature on root system growth of two spring wheat genotypes. The authors found positive interaction effects on root biomass at ambient temperature irrespective of genotype, however when plants were grown under high temperature the positive effect of elevated CO_2 was lowered.

12.2.1.6 Drought and [eCO_2] Interaction

As mentioned previously in this chapter, the single effect of drought and increase atmospheric CO_2 on crop growth and productivity is well documented, however, as in the case of other climate variables, the interaction between them is not well understood. Reasons might be due to the fact that other abiotic as well as biotic factors could have direct and/or indirect influence on their interaction. Recently, efforts have been made to better understand this interaction and its effect on crop growth and quality.

Due to the fact that elevated CO_2, increase temperature and drought are interrelated, the occurrence of one of them can affect the others depending on the time and severity of its presence during growing season. Other factors such as crop management found to affect crop response to these individual or combined (Rosenzweig and Tubiello 2007). Examples from different studied (Tubiello and Ewert 2002; Kimball 1983; Cure and Acock 1986; Idso and Idso 1994) show higher yield response to elevated CO_2 when crop grown under rainfed conditions compared when it is grown under unlimited water supply (Chaudhuri et al. 1990; Kimball et al. 1995). In Australia, the effect of elevated CO_2 on wheat grain yield was 30 %

higher when the crop experienced dry conditions compared to water availability (Reyenga et al. 1999a). Similar findings (Reyenga et al. 1999b) reported higher wheat yield increase in South Australian under drier conditions. Explanations were given by Niinemets (2010), who found that plants are more susceptible to drought stress under lower CO_2 concentrations.

Modeling by Ewert et al. (2002) demonstrated that the yield-increasing effect of higher CO_2 levels was more intensive if it was associated with water deficiency. In the work of Varga et al. (2010), drought was found to cause a reduction in the grain number per plant, which was most sensitive to drought at first node appearance, while the correlation between grain mass reduction and water deficiency was closest when the latter occurred during ripening. They found that the longer period of water withholding at ripening caused a 27.1–27.3 % decrease in the grain number at normal CO_2, but the higher CO_2 level was able to compensate for this. Li et al. (2001) found a close correlation between drought stress and grain mass, with an increase of 2.1–2.3 mg in the grain mass in drought-stressed plants raised at enhanced CO_2 compared with those grown at the ambient level, while carbon dioxide had no effect on the grain mass in the case of normal water supplies.

In the response of the qualitative parameters to drought stress, protein content was found to be in close positive correlation with the potential evapotranspiration and in negative correlation with the rainfall quantity (Mkhabela et al. 2010). The authors demonstrated an increase in the protein content of spring wheat in response to drought. Varga et al. (2010) found that the quantity of protein was modified primarily by drought during the period up to heading; when water was withheld at ripening the duration of drought had no influence on the protein content. Without drought stress, the authors detected lower protein contents at the higher CO_2 level, but the protein contents recorded at enhanced CO_2 after 14 days of water withholding were similar to those obtained at normal CO_2 with normal water supplies.

12.3 Crop Responses to Biotic Stresses – Overview

Population growth in the next few decades will increase the need for food production, however the changing climate and changing threats from pests and pathogens could impact yields of major food crops. Together with abiotic stresses, biotic stresses represent major limiting factors for high crop productivity worldwide. We report here the main biotic stresses and their response to climate variability with reference to wheat production.

12.3.1 Weeds

Within the efforts to understand climate variable effect on weed population, Wittwer (1990) suggested that the rising CO_2 levels will generally favor crop production since the majority of important food crops have the C3 photosynthesis pathway, while a high percentage of the major weed pests are C4 plants that will likely benefit less from CO_2 enrichment (Bazzaz and Garbutt 1988; Patterson and Flint 1990). Reilly et al. (1996) detailed the effect of climate variables on weed community and distribution through collection of information from different studies. Some of them (e.g. Patterson 1993, 1995) found climate change and associated increase in CO_2 levels to affect national and international efforts for controlling weeds. Some of the difficulties that might face and influence the efficiency of biological, chemical as well as mechanical control of weeds include different environmental variables (e.g. precipitation, soil type, humidity and wind). For example, some of these factors can influence the efficiency of chemical herbicides by altering their metabolism and uptake by targeted weeds (Hatzios and Penner 1982). In addition, we reported earlier in this chapter that elevated CO_2 increase starch concentrations in C3 plant leaves. This effect found to have an influence on herbicide efficiency though the impact on its metabolic activity (Wong 1990). Moreover, under elevated CO_2 both chemical as well as mechanical control of weeds with deep rhizomes and tubers would be difficult because the growth of these sophisticated roots is simulated by CO_2 enrichment (Oechel and Strain 1985).

In the response of weeds to increased temperature, different responses can be expected depending on the area of origin or whether weed species belong to C3 or C4 plants. Usually, weeds grown in warm areas found to be more responsive (increase in growth) even to slight increase in day and/or night temperatures (Flint et al. 1984). However, some weed species are not stimulated by warmer climate (Patterson et al. 1986). In addition, the biomass of some C4 weed species showed remarkable increases under small increase in temperature (Flint and Patterson 1983 and Patterson 1993).

12.3.2 Pests

Most evidences suggest that insect pests represent huge threat to crops productivity (Rosenzweig and Hillel 1998; Patterson et al. 1999; Gutierrez 2000). This threat, due to pest damage, found to cause as much as 25 % of food losses worldwide (Sinha et al. 1988). The main effect of climate change on insect pests is on the severity of their outbreaks (Reilly et al. 1996) but this might also include some aspects related to the ability to modify the suitability of some areas, habitat or some crops to host these pests. Therefore, pests response to the changes in climate will vary between different environments and among pest species (Sutherst (1991)) and Sutherst et al. (1995).

Usually, humid conditions together with increase in temperatures represent the most climate factors that are suitable for pests' survival and could cause significant increases in the pest populations. In addition, changing metabolic pathway of some plants due to drought stress, found to increase their vulnerability to the insect pests attack (Mattson and Haack 1987). Other factors (DeBach 1965; Rainey 1989) such as the availability of host plants, natural enemies and the ability of pests to adapt to new climate scenarios under changing climate play also a fundamental role in the widespread of pests. However, the main concern remain is the ability of some pest species for regeneration more than once a year with the availability of suitable conditions (e.g. warmer temperature), which lead to change their geographical allocation and increase their numbers. This was the case for European Corn Borer (Ostrinia nubilalis) when a remarkable northward shift in distribution was observed for each 1 °C increase in temperature (Porter et al. 1991). While expansion in range were also reported for some other pests in Japan (Mochida 1991).

In addition to the factors that affect the survival and distribution of insect pests, the direct effect of elevated CO_2 on insect pests is still not well understood. Modest efforts suggest that elevated CO_2 can directly affect the ability of some insect pests to locate their host plants (Bernklau et al. 2004). The most commonly observed effects of elevated CO_2 on insects are indirect (Bale et al. 2002; Murray et al. 2013) through changes in plant physiology and biochemistry (Berzitis 2013). For example, increase CO_2 concentration thought to increase the carbon content in plant tissues (Deka et al. 2009). This increase will be on the expense of nitrogen (N) availability in plant (Ainsworth et al. 2002; Robinson et al. 2012) making N, as sole nutrient source, not available for insects with enough quantity (Mattson 1980). This alteration in C/N ratio reduces the nutritional quality of the leaves (Ehleringer et al. 2002), which leads to acceleration in food quantity to compensate any lack of N in plant leaves. In addition to decreases in leaf N concentration, changes in plant chemical defences have been documented under elevated CO_2 and these changes could further impact herbivore performance (Robinson et al. 2012).

With reference to the climate factors mentioned before, excessive occurrence of each one of them may have a control role threatened the survival of pests at different stages of their life cycle. Heavy rain during growing season, for example, found to reduce the occurrence and success of oviposition by insects such as the European corn borer (Davidson and Lyon 1987). However, this control role may turn to be negative on crop plants depending on the time. Recently, in-depth understanding of pests behaviour allow the estimation of their potential threats to crop under changing climate using simulation models (Goodenough and McKinion 1992). This was done by the identification of climate variable limitations (variation in temperatures and water limitation) for most widespread agriculture pests (e.g., Davidson and Lyon 1987).

12.3.3 Diseases

Similar to other biotic stresses, wormer conditions, precipitation and other climate factors were found to have great influence on the seasonal phenology, population growth and the distribution of most wheat's common diseases (e.g. Chakraborty et al. 2000; Melloy et al. 2010; Sutherst et al. 2011 and West et al. 2012; Pangga et al. 2012; Ghini et al. 2012; Pautasso et al. 2012). Temperate winter with warmer temperatures present optimal conditions for most common diseases (e.g powdery mildew) (Treharne 1989) and cause stronger outbreaks for some others (Parry et al. 1990). Whereas, the frequency of summer rainfall would increase incidences of some diseases because rainwater represents dynamic means by which disease spores will be diffused (Royle et al. 1986). However, plant resistant to most fungal diseases increases under higher summer temperatures and drier conditions.

In wheat cultivation, leaf and stem rust, was classified to be amongst the most important diseases (Todorovska et al. 2009), which cause substantial yield losses worldwide. Within the international efforts to control this pathogen, the development of new cultivars with resistant genes was the best method to control this disease (Roelfs 1990). Recent work by Mohammadi et al. (2015) using a landrace collection consisting of 380 durum wheat entries originating in several countries along with four check varieties were evaluated for biotic stress (in particular, yellow rust (Puccinia striiformis Westendorf f. sp. tritici)). The authors observed significant changes in reactions of landraces to yellow rust. Moreover, percentage reduction due to the yellow rust was 11.4 % for 1000-kernel weight and 19.9 % for grain yield. Crop losses due to disease invasion may increase to 15 % as effect of doubling CO_2 concentration Bergthorsson et al. (1988).

12.4 Modelling and Simulation

12.4.1 Growth Simulation Models to Understand Climatic Variability with Reference to Wheat

Globally, impact assessment of future climate change is expected to affect wheat yields (Deryng et al. 2011). However, it is still uncertain if the overall change in climate will result in wheat yield increases or decreases (Wilcoxa and Makowskia 2014). General assumption suggests that the severity of the impact on crop yield will depend on climatic and other factors and the interaction between them. This include the combined effects of changed temperature, precipitation, and CO_2 concentrations which have been found to have variable effects on projected future wheat yields across the globe.

In Europe, the effect of climate change on wheat is expected to be more obvious throughout the continent threatened its productivity. However, the severity of the impact will depend on the occurrence and type of climate pattern, and will differ from site to another. Several researches conducted during the last decades over Europe on human driven environmental changes expected increases in temperature and alteration of precipitation patterns. These environmental changes thought to be more pronounced in Pannonian Basin and northern Europe than other parts. In, fact both positive and negative impacts of climate change on European wheat can be expected. Late in the last century, Nonhebel (1996) predicted wheat yield increase in Western Europe. Similar increase was predicted by Harrison and Butterfield (1996) throughout Europe. Whereas Olesen et al. (2011) expected negative climate change impacts on European wheat. In fact, Ozdogan (2011) found that increased

CO_2 concentrations alone had relative positive effect on grain yields when excluding the effects of other climatic conditions. However, remarkable decrease in yields was observed when the crop experienced varied temperatures and precipitation under increased CO_2 concentrations. In another study, both frequent occurrences in climatic stress (e.g. spatial variation in temperature) and changes in crop management (usually leads to changes in leaf area index and differences in maturity period) found to cause remarkable differences between the current and predicted yield in response to future climate change scenarios (Reidsma et al. 2010).

A meta-analysis concerning wheat yields prediction under climate change was carried out to by Wilcoxa and Makowskia (2014) to identify if the combined effects of these climate variables will lead to a yield increase or decrease. In their conclusion, the authors reported increase in yields as a response to the increase above 640 ppm in CO_2 concentrations. This effect found to overwhelm both the effect of temperature increases and lower precipitation. However, results were varied greatly from site to site, likely due to differences in topography, soils and farming practices. Table12.9 shows wheat yield response to climate variables in different parts of the world.

Different studies show that the predicted wheat yield varied between counties and even within the country, among different regions (Table 12.9). This is due to the different circulation models, prediction period, climate change scenarios and simulation model used in study. Among wheat yield prediction results reported across the globe, a wide range of reduction has been reported from 1 % in Iran (Valizadeh et al. 2014) to 40 % in Italy (Tubiello et al. 2000) and Russia (Pavlova et al. 2014) (Table 12.9). The table shows, however, less variability in increase of predicted wheat yield ranged from 3.55 % in Northwestern China (Xiao et al. 2008) to 36 % in Turkey (Yano et al. 2007). Climate variables as well as input information had a strong effect on yield prediction. International institutions such as ICARDA and CIMMYT estimated wheat yield reduction by approximately 20–30 % by 2050 in key production areas of the developing countries due to the forecasting temperature increase.

In the study of climate change related-issues, model-based tools are important to support decision makers and planners especially in agriculture. Their wide use in plant science researches (Sinclair and Muchow 2001; Debaeke and Aboudrare 2004; Porter and Semenov 2005) make it possible to identify and analyze the relationship between climate variables and simulated parameters (e.g. wheat yield) quantitatively (Tardieu 2003; Hansen 2005; Semenov et al. 2007). Since the second half of the last century different crop models have been tested to better understand their efficiency in relation to yield prediction under different climate change scenarios for many crops. This enabled researchers to even highlight the potential threats for future crop production (Jamieson et al. 2000; Richter and Semenov 2005; Olesen et al. 2007). Recently, many model-based studies have been conducted for wheat (Palosuo et al. 2011; Zhao et al. 2015; Webber et al. 2015; Toumi et al. 2016; Huang et al. 2016) to understand how hard predicted changes in climate would be on crop growth (Nonhebel 1994; Semenov and Porter 1995; Asseng et al. 2013). Major results demonstrate that these changes would have a significant impact on

Table 12.9 Predicted wheatyield response to climate variability in different parts of the world

Region/country	Reference	yield trend	%
Denmark	Kristensen et al. (2011)	↓	8
Italy	Tubiello et al. (2000)	↓	10–40
United Kingdom	Ferrara et al. (2010)	↑	–
Finland	Laurila (2001)	↑	30
Switzerland	Torriani et al. (2007)	↑	–
Czech Republic	Trnka et al. (2004)	↑	7.5–25.3
Turkey	Yano et al. (2007)	↑	16–36
Turkey	Ozdogan (2011)	↓	5–35
India	Kaur et al. (2012)	↑	–
Pakistan	Sultana et al. (2009)	↑	–
China	Lin et al. (2005)	↑	5–15
North China Plain	Mo et al. (2009); Guo et al. (2010)	↑	13–19/9.8
Northwestern China	Xiao et al. (2008)	↑	3.55
Northern China	Jia et al. (2011)	↓	–
South-eastern Australia	Anwar et al. (2007)	↓	25
Southern Australia	Luo et al. (2005)	↓	13.5–32
South-east United States	Tsvetsinskaya et al. (2003)	↓	–
Colorado, United States	Ko et al. (2012)	↓	30
Québec, Canada	Brassard and Singh (2008)	↑	14.8
United States	Izaurralde et al. (2003)	↑	–
Midwestern United States	Southworth et al. (2002)	↑	20
Central and eastern Washington state	Thomson et al. (2002)	↑	–
Oklahoma, United States	Zhang (2005)	↑	14
Chile	Meza and Silva (2009)	↓	10
Seville, Spain	Semenov et al. (1996)	↑	–
Khuzestan, Iran	Andarzian et al. (2008)	↑	19
Sweden	Eckersten et al. (2001)	↑	10–20
Southern Russia and north-eastern Kazakhstan	Pavlova et al. (2014)	↓	20–40
Sistan and Baluchestan region, Iran	Valizadeh et al. (2014)	↓	1–37

Modified from Wilcoxa and Makowskia (2014); ↓ = Decrease, ↑ = Increase and – = not indicated

crop growth. However, crop response to certain variables or their combined effect was found to be positive depending on the cultivar and local conditions. Asseng et al. (2011), for example, found predicted increase in average temperatures to cause a significant reduction in wheat yields results in potential threat to global food security. Whereas, a decrease in temperature (by 2 °C) simulated to cause a reduction in durum wheat yield and biomass (Toscano et al. 2012). In the relation to predicted effects of combined increases in temperature and alteration of precipitation pattern, Long (1991) reported reduction in crop productivity. In combination with elevated CO_2, Torriani et al. (2007) found Swiss winter wheat to be the only one respond positively to climate change among three studied crops.

Semenov et al. (2014) simulated wheat yield potential in ten wheat-growing areas of Europe. The authors used a simulation model to test the performance of wheat ideotypes to counteract future climatic changes and found that despite the predicted increase of heat frequency at meiosis and anthesis, yield improvement can be obtained through extending the maturity period. Supit et al. (2010) reported a reduction in European wheat yield due to shortening crop cycle resulted from increased average growing season temperatures. However, Supit et al. (2012) predicted that winter wheat yield will benefits from the temperature increase and elevated CO_2 concentration up to 2050. In Australia, current wheat yields would not be reached by 2080 even if crop management was improved (N levels and crop cultivars changes) (Luo et al. 2009). This variation in yield prediction under future changes in climate scenarios will highly depend on the accuracy and robustness on the model used. In fact, several studies have been conducted to test the ability of different wheat models to simulate crop yield and other crop variables under different climate conditions (e.g. Landau et al. 1998; Meinke et al. 1998 and Mishra et al. 2013). In Table 12.10 we report some of these studies. In Table 12.11, however, we reported a list of the most common, widely used and well-documented wheat models with their reference studies.

Most of crop models reported in Table 12.11 used different daily input of environmental variables to simulate wheat growth and development as a response to variable nitrogen, precipitation, temperature and water conditions. Others (e.g. Southworth et al. 2002) simulate crop yield taking into consideration the CO_2 fertilization affect (this is also applicable for studies reported in Table 12.8). In the frame of understanding the performance of simulation models and their predictability for crop growth and yield under different climate scenarios, Porter et al. (1993) used experimental data to compare the performance of three models (Table 12.10) for potential yield improvement using water and fertilizer as changing variables. Yield predictions varied largely between models when different climate variables and experimental data were used. Another comparison consisted the use of 5 wheat models, Semenov et al. (1996) tested the performance of these models (Table 12.10) in two sites in Europe (Rothamsted, UK, and Seville, Spain) which were considered representative of temperate and Mediterranean climate zone, respectively. The models were run for climate scenarios derived from a number of general circulation models (GCMs) with the corresponding increase in CO_2 concentration to study the effect of climatic variability changes on crop yields simulated by different wheat models. The predicted wheat yields were very different in some of the European locations for scenarios with and without changed variability. Such climatic variability changes, which result in both a large decrease in average grain yield and a large increase in yield variability and thus a high agricultural risk, would make the region around Seville unsuitable for wheat production.

The study of Jamieson et al. (1998b) compared winter wheat yield prediction response to varying water supply at Lincoln, New Zealand in 1991–1992 using five simulation models (Table 12.10). General results showed good grain yield prediction from the models with relative accuracy except for SWHEAT model which predicted yield reduction with lower accuracy as response to drought stress (Fig. 12.10).

Table 12.10 Different simulation models used in different comparison studies

Crop simulation models	AFRCWHEAT2	DSSAT (CERES-Wheat)	SWHEAT	NWHEAT	SIRIUS	SOILN	SUCROS2	APES	CROPSYST	DAISY	FASSET	HERMES	STICS	WOFOST	SSM	APSIM
Comparison study																
van Laar et al. (1992)	+	+	−	−	−	−	+	−	−	−	−	−	−	−	−	−
Porter et al. (1993)	+	+	+	−	−	−	−	−	−	−	−	−	−	−	−	−
Semenov et al. (1996)	+	+	−	+	+	+	−	−	−	−	−	−	−	−	−	−
Jamieson et al. (1998b)	+	+	+	−	+	−	+	−	−	−	−	−	−	−	−	−
Landau et al. (1998)	+	+	−	−	+	−	−	−	−	−	−	−	−	−	−	−
Olesen et al. (2002)	+	+	+	+	+	−	−	−	−	+	+	−	−	−	−	−
Palosuo et al. (2011)	−	+	−	−	−	−	−	+	+	+	+	+	+	+	−	−
Soltani and Sinclair (2015)	−	+	−	−	−	−	−	−	+	−	−	−	−	−	+	+

+ = used in the comparison study and − = not used

Table 12.11 List of the most common and widely used wheat simulation models with their reference studies

Model	Model description	Reference studies[a]
APSIM	(McCown et al. 1996; Keating et al. 2003)	(Zhao et al. 2014; Mohanty et al. 2012; Hammer et al. 2010; Luo et al. 2005, 2009; Asseng et al. 1998, 2000)
EPIC	Williams et al. (1984)	Tubiello et al. (2000)
CERES	(Ritchie and Otter 1985; Otter-Nacke et al. 1987; Ritchie et al. 1988)	(Jamieson et al. 2000; Southworth et al. 2002; Luo et al. 2003; Hoogenboom et al. 2013)
AquaCrop	Steduto et al. (2009)	(Soddu et al. 2013; Andarzian et al. 2011)
Sirius	(Jamieson and Wilson 1988; Jamieson et al. 1998a)	Semenov et al. (2014)
CropSyst	(Stockle et al. 1994, 2003)	(Pannkuk et al. 1998; Benli et al. 2007)
NWHEAT	Groot (1987)	Semenov et al. (1996)
DSSAT	(Jones et al. 2003; Hoogenboom et al. 2012)	Brassard and Singh (2008)
AFRCWHEAT2	Porter (1984, 1993); Weir et al. (1984)	(Porter 1993; Toscano et al. 2012)
APES	Donatelli et al. (2010)	(Palosuo et al. 2011; Therond et al. 2011)
DAISY	Hansen et al. (1990, 1991)	(Ghaley and Porter 2014; Hansen et al. 2012)
HERMES	Kersebaum (1995)	Palosuo et al. (2011)
STICS	Brisson et al. (1998)	(Guillaume et al. 2011; Coucheney et al. 2015)
WOFOST	Van Diepen et al. (1989)	(Huang et al. 2015; Boogaard et al. 2013; Supit et al. 2012, 2010)
SUCROS2	van Laar et al. (1992)	Jamieson et al. (1998b)
SWHEAT	van Keulen and Seligman (1987)	Porter et al. (1993)
SSM	(Amir and Sinclair 1991; Soltani and Sinclair 2012; Soltani et al. 2013)	Soltani and Sinclair (2015)

[a] Studies in which the model has been tested for validation or compared to other models for crop yield prediction under different climate variability

However, biomass accumulation (e.g. aboveground biomass and leaf area index (LAI)) were predicted to vary significantly between models due to the reductions in photosynthetic or light use efficiency. Compared with experimental data, these factors found have relative small effect on the reduction. Across three studied wheat models, van Laar et al. (1992) final biomass was found to be in acceptable range. However, SUCROS2 predicted higher biomass and grain yield compared to other models. In two different climatic areas (using US and European data), the performance of the three models was similar, but prediction results were always higher in Europe (Fig. 12.11) reflecting the effect of different climate conditions on biomass

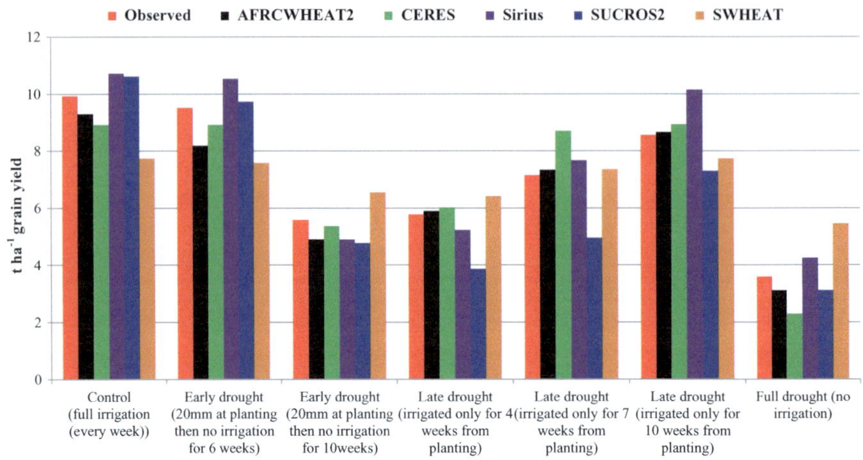

Fig. 12.10 Comparison of final observed and predicted grain biomass from the experiment and simulations under different water supply treatments (Source: Jamieson et al. (1998b))

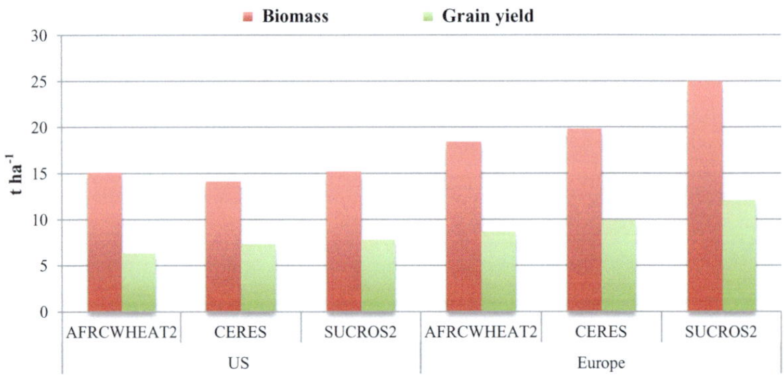

Fig. 12.11 Comparison of predicted grain yield and biomass production in two different environment using simulation models

predictions. In contrast, in the prediction of wheat grain yield, Landau et al. (1998) found differences in yield predictions using simulation models (Table 12.10). The variability in wheat yield predictions was found to be more related to the differences in leaf area index and light use efficiency.

Recently, long-term filed data (including weather data, soil properties, crop phenology and crop and soil management) collected throughout Europe (Palosuo et al. 2011) were used to evaluate the performance of widely used crop models (Table 12.10) in predicting winter wheat yield with reference to variation in date set for model application. Special attention was given to the uncertainty related to the different model prediction and their large-scale applications. Within tested models, high inaccuracy was reported and showed significant variability in terms of yield

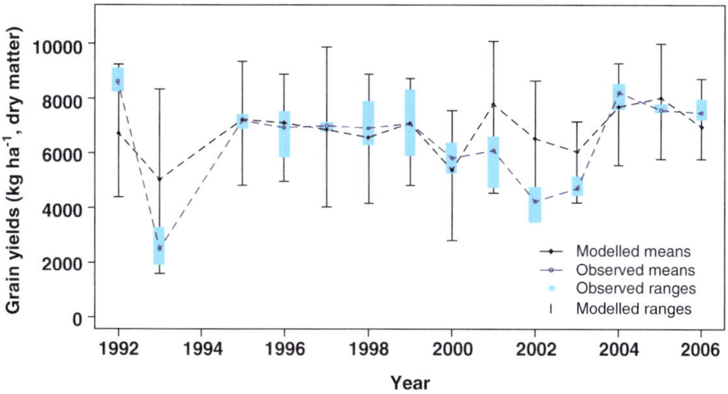

Fig. 12.12 Means and ranges of model-estimated (eight models) and observed (three or four measured plots) yields for the studied growing seasons (14) in Lednice study sites (Source: (Palosuo et al. 2011))

under different environments. Predicted yield was reported to be over or/and under-estimated with close to the observed ones in some case. However, the average predicted yields was consistent with observed once (Fig. 12.12). The results highlighted the importance of calibration in applying crop simulation models to reduce uncertainty. Therefore, the use of different crop models rather than building our conclusion using just one model will be recommended in order to reduce inaccuracy.

To acclimatize wheat cultivar to the different climate change scenarios in order to enhance future production under changing climate, more reasonable results for our future yield prediction should be obtained. To do that, some aspects related to the use of model calibration, complexity (e.g. number of parameters) and characteristics have to be taken into consideration. The accuracy of predication is important element that has to be considered in the simulation studies. In most studies, simulated yields are slightly below or above the observed level including some calibration steps. In the evaluation of crop models accuracy, Wallach et al. (2014) indicated that accuracy in prediction by a model could be obtained when avoid calibration with site-specific and local information (e.g cultivar choice, nitrogen fertilizer and water availability). It is important, however, to evaluate the model across different agro-environmental conditions as well as different output variables. Study by Asseng et al. (2013) indicates that the higher uncertainty in climate change impact projection is due to variation among crop models. Soltani and Sinclair (2015) compared four wheat models (Table 12.10) to tested their transparency and robustness. Although the authors used simpler models (CropSyst (50 parameters) and SSM (55 parameters)) and more complex models (APSIM (292 parameters) and DSSAT (211 parameters)), the simulation results showed no significant relationship between model performance and parameter number. Furthermore, results show that the two simpler models were found to be more robust than the complex models. Therefore, their conclusion highlighted that increasing model complexity does not necessarily lead to model performance improvement. This was in agreement with the conclusion of Sinclair and Seligman (2000) who indicate that models should be kept as simple as possible. To simulate potential yields at large-scale using process-based models, Adam et al. (2011) studied the effect of large data set on predicted yield under different climatic change scenarios in Europe. Comparing simple and detailed approach for simulating dry matter production and LAI changes, the authors found that the simple approach (constant RUE) for dry matter production simulated higher yields in southern Europe, while the detailed approach (the Farquhar approach) simulated higher yields in northern Europe. The results indicate that the effect of location was more pronounced than approach selected on simulated yields. However, for LAI, the authors found that different light interception via LAI approach had resulted in remarkable differences in yield simulation regardless the location. Their conclusion highlighted the importance of well understanding and modeling of the process regarding leaf deterioration with age to decrease model uncertainty in yield simulations.

The results obtained from different simulation models can help to evaluate and improve modeling approaches for better management of the agro-system. This could also help to better assess crop yield potential (e.g. wheat) under climate change in order to overcome potential threats on future crop yield and identify challenges in order to achieve sustainable production.

12.5 Adaptation Options for Wheat to Climate Variability

The unexpected changes in climate event will continue to threat cereal productivity in many part of the world if alternative management options are not adopted. As temperature increase during growing season is usually associated with water scarcity. The frequent occurrence of these climate events will present major threats to future crop productivity. Any adoption efforts to counteract the negative impact of climate change should involve the development of resistant cultivars to both kinds of environmental stress (Tester and Bacic 2005). Recent experimental results agreed that the adoption of resistant cultivars show high efficiency against abiotic as well as some biotic stresses. In addition, agronomical options including water and nitrogen use efficiencies, the use of conservation practices and resource use efficiency can also participate to the mitigation efforts of climate change (Hellin et al. 2012). In most cases, the response of the adapted management options was positive in terms of yield under different studied climate scenarios. However, this positive effect can be influenced depending on cultivars, region and, time and severity of the climate events. Therefore, there is a need for the development of new management option that can be adopted in different geographical areas, to different crops/cultivars and to the wide type of climate changes.

Nowadays, the improvement of agronomical options in order to fight the future changes in climate become of great importance (Howden et al. 2007). Several adoption options were tested and applied to validate their workability in different climate areas across the globe. In Europe, the adoption of these options was assessed under different climate variability Reidsma et al. (2010). We present here some of those that found to have great benefits for wheat cropping system under moderate climate change. However, their effectiveness under more severe climate changes is still questionable.

12.5.1 Agronomic Options

The continuous occurrence of hot and dry summers weather in Europe urged many agricultural researchers to develop options in order to avoid these redundant stresses. Some of them (e.g. Akkaya et al. (2006) and Richards (2006)) suggest that earlier maturation was found to be an option to reduce the negative impact of seasonal variations of light and water on wheat productivity. However, it may result in lower yield. Other studies (e.g. Evans and Fischer 1999) suggest that prolongation of grain filling period is important feature in improving wheat grain yield. By modifying crop development rate, Semenov et al. (2014) found increase grain filling duration to be the key factor to increase wheat yield. But in order to achieve this increase in yield, the authors indicate that both green area index and the number of fertile florets at anthesis should be kept to the maximum. In addition, Richards (1991) found the optimization of flowering time to be the most important aspects in relation to grain yield increase in dry environments such as southern Europe. After anthesis, the improvement of N uptake by the plant was found to be another strategy to increase grain yield. But this will depend very much on the ability of roots to absorb N during grain filling (Andersson et al. 2004; Martre et al. 2006). Under Australian climate change scenarios, Reyenga et al. (1999a) validated the workability for some wheat management options to negate the potential impacts of climate change and increased atmospheric CO_2 on wheat production. Results indicate that nitrous fertilization lead to yield improvement under all studied scenarios, whereas late and early maturity resulted in lower yield.

At field level, agriculture practices represent some other mitigation options, which could help to reduce the negative impact of changing climate on sustainable wheat production. For example, conservation tillage practices and water use efficiency have been tested in different climate and proven to be less costly and viable mitigation options. Water supply, in particular, can cause land degradation through poor water management practices. FAO (1989, 1991) developed some conservation techniques to optimize water efficiency during irrigation. One of these techniques involves the application of conservation tillage practices by which crop residues are incorporated into the soil. Together with the application of green manure and cover crop, conservation tillage provide options to restore soil organic matter in the soil, reduce the risk of erosion and supply the plants with the necessary means to survive (Langdale et al. 1992; Peterson et al. 1993). In addition, conservation tillage consists mainly on the reduction of tillage operations, which can improve the income of rural communities and reduce the environmental loads. In fact, the application of this tillage technique has led to improvement in wheat yield throughout Europe including southern Italy with minimum environmental impacts (Alhajj Ali et al. 2015). However, the applicability of conservation tillage to compensate for the potential future yield losses due to climate change is site-specific. Recently, due to its success in different parts of Europe, the increasing interest of conservation tillage application is becoming more pronounced (Cameron and Oram 1994).

12.5.2 Root Studies

In fact, the complexity of plants' root system and their interaction with the surrounded environments make them difficult to study. However, some efforts to understand the function of wheat root systems under environmental stresses show that they have a great effect on the resistance of environmental stress such as drought due to their sophisticated characteristics (Hurd 1968; Blum 1988). Root length and density in particular are important traits for yield stability especially when water is scarce due to their ability to travel deeper in the soil looking for water. But when adequate amount of rainfall is supplied no yield advantages can be expected even from good plant rooting system. Semenov and Halford (2009) highlighted some root traits including growth and distribution and its role in water stress avoidance. Adaptation options of wheat to climate variability through its rooting system should involve the minimization of stress occurrence through development of a good root system. This include changing the root distribution among soil layers, prolonging the root growth duration, increasing root growth rate and enhancing the absorption ability of water and nutrition from deep soil layers (Wang et al. 2014). These options can, in the case of drought, permits water to be accessed deeper in the soil, whereas, in the case of heat, make a balance between transpiration rates and evaporative demand, which therefore lead to higher carbon fixation rate (Reynolds et al. 2010) and wheat grain yield improvement.

Due to the fact that wheat aboveground and belowground biomasses are intercorrelated (Qin et al. 2012), plant green part exposure to high temperature can influence root development and/or function. While the increases in CO_2 concentrations fount to stimulate root biomass (Benlloch-Gonzalez et al. 2014) and increase total root length, the combined occurrence of high temperature found to negated/reduce this positive effect.

12.5.3 Breeding Approaches

The expected advent of more adverse weather conditions for wheat production as a consequence of changing climate events (Trnka et al. 2014; Yang et al. 2014) will require increasing efforts towards the release of new cultivars. The importance to yield stability to ensure food security (Cattivelli et al. 2008) urged many scientists to develop new cultivars that are more adapted to abiotic and biotic stresses predicted to occur under global climate change. The development of new varieties using wheat landraces that are more adapted to local biotic and abiotic stresses presents a viable strategy to improve and sustain yields, especially under stresses and future changes in climate (De Oliveira et al. 2014; Mohammadi et al. 2015). However, landraces with high genetic diversity should be selected and crossed with locally adapted landraces and varieties to achieve breakthroughs in wheat genetic improvement (Mohammadi et al. 2015) in order to increase tolerance which therefore results in increase yield potential and to respond to climate change (Semenov et al. 2014).

In the effort to increase our understanding of how genes control different traits of importance and become transmitted from one generation to another, plant breeders started to incorporate gene(s) of choice to the desired genetic backgrounds through different hybridization and selection techniques (Talukdar and Talukdar 2013). However, genes are not independent of external and internal environment, and often show their impact through interactions. The understanding of statistics and biometrical genetics helped in unraveling the role of the major and minor genes and the environment, and showed a way to manipulate such traits (i.e. the quantitative traits). The locus that governs such quantitative traits is often called quantitative traits locus (QTL). Using genetic principles in breeding, breeders succeeded in enhancing the yield of crops as well as crops' ability to fight against disease and insect-pest attacks to a great extent. With development of specific alleles of genes, Talukdar and Talukdar (2013) reported that yield of wheat and rice were enhanced to such an extent that it saved millions of poor people from hunger, particularly in Asia and Africa.

To achieve comprehensive understanding of the genetic basis of adaptation among elite cultivars, plant improvement has relied heavily on modifying the phenotype of crops. A very successful intervention has been made to modify phonological patterns of crops to avoid stress (Ludlow and Muchow 1990). This include strategy to avoid water scarcity through modifying root traits, leaf area index and osmotic all of which has been proven to be important mechanism for drought stress tolerance in some crop species. Additional strategies to keep plant metabolism in action during water limitation include stem reserve mobilization after anthesis (Plaut et al. 2004) and functional stay green phenotype (Sanchez et al. 2002). Another intervention involves the reduction of stress occurrence through the development of a good root system that permits water to be accessed deeper in the soil when drought occurs (e.g., Lopes and Reynolds 2010).

The increase interest in molecular breeding allows the identification of QTLs, which helped for the creation of highly stress resistant cultivars. Identified QTLs in wheat show high percentage of phenotype variation were responsible for yield reduction under environmental stress (Pinto et al. 2010). Since the QTLs responsible for yield are linked to other traits, the identification of physiological and genetic components that are responsible for yield responses under drought and heat stresses is fundamental in order to better draw strategic approach to breeding (Reynolds and Tuberosa 2008).

Due to its importance in threatened world's food production, salinity is becoming of great interest to crop breeders. In durum wheat, aspects related to salinity tolerance have been deeply investigated under breeding program, which resulted in the identification of major QTLs responsible for salt-tolerance (James et al. 2006). The creation of this new cultivar found to increase durum wheat yield by 25 % in some cases (Munns et al. 2012).

12.5.4 Molecular Approaches

The discovery of a useful gene is a major scientific challenge determining the success of developing plant varieties suitable for climate-resilient agriculture. The improvement of plant genotypes through breeding programs has led to the creation of resistant cultivars to dives environmental stresses. The development of such cultivars was found to be important strategy to ensure food security. Within the recent breeding initiatives, there is an increasing interest for the development of new cultivars in order to improve crop yield potential and increase crop adaptability to climate changes (Araus et al. 2008).

Crop breeding, both through conventional techniques, and GM assisted breeding could help meet the challenge of adapting wheat to climate variability, if adequately supported by appropriate information on the future climate. With the raise of global population and corresponding increase in eco-environmental problems, the plant breeding techniques demand modernization for enhanced efficiency and environmental stability. This led to the introduction and application of molecular biology tools and techniques of modernization of conventional plant breeding which is called "molecular breeding" (Talukdar and Talukdar 2013). Molecular breeding involves gene discovery and QTL mapping at DNA level in order to improve traits of interest in plant though advanced breeding techniques such as molecular marker-assisted selection, gene manipulation and genomic selection (Jiang 2013). Genomics, for example, is the study of genetics that involves the use of DNA sequencing techniques to sequence, assemble, and analyze the function and structure of genomes with the aim to combat variability in food production (Henry 2014) under varied environmental stresses (Bansal et al. 2013). Molecular genetics technique, however, is applied to better understand the molecular function and interactions among genes to identify their genetic variation through genetic markers in order to classify traits of agronomic and economic interest at a molecular level. There are two categories of genetic markers (DNA and classical markers) (Xu 2010). Their application in plant breeding is becoming widely used because of their ability to transmit specific traits to following generations, which permits the identification of individuals (Jiang 2013).

In plant pathology, in order to increase the accuracy of selection in wheat breeding, important QTLs were mapped (Prasad et al. 1999) using DNA markers. For example, leaf rust resistance genes were tagged in wheat (Roy et al. 1999) in order to improve selection efficiency in wheat breeding (Todorovska et al. 2009).

The use of other biotechnological methods such as haploid and doubled haploid in wheat breeding program present innovative technologies that support the creation and improvement of targeted traits from genetic variations (Germana 2011) able to face future challenges due to climate changes. The use of doubled haploid technology in plant breeding or genetic research has been effective because they reach 100 % homozygosity after one generation after the induction of haploids (instead of several generations of inbreeding through selfing). Recently, both Marker-aided breeding and doubled haploid technology have been used to improve host plant resistance in wheat (Dwivedi et al. 2015). This has found to lead to reduction in farm input through increase input-use efficiency, therefore could present a potential mitigation strategy to the negative impact due to further climate change (Ortiz et al. 2014).

12.5.5 Modelling Approaches

With regard to the information reported previously in this chapter, the projected changes in climate and other environmental factors are interacted in a way that will affect crop productivity. For more than a decade, researches on the potential impacts of changing climate on wheat yield have been pursued worldwide (Luo et al. 2005). The obtained results highlighted the possibility of yield improvement if proper management and adoption option took place especially in most vulnerable areas. However, the main challenges still whether the agroecosystems will adapt to the projected changes in climate.

Wheat is of the most vulnerable crops to environmental stresses such as heat and drought, therefore holistic understanding of the potential impacts on its productivity will help finding a way to increase economical return to farmers and, at the same time, minimize risk. In this regard, modelling techniques through the use of crop growth simulation models are becoming widely applied in agriculture to achieve these goals (Toscano et al. 2012). They are able to deeply investigate the cause behind the potential crop yield reduction resulted from the interaction between different environmental as well as agronomical factors (Pannkuk et al. 1998; Ahmed and Hassan 2011). This will give the possibility for regional assessment to better identify and therefore manage any possible threat to wheat production under different climate change scenarios (Challinor et al. 2010; Ahmed and Hassan 2011; Therond et al. 2011; White et al. 2011). In recent years, they are becoming widely applied at large scale to evaluate crop productivity under climate change impacts (Asseng et al. 2015). However, to be more accurate, crop simulation models will need intensive primary farm data concerning climate conditions, cultivars used, soil properties and crop management. Asseng et al. (2013) highlighted the importance of such information in the accurate prediction of crop productivity. Recent model-based studies have been conducted on wheat to understand crop growth response to foreseen changes in climate. Most results demonstrate that these changes would have a significant impact on crop growth and yield if proper adoption strategies are not applied.

12.6 Conclusion

It is generally accepted that foreseen changes in European climate will cause a substantial crop yield losses. Therefore, the identification of best adoption strategy to the wide variation in future climate will be a vital option to sustain crop productivity. Important strategy will include the modification of phenological pattern to avoid stressful period during plant development. In addition, comprehensive understanding of responses of each plant growth stage to environmental variables such as elevated CO_2 concentration, temperature and drought stress alone or in combination will help to acclimatize crop plant to these changes. Likewise, in the study of the climate variability impact on wheat production, these variables need to be considered in any agenda concerning mitigation strategies to reduce the risk of climate change in crop yield and growth. Information about the time in which climate variable(s) occurred in the field is important as the severity of its effect/their combined effect can vary largely. Example could be made from the effects of drought on wheat yield components which was found to rely heavily on which growth stage the stress occurs. In spite of climate variables, agronomic practices such as cultivar choices, water and nitrogen supply, nutrients availability and growing conditions affect crop responses to different climate variables and, therefore, need to be also taken into account.

In fact, environmental stresses (both abiotic and biotic) were tested in different climate regions across the globe. Their effects were found to be the main factors that significantly limit crop productivity in many parts of the world. Reported information from the literature agreed that the effect of individual variable on crop growth and yield can differ as being the sole variable or in combination with other variable(s). Understanding of the different crop growth stages and their response to varied environmental stresses will be crucial to minimize the negative impacts due to climate change.

Increase population pressure and climate uncertainty will be the major constraints to achieve global food security. the fail to balance between these two factors will lead to remarkable impact on the social, economic, and ecological aspects especially in the most vulnerable regions if proper actions are not taken. Some of these actions are already in place in some regions trying to acclimatize crops to current climate variability and to any potential changes in future climate. Important actions are the use of conservation practices including water harvesting technique and input-use efficiency, modifying crop cycle through early or late sowing date according to the accuracy seasonal weather forecasting and improved the selection of crop cultivars that are well adapted to different environmental conditions.

To some up with, climatic variability is and will be an issue regardless of the current understanding of global temperature trends. The failure of agriculture to adapt to climatic variability will impact global food, especially wheat production. A holistic approach will be paramount to sustaining agriculture and the vitality of the world in the face of climate change.

References

Abayomi, Y., and D. Wright. 1999. Effects of water stress on growth and yield of spring wheat (Triticum aestivum L.) cultivars. *Tropical Agriculture* 76: 120–125.

Abraha, M.G., and M.J. Savage. 2006. Potential impacts of climate change on the grain yield of maize for the midlands of KwaZulu-Natal, South Africa. *Agriculture Ecosystems and Environment* 115: 150–160.

Adam, M., L.G.J. Van Bussel, P.A. Leffelaar, H. Van Keulen, and F. Ewert. 2011. Effects of modelling detail on simulated potential crop yields under a wide range of climatic conditions. *Ecological Modelling* 222: 131–143.

Aggarwal, P.K., and R.K. Mall. 2002. Climate change and rice yields in diverse agro-environments of India. II. Effect of uncertainties in scenarios and crop models on impact assessment. *Climatic Change* 52(3): 331–343.

Ahmed, M., and Fayyaz-ul-Hassan. 2011. APSIM and DSSAT models as decision support tools. The 19th international congress on modelling and simulation, Perth, Australia. http://mssanz.org.au/modsim2011.

Ainsworth, E.A., and S.P. Long. 2005. What have we learned from 15 years of Free-Air CO2 Enrichment (FACE)? A meta-analytic review of the responses of photosynthesis, canopy. and plant production to rising CO2. *The New Phytologist* 165: 351–371.

Ainsworth, E.A., P.A. Davey, C.J. Bernacchi, O.C. Dermody, E.A. Heaton, D.J. Moore, P.B. Morgan, S.L. Naidu, H.S.Y. Ra, X.G. Zhu, P.S. Curtis, and S.P. Long. 2002. A meta-analysis of elevated CO2 effects on soybean (Glycine max) physiology, growth and yield. *Glob Change Biology* 8(8): 695–709.

Akkaya, A., T. Dokuyucu, R. Kara, and M. Akçura. 2006. Harmonization ratio of post- to pre-anthesis durations by thermal times for durum wheat cultivars in a Mediterranean environment. *European Journal of Agronomy* 24: 404–408.

Akram, M. 2011. Growth and yield components of wheat under water stress of different growth stages. *Bangladesh Journal of Agricultural Research* 36(3): 455–468.

Albert, K.R., T.N. Mikkelsen, A. Michelsen, H. Ro-Poulsen, and L. van der Linden. 2011. Interactive effects of drought, elevated CO_2 and warming on photosynthetic capacity and photosystem performance in temperate heath plants. *Journal of Plant Physiology* 168: 1550–1561.

Alcamo, J., G.J.J. Kreileman, J.C. Bollen, G.J. van den Born, R. Gerlagh, M.S. Krol, A.M.C. Toet, and H.J.M. Vries. 1996. Baseline scenarios of global environmental change. *Global Environmental Change* 6: 261–303.

Alexander, B., P. Hayman, G. McDonald, A. Talukder, G. Gill. 2010. Characterizing the risk of heat stress on wheat in South Australia: Meteorology, climatology and the design of a field heating system. In: 15th Australian agronomy conference, Christchurch, New Zealand. http://www.agronomy.org.au/.

Alhajj Ali, S., L. Tedone, and G. De Mastro. 2015. Optimization of the environmental performance of rainfed durum wheat by adjusting the management practices. *Journal of Cleaner Production* 87: 105–118.

Al-Issa, T.A., and N.H. Samarah. 2006. Tillage practices in wheat production under rainfed conditions in Jordan: An economic comparison. *World Journal of Agricultural Research* 2(3): 322–325.

Allen Jr., L.H. 1990. Plant responses to rising carbon dioxide and potential interactions with air pollutants. *Journal of Environmental Quality* 19: 15–34.

Altenbach, S.B., F.M. DuPont, K.M. Kothari, R. Chan, E.L. Johnson, and D. Lieu. 2003. Temperature, water and fertilizer influence the timing of key events during grain development in a US spring wheat. *Journal of Cereal Science* 37: 9–20.

Amir, J., and T.R. Sinclair. 1991. A model of water limitation on spring wheat growth and yield. *Field Crops Research* 28(1–2): 59–69.

Amthor, J.S. 2001. Effects of atmospheric CO_2 concentration on wheat yield: Review of results from experiments using various approaches to control CO_2 concentration. *Field Crops Research* 73: 1–34.
Andarzian, B., A.M. Bakhshandeh, M. Bannayan, Y. Emam, G. Fathi, and K.A. Saeed. 2008. WheatPot: A simple model for spring wheat yield potential using monthly weather data. *Biosystems Engineering* 99(4): 487–495.
Andarzian, B., M. Bannayan, P. Steduto, H. Mazraeh, M.E. Barati, M.A. Barati, and A. Rahnama. 2011. Validation and testing of the AquaCrop model under full and deficit irrigated wheat production in Iran. *Agricultural Water Management* 100: 1–8.
Andersson, A., E. Johansson, and P. Oscarson. 2004. Post-anthesis nitrogen accumulation and distribution among grains in spring wheat spikes. *The Journal of Agricultural Science* 142: 525–533.
Annicchiarico, P., and L. Pecetti. 1993. Contribution of some agronomic traits to durum wheat performance in a dry Mediterranean region of Northern Syria. *Agronomie, EDP Sciences* 13(1): 25–34.
Annicchiarico, P., L. Pecetti, and A.B. Damania. 1995. Relationships between phenotypic variation and climatic factors at collecting sites in durum wheat landraces. *Hereditas* 122(2): 163–167.
Anonymous. 2009. *How to feed the world in 2050. High-level experts forum*, 35. Rome: FAO.
Anwar, M.R., G. O'Leary, D. McNeil, H. Hossain, and R. Nelson. 2007. Climate change impact on rainfed wheat in south-eastern Australia. *Field Crops Research* 104: 139–147.
Araus, J.L., G.A. Slafer, M.P. Reynolds, and C. Royo. 2002. Plant breeding and water relations in C_3 cereals: What to breed for? *Annals of Botany* 89: 925–940.
Araus, J.L., G.A. Slafer, C. Royo, and M.D. Serret. 2008. Breeding for yield potential and stress adaptation in cereals. *Critical Reviews in Plant Science* 27: 377–412.
Ashraf, M.Y. 1998. Yield and yield components response of wheat (*Triticum aestivum* L.) genotypes tinder different soil waler deficit conditions. *Acta Agronomica Hungarica* 46: 45–51.
Asseng, S., B.A. Keatingb, I.R.P. Fillerya, P.J. Gregoryd, J.W. Bowdenc, N.C. Turnera, J.A. Paltaa, and D.G. Abrechte. 1998. Performance of the APSIM-wheat model in Western Australia. *Field Crops Research* 57(2): 163–179.
Asseng, S., H. van Keulen, and W. Stol. 2000. Performance and application of the APSIM N wheat model in the Netherlands. *European Journal of Agronomy* 12: 37–54.
Asseng, S., L. Foster, and N.C. Turner. 2011. The impact of temperature variability on wheat yield. *Global Change Biology* 17: 997–1012.
Asseng, S., F. Ewert, C. Rosenzweig, J.W. Jones, J.L. Hatfield, A.C. Ruane, K.J. Boote, P.J. Thorburn, R.P. Rotter, D. Cammarano, N. Brisson, B. Basso, P. Martre, P.K. Aggarwal, C. Angulo, P. Bertuzzi, C. Biernath, A.J. Challinor, J. Doltra, S. Gayler, R. Goldberg, R. Grant, L. Heng, J. Hooker, L.A. Hunt, J. Ingwersen, R.C. Izaurralde, K.C. Kersebaum, C. Müller, S. Naresh Kumar, C. Nendel, G. O'Leary, J.E. Olesen, T.M. Osborne, T. Palosuo, E. Priesack, D. Ripoche, M.A. Semenov, I. Shcherbak, P. Steduto, C. Stockle, P. Stratonovitch, T. Streck, I. Supit, F. Tao, M. Travasso, K. Waha, D. Wallach, J.W. White, J.R. Williams, and J. Wolf. 2013. Uncertainty in simulating wheat yields under climate change. *Nature Climate Change* 3: 827–832.
Asseng, S., F. Ewert, P. Martre, R.P. Rötter, D.B. Lobell, D. Cammarano, B.A. Kimball, M.J. Ottman, G.W. Wall, J.W. White, M.P. Reynolds, P.D. Alderman, P.V.V. Prasad, P.K. Aggarwal, J. Anothai, B. Basso, C. Biernath, A.J. Challinor, G. De Sanctis, J. Doltra, E. Fereres, M. Garcia-Vila, S. Gayler, G. Hoogenboom, L.A. Hunt, R.C. Izaurralde, M. Jabloun, C.D. Jones, K.C. Kersebaum, C. Müller, A.K. Koehler, S. Naresh Kumar, C. Nendel, G. O'Leary, J.E. Olesen, T. Palosuo, E. Priesack, E. Eyshi Rezaei, A.C. Ruane, M.A. Semenov, I. Shcherbak, C. Stöckle, P. Stratonovitch, T. Streck, I. Supit, F. Tao, P. Thorburn, K. Waha, E. Wang, D. Wallach, J. Wolf, Z. Zhao, and Y. Zhu. 2015. Rising temperatures reduce global wheat production. *Nature Climate Change* 5: 143–147. doi:10.1038/nclimate2470.
Baker, J.T., and L.H. Allen Jr. 1993. Contrasting crop responses to CO_2 and temperature: Rice, soybean and citrus. *Vegetatio* 104/105: 239–260.

Bale, J.S., G.J. Masters, I.D. Hodkinson, C. Awmack, T.M. Bezemer, V.K. Brown, J. Butterfield, A. Buse, J.C. Coulson, J. Farrar, J.E.G. Good, R. Harrington, S. Hartley, T.H. Jones, R.L. Lindroth, M.C. Press, I. Symrnioudis, A.D. Watt, and J.B. Whittaker. 2002. Herbivory in global climate change research: Direct effects of rising temperature on insect herbivores. *Global Change Biology* 8: 1–16.

Balla, K., M. Rakszegi, Z. Li, F. Békés, S. Bencze, and O. Veisz. 2011. Quality of winter wheat in relation to heat and drought shock after anthesis. *Czech Journal of Food Sciences* 29(2): 117–128.

Bansal, K.C., S.K. Lenaka, and T.K. Mondal. 2013. Genomic resources for breeding crops with enhanced abiotic stress tolerance. *Plant Breeding*. doi:10.1111/pbr.12117.

Barnabás, B., K. Jäger, and A. Fehér. 2008. The effect of drought and heat stress on reproductive processes in cereals. *Plant, Cell and Environment* 31: 11–38.

Barneix, A.J., and M.R. Guitman. 1993. Leaf regulation of the nitrogen concentration in the grain of wheat plants. *Journal of Experimental Botany* 44: 1607–1612.

Battisti, D.S., and R.L. Naylor. 2009. Historical warnings of future food insecurity with unprecedented seasonal heat. *Science* 323: 240–244.

Batts, G.R., J.I.L. Morison, R.H. Ellis, P. Hadley, and T.R. Wheeler. 1997. Effects of CO2 and temperature on growth and yield of crops of winter wheat over four seasons. *European Journal of Agronomy* 7: 43–52.

Bazzaz, F.A., and K. Garbutt. 1988. The response of annuals in competitive neighborhoods: Effects of elevated CO2. *Ecology* 69: 937–946.

Bender, J., U. Hertstein, and C.R. Black. 1999. Growth and yield responses of spring wheat to increasing carbon dioxide, ozone and physiological stresses: A statistical analysis of 'ESPACE-Wheat' results. *European Journal of Agronomy* 10: 185–195.

Benli, B., M. Pala, C. Stockle, and T. Oweis. 2007. Assessment of winter wheat production under early sowing with supplemental irrigation in a cold highland environment using CropSyst simulation model. *Agricultural Water Management* 93(1–2): 45–53.

Benlloch-Gonzalez, M., R. Bochicchio, J. Berger, H. Bramley, and J.A. Palta. 2014. High temperature reduces the positive effect of elevated CO_2 on wheat root system growth. *Field Crops Research* 165: 71–79.

Bergthorsson, P., H. Bjornsson, O. Dyrmundsson, B. Gudmundsson, A. Helgadottir, and J.V. Jonmundsson. 1988. The effects of climatic variations on agriculture in Iceland. In *The impact of climatic variations on agriculture, volume 1, Assessments in cool temperate and cold regions*, ed. M.L. Parry, T.R. Carter, and N.T. Konijn, 383–509. Dordrecht: Kluwer.

Bernklau, E.J., E.A. Fromm, and L.B. Bjostad. 2004. Disruption of host location of western corn rootworm larvae (Coleoptera: Chrysomelidae) with carbon dioxide. *Journal of Economic Entomology* 97: 330–339.

Berntsen, J., B.M. Petersen, B.H. Jacobsen, J.E. Olesen, and N.J. Hutchings. 2003. Evaluating nitrogen taxation scenarios using the dynamic whole farm simulation model FASSET. *Agricultural Systems* 76(3): 817–839.

Berzitis, E. 2013. Climate change effects on the pest status and distribution of the bean leaf beetle (Cerotoma trifurcata). A PhD thesis, University of Guelph, Ontario, Canada.

Bita, C.E., and T. Gerats. 2013. Plant tolerance to high temperature in a changing environment: Scientific fundamentals and production of heat stress-tolerant crops. *Frontiers in Plant Science* 4: 273. doi:10.3389/fpls.2013.00273.

Blum, A. 1988. *Plant breeding for stress environment*. Boca Raton: CRC Press.

Blum, A. 1998. Improving wheat grain filling under stress by stem reserve mobilisation. *Euphytica* 100: 77–83.

Blumenthal, C.S., E.W.R. Barlow, and C.W. Wrigley. 1993. Growth environment and wheat quality: The effect of heat stress on dough properties and gluten proteins. *Journal of Cereal Science* 18: 3–21.

Boogaard, H., J. Wolf, I. Supit, S. Niemeyer, and M. van Ittersum. 2013. A regional implementation of WOFOST for calculating yield gaps of autumn-sown wheat across the European Union. *Field Crops Research* 143: 130–142.

Borghi, B., M. Corbellini, C. Minoia, M. Palumbo, N. Di Fonzo, and M. Perenzin. 1997. Effects of Mediterranean climate on wheat bread-making quality. *European Journal of Agronomy* 6: 145–154.

Borjigidai, A., K. Hikosaka, T. Hirose, T. Hasegawa, M. Okada, and K. Kobayashi. 2006. Seasonal changes in temperature dependence of photosynthetic rate in rice under a free-air CO2 enrichment. *Annals of Botany* 97: 549–557.

Borlaug, N.E., and C.R. Dowswell. 1996. The acid lands: One of agriculture's last frontiers. In *Plant-soil interactions at low pH. Proceedgins of the 4th international symposium*, ed. A.C. Moniz, A.M.C. Furlani, R.E. Schaffert, N.K. Fageria, C.A. Rosolem, and H. Cantorella. Belo Horizonte: The Brazilian Soil Science Society.

Bosello, F., and Zhang, J. 2005. Assessing climate change impacts: Agriculture. CIP – Climate impacts and policy division, working paper n. 02.2007.

Bowes, G. 1993. Facing the inevitable: Plants and increasing atmospheric CO_2. *Annual Review of Plant Physiology and Plant Molecular Biology* 44: 309–332.

Brassard, J.P., and B. Singh. 2008. Impacts of climate change and CO(2) increase on agricultural production and adaptation options for Southern Quebec, Canada. *Mitigation and Adaptation Strategies for Global Change* 13: 241–265.

Braun, H.J., G. Atlin, and T. Payne. 2010. Multi-location testing as a tool to identify plant response to global climate change. In *Climate change and crop production*, ed. M. Reynolds. London: CABI.

Bray, S., and D.M. Reid. 2002. The effect of salinity and CO_2 enrichment on the growth and anatomy of the second trifoliate leaf of *Phaseolus vulgaris. Canadian Journal of Botany* 80: 349–359.

Briggle, L.W. 1980. Origin and botany of wheat. In *Wheat documenta cibageigy*, ed. E. Häfliger, 6–13. Switzerland: Basle.

Brisson, N., B. Mary, D. Ripoche, M.H. Jeuffroy, F. Ruget, B. Nicoullaud, P. Gate, F. Devienne-Barret, R. Antonioletti, and C. Durr. 1998. STICS: A generic model for the simulation of crops and their water and nitrogen balances. I. Theory and parameterization applied to wheat and corn. *Agronomie* 18: 311–346.

Buchner, P., M. Tausz, R. Ford, A. Leo, G.J. Fitzgerald, M.J. Hawkesford, and S. Tausz-Posch. 2015. Expression patterns of C- and N-metabolism related genes in wheatare changed during senescence under elevated CO2 in dry-land agriculture. *Plant Science* 236: 239–249.

Bunce, J.A. 1998. The temperature dependence of the stimulation of photosynthesis by elevated carbon dioxide in wheat and barley. *Journal of Experimental Botany* 49: 1555–1561.

Cameron, D., and P. Oram. 1994. *Minimum and reduced tillage: Its use in North America and Western Europe and its potential application in Eastern Europe, Russia, and Central Asia*. Washington, DC: International Food Policy Research Institute. 121 pp.

Campbell, G.S., and J.M. Norman. 1989. The description and measurement of plant canopy structure. In *Plant canopies: Their growth, form and function*, ed. G. Russell, B. Marshall, and P.G. Jarvis, 1–19. Cambridge: Cambridge University Press.

Cardoso-Vilhena, J., and J. Barnes. 2001. Does nitrogen supply affect the response of wheat (Triticum aestivum cv. Hanno) to the combination of elevated CO_2 and O_3? *Journal of Experimental Botany* 52: 901–1911.

Carter, T.R., R.N. Jones, and X.L. Lu. 2007. New assessment methods and the characterization of future conditions. In *Climate change 2007: Impacts, adaptation and vulnerability, contribution of working group II to the fourth assessment report of the Intergovernmental Panel on Climate Change*, ed. IPCC, 133–171. Cambridge: Cambridge University Press.

Cattivelli, L., F. Rizza, F.-W. Badeck, E. Mazzucotelli, A.M. Mastrangelo, E. Francia, C. Marè, A. Tondelli, and M.A. Stanca. 2008. Drought tolerance improvement in crop plants: An intergrated view from breeding to economics. *Field Crops Research* 105: 1–14.

Chakraborty, S., A. von Tiedemann, and P.S. Teng. 2000. Climate change: Potential impact on plant diseases. *Environmental Pollution* 108: 317–326.

Challinor, A.J., E.S. Simelton, E.D.G. Fraser, D. Hemming, and M. Collins. 2010. Increased crop failure due to climate change: Assessing adaptation options using models and socio-economic data for wheat in China. *Environmental Research Letters* 5: 034012–034012.

Chaudhuri, U.N., M.B. Kirkam, and E.T. Kanemasu. 1990. Root growth of winter wheat under elevated carbon dioxide and drought. *Crop Science* 30: 853–857.

Chaves, M.M., J.P. Maroco, and J.S. Pereira. 2003. Understanding plant responses to drought—from genes to the whole plant. *Functional Plant Biology* 30(3): 239–264.

Chen, De.X., H.W. Hunt, and J.A Morgan. 1996. Responses of a C3 and C4 perennial grass to CO_2 enrichment and climate change: Comparison between model predictions and experimental data. *Ecological Modeling* 87, 11–27.

Christensen, J.H., B. Hewitson, A. Busuioc, A. Chen, X. Gao, I. Held, R. Jones, R.K. Kolli, W.-T. Kwon, R. Laprise, V. Magaña Rueda, L. Mearns, C.G. Menéndez, J. Räisänen, A. Rinke, A. Sarr, and P. Whetton. 2007. Regional climate projections. In *Climate change 2007: The physical science basis. Contribution of working group I to the fourth assessment report of the Intergovernmental Panel on Climate Change*, ed. S. Solomon, D. Qin, M. Manning, Z. Chen, M. Marquis, K.B. Averyt, M. Tignor, and H.L. Miller. Cambridge/New York: Cambridge University Press.

CIMMYT, International Maize and Wheat Improvement Centre. 2014 Wheat improvement – the mandate of CIMMYT's global wheat program. Wheat Research, CIMMYT website.

Conroy, J. 1992. Influence of elevated atmospheric CO_2 concentration on plant nutrition. *Australian Journal of Botany* 40: 445–456.

Conroy, J.P., S. Seneweera, A.S. Basra, G. Rogers, and B. Nissen-Wooller. 1994. influence of rising atmospheric concentrations and temperature on growth, yield and grain quality of cereal crops. *Australian Journal of Plant Physiology* 21: 741–758.

Coucheney, E., S. Buis, M. Launay, J. Constantin, B. Mary, I. García de Cortazar-Atauri, D. Ripoche, N. Beaudoin, F. Ruget, K.S. Andrianarisoa, C. Le Bas, E. Justes, and J. Leonard. 2015. Accuracy, robustness and behavior of the STICS soil-crop model for plant, water and nitrogen outputs: Evaluation over a wide range of agro-environmental conditions in France. *Environmental Modelling & Software* 64: 177–190.

Crafts-Brandner, S.J., and R.D. Law. 2000. Effect of heat stress on the inhibition and the recovery of the ribulose-1,5-bisphosphate carboxylase/oxygenase activation state. *Planta* 212: 67–74.

Cure, J.D., and B. Acock. 1986. Crop responses to carbon dioxide doubling: A literature survey. *Agricultural and Forest Meteorology* 38: 127–145.

Curtis, B.C. 1982. Potential for a yield increase in wheat. In Proceedings of the national wheat research conference, Beltsville, MD, USA, 26–28 October, p. 5–19. Washington, DC: National Association of Wheat Growers Foundation.

Dahal, K., V.L. Knowles, W.C. Plaxton, and N.P.A. Hüner. 2014. Enhancement of photosynthetic performance, water use efficiency and grain yield during long-term growth under elevated CO2 in wheat and rye is growth temperature and cultivar dependent. *Environmental and Experimental Botany* 106: 207–220.

Dalrymple, D.G. 1986. *Development and spread of high yielding varieties of wheat in developing countries*. Washington, DC: Bureau for Science and Technology, Agency for International Development, US Government Printing Office.

Davidson, R.H., and W.F. Lyon. 1987. *Insect pests of farm, garden and orchard*. New York: Wiley. 640 pp.

Day, A.D., and S. Intalap. 1970. Some effects of soil moisture stress on growth of wheat (Triticum aestiuum). *Agronomy Journal* 62: 27–29.

De Oliveira, A.C., N. Marini, and D.R. Farias. 2014. Climate change: New breeding pressures and goals. In *Encyclopedia of agriculture and food systems*, 284–293. Amsterdam: Elsevier/Academic Press.

DeBach, P. 1965. Some biological and ecological phenomena associated with colonizing entomophagous insects. In *The genetics of colonizing species*, ed. H.G. Baker and G.L. Stebbins, 287–303. New York: Academic.

Debaeke, P., and A. Aboudrare. 2004. Adaptation of crop management to water-limited environments. *European Journal of Agronomy* 21: 433–446.

Deka, S., K. Byjesh, U. Kumar, and R. Choudhary. 2009. Climate change and impacts on crop pests — A critique. In *Impact of climate change on agriculture*, ed. S. Panigrahy, S.S. Ray, and J.S. Parihar. Ahmedabad: ISPRS Workshop Proceedings XXXVIII-8/W3.

Denby, K., and C. Gehring. 2005. Engineering drought and salinity tolerance in plants: Lessons from genome-wide expression profiling in Arabidopsis. *Trends in Biotechnology* 23(11): 547–552.

Denčić, S., R. Kastori, B. Kobiljski, and B. Duggan. 2000. Evaluation of grain yield and its components in wheat cultivars and landraces under near optimal and drought conditions. *Euphytica* 113(1): 43–52.

Denyer, K., C.M. Hylton, and A.M. Smith. 1994. The effect of high temperature on starch synthesis and the activity of starch synthase. *Australian Journal of Plant Physiology* 21: 783–789.

Deryng, D., W.J. Sacks, C.C. Barford, and N. Ramankutty. 2011. Simulating the effects of climate and agricultural management practices on global crop yield. *Global Biogeochemical Cycles* 25, GB2006.

Dhakhwa, G.B., C.L. Campbell, S.K. LeDuc, and E.J. Cooter. 1997. Maize growth: Assessing the effect of global warming and CO_2 fertilization with crop models. *Agricultural and Forest Meteorology* 87: 253–272.

Dickin, E., and D. Wright. 2008. The effects of winter waterlogging and summer drought on the growth and yield of winter wheat. *European Journal of Agronomy* 28: 234–244.

Doheny-Adams, T., L. Hunt, P.J. Franks, D.J. Beerling, and J.E. Gray. 2012. Genetic manipulation of stomatal density influences stomatal size, plant growth and tolerance to restricted water supply across a growth carbon dioxide gradient. *Philosophical Transactions of the Royal Society* 367: 547–555.

Donatelli, M., G. Russell, A.E. Rizzoli, M. Acutis, M. Adam, I.N. Athanasiadis, et al. 2010. A component-based framework for simulating agricultural production and externalities. In *Environmental and agricultural modeling-integrated approaches for policy impact assessment*, ed. M.K. van Ittersum and F.M. Brouwer, 63–108. Dordrecht/New York: Springer.

Dupont, F., and S. Altenbach. 2003. Molecular and biochemical impacts of environmental factors on wheat grain development and protein synthesis. *Journal of Cereal Science* 38: 133–146.

Dwivedi, S.L., A.B. Britt, L. Tripathi, S. Sharma, H.D. Upadhyaya, and R. Ortiz. 2015. Haploids: Constraints and opportunities in plant breeding. *Biotechnology Advances*. doi:10.1016/j.biotechadv.2015.07.001.

Eckersten, H., K. Blombäck, T. Kätterer, and P. Nyman. 2001. Modelling C, N, water and heat dynamics in winter wheat under climate change in southern Sweden. *Agriculture Ecosystems and Environment* 86: 221–235.

Ehleringer, J.R., T. Cerling, and M. Dearing. 2002. Atmospheric CO2 as a global change driver influencing plant-animal interactions. *Integrative and Comparative Biology* 42: 424–430.

Ekman, A., L. Bülow, and S. Stymne. 2007. Elevated atmospheric CO2 concentration and diurnal cycle induce changes in lipid composition in Arabidopsis thaliana. *The New Phytologist* 174(3): 591–599.

Erbs, M., R. Manderscheid, G. Jansen, S. Seddig, A. Pacholski, and H.J. Weigel. 2010. Effects of free-air CO2 enrichment and nitrogen supply on grain quality parameters and elemental composition of wheat and barley grown in a crop rotation. *Agriculture Ecosystems and Environment* 136: 59–68.

EUROSTAT. 2015. http://ec.europa.eu/eurostat/data/database. Last Accessed July, 2015.

Evans, L.T., and R.A. Fischer. 1999. Yield potential: Its definition, measurement and significance. *Crop Science* 39: 1544–1551.

Ewert, F., D. Rodriguez, P. Jamieson, M.A. Semenov, R.A.C. Mitchell, J. Goudriaan, J.R. Porter, B.A. Kimball, P.J. Pinter Jr., R. Manderscheid, H.J. Weigel, A. Fangmeier, E. Fereres, and F. Villalobos. 2002. Effects of elevated CO_2 and drought on wheat: Testing crop simulation models for different experimental and climatic conditions. *Agriculture Ecosystems and Environment* 93: 249–266.

Ewert, F., M.D.A. Rounsevell, I. Reginster, M.J. Metzger, and R. Leemans. 2005. Future scenarios of European agricultural land use I. Estimating changes in crop productivity. *Agriculture Ecosystems and Environment* 107: 101–116.

Ewert, F., J.R. Porter, and M.D.A. Rounsevell. 2007. Crop models CO2, and climate change. *Science* 315: 459.

Falloon, P., D. Bebber, J. Bryant, M. Bushell, A.J. Challinor, S. Dessai, S. Gurr, and A.-K. Koehler. 2015. Using climate information to support crop breeding decisions and adaptation in agriculture. *World Agriculture* 5(1): 25–43.

FAO. 1989. *Guidelines for designing and evaluating surface irrigation systems*, Irrigation and drainage papers, 45. Rome: FAO, 137 pp.

FAO. 1991. *Water harvesting*, AGL miscellaneous papers, 17. Rome: FAO. 133 pp.

FAO (Food and agriculture organization of the united nation). 2009. The second report on the state of the world's plant genetic resources for food and agriculture. Intergovernmental technical working group on plant genetic resources for food and agriculture. Commission on genetic resources for food and agriculture. Rome, 15–17 July.

FAO (Food and Agriculture Organization of the United Nations). 2013. FAOSTAT: Production-crops, 2013 data.

FAOSTAT Data. 2013. *Food and Agriculture Organization of the United Nations. Statistical database*. http://faostat.fao.org/site/291/default.aspx

FAOSTAT Data. 2014. *Food and Agriculture Organization of the United Nations. Statistical database*. http://faostat.fao.org/site/291/default.aspx

FAOSTAT Data. 2015. *Food and Agriculture Organization of the United Nations. Statistical database*. http://faostat.fao.org/site/291/default.aspx

Farooq, M., H. Bramley, J.A. Palta, and K.H.M. Siddique. 2011. Heat stress in wheat during reproductive and grain-filling phases. *Critical Reviews in Plant Sciences* 30: 491–507.

Farooq, M., M. Hussain, and K.H.M. Siddique. 2014. Drought stress in wheat during flowering and frain-filling periods. *Critical Reviews in Plant Sciences* 33: 331–349.

Farrar, J.F., and M.L. Williams. 1991. The effects of increased atmospheric carbon-dioxide and temperature on carbon partitioning, source-sink relations and respiration. *Plant, Cell and Environment* 14: 819–830.

Fedoroff, N.V., D.S. Battisti, R.N. Beachy, P.J.M. Cooper, D.A. Fischhoff, C.N. Hodges, V.C. Knauf, D. Lobell, B.J. Mazur, D. Molden, M.P. Reynolds, P.C. Ronald, M.W. Rosegrant, P.A. Sanchez, A. Vonshak, and J.-K. Zhu. 2010. Radically rethinking agriculture for the 21st century. *Science* 327: 833–834.

Fernando, N., J. Panozzo, M. Tausz, R.M. Norton, G.J. Fitzgerald, S. Myers, M.E. Nicolas, and S. Seneweer. 2014. Intra-specific variation of wheat grain quality in response to elevated [CO_2] at two sowing times under rain-fed and irrigation treatments. *Journal of Cereal Science* 59: 137–144.

Ferrara, R.M., P. Trevisiol, M. Acutis, G. Rana, G.M. Richter, and N. Baggaley. 2010. Topographic impacts on wheat yields under climate change: Two contrasted case studies in Europe. *Theoretical and Applied Climatology* 99: 53–65.

Ferris, R., R.H. Ellis, T.R. Wheeler, and P. Hadley. 1998. Effect of high temperature stress at anthesis on grain yield and biomass of field-grown crops of wheat. *Annals of Botany* 82: 631–639.

Ferris, R., T.R. Wheeler, R.H. Ellis, P. Hadley, B. Wollenweber, J.R. Porter, T.S. Karacostas, M.N. Papadopoulos, and J. Schellberg. 2000. Effects of high temperature extremes on wheat. In *Final report of CLIVARA project (climate change, climatic variability and agriculture in Europe: an integrated assessment)*, Research report no. 21, ed. T.E. Downing, P.A. Harrison, R.E. Butterfield, and K.G. Lonsdale, 31–55. Oxford: Environmental Change Institute, University of Oxford.

Flagella, Z., M.M. Giuliani, L. Giuzio, C. Volpi, and S. Masci. 2010. Influence of water deficit on durum wheat storage protein composition and technological quality. *European Journal of Agronomy* 33: 197–207.

Flexas, J., J. Bota, F. Loreto, G. Cornic, and T.D. Sharkey. 2004. Diffusive and metabolic limitations to photosynthesis under drought and salinity in C3 plants. *Plant Biology* 6(3): 269–279.

Flint, E.P., and D.T. Patterson. 1983. Interference and temperature effects on growth in soybean (Glycine max) and associated C3 and C4 weeds. *Weed Science* 31: 193–199.

Flint, E.P., D.T. Patterson, D.A. Mortensen, G.H. Riechers, and J.L. Beyers. 1984. Temperature effects on growth and leaf production in three weed species. *Weed Science* 32: 655–663.

Foulkes, M.J., R. Sylvester-Bradley, R. Weightman, and J.W. Snape. 2007. Identifying physiological traits associated with improved drought resistance in winter wheat. *Field Crops Research* 103: 11–24.

Fuhrer, J. 2003. Agroecosystem responses to combinations of elevated CO_2, ozone, and global climate change – A review. *Agriculture, Ecosystem & Environment* 97(1–3): 1–20.

Garrido-Lestache, E., R.J. López-Bellido, and L. López-Bellido. 2005. Durum wheat quality under Mediterranean conditions as affected by N rate, timing and splitting, N form and S fertilization. *European Journal of Agronomy* 23: 265–278.

Germana, M.A. 2011. Gametic embryogenesis and haploid technology as valuable support to plant breeding. *Plant Cell Reports* 30: 839–857. doi:10.1007/s00299-011-1061-7.

Ghaley, B.B., and J.R. Porter. 2014. Ecosystem function and service quantification and valuation in a conventional winter wheat production system with DAISY model in Denmark. *Ecosystem Services* 10: 79–83.

Ghini, R., E. Hamada, F. Angelotti, L.B. Costa, and W. Bettiol. 2012. Research approaches, adaptation strategies, and knowledge gaps concerning the impacts of climate change on plant diseases. *Tropical Plant Pathology* 37: 5–24.

Giannakopoulos, C., P. Le Sager, M. Bindi, E. Moriondo Kostopoulou, and C.M. Goodess. 2009. Climatic changes and associated impacts in the Mediterranean resulting from global warming. *Global and Planetary Change* 68: 209–224.

Goodenough, J.L., and J.M. Mckinion (eds.). 1992. *Basics of insect modeling*, ASAE monograph no. 10. St. Joseph: American Society Agricultural Engineers. 221 pp.

Gooding, M.J., R.H. Ellis, P.R. Shewry, and J.D. Schofield. 2003. Effects of restricted water availability and increased temperature on the grain filling, drying and quality of winter wheat. *Journal of Cereal Science* 37: 295–309.

Grace, J. 1988. Temperature as a determinant of plant productivity. In *Plants and temperature*, ed. S.P. Long and F.I. Woodward, 91–107. Cambridge: The Company of Biologists.

Grasty, S. 1999. Agriculture and climate change: In *TDRI quarterly review*, ed. P. Auger,and R. Suwanraks, vol. 14, no. 2, pp. 12–16.

Graybosch, R.A., C.J. Peterson, P.S. Baenziger, and D.R. Shelton. 1995. Environmental modification of hard red winter wheat flour protein composition. *Journal of Cereal Science* 22: 45–51.

Griffin, J.J., T.G. Ranney, and D.M. Pharr. 2004. Heat and drought influence photosynthesis and water relations, and soluble carbohydrates of two ecotypes of redbud (Cercis canadensis). *Journal of American Society for Horticultural Science* 129: 497–502.

Groot, J.J.R. 1987. *Simulation of nitrogen balance in a system of winter wheat and soil*, Simulation reports CABO-TT no. 13. Wageningen: Centre for Agrobiological Research and Department of Theoretical Production Ecology, Agricultural University.

Guillaume, S., J.E. Bergez, D. Wallach, and E. Justes. 2011. Methodological comparison of calibration procedures for durum wheat parameters in the STICS model. *European Journal of Agronomy* 35: 15–126.

Guo, R.P., Z.H. Lin, X.G. Mo, and C.L. Yang. 2010. Responses of crop yield and water use efficiency to climate change in the North China Plain. *Agricultural Water Management* 97: 1185–1194.

Gutierrez, A.P. 2000. Climate change: Effects on pest dynamics. In *Climate change and global crop productivity*, ed. K.R. Reddy and H.F. Hodges. New York: CAB International.

Guttieri, M.J., R. McLean, J.C. Stark, and E. Souza. 2005. Managing irrigation and nitrogen fertility of hard spring wheats for optimum bread and noodle quality. *Crop Science* 45: 2049–2059.

Hammer, G.L., E. van Oosterom, G. McLean, S.C. Chapman, I. Broad, P. Harland, and R.C. Muchow. 2010. Adapting APSIM to model the physiology and genetics ofcomplex adaptive traits in field crops. *Journal of Experimental Botany* 61: 2185–2202.

Han, X., X. Hao, S.K. Lam, H. Wang, Y. Li, T. Wheeler, H. Ju, and E. Lin. 2015. Yield and nitrogen accumulation and partitioning in winter wheat under elevated CO2: A 3-year free-air CO_2 enrichment experiment. *Agriculture Ecosystems and Environment* 209: 132–137.

Hansen, J.W. 2005. Integrating seasonal climate prediction and agricultural models for insights into agricultural practice. *Philosophical Transactions of the Royal Society Series B, Biological Sciences* 360: 2037–2047.

Hansen, J., H.E. Jensen, N.E. Nielsen, and H. Svendsen. 1990. *DAISY—A soil plant system model. Danish simulation model for transformation and transport of energy and matter in the soil–plant–atmosphere system*. Copenhagen: National Agency for Environmental Protection.

Hansen, S., H.E. Jensen, N.E. Nielsen, and H. Svendsen. 1991. Simulation of nitrogen dynamics and biomass production in winter wheat using the Danish simulation model Daisy. *Fertiliser Research* 27: 245–259.

Hansen, S., P. Abrahamsen, C.T. Petersen, M. Styczen. 2012. Daisy: Model use, calibration, and validation. Trans. ASABE55, 1315–1333.

Harrison, P.A., and R.E. Butterfield. 1996. Effects of climate change on Europe-wide winter wheat and sunflower productivity. *Climate Research* 7: 225–241.

Hatzios, K.K., and D. Penner. 1982. *Metabolism of herbicides in higher plants*. Edina: CEPCO iv., Burgess Publ.

Hays, D.B., J.H. Do, R.E. Mason, G. Morgan, and S.A. Finlayson. 2007. Heat stress induced ethylene production in developing wheat grains induces kernel abortion and increased maturation in a susceptible cultivar. *Journal of Plant Sciences* 172: 1113–1123.

Hellin, J., B. Shiferaw, J.E. Cairns, M. Reynolds, I. Ortiz-Monasterio, M. Banzige, K. Sonder, and R. La Rovere. 2012. Climate change and food security in the developing world: Potential of maize and wheat research to expand options for adaptation and mitigation. *Journal of Development and Agricultural Economics* 4(12): 311–321.

Henry, R.J. 2014. Genomics strategies for germplasm characterization and the development of climate resilient crops. *Frontiers in Plant Science* 5: 68.

Hisas, S. 2011. *The food gap. The impacts of climate change in food production: A 2020 perspective*. Alexandria: Universal Ecological Fund.

Hochman, Z.V.I. 1982. Effect of water stress with phasic development on yield of wheat grown in a semi-arid environment. *Field Crops Research* 5: 55–67.

Hocking, P., and C.P. Meyer. 1991. Effects of CO_2 enrichment and nitrogen stress on growth and partitioning of dry matter and nitrogen in wheat and maize. *Australian Journal of Plant Physiology* 18: 339–356.

Högy, P., and A. Fangmeier. 2008. Effects of elevated atmospheric CO_2 on grain quality of wheat. *Journal of Cereal Science* 48: 580–591.

Högy, P., M. Keck, K. Niehaus, J. Franzaring, and A. Fangmeier. 2010. Effects of atmospheric CO_2 enrichment on biomass, yield and low molecular weight metabolites in wheat grain. *Journal of Cereal Science* 52: 215–220.

Hoogenboom, G., J.W. Jones, P.W. Wilkens, C.H. Porter, K.J. Boote, L.A. Hunt, U. Singh, J.L. Lizaso, J.W. White, O. Uryasev, F.S. Royce, R. Ogoshi, A.J. Gijsman, G.Y. Tsuji, J. Koo. 2012. Decision Support System for Agrotechnology Transfer (DSSAT) Version 4.5. University of Hawaii, Honolulu, Hawaii (CD-ROM).

Hoogenboom, G., J.T. Ritchie, L.A. Hunt, J.W. White, and J. Anothai. 2013. The Cropping System Model (CSM)-CERES-Wheat. In *Proceedings of the workshop on modeling wheat response to high temperature*, CIMMYT, El Batán, Mexico, 19–21 June 2013, ed. P.D. Alderman, E. Quilligan, S. Asseng, F. Ewert, and M.P. Reynolds. Mexico: CIMMYT.

Howden, S.M., J.-F. Soussana, F.N. Tubiello, N. Chhetri, M. Dunlop, and H. Meinke. 2007. Adapting agriculture to climate change. *PNAS* 104(50): 19691–19696.

Huang, J., L. Tian, S. Liang, H. Ma, I. Becker-Reshef, Y. Huang, W. Su, X. Zhang, D. Zhu, and W. Wu. 2015. Improving winter wheat yield estimation by assimilation of the leaf area index from Landsat TM and MODIS data into the WOFOST model. *Agricultural and Forest Meteorology* 204: 106–121.

Huang, J., F. Sedano, Y. Huang, H. Ma, X. Li, S. Liang, L. Tian, X. Zhang, J. Fan, and W. Wu. 2016. Assimilating a synthetic Kalman filter leaf area index series into the WOFOST model to improve regional winter wheat yield estimation. *Agricultural and Forest Meteorology* 216: 188–202.

Hurd, E.A. 1968. Growth of roots of seven cultivars of spring wheat at high and low moisture levels. *Agronomy Journal* 60: 201–205.

Hurkman, W.J., K.F. McCue, S.B. Altenbach, A. Korn, C.K. Tanaka, K.M. Kotharia, E.L. Johnson, D.B. Bechtel, J.D. Wilson, O.D. Anderson, and F.M. DuPont. 2003. Effect of temperature on expression of genes encoding enzymes for starch biosynthesis in developing wheat endosperm. *Plant Science* 164: 873–881.

ICARDA. 2014. *ICARDA annual report 2013*. Beirut: International Center for Agricultural Research in the Dry Areas. 68 pp.

Idso, K.E., and S.B. Idso. 1994. Plant responses to atmospheric carbon dioxide enrichment in the face of environmental constraints-a review of the past 10 years of research. *Agricultural and Forest Meteorology* 69: 152–203.

Idso, S.B., B.A. Kimball, M.G. Anderson, and J.R. Mauney. 1987. Effects of atmospheric CO2 enrichment on plant growth: The interactive role of air temperature. *Agriculture, Ecosystem & Environment* 20(1): 1–10.

Ingram, J., and D. Bartels. 1996. The molecular basis of dehydration tolerance in plants. *Annual Review of Plant Physiology and Plant Molecular Biology* 47(1): 377–403.

IPCC (Intergovernmental Panel on Climate Change). 1996. In *Climate change 1995, summary for policy makers*, ed. J.T. Houghton et al., 1–7. New York: Cambridge University Press.

IPCC. 2007a. *Contribution of working group I to the fourth assessment report of the Intergovernmental Panel on Climate Change*. New York: Cambridge University Press.

IPCC. 2007b. *Contribution of working group II to the fourth assessment report of the Intergovernmental Panel on Climate Change*. New York: Cambridge University Press.

Irmak, S., H.A. Naeem, G.L. Lookhart, and F. MacRitchle. 2008. Effect of heat stress on wheat proteins during kernel development in wheat near-isogenic lines differing at Glu-D1. *Journal of Cereal Science* 48: 513–516.

Izaurralde, R.C., N.J. Rosenberg, R.A. Brown, and A.M. Thomson. 2003. Integrated assessment of Hadley Center (HadCM2) climate-change impacts on agricultural productivity and irrigation water supply in the conterminous United States—Part II. Regional agricultural production in 2030 and 2095. *Agricultural and Forest Meteorology* 117: 97–122.

Jäger, K. 2010. Simultaneous water withholding and elevated temperature alters embryo and endosperm development in wheat. *Acta Agronomica Hungarica* 58: 91–95.

James, R.A., R.J. Davenport, and R. Munns. 2006. Physiological characterization of two genes for Na+exclusion in durum wheat, Nax1 and Nax2. *Plant Physiology* 142: 1537–1547.

Jamieson, P.D., and D.R. Wilson. 1988. Agronomic uses of a model of wheat growth, development and water use. *Proceedgins of Agronomy Society NZ* 18: 7–10.

Jamieson, P.D., J.R. Porter, J. Goudriaan, J.T. Ritchie, H. van Keulen, and W. Stol. 1998a. A comparison of the models AFRCWHEAT2, CERES-Wheat, Sirius, SUCROS2 and SWHEAT with measurements from wheat grown under drought. *Field Crops Research* 55: 23–44.

Jamieson, P.D., M.A. Semenov, I.R. Brooking, and G.S. Francis. 1998b. SIRIUS: A mechanistic model of wheat response to environmental variation. *European Journal of Agronomy* 8: 161–179.

Jamieson, P.D., J. Berntsen, F. Ewert, B.A. Kimball, J.E. Olesen, P.J.J. Pinter, J.R. Porter, and M.A. Semenov. 2000. Modelling CO_2 effects on wheat with varying nitrogen supplies. *Agriculture Ecosystems and Environment* 82: 27–37.

Jenner, C.F. 1994. Starch synthesis in the kernel of wheat under high temperature conditions. *Australian Journal of Plant Physiology* 21: 791–806.

Ji, X., B. Shiran, J. Wan, D.C. Lewis, C.L. Jenkins, A.G. Condon, R.A. Richards, and R. Dolferus. 2010. Importance of pre-anthesis anther sink strength for maintenance of grain number during reproductive stage water stress in wheat. *Plant, Cell and Environment* 33(6): 926–942.

Jia, Y.W., S.H. Shen, C.W. Niu, Y.Q. Qiu, H. Wang, and Y. Liu. 2011. Coupling crop growth and hydrologic models to predict crop yield with spatial analysis technologies. *Journal of Applied Remote Sensing* 5(1): 053537.

Jiang, G.L. 2013. Molecular markers and marker-assisted breeding in plants. In *Plant breeding from laboratories to fields*, ed. S.B. Anderson, 45–83. Croatia: InTech.

Jones, J.W., G. Hoogenboom, C.H. Porter, K.J. Boote, W.D. Batchelor, L.A. Hunt, P.W. Wilkens, U. Singh, A.J. Gijsman, and J.T. Ritchie. 2003. The DSSAT cropping system model. *European Journal of Agronomy* 18: 235–265.

Kang, Y., S. Khan, and X. Ma. 2009. Climate change impacts on crop yield, crop water productivity and food security – a review. *Progress in Natural Science* 19(12): 1665–1674.

Kaur, H., S.K. Jalota, R. Kanwar, and B.B. Vashisht. 2012. Climate change impacts on yield, evapotranspiration and nitrogen uptake in irrigated maize (Zea mays)-wheat (Triticum aestivum) cropping system: A simulation analysis. *The Indian Journal of Agricultural Sciences* 82: 213–219.

Keating, B.A., P.S. Carberry, G.L. Hammer, M.E. Probert, M.J. Robertson, D. Holzworth, N.I. Huth, J.N.G. Hargreaves, H. Meinke, Z. Hochman, G. McLean, K. Verburg, V. Snow, J.P. Dimes, M. Silburn, E. Wang, S. Brown, K.L. Bristow, S. Asseng, S. Chapman, R.L. McCown, D.M. Freebairn, and C.J. Smith. 2003. An overview of APSIM, a model designed for farming systems simulation. *European Journal of Agronomy* 18: 267–288.

Keeling, P.L., P.J. Bacon, and D.C. Holt. 1993. Elevated temperature reduces starch deposition in wheat endosperm by reducing the activity of soluble starch synthase. *Planta* 191: 342–348.

Kersebaum, K.C. 1995. Application of a simple management model to simulate water and nitrogen dynamics. *Ecological Modelling* 81: 145–156.

Kimball, B.A. 1983. Carbon dioxide and agricultural yield: An assemblage and analysis of 430 prior observations. *Agronomy Journal* 75: 779–788.

Kimball, B.A., and C.J. Bernacchi. 2006. Evapotranspiration, canopy temperature, and plant water relations. In *Managed ecosystems and* CO_2, ed. J. Nösberger, S.P. Long, R.J. Norby, M. Stitt, G.R. Hendrey, and H. Blum, 311–324. Berlin/Heidelberg: Springer.

Kimball, B.A., P.J. Pinter Jr., R.L. Garcia, R.L. LaMorte, G.W. Wall, D.J. Hunsaker, G. Wechsung, F. Wechsung, and Th Kartschall. 1995. Productivity and water use of wheat under free-air CO_2 enrichment. *Global Change Biology* 1(6): 429–442.

Kimball, B.A., K. Kobayashi, and M. Bindi. 2002. Responses of agricultural crops to free air CO_2 enrichment. *Advances in Agronomy* 77: 293–368.

Ko, J.H., L.R. Ahuja, S.A. Saseendran, T.R. Green, L.W. Ma, D.C. Nielsen, and C.L. Walthall. 2012. Climate change impacts on dryland cropping systems in the Central Great Plains, USA. *Climatic Change* 111: 445–472.

Kristensen, K., K. Schelde, and J.E. Olesen. 2011. Winter wheat yield response to climate variability in Denmark. *The Journal of Agricultural Science* 149: 33–47.

Lal, R. 2005. Climate change, soil carbon dynamics, and global food security. In *Climate change and global food security*, ed. R. Lal, B. Stewart, N. Uphoff, et al., 113–143. Boca Raton: CRC Press.

Lam, S.K., D. Chen, and R. Norton. 2012. Nitrogen demand and the recovery of ^{15}N labelled fertilizer in wheat grown under elevated carbon dioxide in southern Australia. *Nutrient Cycling in Agroecosystems* 92(2): 133–144.

Landau, S., R.A.C. Mitchell, V. Barnett, J.J. Coils, J. Craigon, K.L. Moore, and R.W. Payne. 1998. Testing winter wheat simulation models' predictions against observed UK grain yields. *Agricultural and Forest Meteorology* 89: 85–99.

Langdale, G.W., L.T. West, and R.R. Bruce. 1992. Restoration of eroded soil with conservation tillage. *Soil Technology* 5: 81–90.

Larcher, W. 2003. *Physiological plant ecology*, 4th ed. Berlin/Heidelberg: Springer.

Laurila, H. 2001. Simulation of spring wheat responses to elevated CO_2 and temperature by using CERES-wheat crop model. *Agricultural and Food Science Finland* 10: 175–196.

Lawlor, D.W. 1995. The effect of water deficit on photosynthesis. In *Environment and plant metabolism. Flexibility and acclimation*, ed. N. Smirnoff, 129–160. Oxford: BIOS Scientific Publisher.

Lawlor, D.W., and G. Cornic. 2002. Photosynthetic carbon assimilation and associated metabolism in relation to water deficits in higher plants. *Plant, Cell and Environment* 25(2): 275–294.

Lee-Ho, E., L.J. Walton, D.M. Reid, E.C. Yeung, and L.V. Kurepin. 2007. Effects of elevated carbon dioxide and sucrose concentrations on *Arabidopsis thaliana* root architecture and anatomy. *Canadian Journal of Botany* 85: 324–330.

Leemans, R., and A. Solomon. 1993. Modeling the potential change in yield and distribution of the earth's crops under a warmed climate. *Climate Research* 3: 79–96.
Leonard, W.H., and J.H. Martin. 1963. *Cereal crops*. New York: MacMillan Publishing.
Lepetz, V., M. Massot, D.S. Schmeller, and J. Clobert. 2009. Biodiversity monitoring: Some proposals to adequately study species' responses to climate change. *Biodiversity and Conservation* 18: 3185–3203.
Li, A., Y. Hou, G.W. Wall, A. Trent, B.A. Kimball, and P.J. Pinter Jr. 2000. Free-air CO_2 enrichment and drought stress effects on grain filling rate and duration in spring wheat. *Crop Science* 40: 1263–1270.
Li, A., Y. Hou, and A. Trent. 2001. Effects of elevated atmospheric CO_2 and drought stress on individual grain filling rates and duration of the main stem in spring wheat. *Agricultural and Forest Meteorology* 106: 289–301.
Li, Y.-F., Y. Wu, N. Hernandez-Espinosa, and R.J. Peña. 2013. Heat and drought stress on durum wheat: Responses of genotypes, yield, and quality parameters. *Journal of Cereal Science* 57: 398–404.
Lin, E.D., W. Xiong, H. Ju, Y.L. Xu, Y. Li, L.P. Bai, and L.Y. Xie. 2005. Climate change impacts on crop yield and quality with CO2 fertilization in China. *Philosophical Transactions of the Royal Society B* 360: 2149–2154.
Lobell, D.B., M.B. Burke, C. Tebaldi, M.D. Mastrandrea, W.P. Falcon, and R.L. Naylor. 2008. Prioritizing climate change adaptation needs for food security in 2030. *Science* 319: 607–610.
Lobell, D.B., G.L. Hammer, G. McLean, C. Messina, M.J. Roberts, and W. Schlenker. 2013. The critical role of extreme heat for maize production in the United States. *Nature Climate Change* 3: 497–501.
Lonbani, M., and A. Arzani. 2011. Morpho-physiological traits associated with terminal drought stress tolerance in triticale and wheat. *Agronomy Research* 9(1–2): 315–329.
Long, S.P. 1991. Modification of the response of photosynthetic productivity to rising temperature by atmospheric CO_2 concentrations: Has its importance been underestimated? *Plant, Cell & Environment* 14(8): 729–740.
Lopes, M.S., and M.P. Reynolds. 2010. Partitioning of assimilates to deeper roots is associated with cooler canopies and increased yield under drought in wheat. *Functional Plant Biology* 37: 147–156.
Ludlow, M.M., and R.C. Muchow. 1990. A critical evaluation of traits for improving crop yields in water-limited environments. *Advances in Agronomy* 43: 107–153.
Luo, Q., M.A.J. Williamsa, W. Bellottib, and B. Bryana. 2003. Quantitative and visual assessments of climate change impacts on South Australian wheat production. *Agricultural Systems* 77: 173–186.
Luo, Q., W. Bellotti, M. Williams, and B. Bryan. 2005. Potential impact of climate change on wheat yield in South Australia. *Agricultural and Forest Meteorology* 132: 273–285.
Luo, Q., W. Bellotti, M. Williams, and E. Wang. 2009. Adaptation to climate change of wheat growing in South Australia: Analysis of management and breeding strategies. *Agriculture Ecosystems & Environment* 129(1–3): 261–267.
Ma, H., J. Zhu, Z. Xie, G. Liu, Q. Zeng, and Y. Han. 2007. Responses of rice and winterwheat to free air CO_2 enrichment (China FACE) at rice/wheat rotation system. *Plant and Soil* 294: 137–146.
Manderscheid, R., J. Bender, H.J. Jager, and H.J. Weigel. 1995. Effects of season long CO2 enrichment on cereals. II. Nutrient concentrations and grain quality. *Agriculture, Ecosystem and Environment* 54: 175–185.
Manschadi, A.M., J. Christopher, P. Devoil, and G.L. Hammer. 2006. The role of root rchitectural traits in adaptation of wheat to water-limited environments. *Functional Plant Biology* 33: 823–837.
Martre, P., P.D. Jamieson, M.A. Semenov, R.F. Zyskowski, J.R. Porter, and E. Triboi. 2006. Modelling protein content and composition in relation to crop nitrogen dynamics for wheat. *European Journal of Agronomy* 25: 138–154.

Masters, G., P. Baker, and J. Flood. 2010. Climate change and agricultural commodities. CABI working paper 02, Europe-UK.
Mattson, W.J. 1980. Herbivory in relation to plant nitrogen content. *Annual Review of Ecology and Systematics* 11: 119–161.
Mattson, W.J., and R.A. Haack. 1987. The role of drought in outbreaks of plant-eating insects. *BioScience* 37(2): 110–118.
McCown, R.L., G.L. Hammer, J.N.G. Hargreaves, D.P. Holzworth, and D.M. Freebairn. 1996. APSIM: A novel software system for model development, model testing and simulation in agricultural system research. *Agricultural Systems* 50: 255–271.
McDonald, A.J.S., and W.J. Davies. 1996. Keeping in touch: Responses of the whole plant to deficits in water and nitrogen supply. *Advances in Botanical Research* 22: 229–300.
Meinke, H., R. Rabbinge, G. Hammer, H. van Keulen, and P. Jamieson. 1998. Improving wheat simulation capabilities in Australia from a cropping systems perspective. II. Testing simulation capabilities of wheat growth. *European Journal of Agronomy* 8: 83–99.
Melloy, P., G. Hollaway, J. Luck, R. Norton, E. Aitken, and S. Chakraborty. 2010. Production and fitness of Fusarium pseudograminearum inoculum at elevated carbon dioxide in FACE. *Global Change Biology* 16: 3363–3373.
Meza, F.J., and D. Silva. 2009. Dynamic adaptation of maize and wheat production to climate change. *Climatic Change* 94: 143–156.
Milad, S.I., L.E. Wahba, and M.N. Barakat. 2011. Identification of RAPD and ISSR markers associated with flag leaf senescence under water-stressed conditions in wheat (Triticum aestivum L.). *Australian Journal of Crop Science* 5(3): 337–343.
Mishra, A., R. Singh, N.S. Raghuwanshi, C. Chatterjee, and J. Froebrich. 2013. Spatial variability of climate change impacts on yield of rice and wheat in the Indian Ganga Basin. *Science of the Total Environment* 468–469: S132–S138.
Mkhabela, M., P. Bullock, M. Gervais, G. Finlay, and H. Sapirstein. 2010. Assessing indicators of agricultural drought impacts on spring wheat and quality on the Canadian prairies. *Agricultural and Forest Meteorology* 150: 399–410.
Mo, X.G., S.X. Liu, Z.H. Lin, and R.P. Guo. 2009. Regional crop yield, water consumption and water use efficiency and their responses to climate change in the North China Plain. *Agriculture Ecosystems and Environment* 134: 67–78.
Mochida, O. 1991. Impact of CO_2-climate change on pests distribution. *Agriculture and Horticulture* 66: 128–136.
Moffat, A.S. 2002. Plant genetics. Finding new ways to protect drought-stricken plants. *Science* 296: 1226–1229.
Mohammadi, R., B. Sadeghzadeh, H. Ahmadi, N. Bahrami, and A. Amri. 2015. Field evaluation of durum wheat landraces for prevailing abiotic and biotic stresses in highland rainfed regions of Iran. *The Crop Journal*. doi:10.1016/j.cj.2015.03.008.
Mohanty, M., M.E. Probert, K.S. Reddy, R.C. Dalal, A.K. Mishra, A.S. Rao, M. Singh, and N.W. Menzies. 2012. Simulating soybean–wheat cropping system: APSIM model parameterization and validation. *Agriculture Ecosystems and Environment* 152: 68–78.
Morison, J.I.L. 1985. Sensitivity of stomata and water use efficiency to high CO_2. *Plant, Cell and Environment* 8: 467–474.
Mullarkey, M., and P. Jones. 2000. Isolation and analysis of thermotolerant mutants of wheat. *Journal of Experimental Botany* 51: 139–146.
Munns, R., R.A. James, B. Xu, A. Athman, S.J. Conn, C. Jordans, C.S. Byrt, R.A. Hare, S.D. Tyerman, M. Tester, D. Plett, and M. Gilliham. 2012. Wheat grain yield on saline soils is improved by an ancestral Na(+) transporter gene. *Nature Biotechnology* 30: 360–364.
Murray, T.J., D.S. Ellsworth, D.T. Tissue, and M. Riegler. 2013. Interactive direct and plant-mediated effects of elevated atmospheric CO2 and temperature on a eucalypt-feeding insect herbivore. *Global Change Biology* 19: 1407–1416.
Natural Resources Institute Finland (NRIF). 2015. Global warming reduces wheat production markedly if no adaptation takes place. ScienceDaily. Retrieved 19 Feb, 2015 from www.sciencedaily.com/releases/2015/01/150112082944.htm.

Nawaz, A., M. Farooq, S.A. Cheema, and A. Wahid. 2013. Differential response of wheat cultivars to terminal heat stress. *International Journal of Agriculture and Biology* 15: 1354–1358.

Nezhadahmadi, A., Z.H. Prodhan, and G. Faruq. 2013. Drought tolerance in wheat. *Review Article* 610721: 12. http://dx.doi.org/10.1155/2013/610721.

NIC, The National Intelligence Council. 2012. *Global trends 2030: Alternative worlds*. Washington, DC: NIC.

Nicolas, M.E., R.M. Gleadow, and M.J. Dalling. 1985. Effect of post-anthesis drought on cell-division and starch accumulation in developing wheat grains. *Annals of Botany* 55: 433–444.

Niinemets, Ü. 2010. Responses of forest trees to single and multiple environmental stresses from seedlings to mature plants: Past stress history, stress interactions, tolerance and acclimation. *Forest Ecology and Management* 260: 1623–1639.

Nonhebel, S. 1994. Inaccuracies in weather data and their effects on crop growth simulation results. II. Water-limited production. *Climate Research* 4: 61–74.

Nonhebel, S. 1996. Effects of temperature rise and increase in CO_2 concentration on simulated wheat yields in Europe. *Climatic Change* 34: 73–90.

Oechel, W.C., and B.R. Strain. 1985. Native species responses to increased atmospheric carbon dioxide concentration. In *Direct effects of increasing carbon dioxide on vegetation*, ed. B.R. Strain and J.D. Cure, 117–154. Washington, DC: DOE/ER-0238, U.S. Department of Energy.

Olesen, J.E., B.M. Petersen, J. Berntsen, S. Hansen, P.D. Jamieson, and A.G. Thomsen. 2002. Comparison of methods for simulating effects of nitrogen on green area index and dry matter growth in winter wheat. *Field Crops Research* 74: 131–149.

Olesen, J.E., T.R. Carter, C.H. Diaz-Ambrona, et al. 2007. Uncertainties in projected impacts of climate change on European agriculture and terrestrial ecosystems based on scenarios from regional climate models. *Climatic Change* 81: 123–143.

Olesen, J.E., M. Trnka, K.C. Kersebaum, A.O. Skjelvag, B. Seguin, P. Peltonen-Sainio, F. Rossi, J. Kozyra, and F. Micale. 2011. Impacts and adaptation of European crop production systems to climate change. *European Journal of Agronomy* 34: 96–112.

Ortiz, R., K.D. Sayre, B. Govaerts, R. Gupta, G.V. Subbarao, T. Ban, D. Hodson, J.M. Dixon, J.I. Ortiz-Monasterio, and M. Reynolds. 2008. Climate change: Can wheat beat the heat? *Agriculture Ecosystems and Environment* 126: 46–58.

Ortiz, R., D. Edwards, S. Mayes, F.C. Ogbonnaya, and C. Kole. 2014. Agro-biodiversity, climate change challenges and potential genomics-led plant breeding solutions, pp 3–12. In 'Application of Genomics to the Production of Climate Resilient Crops: Challenges and opportunities'. International Climate Resilient Crops Genomics Consortium (ICRCGC) White Paper, ed. D. Edwards, R. Henry, and C. Kole. Available for download at http://www.climatechangegenomics.org/.

Otter-Nacke, S., D.C. Godwin, and J.T. Ritchie. 1987. *Testing and validating the CERES-Wheat model in diverse environments*, AgRISTARS Publication no. YM-15-00407. Springfield: NTIS.

Ozdogan, M. 2011. Modeling the impacts of climate change on wheat yields in Northwestern Turkey. *Agriculture Ecosystems and Environment* 141: 1–12.

Palosuo, T., K.C. Kersebaum, C. Angulo, P. Hlavinka, M. Moriondo, J.E. Olesen, R.H. Patil, F. Ruget, C. Rumbaur, J. Takác, M. Trnka, M. Bindi, B. Caldag, F. Ewert, R. Ferrise, W. Mirschel, L. Saylan, B. Siskak, and R. Rötter. 2011. Simulation of winter wheat yield and its variability in different climates of Europe: A comparison of eight crop growth models. *European Journal of Agronomy* 35: 103–114.

Pangga, I.B., J. Hanan, and S. Chakraborty. 2012. Climate change impacts on plant canopy architecture: Implications for pest and pathogen management. *European Journal of Plant Pathology* 135(3): 595–610.

Pannkuk, C.D., C.O. Stockle, and R.I. Papendick. 1998. Evaluating CropSyst simulations of wheat management in a wheat-fallow region of the US pacific northwest. *Agricultural Systems* 57(2): 121–134.

Parry, M.L., J.H. Porter, and T.R. Carter. 1990. Agriculture: Climatic change and its implications. *Trends in Ecology and Evolution* 5: 318–322.

Passioura, J.B. 1996. Drought and drought tolerance. In *Drought tolerance in higher plants: Genetical, physiological and molecular biological analysis*, ed. E. Belhassen, 3–12. Dordrecht: Kluwer.
Patterson, D.T. 1993. Implications of global climate change for impact of weeds, insects, and plant diseases. In *International crop science*, vol. I, 273–280. Madison: Crop Science Society of America.
Patterson, D.T. 1995. Weeds in a changing climate. *Review of Weed Science* 43(4): 685–701.
Patterson, D.T., and E.P. Flint. 1990. Implications of increasing carbon dioxide and climate change for plant communities and competition in natural and managed ecosystems. In *Impact of CO_2, trace gases, and climate change on global agriculture*, Special publication 53, ed. B.A. Kimball, N.J. Rosenberg, and L.H. Allen Jr., 83–110. Madison: American Society of Agronomy.
Patterson, D.T., A.E. Russell, D.A. Mortensen, R.D. Coffin, and E.P. Flint. 1986. Effects of temperature and photoperiod on Texas panicum (Panicum texanum) and wild proso millet (Panicum miliaceum). *Weed Science* 34: 876–882.
Patterson, D.T., J.K. Westbrook, R.J.V. Joyce, P.D. Lingren, and J. Rogasik. 1999. Weeds, insects, and disease. *Climatic Change* 43: 711–727.
Pautasso, M., T.F. Döring, M. Garbelotto, L. Pellis, and M.J. Jeger. 2012. Impacts of climate change on plant diseases-opinions and trends. *European Journal of Plant Pathology* 133: 295–313.
Pavlova, V.N., S.E. Varcheva, R. Bokusheva, and P. Calanca. 2014. Modelling the effects of climate variability on spring wheat productivity in the steppe zone of Russia and Kazakhstan. *Ecological Modelling* 277: 57–67.
Peterson, G.A., D.G. Westfall, and C.V. Cole. 1993. Agroecosystem approach to soil and crop management research. *Soil Science Society of America Journal* 57: 1354–1360.
Pinto, R.S., M.P. Reynolds, K.L. Mathews, C.L. McIntyre, J.J. Olivares-Villegas, and S.C. Chapman. 2010. Heat and drought adaptive QTL in a wheat population designed to minimize confounding agronomic effects. *Theoretical and Applied Genetics* 121: 1001–1021.
Plaut, Z., B.J. Butow, C.S. Blumenthal, and C.W. Wrigley. 2004. Transport of dry matter into developing wheat kernels and its contribution to grain yield under post-anthesis water deficit and elevated temperature. *Field Crops Research* 86: 185–298.
Porter, J.R. 1984. A model of canopy development in winter wheat. *Journal of Agricultural Science* 102: 383–392.
Porter, J.R. 1993. AFRCWHEAT2: A model of the growth and development of wheat incorporating responses to water and nitrogen. *European Journal of Agronomy* 2: 69–82.
Porter, J.R., and D.J. Moot. 1998. Research beyond the means climatic variability and plant growth. COST Symposium on Applied Agrometeorology and Agroclimatology.
Porter, J.R., P.D. Jamieson, and D.R. Wilson. 1993. Comparison of the wheat simulation models AFRCWHEAT2, CERES-wheat and SWHEAT for non-limiting conditions of crop growth. *Field Crops Research* 33(1–2): 131–157.
Porter, J.R., and M.A. Semenov. 2005. Crop responses to climatic variation. *Philosophical Transactions of the Royal Society B* 360: 2021–2035.
Porter, J.H., M.L. Parry, and T.R. Carter. 1991. The potential effects of climatic change on agricultural insect pests. *Agricultural Forest Meteorology* 57: 221–240.
Pradhan, G.P., P.V.V. Prasad, A.K. Fritz, M.B. Kirkham, and B.S. Gill. 2012. Effects of drought and high temperature stress on synthetic hexaploid wheat. *Functional Plant Biology* 39: 90–198.
Prasad, M., R.K. Varshney, A. Kumar, H.S. Balyan, P.C. Sharma, K.J. Edwards, H. Singh, H.S. Dhaliwal, J.K. Roy, and P.K. Gupta. 1999. A microsatellite marker associated with a QTL for grain protein content on chromosome arm 2DL of bread wheat. *Theoretical and Applied Genetics* 99: 341–345.
Pritchard, S.G., and H.H. Rogers. 2000. Spatial and temporal deployment of crop roots in CO_2-enriched environments. *The New Phytologist* 147: 55–71.
Qaderi, M.M., and D.M. Reid. 2009. Crop responses to elevated carbon dioxide and temperature. In *Climate change and crop*, ed. S.N. Singh. Berlin: Springer. doi:10.1007/978-3-540-88246-61.

Qin, Z., Q. Zhuang, and M. Chen. 2012. Impacts of land use change due to biofuel crops on carbon balance, bioenergy production, and agricultural yield, in the conterminous United States. *GCB Bioenergy* 4: 277–288.

Rainey, R.C. 1989. *Migration and meteorology: Flight behavior and the atmospheric environment of migrant pests*. New York: Oxford University Press. 314 pp.

Rawson, H.M. 1992. Plant responses to temperature under conditions of elevated CO_2. *Australian Journal of Botany* 40: 473–490.

Rawson, H.M. 1995. Yield responses of two wheat genotypes to carbon dioxide and temperature in field studies using temperature gradient tunnels. *Australian Journal of Plant Physiology* 22: 23–32.

Reddy, V.R., and Y.A. Pachepsky. 2000. Predicting crop yields under climate change conditions from monthly GCM weather projections. *Environmental Modelling & Software* 15: 79–86.

Reddy, A.R., K.R. Reddy, and H.F. Hodges. 1998. Interactive effects of elevated carbon dioxide and growth temperature on photosynthesis in cotton leaves. *Plant Growth Regulation* 26: 33–40.

Reidsma, P., F. Ewert, Lansink A. Oude, and R. Leemans. 2010. Adaptation to climate change and climate variability in European agriculture: The importance of farm level responses. *European Journal of Agronomy* 32: 91–102.

Reilly, J., W. Baethgen, F.E. Chege, S.C. van de Geijn, L. Erda, A. Iglesias, G. Kenny, D. Patterson, J. Rogasik, R. Rötter, C. Rosenzweig, W. Sombroek, and J. Westbrook. 1996. Agriculture in a changing climate: Impacts and adaptation. In *Climate change 1995: Impacts, adoption, and mitigation of climate change: Scientific and technical analysis*, ed. R.T. Watson, M.C. Zinyowera, and R.H. Moss, 427–467. Cambridge: Cambridge University Press.

Reyenga, P.J., S.M. Howden, H. Meinke, and G.M. Mckeon. 1999a. Modelling global change impacts on wheat cropping in south-east Queensland, Australia. *Environmental Modelling and Software* 14: 297–306.

Reyenga, P.J., S.M. Howden, H. Meinke, and W.B. Hall. 1999b. Global change impacts on the wheat production along an environmental gradient in South Australia. In *MODSIM 99 Proceedings of the international congress on modeling and simulation, 6–9 December 1997, Hamilton, New Zealand*, ed. L. Oxley and F. Scrimgeour. Perth: Modelling and Simulation Society of Australia and New Zealand.

Reynolds, M., and R. Tuberosa. 2008. Translational research impacting on crop productivity in drought-prone environments. *Current Opinion in Plant Biology* 11: 171–179.

Reynolds, M.P., J. Pietragalla, and H.-J. Braun. 2008. *International symposium on wheat yield potential: Challenges to international wheat breeding*. Mexico: CIMMYT.

Reynolds, M., Y. Manes, A. Izanloo, and P. Langridge. 2009. Phenotyping approaches for physiological breeding and gene discovery in wheat. *The Annals of Applied Biology* 155: 309–320.

Reynolds, M.P., D. Hays, and S. Chapman. 2010. Breeding for adaptation to heat and drought stress. In *Climate change and crop production*, ed. M.P. Reynolds, 447–454. London: CABI.

Rharrabtia, Y., C. Royo, D. Villegas, N. Aparicio, and L.F. García del Moral. 2003a. Durum wheat quality in Mediterranean environments I. Quality expression under different zones, latitudes and water regimes across Spain. *Field Crops Research* 80: 123–131.

Rharrabtia, Y., D. Villegas, C. Royo, V. Martos-Nunez, and L.F. García del Moral. 2003b. Durum wheat quality in Mediterranean environments II. Influence of climatic variables and relationships between quality parameters. *Field Crops Research* 80: 133–140.

Ribas-Carbo, M., N.L. Taylor, L. Giles, et al. 2005. Effects of water stress on respiration in soybean leaves. *Plant Physiology* 139(1): 466–473.

Richards, R.A. 1991. Crop improvement for temperate Australia: Future opportunities. *Field Crops Research* 26: 141–169.

Richards, R.A. 2006. Physiological traits used in the breeding of new cultivars for water-scarce environments. *Agricultural Water Management* 80: 197–211.

Richter, G.M., and M.A. Semenov. 2005. Modelling impacts of climate change on wheat yields in England and Wales: Assessing drought risks. *Agricultural Systems* 84: 77–97.

Ritchie, J.T., and S. Otter. 1985. Description and performance of CERES-Wheat: A user-riented wheat yield model. In *ARS wheat yield project*, ARS-38, ed. W.O. Willis, 159–175. Washington, DC: Department of Agriculture, Agricultural Research Service.

Ritchie, J.T., D.C. Godwin, and S. Otter-Nacke. 1988. *CERES-Wheat: A simulation model of wheat growth and development*. College Station: Texas A&M University Press.

Rizhsky, L., H. Liang, and R. Mittler. 2002. The combined effect of drought stress and heat shock on gene expression in tobacco. *Plant Physiology* 130(3): 1143–1151.

Rizza, F., F.W. Badeck, L. Cattivelli, O. Lidestri, N. di Fonzo, and A.M. Stanca. 2004. Use of a water stress index to identify barley genotypes adapted to rainfed and irrigated conditions. *Crop Science* 44(6): 2127–2137.

Robinson, E.A., G.D. Ryan, and J.A. Newman. 2012. A meta-analytical review of the effects of elevated CO2 on plant-arthropod interactions highlights the importance of interacting environmental and biological variables. *New Phytologist* 194: 321–336.

Roelfs, A.P. 1990. Epidemiology of the cereal rusts in north America. *Canadian Journal of Plant Pathology* 11: 86–90.

Rosenzweig, C., and D. Hillel. 1998. *Climate change and the global harvest*. New York: Oxford University Press.

Rosenzweig, C., and F.N. Tubiello. 2007. Adaptation and mitigation strategies in agriculture: An analysis of potential synergies. *Mitigation and Adaptation Strategies for Global Change* 12: 855–873. doi:10.1007/s11027-007-9103-8.

Roy, J.K., M. Prasad, R.K. Varshney, H.S. Balyan, T.K. Blake, H.S. Dhaliwal, H. Singh, K.J. Edwards, and P.K. Gupta. 1999. Identification of a microsatellite on chromosomes 6B and a STS on 7D of bread wheat showing an association with pre-harvest sprouting tolerance. *Theoretical and Applied Genetics* 99: 336–340.

Royle, D.J., M.W. Shaw, and R.J. Cook. 1986. Patterns of development of Septoria nodorume and S. tritici in some winter wheat crops in western Europe, 1981-1983. *Plant Pathology* 35: 466–476.

Rucker, K.S., C.K. Kevin, C.C. Holbrook, and J.E. Hook. 1995. Identification of peanut genotypes with improved drought avoidance traits. *Peanut Science* 22: 14–18.

Saini, H.S., and D. Aspinall. 1982. Abnormal sporogenesis in wheat (Triticum aestivum L.) induced by short periods of high temperature. *Annals of Botany* 49: 835–846.

Saini, H.S., and M.E. Westgate. 1999. Reproductive development in grain crops during drought. *Advances in Agronomy* 68: 59–96.

Saini, H.S., M. Sedgley, and D. Aspinall. 2010. Effect of Heat stress during floral development on pollen tube growth and ovary anatomy in wheat (Triticum aestivum L.). *Australian Journal of Plant Physiology* 10: 137–144.

Salekdeh, G.H., H.J. Siopongco, L.J. Wade, B. Ghareyazie, and J. Bennett. 2002. A proteomic approach to analyzing drought- and salt-responsiveness in rice. *Field Crops Research* 76(2-3): 199–219.

Sanchez, A.C., P.K. Subudhi, D.T. Rosenow, and H.T. Nguyen. 2002. Mapping of QTLs associated with drought resistance in sorghum (Sorghum bicolor L. Moench). *Plant Molecular Biology* 48: 713–726.

Savin, R., and M.E. Nicolas. 1996. Effects of short periods of drought and high temperature on grain growth and starch accumulation of two malting barley cultivars. *Australian Journal of Plant Physiology* 23: 201–210.

Schapendonk, A.H.C.M., H.Y. Xu, P.E.L.V.D. Putten, and J.H.J. Spiertz. 2007. Heat-shock effects on photosynthesis and sink-source dynamics in wheat (Triticum aestivum L.). *NJAS-Wageningen Journal of Life Sciences* 55: 37–54.

Schneekloth, J., T. Bauder, and N. Hansen. 2012. Limited irrigation management: Principles and practices. Colorado State University Extension. 4/03. Fact Sheet No. 4.720. http://www.ext.colostate.edu/pubs/crops/04720.html.

Semenov, M.A., and N.G. Halford. 2009. Identifying target traits and molecular mechanisms for wheat breeding under a changing climate. *Journal of Experimental Botany* 60(10): 2791–2804.

Semenov, M.A., and J.R. Porter. 1995. Climatic variability and the modelling of crop yields. *Agricultural and Forest Meteorology* 73: 265–283.
Semenov, M.A., and P.R. Shewry. 2011. Modelling predicts that heat stress, not drought, will increase vulnerability of wheat in Europe. *Science Reports* 1: 66.
Semenov, M.A., J. Wolf, L.G. Evans, H. Eckerstern, and A. Iglesias. 1996. Comparison of wheat simulation models under climate change. II. Application of climate change scenarios. *Climate Research* 7: 271–281.
Semenov, M.A., P.D. Jamieson, and P. Martre. 2007. Deconvoluting nitrogen use efficiency in wheat: A simulation study. *European Journal of Agronomy* 26: 283–294.
Semenov, M.A., R.A.C. Mitchell, A.P. Whitmore, M.J. Hawkesford, M.A.J. Parry, and P.R. Shewry. 2012. Shortcomings in wheat yield predictions. *Nature Climate Change* 2: 380–382.
Semenov, M.A., P. Stratonovitch, F. Alghabari, and M.J. Gooding. 2014. Adapting wheat in Europe for climate change, review. *Journal of Cereal Science* 59: 245–256.
Sinclair, T.R., and R.C. Muchow. 2001. System analysis of plant traits to increase grain yield on limited water supplies. *Agronomy Journal* 93: 263–270.
Sinclair, T.R., and N. Seligman. 2000. Criteria for publishing papers on crop modeling. *Field Crops Research* 68: 165–172.
Sinha, S.K., N. H. Rao, and M.S. Swaminathan. 1988. Food security in the changing global climate. In the conference proceedings for the changing atmosphere: Implications for global security, 27–30 June 1988, in Toronto, Canada, 167-192. WMO-No. 170. Geneva: World Meteorological Organization.
Sivamani, E., A. Bahieldin, J.M. Wraith, et al. 2000. Improved biomass productivity and water use efficiency under water deficit conditions in transgenic wheat constitutively expressing the barley HVA1 gene. *Plant Science* 155(1): 1–9.
Smale, M., P. Aquino, J. Crossa, E. del Toro, J. Dubin, T. Fischer, P. Fox, M. Khairallah, A. Mujeeb-Kazi, K.J. Nightingale, I. Ortiz-Monasterio, S. Rajaram, R. Singh, B. Skovmand, M. van Ginkel, G. Varughese, and R. Ward. 1996. *Understanding global trends in the use of wheat diversity and international flows of wheat genetic resources*, Economics working paper 96-02. Mexico: CIMMYT.
Smirnoff, N. 1993. The role of active oxygen in the response of plants to water deficit and desiccation. *New Phytologist* 125: 27–58.
Sobhkhizi, A., Mf Rayni, H.B. Barzin, and M. Noori. 2014. Influence of drought stress on photosynthetic enzymes, chlorophyll, protein and relative water content in crop plants. *International Journal of Biosciences* 5(7): 89–100.
Soddu, A., R. Deidda, M. Marrocu, R. Meloni, C. Paniconi, R. Ludwig, M. Sodde, G. Mascaro, and E. Perra. 2013. Climate variability and durum wheat adaptation using the AquaCrop model in southern Sardinia. *Procedia Environmental Sciences* 19: 830–835.
Soltani, A., and T.R. Sinclair. 2012. *Modeling physiology of crop development, growth and yield.* Wallingford: CABI Publication. 322 pp.
Soltani, A., and T.R. Sinclair. 2015. A comparison of four wheat models with respect to robustness and transparency: Simulation in a temperate, sub-humid environment. *Field Crops Research* 175: 37–46.
Soltani, A., V. Maddah, and T.R. Sinclair. 2013. SSM-wheat: A simulation model for wheatdevelopment, growth and yield. *International Journal of Plant Production* 7: 711–740.
Sommer, R., M. Glazirina, T. Yuldashev, A. Otarov, M. Ibraev, L. Martynovac, M. Bekenov, B. Kholov, Kobilov R. Ibragimov, S. Karaev, M. Sultonov, F. Khasanova, M. Esanbekov, D. Mavlyanov, S. Isaev, S. Abdurahimov, R. Ikramov, L. Shezdyukova, and E. de Pauw. 2013. Impact of climate change on wheat productivity in Central Asia. *Agriculture Ecosystems and Environment* 178: 78–99.
Southworth, J., R.A. Pfeifer, M. Habeck, J.C. Randolph, O.C. Doering, and D.G. Rao. 2002. Sensitivity of winter wheat yields in the Midwestern United States to future changes in climate, climate variability, and CO2 fertilization. *Climate Research* 22: 73–86.
Steduto, P., T.C. Hsiao, D. Raes, and E. Fereres. 2009. AquaCrop: The FAO crop model to simulate yield response to water: I. Concepts and underlying principles. *Agronomy Journal* 112: 426–437.

Stitt, M. 1991. Rising CO2 levels and their potential significance for carbon flow in photosynthetic cells. *Plant, Cell and Environment* 14: 311–317.

Stockle, C.O., S. Martin, and G.S. Campbell. 1994. CropSyst, a cropping systems model: Water/nitrogen budgets and crop yield. *Agricultural Systems* 46: 335–359.

Stockle, C.O., M. Donatellib, and R. Nelsona. 2003. CropSyst, a cropping systems simulation model. *European Journal of Agronomy* 18(3–4): 289–307.

Sultana, H., N. Ali, M.M. Iqbal, and A.M. Khan. 2009. Vulnerability and adaptability of wheat production in different climatic zones of Pakistan under climate change scenarios. *Climatic Change* 94: 123–142.

Supit, I., C.A. Van Diepen, A.J.W. De Wit, P. Kabat, B. Baruth, and F. Ludwig. 2010. Recent changes in the climatic yield potential of various crops in Europe. *Agricultural Systems* 103: 683–694.

Supit, I., C.A. van Diepen, A.J.W. de Wit, J. Wolf, P. Kabat, B. Baruth, and F. Ludwig. 2012. Assessing climate change effects on European crop yields using the crop growth monitoring system and a weather generator. *Agricultural and Forest Meteorology* 164: 96–111.

Sutherst, R.W. 1991. Pest risk analysis and the greenhouse effect. *Review of Agricultural Entomology* 79: 1177–1187.

Sutherst, R.W., G.F. Maywald, and D.B. Skarrate. 1995. Predicting insect distributions in a changed climate. In *Insects in a changing environment*, ed. R. Harrington and N.E. Stork, 59–91. London: Academic.

Sutherst, R.W., F. Constable, K.J. Finlay, R. Harrington, J. Luck, and M.P. Zalucki. 2011. Adapting to crop pest and pathogen risks under a changing climate. *Wiley Interdisciplinary Reviews – Climate Change* 2: 220–237.

Szegletes, Z.S., L. Erdei, I. Tari, and L. Cseuz. 2000. Accumulation of osmoprotectants in wheat cultivars of different drought tolerance. *Cereal Research Communications* 28(4): 403–410.

Talukdar, A., and P. Talukdar. 2013. Emerging concepts and strategies for genomics and breeding. In *Genomics and breeding for climate-resilient crops, vol. I: Concepts and strategies*, ed. C. Kole. Berlin/Heidelberg: Springer. doi:10.1007/978-3-642-37045-8.

Talukder, A., G. Gill, G. McDonald, P. Hayman, B. Alexander. 2010. Field evaluation of sensitivity of wheat to high temperature stress near flowering and early grain set. In 15th Australian agronomy conference, Christchurch, New Zealand. http://www.agronomy.org.au/.

Tardieu, F. 2003. Virtual plants: Modelling as a tool for genomics of tolerance to water deficit. *Trends in Plant Science* 8: 9–14.

Tashiro, T., and I.F. Wardlaw. 1989. A comparison of the effect of high temperature on grain development in wheat and rice. *Annals of Botany* 64: 59–65.

Tester, M., and M. Bacic. 2005. Abiotic stress tolerance in grasses. From model plants to crop plants. *Plant Physiology* 137: 791–793.

Tewolde, H., C.J. Fernandez, and C.A. Erickson. 2006. Wheat cultivars adapted to post-heading high temperature stress. *Journal of Agronomy and Crop Science* 192: 111–120.

Therond, O., H. Hengsdijk, E. Casellas, D. Wallach, M. Adam, H. Belhouchette, R. Oomen, G. Russell, F. Ewert, J.-E. Bergez, S. Janssen, J. Wery, and M.K. Van Ittersum. 2011. Using a cropping system model at regional scale: Low-data approaches for crop management information and model calibration. *Agriculture Ecosystems and Environment* 142(1-2): 85–94.

Thompson, G.B., and F.J. Woodward. 1994. Some influences of CO_2 enrichment, nitrogen nutrition and competition on grain yield and quality in spring wheat and barley. *Journal of Experimental Botany* 45: 937–942.

Thomson, A.M., R.A. Brown, S.J. Ghan, R.C. Izaurralde, N.J. Rosenberg, and L.R. Leung. 2002. Elevation dependence of winter wheat production in Eastern Washington State with climate change: A methodological study. *Climatic Change* 54: 141–164.

Todorovska, E., N. Christov, S. Slavov, P. Christova, and D. Vassilev. 2009. Biotic stress resistance in wheat—Breeding and genomic selection implications. *Biotechnology & Biotechnological Equipment* 23(4): 1417–1426. doi:10.2478/V10133-009-0006-6.

Torriani, D.S., P. Calanca, S. Schmid, M. Beniston, and J. Fuhrer. 2007. Potential effects of changes in mean climate and climate variability on the yield of winter and spring crops in Switzerland. *Climate Research* 34: 59–69.

Toscano, P., R. Ranieri, A. Matese, F.P. Vaccari, B. Gioli, A. Zaldei, M. Silvestri, C. Ronchi, P. La Cava, J.R. Porter, and F. Miglietta. 2012. Durum wheat modeling: The Delphi system, 11 years of observations in Italy. *European Journal of Agronomy* 43: 108–118.

Toumi, J., S. Er-Raki, J. Ezzaharc, S. Khabba, L. Jarland, and A. Chehbounid. 2016. Performance assessment of AquaCrop model for estimating evapotranspiration, soil water content and grain yield of winter wheat in Tensift Al Haouz (Morocco): Application to irrigation management. *Agricultural Water Management* 163: 219–235.

Treharne, K. 1989. The implications of the "greenhouse effect" for fertilizers and agrochemicals. In *The "Greenhouse Effect" and UK Agriculture*, No. 19, ed. R.M. Bennet, 67–78. Reading: Center for Agricultural Strategy, University of Reading.

Trethowan Mvan Ginkel, M., and S. Rajaram. 2002. Progress in breeding wheat for yield and adaptation in global drought affected environments. *Crop Science* 42: 1441–1446.

Trnka, M., M. Dubrovsky, D. Semeradova, and Z. Zalud. 2004. Projections of uncertainties in climate change scenarios into expected winter wheat yields. *Theoretical and Applied Climatology* 77: 229–249.

Trnka, M., R.P. Rötter, M. Ruiz-Ramos, K.C. Kersebaum, J.E. Olesen, Z. Žalud, and M.A. Semenov. 2014. Adverse weather conditions for European wheat production will become more frequent with climate change. *Nature Climate Change* 4: 637–643.

Tsvetsinskaya, E.A., L.O. Mearns, T. Mavromatis, W. Gao, L. McDaniel, and M.W. Downton. 2003. The effect of spatial scale of climatic change scenarios on simulated maize, winter wheat, and rice production in the southeastern United States. *Climatic Change* 60: 37–71.

Tuba, Z., K. Szente, and J. Koch. 1994. Response of photosynthesis, stomatal conductance, water use efficiency and production to long-term elevated CO_2 in winter wheat. *Journal of Plant Physiology* 144: 661–668.

Tuba, Z., K. Szente, Z. Nagy, Z. Csintalan, and J. Koch. 1996. Responses of CO_2 assimilation, transpiration and water use efficiency to long term elevated CO_2 in perennial C3 xeric loess steppe species. J. *Plant Physiology* 148: 356–361.

Tuberosa, R., and S. Salvi. 2006. Genomics-based approaches to improve drought tolerance of crops. *Trends in Plant Science* 11(8): 405–412.

Tubiello, F.N., and F. Ewert. 2002. Modeling the effects of elevated CO_2 on crop growth and yield: A review. *European Journal of Agronomy* 18(1–2): 57–74.

Tubiello, F.N., M. Donatelli, C. Rosenzweig, and C.O. Stockle. 2000. Effects of climate change and elevated CO2 on cropping systems: Model predictions at two Italian locations. *European Journal of Agronomy* 13: 179–189.

Turner, N.C. 2004. Sustainable production of crops and pastures under drought in a Mediterranean environment. *The Annals of Applied Biology* 144: 139–147.

UN, United Nations. 2013. *World population prospects: The 2012 revision, highlights and advance tables*, ESA/P/WP.228. New York: Department of Economic and Social Affairs, Population Division.

USDA, United States Department of Agriculture. 2015. *World agricultural production*, Foreign agricultural service circular series WAP 7-15. Washington, DC: USDA.

Valizadeh, J., S.M. Ziaei, and S.M. Mazloumzadeh. 2014. Assessing climate change impacts on wheat production (a case study). *Journal of the Saudi Society of Agricultural Sciences* 13(2): 107–115.

Van Diepen, C.A., J. Wolf, H. van Keulen, and C. Rappoldt. 1989. WOFOST: A simulation model of crop production. *Soil Use Manage* 5: 16–24.

Van Keulen, H., and N.G. Seligman. 1987. *Simulation of water use, nitrogen nutrition and growth of a spring wheat crop*, Simulation monographs. Wageningen: Pudoc. 310 pp.

van Laar, H.H., J. Goudriaan, and H. van Keulen. 1992. Simulation of crop growth for potential and water limited production situations (as applied to spring wheat). Simulation Reports CABO-TT, 27, CABO-DLO/TPE-WAU, Wageningen, 78 pp.

Vanaja, M., S.K. Yadav, G. Archana, N.J. Lakshmi, P.R. Ram Reddy, P. Vagheera, S.K. Abdul Razak, M. Maheswari, and B. Venkateswarlu. 2011. Response of C4 (maize) and C3 (sun-

flower) crop plants to drought stress and enhanced carbon dioxide concentration. *Plant, Soil and Environment* 57(5): 207–215.
Varga, B., K. Balla, S. Bencze, and O. Veisz. 2010. Combined effect of the drought duration and elevated atmospheric co2 level on physiological and yield parameters of winter wheat. *Acta Agronomica Hungarica* 58(4): 323–331. doi:10.1556/AAgr.58.2010.4.1.
Volaire, F. 2003. Seedling survival under drought differs between an annual (*Hordeum vulgare*) and a perennial grass (*Dactylis glomerata*). *New Phytologist* 160: 501–510.
Vu, J.C.V. 2005. Acclimation of peanut (Arachis hypogaea L.) leaf photosynthesis to elevated growth CO2 and temperature. *Environmental and Experimental Botany* 53: 85–95.
Wahid, A., S. Gelani, M. Ashraf, and R. Foolad. 2007. Heat tolerance in plants: An overview. *Environmental and Experimental Botany* 61: 199–223.
Wall, G.W. 2001. Elevated atmospheric CO_2 alleviates drought stress in wheat. *Agriculture Ecosystems and Environment* 87: 261–271.
Wall, G.W., N.R. Adam, T.J. Brooks, B.A. Kimball, P.J. Pinter Jr., R.L. LaMorte, F.J. Adamsen, D.J. Hunsaker, G. Wechsung, F. Wechsung, S. Grossman-Clarke, S.W. Leavitt, A.D. Matthias, and A.N. Webber. 2000. Acclimation response of spring wheat in a free-air CO_2 enrichment (FACE) atmosphere with variable soil nitrogen regimes. 2. Net assimilation and stomatal conductance of leaves. *Photosynthesis Research* 66: 79–95.
Wallach, D., D. Makowski, J.W. Jones, and F. Brun. 2014. *Working with dynamic crop models*, 2nd ed, 394. London: Academic.
Wang, Z., and B. Huang. 2004. Physiological recovery of Kentucky bluegrass from simultaneous drought and heat stress. *Crop Science* 44: 1729–1736.
Wang, X., J. Cai, D. Jiang, F. Liu, T. Dai, and W. Cao. 2011. Pre-anthesis high-temperature acclimation alleviates damage to the flag leaf caused by post-anthesis heat stress in wheat. *Journal of Plant Physiology* 168: 585–593.
Wang, L., Z. Feng, and J.K. Schjoerring. 2013. Effects of elevated atmospheric CO_2 on physiology and yield of wheat (Triticum aestivum L.): A meta-analytic test of current hypotheses. *Agriculture Ecosystems and Environment* 178: 57–63.
Wang, Y., W. Hu, X. Zhang, L. Li, G. Kang, W. Feng, Y. Zhu, C. Wang, and T. Guo. 2014. Effects of cultivation patterns on winter wheat root growth parameters and grain yield. *Field Crops Research* 156: 208–218.
Watson, R.T., M.C. Zinyowera, and R.H. Moss. 1996. *Climate change 1995: Impacts, adaptations and mitigation of climate change scientific-technical analyses*. Cambridge: Cambridge University Press.
Webber, H., P. Martre, S. Asseng, B. Kimball, J. White, M. Ottman, G.W. Wall, G. De Sanctis, J. Doltra, R. Grant, B. Kassie, A. Maiorano, J.E. Olesen, D. Ripoche, E.E. Rezaei, M.A. Semenov, P. Stratonovitch, F. Ewert. 2015. Canopy temperature for simulation of heat stress in irrigated wheat in a semi-arid environment: A multi-model comparison. Field Crops Research In Press.
Weir, A.H., P.L. Bragg, J.R. Porter, and J.H. Rayner. 1984. A winter wheat crop simulation model without water or nutrient limitations. *Journal of Agricultural Science* 102: 371–382.
West, J.S., S. Holdgate, J.A. Townsend, S.G. Edwards, P. Jennings, and B.D.L. Fitt. 2012. Impacts of changing climate and agronomic factors on fusarium ear blight of wheat in the UK. *Fungal Ecology* 5: 53–61.
Wheeler, T.R., G.R. Batts, R.H. Ellis, P. Hardley, and J.I.L. Mortison. 1996a. Growth and yield of winter wheat (Triticum aestivum) crops in response to CO_2 and temperature. *The Journal of Agricultural Science* 127: 37–48.
Wheeler, T.R., T.D. Hong, R.H. Ellis, G.R. Batts, J.I.L. Morison, and P. Hadley. 1996b. The duration and rate of grain growth, and harvest index, of wheat (Triticum aestivum L) in response to temperature and CO_2. *Journal of Experimental Botany* 47: 623–630.
White, J.W., G. Hoogenboom, B.A. Kimball, and G.W. Wall. 2011. Methodologies for simulating impacts of climate change on crop production. *Field Crops Research* 124(3): 357–368.
Wilcoxa, J., and D. Makowskia. 2014. A meta-analysis of the predicted effects of climate change on wheat yields using simulation studies. *Field Crops Research* 156: 180–190.

Wilks, D.S., and S.J. Riha. 1996. High-frequency climate variability and crop yields. *Climatic Change* 32(3): 231–235.
Williams, J.R., C.A. Jones, and P.T. Dyke. 1984. A modeling approach to determining the relationship between erosion and soil productivity. *Transactions of ASAE* 27: 129–144.
Williams, M., P.R. Shewry, and J.L. Harwood. 1994. The influence of the 'greenhouse effect' on wheat (Triticum aestivum L.) grain lipids. *Journal of Experimental Botany* 45: 1379–1385.
Witcombe, J.R., P.A. Hollington, C.J. Howarth, S. Reader, and K.A. Steele. 2008. Breeding for abiotic stresses for sustainable agriculture. *Philosophical Transactions of the Royal Society, Series B, Biological Sciences* 363: 703–716.
Wittwer, S.H. 1990. Implications of greenhouse effect on crop productivity. *HortScience* 25: 1560–1567.
Wolf, J., and C. Kempenaar. 1998. Results from Wageningen. In Model evaluation of experimental variability to improve predictability of crop yields under climate change, ed. J. Wolf (Compilator). Final report of the EU MODEXCROP project, pp. 3.1–3.106. Essex/Copenhagen/Wageningen: University of Essex/Royal Veterinary and Agricultural University/Wageningen University and Research Center.
Wolf, J., M. van Oijen, and C. Kempenaar. 2002. Analysis of the experimental variability in wheat responses to elevated CO_2 and temperature Agriculture. *Ecosystems and Environment* 93: 227–247.
Wolfe, D.W., and J.D. Erickson. 1993. Carbon dioxide effects on plants: Uncertainties and implications for modelling crop response to climate change. In *Agricultural dimensions of global climate change*, ed. H.M. Kaiser and T.E. Drennen, 153–178. Delray Beach: St. Lucie Press.
Wollenweber, B., J.R. Porter, and J. Schellberg. 2003. Lack of interaction between extreme high-temperature events at vegetative and reproductive growth stages in wheat. *Journal of Agronomy and Crop Science* 189: 142–150.
Wong, S.C. 1990. Elevated atmospheric partial pressure of CO2 and plant growth. II. Non-structural carbohydrate content in cotton plants and its effect on growth parameters. *Photosynthesis Research* 23: 171–180.
Woodward, F.I., and F.A. Bazzaz. 1988. The responses of stomatal density to CO_2 partial pressure. *Journal of Experimental Botany* 39: 1771–1781.
World Bank. 2008. *World development report 2008: Agriculture for development*. Washington, DC: The world Bank. 365pp.
Xiao, G.J., Q. Zhang, Y.B. Yao, H. Zhao, R.Y. Wang, H.Z. Bai, and F.J. Zhang. 2008. Impact of recent climatic change on the yield of winter wheat at low and high altitudes in semi-arid northwestern China. *Agriculture Ecosystems and Environment* 127: 37–42.
Xu, Y. 2010. *Molecular plant breeding*. Wallingford/Cambridge, MA: CAB International.
Yang, J., and J. Zhang. 2006. Grain filling of cereals under soil drying. *New Phytologist* 169: 223–236.
Yang, J.C., J.H. Zhang, Z.Q. Wang, G.W. Xu, and Q.S. Zhu. 2004. Activities of key enzymes in sucrose-to-starch conversion in wheat grains subjected to water deficit during grain filling. *Plant Physiology* 135: 1621–1629.
Yang, Y., D.L. Liu, M.R. Anwar, H. Zuo, and Y. Yang. 2014. Impact of future climate change on wheat production in relation to plant-available water capacity in a semiarid environment. *Theoretical and Applied Climatology* 115: 391–410.
Yano, T., M. Aydin, and T. Haraguchi. 2007. Impact of climate change on irrigation demand and crop growth in a Mediterranean environment of Turkey. *Sensors* 7: 2297–2315.
Yordanov, I., V. Velikova, and T. Tsonev. 2003. Plant responses to drought and stress tolerance. *Bulgarian Journal of Plant Physiology* Special Issue: 187–206.
Zhang, X.C. 2005. Spatial downscaling of global climate model output for site specific assessment of crop production and soil erosion. *Agricultural and Forest Meteorology* 135: 215–229.
Zhang, H.P., and T. Oweis. 1998. Water yield relation and optimal irrigation scheduling of wheat in Mediterranean regions. *Agriculture Water Management* 3: 195–211.

Zhao, G., B.A. Bryana, and X. Songa. 2014. Sensitivity and uncertainty analysis of the APSIM-wheat model: Interactions between cultivar, environmental, and management parameters. *Ecological Modelling* 279: 1–11.

Zhao, Z., X. Qin, E. Wang, P. Carberry, Y. Zhang, S. Zhou, X. Zhang, C. Hu, and Z. Wang. 2015. Modelling to increase the eco-efficiency of a wheat–maize double cropping system. *Agriculture Ecosystems & Environment* 210: 36–46.

Zhu, J.K. 2002. Salt and drought stress signal transduction in plants. *Annual Review of Plant Biology* 53: 243–273.

Ziska, L.H., C.F. Morris, and E.W. Goins. 2004. Quantitative and qualitative evaluation of selected wheat varieties released since 1903 to increasing atmospheric carbon dioxide: Can yield sensitivity to carbon dioxide be a factor in wheat performance? *Global Change Biology* 10: 1810–1819.

Chapter 13
Quantification of Climate Change and Variability Impacts on Maize Production at Farm Level in the Wami River Sub-Basin, Tanzania

Sixbert K. Mourice, Winfred Mbungu, and Siza D. Tumbo

Contents

S.K. Mourice
Department of Crop Science and Production, Sokoine University of Agriculture, P. O. Box 3005, Morogoro, Tanzania

W. Mbungu (✉)
Department of Agricultural Engineering and Land use planning, Sokone University of Agriculture, P. O. Box 3003, Morogoro, Tanzania

Department of Biological Systems Engineering, Virginia Tech, Blacksburg, VA 24060, USA
e-mail: winfred@vt.edu; winfried@suanet.ac.tz

S.D. Tumbo
Department of Agricultural Engineering and Land use planning, Sokone University of Agriculture, P. O. Box 3003, Morogoro, Tanzania

M. Ahmed, C.O. Stockle (eds.), *Quantification of Climate Variability, Adaptation and Mitigation for Agricultural Sustainability*,
DOI 10.1007/978-3-319-32059-5_13

Abstract Up to 95 % of food production in Tanzania depends on rainfall, whose timing, quantity and distribution is highly affected by climate variability and will highly likely change as a result of global warming. Several analyses have been done on the response of several crops in different agro-ecological zones and cropping systems to the impacts of changing climate and interactions of several climate variables have been highlighted. Many of the previous efforts have based on aggregations at sub-national and national scales, and have not considered the impacts and adaptation initiatives on individual fields, where the impacts will be directly felt. In this study, we quantify climate change impacts and adaptation for coping with future climate by individual farm fields in the Wami River sub-basin in Tanzania. The assessment was based on two RCPS and five downscaled GCMs with two time periods up to year 2100, involving a total of 168 farm fields. Maize yield change was projected to be in the negative direction for all the GCMs in both RCPs and periods. Organic matter application was an important climate change adaptation option whereas nitrogen fertilizer would only be suitable in more humid, rather than in semi-arid sections of the study area. We conclude that farm level climate change impacts quantification and adaptation assessment is key in designing sound adaptation strategies based on biophysical and socioeconomic endowments.

Keywords AgMIP • CERES • Quantification • Nitrogen • RCPs • Yield • Climate variability • Climate change

13.1 Introduction

Food availability, access, stability and utilization are considered to be key dimensions of food security that forms the essential component of the human survival and well-being of the global population (Schmidhuber and Tubiello 2007; Rowhani et al. 2011). Thus, food security of the world population is becoming a major concern not only in developing countries including sub-Saharan Africa, Tanzania in particular, where up to 95 % of crop production by smallholder farmers relies on rainfall, but also at regional and global levels. Smallholder farmers in Tanzania are more vulnerable to the impacts because the majority of them are subsistence farmers and directly derive their livelihood and food from agriculture related activities, both on and off-farm (Ito and Kurosaki 2009), with agriculture playing a significant role in efforts to reduce rural as well as aggregate poverty (Sarris et al. 2006). On the other hand, food security is considered as a development issue and needs to be streamlined in the development agenda for a country to forge ahead (URT 2012b). In general, agriculture is the major employer of the majority of the population, employing more than 75 % of the working population (Tumbo et al. 2012; Mbungu et al. 2015; URT 2012b, c), and the sector accounts for about 50 % of national GDP and about 75 % of export earnings (Leyaro and Morrissey 2013; World_Bank 2009; URT 2008a). Apart from the many non-climatic factors such as use of low inputs (such as improved seeds and fertilizer) and application of low levels of technology (Sarris et al. 2006; Leyaro and Morrissey 2013; Lokina et al. 2011) resulting in low yields (URT 2008a), the agriculture and overall food availability in Tanzania is highly vulnerable to the impacts of changes in climate (Rowhani et al. 2011; Morton 2007; Paavola 2008; URT 2008b, 2012b, c) because of high reliance on weather-sensitive rainfed agriculture(Ahmed et al. 2011) and low adaptive capacity (URT 2012b, c). According to the World Bank (2012), the percentage of population who lived on less than $1.25 at 2005 international prices was about 43.5 % and there are fears that this could get worse with climate variability and change (Ahmed et al. 2009; Rowhani et al. 2011). The total area currently under irrigation is less than 0.5 Mha, of which only 0.4 Mha (1.2 % of the total irrigation potential area) has good irrigation infrastructure, while another 0.1 Mha is still under traditional irrigation practice (URT 2012a). Estimates show that there are 2.3 Mha of high potential, 4.8 Mha and 22.3 of medium and low irrigation potential land respectively (URT 2002).

Rainfall timing, quantity, distribution and reliability are changing and likely to continue to change and affect rain-fed agriculture as a result of global warming. There is ample evidence suggesting that, changes in weather extremes, especially rainfall, minimum and maximum temperatures are already occurring at local (Mbungu et al. 2012), regional and global scales (Lobell and Gourdji 2012; IPCC 2014a) These changes have negative impacts on crop sector, along natural systems, in all regions of the world (IPCC 2014b; Morton 2007; Moore et al. 2012). As a result, rain fed cropping systems have become or will be highly vulnerable to these global climate changes (Rowhani et al. 2011; Ahmed et al. 2011). Smallholder farmers at subsistence level are more likely to be hit by hunger and poverty as destructive impacts of changing climate continue to be realized and lack of investments and clear policy to circumvent the situation at the local and national levels (Rowhani et al. 2011; Schlenker and Lobell 2010).

The National Adaptation Action Plan (NAPA) for Tanzania (URT 2007) acknowledges the frequent and severe droughts that have hit many parts of the country and their consequences continue to be felt on food production and water sectors and ranked agriculture as the most vulnerable sector. And there are calls for assessments and studies on the impact of climate variability and change in crop production at farm, landscape, regional and national levels.

Assessments on what the climate change would bring on Tanzanian agricultural sector are generally available, with mixed results and most pointing to negative climate change impacts. Mwandosya et al. (1998) point out that central Tanzania would experience temperature increase of up to 4 °C, associated with decline in rainfall towards the end of century under doubled carbon dioxide concentration scenario. Arndt et al. (2012) reported that food security will generally deteriorate by the mid-century, mainly due to increase in temperature and changes in rainfall patterns. Rowhani et al. (2011) indicated that in Tanzania by the mid-century, an increase in 2 °C projected temperature, maize, sorghum and rice yields would decline by 13 %, 8.8 % and 7.6 % respectively. This study concludes that climate change impacts on crop yields would be under estimated if the focus is on climatic or seasonal means rather than climatic variability or intra-seasonal variability. (Tumbo et al. 2012) in their study, focused on the significance of agronomic practices in reducing climate change impacts on maize yields in Same District, Tanzania. They concluded that maize yields would increase under conventional (no manure or fertilizers) and recommended (40 kg N/ha) agronomic practices towards the mid-century time period if maize was planted during the months of October –December/ January (short rains) rather that during the long rains (March–June) season. In another study Mbungu et al. (2015) evaluated the performance of different practices on maize cultivars under changing climate noted a decrease in maize yield of some maize cultivars due to 2 °C rise in temperature in the long rainy season, and recommended more site-specific climate change studies to evaluate other crop varieties in the area and other areas.

However, these and other studies linking climate change and food security have had some shortfalls. First, since climate change is expected to vary widely across the country, so will be its affects. At farm level, climate change impacts are rare in global change discourses. Although small scale farmers under rain fed cropping systems share common characteristics in terms of the type of agricultural enterprises, they differ in the ways they are prepared to dealing with climatic shocks (Lyimo and Kangalawe 2010; Mongi et al. 2010). Due to the diversity existing among the farms, farmers and their locations, farm- or -location specific impacts assessment is significant for two reasons; to develop an understanding how available biophysical resources shield or expose the farmers to climate change impacts; and to develop farm tailored adaptation measures. Mwandosya et al. (1998) reported on the differences in temperature and rainfall projections between major agro-ecological zones. National level assessment seems to be wide yet, since adaptation strategies need high resolution information possibly at subnational level (Lobell and Field 2007; Lobell et al. 2008). Second, agricultural crop models used in the assessment are at most calibrated to crop parameters only, assuming uniform crop management and initial field conditions across the area of interest. However, the variation in planting dates, soil initial conditions, use of organic matter and fertilizers as well as the soil water conditions at planting, among farms, districts or river basins accounts for yield difference which is not normally explained by climate change assessment using crop models. Third, the representative concentration pathways (RCPs) (van Vuuren et al. 2011) which are based on anthropogenic greenhouse gas (GHG) emissions as driven by population size, economic activity, lifestyle, energy use, land use patterns, technology and climate policy has not been widely incorporated in climate change studies in Tanzania. RCPs are set of emission scenarios spanning the range of year 2100 radiative forcing. The RCPs include four scenarios,

RCP 2.6 – (strict mitigation scenario) RCP4.5 and RCP6.0 (two intermediate scenarios) and RCP8.5, a scenario with very high GHG emissions (International Panel on Climate Change-IPCC 2014a). A number of climate change assessments have employed Coupled Model Inter-comparison Project Phase Three (CMIP3) scenarios (Rowhani et al. 2011; Tumbo et al. 2012; Mbungu et al. 2015) although we do not intend to make a comparison between CMIP3 and CMIP5 projections as they use different scenarios.

13.2 Agro-Ecological Zones, Livelihood and Cropping Systems/Crops of Tanzania

Growth, development and sustainability of the agriculture sector is dependent on environmental resources such as land forest, air and water resources. Based on weather, altitude dependable growing seasons and average water holding capacity of the soils and physiographic features, Tanzania is divided into seven main agro-ecological zones (AEZs) ranging from higher to lower rainfall areas as summarized in Fig. 13.1.

There are a lot of variations in the cropping systems in Tanzania and mostly associated with climatic and agro-ecological conditions (as in Fig. 13.1). Major subsistence food crops such as maize have wide coverage throughout the country, with the economic base of rural livelihoods varying among and within the AEZs (Fig. 13.2 and Table 13.1) (URT 2014). Areas with higher rainfall such as those in

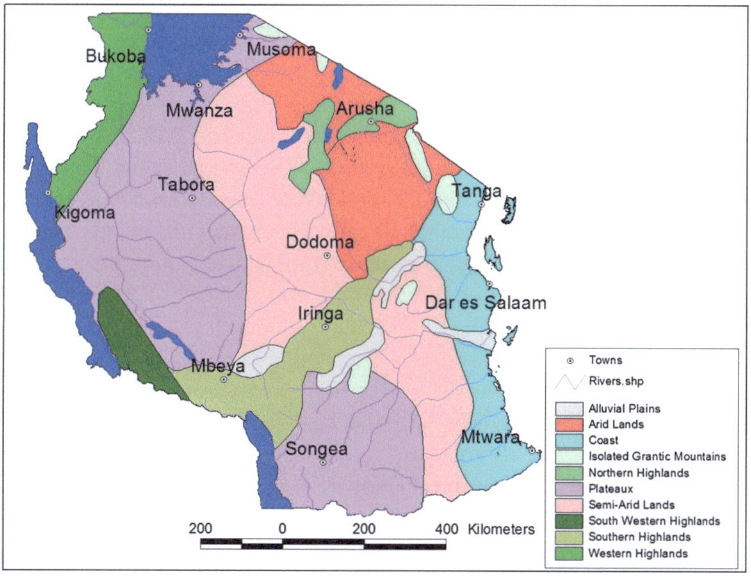

Fig. 13.1 Agro-ecological zones of Tanzania (Source: SUA (2010))

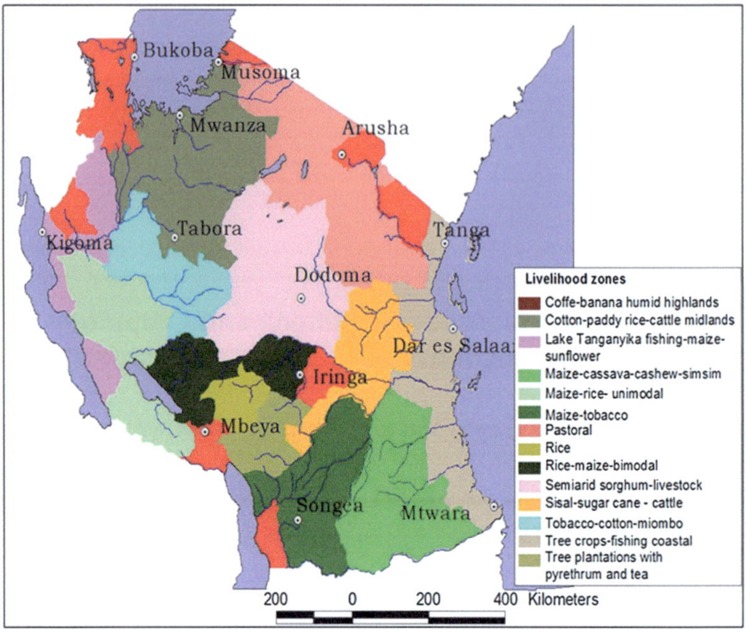

Fig. 13.2 Livelihood zones of Tanzania with dominant crops (Source: URT 2014)

the southern highlands in the Southern Agricultural Growth Corridor of Tanzania (SAGCOT) have diversity of crop livelihoods and areas in the lake zones have a mix of fishing and food crops. However, arid and semi-arid areas are dominantly pastoralists and have high dependence on drought-tolerant crops such as sorghum (URT 2014). On average Tanzania receives about 1071 mm of rainfall, while some areas such as the Lake Tanganyika basin and the southern highlands receiving up to 3000 mm annually, and about half the country receives less than 762 mm annually. Two rainfall regimes are common in Tanzania, the bimodal rainfall regime which are characterized by long rains from March to May (locally known as masika) and short rains from October to December (locally known as vuli) mostly dominant in the northern part of the country and to areas extending up to some parts of Morogoro, while the rest of the country is unimodal with the majority of the rainfall falling between December and April. The El Niño/La Niña South Oscillation (ENSO) phenomenon can also result in substantial impacts on intra-seasonal rainfall variability. Tanzania experiences a lot of rainfall variability in most areas with extreme dry and wet conditions throughout the year, with less changes and variations in annual temperature (URT 2014).

As noted earlier, agricultural production is dominated by smallholder farmers with the use of low technology and low use of inputs farming an average of less than an acre to 3 acres. Maize, a staple food in Tanzania is produced by the majority of smallholder farmers, with Mbeya, Rukwa, Ruvuma, Iringa and Njombe being the main producing regions. Other important regions include Arusha, Dodoma and

Table 13.1 Major cropping systems in Tanzania

Cropping system	Where found	Description
Maize/legume system	Rukwa, Ruvuma, Arusha, Kagera, Shinyanga, Iringa, Mbeya, Kigoma, Tabora, Tanga, Morogoro, Kahama, Biharamulo	Shifting cultivation, maize & legumes, beans and groundnuts intercropped, Arabic coffee
Banana/coffee/horticulture system	Kagera, Kilimanjaro, Arusha, Kigoma, and Mbeya Regions	Tree crops, high intensive land use, volcanic soils with high fertility, land scarcity
Cashew/coconut/cassava system	Coast region; eastern Lindi and Mtwara	Low rainfall, low soil fertility, cassava, coconut and cashew, land is not scarce, shifting cultivation
Rice/sugar cane system	Alluvial river valleys	Rice and sugarcanes
Sorghum/bulrush millet/ livestock system	Sukumaland; Shinyanga and rural Mwanza	Sorghum, millet, maize and cotton, oilseeds and rice, intense population pressure, declining soil fertility
Tea/maize/pyrethrum system	Njombe and Mufindi districts in Iringa region	Tea, Maize, Irish potatoes, beans, wheat, pyrethrum, wattle trees and sunflower
Cotton/maize system	Mwanza, Shinyanga Kagera, Mara, Singida, Tabora and Kigoma, Morogoro, Coast, Mbeya, Tanga, Kilimanjaro and Arusha	cotton, sweet potatoes, maize, sorghum and groundnuts, intensive cultivation, livestock kept
Horticulture based system	Lushoto district; Tanga region, Morogoro rural; Morogoro region and Iringa rural in Iringa region	Vegetables, (cabbages, tomatoes, sweet pepper, cauliflower lettuce and indigenous vegetables) and fruits, (pears, apples, plums, passion fruits and avocado), Maize, coffee, Irish potatoes, tea and beans
Paddy rice and irrigated system	River valleys and alluvial plains, Kilombero, Wami Valleys, Kilosa, Lower Kilimanjaro, Ulanga, Kyela, Usangu and Rufiji	
Pastoralists and agropastoralist system	Semi-arid areas i.e. Dodoma, Singida, parts of Mara and Arusha; Chunya districts, Mbeya and Igunga district in Tabora	Livestock and simple cropping system, shifting cultivation of sorghum millet, moderate population density 30 per km^2, limited resource base and poor and variable rainfall

Source, URT 2014

Morogoro. Other important cereals include paddy rice, sorghum, wheat, and millet. According to statistics from FAO (2015) yields of the important food crops are variable and hardly go beyond 3 tones/ha (as shown in Fig. 13.2) using data from 1961 to 2012. The variability in yields apart from other non-climatic factors are mainly attributed to the variability of climate and climate change. Changing climate has resulted in a general decline in agricultural productivity, shifting agro-ecological zones (AEZ), prevalence of crop pests and diseases and increasing rainfall unreliability (URT 2014), posing more challenge to agriculture. Studies and projections indicate that some of the previous highly productive areas such as the southern and northern highlands will continue to be affected by declining rainfall, frequent droughts and significant increase in spatial and temporal variability of rainfall with long term implications in the agricultural sector planning and resources allocation such as seeds, pesticides and even shift in types of agricultural produce (URT 2009b, 2014). Alternating dry conditions with heavy rainfall combine with inadequate land management in many areas that exacerbates land degradation and increase vulnerability to weather-related shocks (Enfors and Gordon 2008).

It is highly likely also that the uncertainty in rainfall may encourage land degradation and landscape fragmentation and ultimately affect agro-ecosystems (Soini 2005) as shifting cultivation in search for more productive soils and land increase. Fischer et al. (2005) and Cassman et al. (2003) showed that land expansion will likely increase especially in sub-Saharan Africa and Latin America. There is compelling evidence to suggest that agricultural productivity gains in Tanzania has basically come from land expansion (Salami et al. 2010) and not increase in crop yields (URT 2014). On the other hand, studies have indicated that land expansion and land use practices have played a role in changing the global climate and have degraded the ecosystems and services upon which we depend (Foley et al. 2005; Gibbs et al. 2010; URT 2012b) (Fig. 13.3).

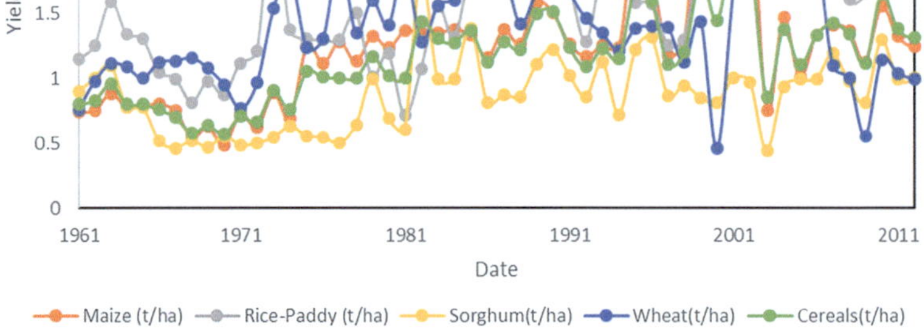

Fig. 13.3 Yield trends for selected food crops in Tanzania (Raw data from FAO 2015)

13.3 Quantification of Climate Change and Variability Impacts on Crops Productivity

The response of crops to the impacts of climate change and climate variability are likely to increase crop losses, and these have been assessed through experiments that are used for quantification of direct impacts of the elevated atmospheric CO_2, management and change in such climate variables as higher temperatures, altered precipitation and transpiration regimes, increased frequency of extreme temperature and precipitation events, and weeds, pest and pathogen pressure or by crop simulation models (Kang et al. 2009; Tubiello et al. 2007). Research in recent years has made progress towards quantification of the potential effects of the key interactions.

13.3.1 High Temperature and Elevated CO_2 Concentration Interaction

Crops productivity at local and global levels are and are likely going to be affected by the elevated concentrations of atmospheric CO_2 and changes in related climatic variables (Abraha and Savage 2006; Kang et al. 2009). Crop development, growth and yield and total crop production will respond to the increases in atmospheric CO_2 concentration (Tubiello et al. 2007). Quantification has been done through research conducted in several places around the globe. Studies have confirmed that plant biomass and yield tend to increase as CO_2 concentrations increase above certain levels. Various experiments mostly in controlled environment ranging from closed chambers, greenhouses, open and closed field top chambers and free-air carbon dioxide enrichment (FACE) experiments have provided robust results over the years for quantification of effects of elevated CO_2 on crops. It has been shown that photosynthetic activities are simulated with CO_2 concentrations that may lead to increased plant productivity and modified water nutrient cycles (Nowak et al. 2004; Kimball et al. 2002). Maize is one of the crops that has been experimented for its response to projected CO_2 fertilization (Long et al. 2006).

Like other C_4 plants, maize possesses CO_2 concentrating mechanisms in its bundle sheath cells (Long et al. 2006; Leakey 2009). The mechanism may increase CO_2 concentration six-times above that of ambient temperature (von Caemmerer and Furbank 2003). Enhanced atmospheric CO_2 concentration increases the rates of net photosynthesis, and also reduces stomata conductance and thus reducing transpiration per unit leaf area. Yield of C4 crops has been reported to increase between 0 and 10 % for concentrations of atmospheric CO_2 concentrations of 550 ppm from the current 350 ppm (Ainsworth and Long 2005; Long et al. 2004; Leakey 2009). Increases were also observed in above ground biomass in the range of 0–30 %. Sicher and Barnaby (2012) reported that CO_2 enrichment delayed the onset of the effects of water stress by 2 days when CO_2 concentration was increased from

380 ppm (ambient) to 700 ppm in growth chambers. However, under climate change scenarios, gains obtained from CO_2 increase may be offset by rise in temperature, since higher temperatures shorten lifecycle thus reducing available time to accumulate photosynthates for the crop (Jat et al. 2016).

13.3.2 High Temperature and Drought Interaction

High temperature and drought are considered as key factors with high potential impact on crop yield (Barnabás et al. 2008). Several regions and parts of the world are projected to experience water scarcity for crop production, and this is going to have detrimental effects to the growth of cereals such as maize. Heat-related include in cereals such as maize include the high-temperature-induced shortening of development phases, reduced light perception over the shortened life cycle and perturbation of the process associated with carbon assimilation (Stone 2001). Increased respiration requires greater carbon fixation for sustained growth and survival (Barnabás et al. 2008). Griffin et al. (2004) reported that plant photosynthesis was limited by the decreasing the activity of Rubisco as temperature rise above 35 °C. The response of plants to heat and drought related stress very much depends on the different stages of the plant as the response occurs at the molecular, cellular and physiological levels (Barnabás et al. 2008). Stress that reduces plant water status and photosynthesis during grain filling induces the conversion of stem reserves into soluble sugars and the mobilization of sugars into the grains (Blum 2005). Some of the effects of heat stress include a reduced synthesis of normal proteins and the accelerated transcription and translation of heat shock proteins, the production of phytohormones and antioxidants (Maestri et al. 2002). Other studies indicate that high temperature during critical flowering period of a crop may lower positive CO_2 effects on yield by reducing grain number, size and quality (Caldwell et al. 2005; Thomas et al. 2003). Increased temperatures during the growing period may also reduce CO2 effects indirectly, by increasing water demand (Tubiello et al. 2007).

It is clear that water plays an important role in the growth of plants and therefore climate impacts on crops will depend on the precipitation scenarios assumed. As most of the crop systems in sub-Saharan Africa are rainfed, the general circulation model-projected changes in precipitation will determine the direction and magnitude of the overall impacts (Reilly et al. 2003; Tubiello et al. 2007). It is understood that ecosystem productivity and function will be highly modified by the evapotranspiration to precipitation ratio, and higher water use efficiency caused by stomatal closure and greater root densities under elevated atmospheric CO2 may help in reducing the drought pressures (Morgan et al. 2004).

13.3.3 Interactions of Elevated CO_2 with Soil Nutrients

Experiments such as FACE have enabled us to understand the interaction of soil nutrients with elevated CO2 concentration. It was confirmed that high N soil contents increase the relative response to elevated atmospheric CO2 concentrations (Nowak et al. 2004; Tubiello et al. 2007). Under high N supply through application of N fertilizer, yield of C3 plants was reported to increase in the 10 years of experiments. Studies have indicated N availability may be maintained or restored through increase in biological N2 fixation under elevated CO_2 concentrations. However, it is realized that legumes may benefit more from elevated atmospheric CO2 concentrations than non-fixing species (Teyssonneyre et al. 2002; Tubiello et al. 2007).

13.3.4 Overview of Biotic Stress (Weeds and Pests)

Interactions of weeds and insect pest as well as diseases with climate change, including CO_2 concentrations has not been adequately quantified, unlike other experiments that manipulate climate management variables. However, it is equally important to understand the role and importance of the roles played by the interactions amid changes in climate. The understanding is that the rise in carbon dioxide concentration is going to favor the growth of wanted and unwanted plants. As a result proliferation of different species of noxious plants may have a damaging effect on the current crops, and there is currently little understanding as to whether CO_2 concentrations selects such noxious plants over the others within ecosystems (Ziska and George 2004). CO2–temperature interactions are recognized as a key factor determining plant damage from pests in future decades; CO_2/precipitation interactions will be likewise important (Tubiello et al. 2007). Studies continue to investigate the damage by pests to crops as part of the increased CO2 rise and changes in climate especially temperature (Agrell et al. 2004).

13.4 Modeling and Simulation

Addressing long term climate change impacts on crop production is important; not only for enabling development of adaptation policies but also understanding the exact process in plant growth that will be affected and enable researchers innovate new mitigation technologies such as breeding of future climate-adapted crop varieties or new, tailored farm practices. Adaptation to the climate change impacts has a bearing on the farmers' ability to sustain food security. Also is important for the policy machinery to develop programs that would enable effective use of biophysical resources for enhanced food production despite the negative effects of climate change.

In this study, we employ Agricultural model inter-comparison and Improvement Project (AgMIP) (Rosenzweig et al. 2014) modeling framework to evaluate the extent to which climate change affects maize (*Zea mays L.)* productivity at individual farm fields in the Wami River sub basin, Tanzania. Specifically, we determine the yield change between baseline and climate change projections at mid-century and end century time periods; and evaluate effects of enhanced soil fertility as an adaptation measure for improved maize yield future emission scenarios.

13.5 Materials and Methods

13.5.1 Study Area Description

Wami river sub basin is located between 5–7°S and 36–39°E, covering an area of 43,000 km² (WRBWO 2007), with an altitudinal gradient of approximately 2260 m (Fig. 13.4). Wami sub basin receives annual rainfall between 550 and 700 mm in the highlands near Dodoma and 900–1000 mm in the lowlands near Dakawa and at the river estuary. Generally, dry periods occur from July to October and wet periods from November to December (*vuli* rains) and from March to June (*masika* rains)

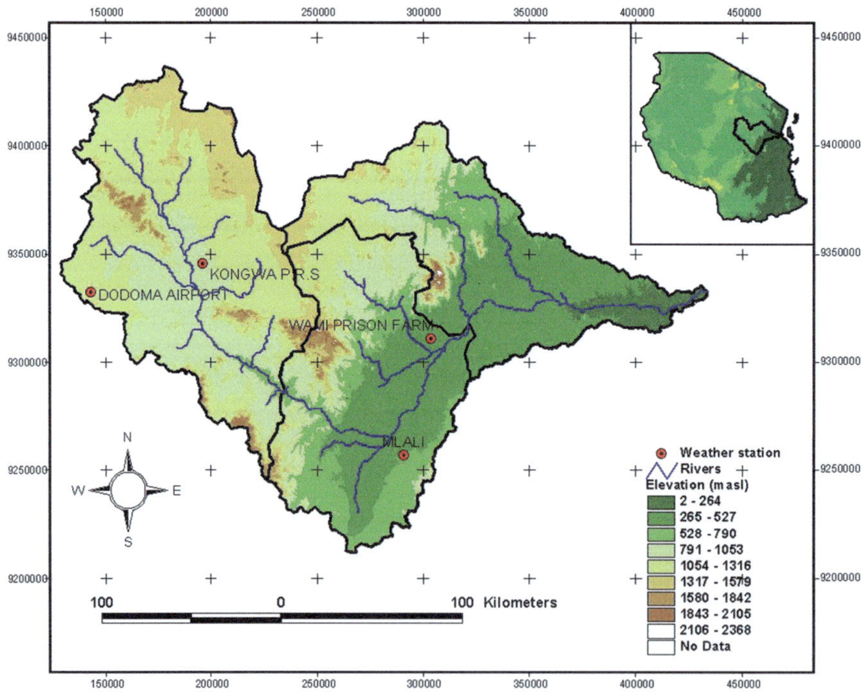

Fig. 13.4 Map of the Wami River sub-basin, Tanzania

(Ngana et al. 2010). The annual mean temperature is approximately 26 °C, coolest month being August with average temperature of 18 °C and hottest month is February with 32 °C.

13.5.2 Farming Systems of the Study Area

Agriculture is the main economic activity in the study area. Such crops as maize, sorghum, sesame, sunflower, and pear millet are grown under rain fed conditions in well drained areas. Rice is also cultivated under irrigation, mostly in river valleys (Ngana et al. 2010). Maize is the major food crop in the Wami River sub basin, accounting for over 65 % of total cereal output. All maize is grown under rain fed conditions, normally during long rains, and occasionally in short rains. Both improved maize cultivars and local landraces are grown in the basin. Such cultivars as *Staha, Situka, TMV1, Kilima, SeedCo* series. Nevertheless *Staha* and *Situka* are the most popular varieties (Mourice et al. 2014a). In this study, maize cultivar *Situka* was selected because it is a short term variety, maturing between 100 and 120 days, good yielding (6 t ha^{-1}) and versatile, i.e. it grows across the study area. Main cropping systems involving maize are practiced in the study area. They include mixed cropping – maize and other crop/s grown without regular arrangement in the same field; sole cropping – only maize crop grown in the field, normally in rows; intercropping- maize and other crop/s are grown in alternating rows in the same field; and relay cropping- pigeon peas planted just before maize anthesis.

13.5.3 CERES- Maize Model

CERES (Crop–Environment–Resource– Synthesis) – Maize module (Jones et al. 1986) within the DSSAT v (4.5) uses simplified functions to predict the growth of maize crop as influenced by major factors that affect yield. These factors include genetics, climate (daily minimum and maximum temperatures, solar radiation and rainfall), soils and management (Hunt and Boote 1998) The rationale of using this crop system model is that, amongst an array of dynamic crop models, CERES-Maize has been used in a wide range of environments. Moreover, this model and other models of the DSSAT suite requires minimum data sets (MDS) (Hunt and Boote 1998) some of which can easily be estimated or obtained from field measurements or observations. Moreover, CERES-Maize model has been fitted with cultivar specific parameters for four locally adapted maize cultivars (Mourice et al. 2014a) CERES-Maize model have the ability to simulate the effects of atmospheric carbon dioxide concentration and the effects of nitrogen deficiency and soil water deficit on the photosynthesis and pathways of carbohydrate movement in plants (Rosenzweig and Iglesias 1998) CERES-Maize CSM uses the radiation use efficiency (RUE) to calculate the total biomass production, given the amount of daily solar radiation available for interception (PAR). The equation used for potential biomass production (PCARB) for a day is (Ritchie et al. 1998)

$$\mathrm{PCARB} = \mathrm{RUE} \times \mathrm{IPAR} \times \mathrm{CO_2} \tag{13.1}$$

where IPAR is the fraction PAR intercepted by the plants and CO_2 is a carbon dioxide modification factor.

$$\mathrm{IPAR} = \mathrm{PAR} \times [1 - \exp(-\mathrm{k} \times \mathrm{LAI})] \tag{13.2}$$

where *k* is an extinction coefficient and *LAI* is the green leaf area index of the plant canopy. Under non-optimal conditions of low or higher temperature, water or nitrogen stress, the actual daily biomass production (*CARBO*) may be less than the *PCARB*. The equation used to reduce biomass due to non-optimal temperature defines the temperature reduction factor (*PRFT*) in terms of the weighted daytime temperature (*TDAY*)

$$\mathrm{PRFT} = 1 - \mathrm{T_C} \times (\mathrm{TDAY} - \mathrm{T_O})^2 \tag{13.3}$$

Where T_C is a constant and T_O is the optimum temperature for photosynthesis. Optimum temperature for photosynthesis in maize is 26 °C (Ritchie et al. 1998). Weighted daytime temperature is given as

$$\mathrm{TDAY} = 0.75 \times \mathrm{TMAX} + 0.25\mathrm{TMIN} \tag{13.4}$$

where *TMAX* and *TMIN* are respectively daily maximum and minimum temperatures. Therefore, non-optimal temperature effects on reduction of biomass production is defined as

$$\mathrm{CARBO} = \mathrm{PCARB} \times \min(\mathrm{PRFT}, 1) \tag{13.5}$$

where *min* indicates the minimum value of PRFT. The PRFT varies between 0 and 1, where 1 is non-limiting and 0 is the maximum deficit (Ritchie et al. 1998).

13.5.4 CERES-Maize Model Calibration and Assumptions

CERES-Maize model was calibrated according to Mourice et al. (2015). Due to the fact that crop residues from previous crop were not measured for each farm, a minimum of 500 kg ha was assumed to be the initial surface residue biomass from the previous crop for all fields. For fields which reported use of inorganic fertilizers, two rounds of application were assumed, the first at 14 days after sowing (DAS) (33 %) and the second at 45 DAS (67 %). For other crop growth limitations which could not be explicit in the survey data, eg insects, weeds or crop diseases, a soil fertility factor (SPLF) of 1.0 was assumed for all farms across the study area. Start and end of model simulation was set to respectively 30 and 200 days before and after sowing.

13.5.5 Future Climate Scenarios

Representative concentration pathways (RCPs) describe climate futures that are likely to occur depending on the extent of greenhouse gas emissions in the future years. There are four pathways with radiative forcing values of 2.6, 4.5, 6 and 8.5 Wm^{-2} spanning the range of year 2100 (van Vuuren et al. 2011). They provide information on possible trajectories for the main forcing agents of climate change. Of the four RCPs, two RCPs (4.5 and 8.5) were selected for this study. For each RCP, five CMIP5 global circulation models (GCMs) with daily weather data for mid-century (2040–2069) and end-century (2070–2099) time periods were used. The five GCMS include CCSM4, GFDL-ESM2G, HadGEM2-ES, MIROC5 and MPI-ESM-LR. The rationale for selecting the two RCPs were that; on one hand, RCP4.5 scenario provides a common platform for climate models to explore the climate system response to stabilizing the anthropogenic components of radiative forcing. Under this scenario, it is assumed that high efficient energy systems, for example, shifting to use of electricity and low emission energy technologies and deployment of carbon capture and geologic storage technology are will be deployed (Thomson et al. 2011). On the other hand, RCP 8.5 corresponds to the scenario with highest GHG emission due to intensified energy demand and absence of climate change policies (Riahi et al. 2011). Therefore, the two RCPs were selected to understand on the impacts of climate change in situations where there is respectively, – mitigation and the emissions have plateaued; and no climate change consciousness (van Vuuren et al. 2011). For RCP4.5 mid-century and end-century, atmospheric CO_2 concentration was respectively 499 and 532 ppm, whereas for RCP 8.5 mid-century and end century CO_2 concentration value was 571 and 801 ppm respectively (Meinshausen et al. 2011).

13.5.6 Soil, Crop, Weather and Management Data

CERES-Maize CSM requires detailed soil information from standard soil profile description. A total of 20 soil profiles were identified across the sub basin, 12 of which were obtained from Africa Soil Information Service (AfSIS) database (Leenaars 2013) and the remaining eight were newly opened, particularly in locations where soil information was not available. Soil hydrological properties for each layer in each soil profile were estimated using soil water properties calculator (Saxton and Rawls 2009). Inputs to the calculator were soil texture, (sand, silt and clay) and organic matter. For each soil profile horizon, drained lower limit (SLLL), drained upper limit (SLDUL) saturation (SLSAT and available water capacity was estimated. In order to allocated each of the farms, soil profiles were sub divided into; good soil, where soil hydrologic properties and organic matter content were increased by 20 % from the measured; average soil where it was maintained to measured values and poor soils; where the measured values were reduced by 20 %. From the respondents ranking for each farm field, soil profiles were accordingly placed to reflect respondents' assessment. Information regarding planting dates, plant population, fertilizer and organic matter applications and crop maturity dates were obtained from key informants' interviews in which 50 farmers 12 extension workers and 3 agricultural officers were involved across the study area. This kind of information was not captured in the panel survey. Daily time series weather data (1981–2010) for 11 assessment weather stations comprising solar radiation, maximum and minimum temperature and precipitation were obtained from AgMERRA datasets.

13.5.7 Climate Change Impacts Assessment Simulations

CERES-Maize model input files and simulations were organized using AgMIP framework (Rosenzweig et al. 2014). Under this framework, model input data were organized using data overlay for multi-model export (DOME) format. AgMIP framework has three DOME files, namely field survey DOME which comprises survey data such as yield as reported by farmers, plant population, plant population, and fertilizer application. Field overlay DOME contains datasets which were not measured from specific spot in the study area but are based on the best agronomic knowledge of cultural practices in the study area. Seasonal strategy DOME dataset contains baseline and future management and climate inputs for modifying existing sites data for analysis of hypothetical scenarios (Rosenzweig et al. 2014). Observed yields for each farm were obtained from National Bureau of Statistics (NBS) panel survey databases (NBS 2012). A QUADUI tool which translates survey, soil, and weather and DOME files into a model-ready format was used to interface the DSSAT model and data. Simulations were carried out for each farm, over each emission scenario and three time periods namely baseline, mid-century and end-century time periods.

13.5.8 Climate Change Adaptation Simulations

Organic matter, 1000 kg/ha and nitrogen fertilizer at a rate of 60 Kg N/ha were assessed as potential adaptation measures to climate change impacts in the study area for both emission scenarios (RCP 4.5 and RCP 8.5) and time periods (mid-century and end century).

13.5.9 *Statistical Analysis*

Empirical cumulative distribution functions (ECDF) were employed to visualize climate change impacts and the effects of adaptation strategies on maize productivity in the study area. Excel 2010 and "ggplot2" and "plyr" packages of the R-Software (R_Core_Team 2013) were used in data analyses.

13.6 Results and Discussion

13.6.1 *Results*

13.6.1.1 Climate Change Impacts on Maize Yields

CERES-Maize simulations of maize yields with respect to five GCMs and baseline indicate yield decline in both mid-century and end-century time periods under RCP 4.5 emission scenario (Fig. 13.5). During the mid-century time period, all four GCMs projected more or less similar yield reduction versus the baseline climate. In the end-century time period, HadGEM2-EC projected substantial yield decline from the baseline yields in the study area. (Fig. 13.5b).

Under the RCP 8.5 emission scenario, climate change projections indicate maize yield decline with respect to the baseline simulations in the study area (Fig. 13.6). GCMs projections were less variable in the mid-century period, unlike in the end century time period where HaDGEM-EC GCM projected significant yield decline as compared to the baseline yields (Fig. 13.6a, b).

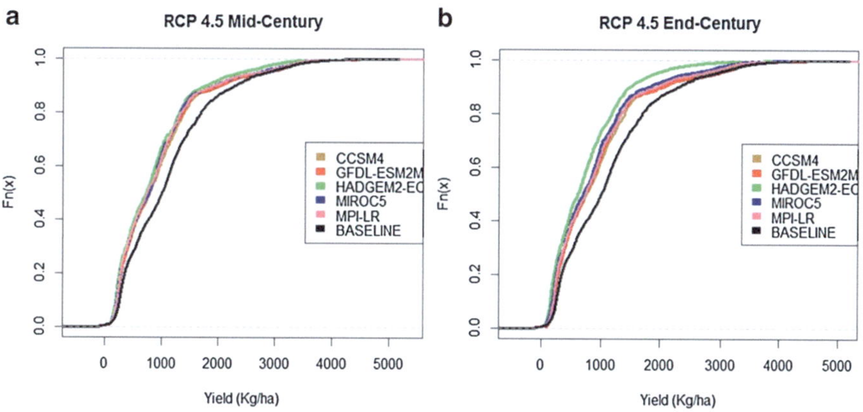

Fig. 13.5 Climate change impacts projections on maize yields for the RCP 4.5 emission scenario (**a**) in the mid-century (2040–2069) and (**b**) end-century (2070–2099) time periods

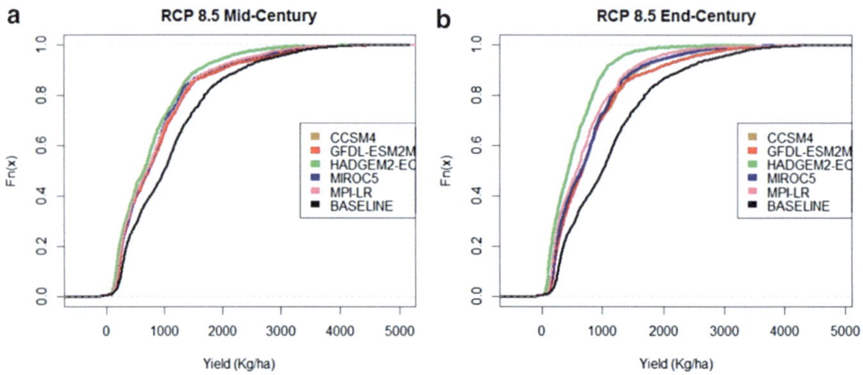

Fig. 13.6 Climate change impacts projections on maize yields under RCP 8.5 emission scenario (**a**) in the mid-century (2040–2069) and (**b**) end-century (2070–2099) time periods

13.6.1.2 Climate ChangeAdaptation Options

In response to changes, smallholder farmers have developed different farming systems finely tuned to many aspects of their environment to cope and adapt to the same. Adaptation and coping mechanism include among others adjustments to planting dates, rainwater harvesting, introduction of legumes in the cropping system, different aspects of conservation of agriculture for increased organic matter and sustainability of the cropping systems and selection of animal species. Effective and sustainable adaptation to climate change and variability depends on, among other things, our ability to assess and understand the impacts and potential opportunities in relation to and help develop functional strategies to address them. It is equally important to emphasize on the importance of improving weather forecasting, its access and reliability in compatibility with local indigenous knowledge (Mahoo et al. 2015). Different challenges exist, but opportunities and possibility for direct observable benefits of the different practices are driving factors for the adoption by many smallholder farmers (Kahimba et al. 2014; Tumbo et al. 2011; Shetto and Owenya 2007). We explore the adaptation options based on model scenarios for quantifying their efficiency in improving smallholder farmers' crop productivity.

13.6.1.3 Organic Matter Application

Adding organic matter to the farms would cause maize yield increase from an average of 1 t/ha to over 1.75 t/ha using base climate. Yield increase was projected with all GCMs if there would be organic matter application practice as an adaptation option under RCP 4.5 climate change scenario during mid-century time period. However, during end-century time period, GCM HadGEM2-EC projects yield decline despite organic matter application as an adaptation option (Fig. 13.7a, b).

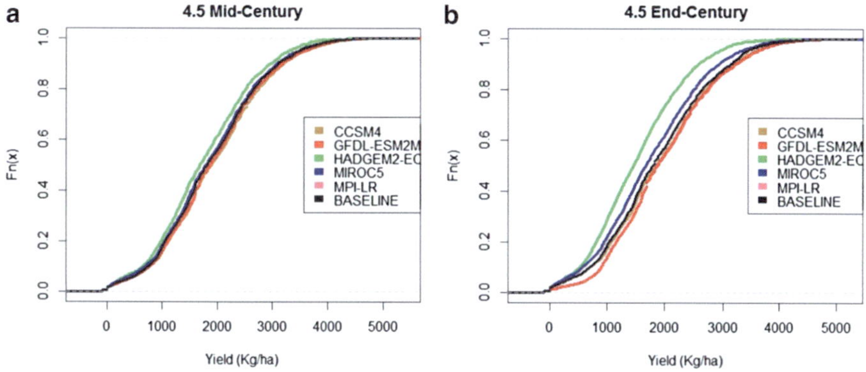

Fig. 13.7 Effects of organic matter application on maize yield as an adaptation measure to climate change under RCP 4.5 scenario during (**a**) mid-century (2040–2069) and (**b**) end-century (2070–2099) time periods

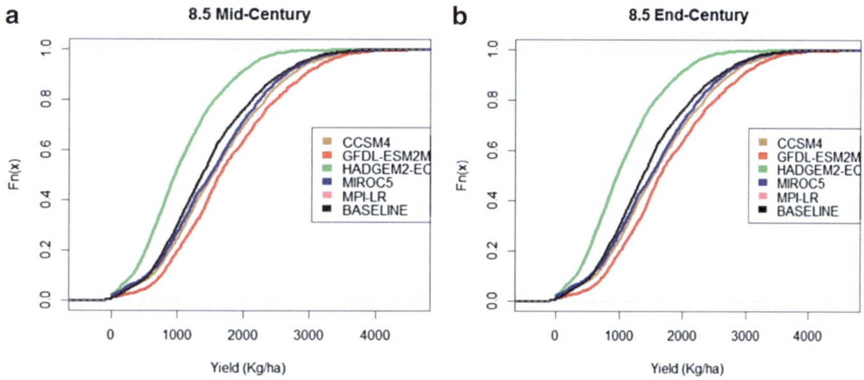

Fig. 13.8 Effects of organic matter application on maize yield as an adaptation measure to climate change under RCP 8.5 scenario during (**a**) mid-century (2040–2069) and (**b**) end-century (2070–2099) time periods

GFDL-ESM2M projections indicate that maize yields would be maintained at or slightly higher than the baseline yield level when application of organic matter is opted as an adaptation strategy during both mid and end-century time periods (Fig. 13.7a, b).

Under RCP 8.5 climate change scenario, organic matter application would result into yield increase above baseline yield for all GCMs except HaDGEM2-EC during both mid- and end-century time periods (Fig. 13.8a, b). HadGEM2-EC projections indicate yield reduction if organic matter application would be adapted in this climate change scenario, both during mid- and end-century time periods.

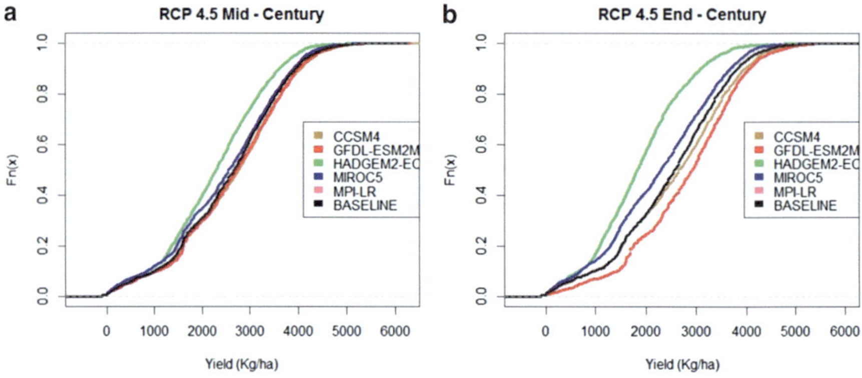

Fig. 13.9 Effects of Nitrogen fertilizer application on maize yield as an adaptation measure to climate change under RCP 4.5 scenario during (**a**) mid-century (2040–2069) and (**b**) end-century (2070–2099) time periods

13.6.1.4 Effects of Nitrogen Fertilizer Application

Model projections indicate that nitrogen fertilizer application to the maize fields would be an important strategy in enhancing yields in the face of climate change. Future maize yields would be higher than the base yields if N fertilizers was generally adopted in the study area. During the mid-century time period, yield projections for most GCMs would exceed that of base climate except that of HadGEM2-EC under RCP 4.5 scenario (Fig. 13.9a). As regards to the end-century time period, projected maize yield for GCMs GFDL- ESM2M and CCSM4 would be higher than that of the base climate (Fig. 13.9b).

Like in RCP 4.5 climate change scenario, yield increase was projected when N fertilizer would be incorporated in GCM projections as compared to the baseline. HadGEM2-EC projected yields below the baseline climate yields despite nitrogen fertilizer application, whereas such GCMs as GFDL-ESM2M, CCSM4, MPI-LR and MIROC5 projected higher yields in both mid- and end-century time periods under RCP 8.5 climate scenario (Fig. 13.10).

On disaggregating fields per geographical area, organic matter application as an adaptation option gave varying response across the study area and also among the GCMs. Effects of organic matter on yield change are presented for RCP 4.5 during the mid-century period (Fig. 13.11). When compared to an no-adaptation scenario, yield increase was projected, ranged from 23 to 495 % for farms in Kibakwe and Mvumi divisions respectively (Fig. 13.11). Significant yield increase due to organic matter adaptation strategy was projected for farms in all locations except those in Kibakwe division, where it ranged between 23 and 47 %.

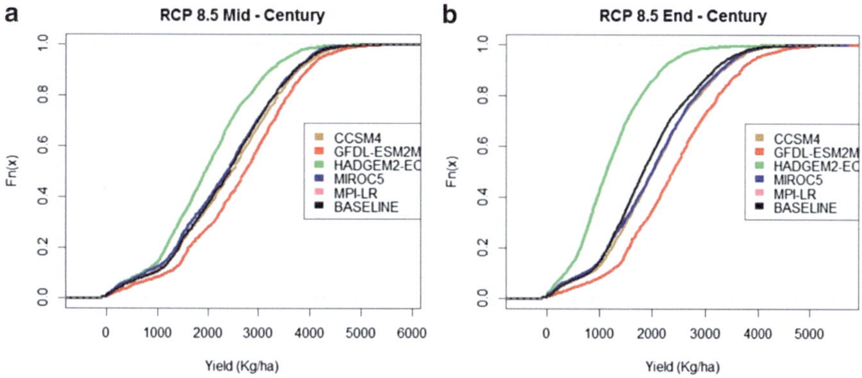

Fig. 13.10 Effects of Nitrogen fertilizer application on maize yield as an adaptation measure to climate change under RCP 4.5 scenario during (**a**) mid-century (2040–2069) and (**b**) end-century (2070–2099) time periods

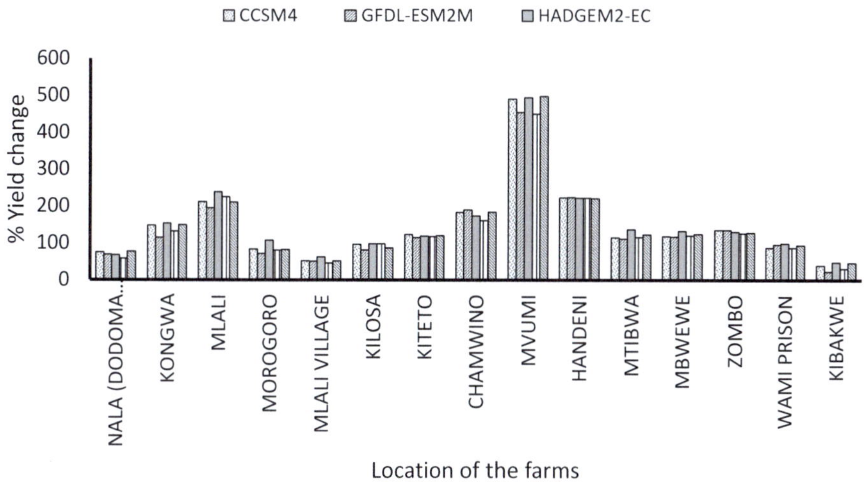

Fig. 13.11 Yield change with matter adaptation option RCP 4.5 scenario during mid-century time period

On the other hand, when nitrogen was used as an adaptation option, maize yield projections ranged from −18 to 106 % across the study area. Farms in such locations as Nala Ward (Dodoma Urban) Mlali village and Kibakwe would register negative maize yields if 60 Kg N was applied as an adaptation option in production (Fig. 13.12). Farms in Mlali Ward, Mvumi, Morogoro, Handeni and Wami Prison are projected to benefit from nitrogen fertilizer application as a climate change adaptation strategy in the mid-century time period.

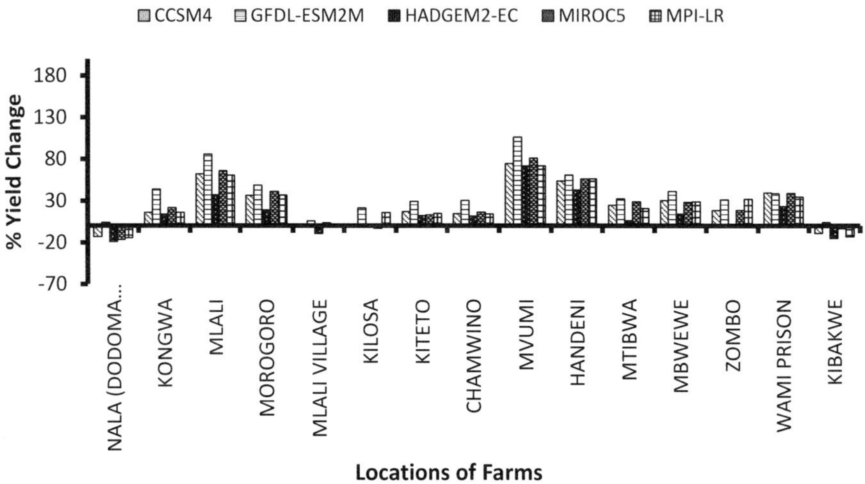

Fig. 13.12 Nitrogen fertilizer application (60 Kg N/ha) as an adaptation strategy under RCP 4.5 mid-century time period

13.7 Discussion

Our results indicate varying effects of climate change based on emission scenarios and the time periods of the projections. However, model projections points to the declining maize yield, in both emission scenarios, as compared to the base climate. Similar trends were pointed out by Rosenzweig et al. (2013) when comparing yield projections under multiple global gridded crop models (GGCMs). They observed model agreement on the direction of yield changes in many agricultural regions at both low and high latitudes but uncertainties related to the representation of CO_2nitrogen and high temperature effects constrain the better understanding of the effects of climate change. The difference in maize yield projections between GCMS may be attributed to the latter's behavior as regards to the seasonal temperature and precipitation magnitude and distribution. For example, HaDGEM2-EC consistently projected high yield decline because it projects high maximum temperature at all locations in the study area.

Climate change is associated with mean temperature increase (IPCC 2014a, b), this would result in shorter crop duration, and thus reduce yield (Haefele et al. 2016). Warming increases vapor pressure deficit, leading to reduced water use efficiency because more water has to be transpired per unit carbon gain (Ray et al. 2002). In low input, rain-fed agricultural systems, crop yields are bound to decline as a result of high temperatures. Rosenzweig and Iglesias (1998) point out a negative yield change in low latitudes (warmer) and positive yield change in higher or mid latitudes (cooler) as a result of climate change. Although CO_2 was enhanced in both emission scenarios (Thomson et al. 2011; Riahi et al. 2011), its effect on maize yield was not obvious. This suggests that maize yield gains due to enhanced CO_2 concentration might have been offset by high temperatures associated with the emission scenarios, especially during end-century time periods. Even though maize is a C4 crop, the interaction of enhanced CO_2 concentration and stresses will likely modulate crop responses. Low nitrogen levels conditions used in simulating crop yields may also have prevented maize the maize crop from taking full advantage of enhanced CO_2 concentration (Kimball 1983).

Adapting is a very fundamental process to ameliorate the negative effects associated with climate change. Common farm practices adaptation approaches in crop production include adopting new crop species/varieties adapted to harsh conditions, use of irrigation technologies, change of planting dates, etc. From this study, organic matter is a farm-level practice which indicate substantial yield gain when compared to unmitigated climate change impacts. Soil organic matter is very important especially in tropical systems where replenishment is often far less than its removal through erosion or degradation (Lal 2009). Organic matter improves soil water and nutrient holding capacity, water infiltration, and prevent soil erosion by enhancing soil aggregate stability (Skidmore et al. 1986). The fact that model projections point to maize yield improvement in all scenarios, it is an indication that adopting organic matter application would minimize risks of crop failure associated with high air temperatures. For HadGEM2-EC GCM, it is apparent that most farms would still experience lower yields than the base climate. As pointed out earlier on, the benefits gained through organic matter application may be offset by unfavorable weather projected by the GCM. On per farm basis, farms in semi-arid locations such as Mvumi, Kongwa, Chamwino, Nala would benefit by adopting organic matter application as a climate impact mitigation option. Solomon et al. (2000) reported that in semi-arid areas of Tanzania, soil organic matter declines rapidly after natural woodlands are opened up for crop cultivation. This rapid decline of soil organic matter content has a bearing on sustainable food security and if no replenishment, the soil becomes less productive and prone to severe erosion, a characteristic typical to most arid areas in Tanzania.

Although nitrogen is important in plant growth and development, it is the most limiting in most cropping systems in Tanzania (Amuri 2015). However, plants response to soil N availability depends on soil available moisture (Mourice et al. 2014b). Using Nitrogen fertilizer as an adaptation measure to climate change may work and may not work for some farms. Farms in semi-arid locations suffer yield loss. This is because, in situations where soil water is not sufficiently available, N supply would cause increased vegetative growth when there is ample water supply. In event of deficit soil moisture, the water demand for the plants due to increased transpiring surface become higher than the soil can supply, leading to closure of stomata and consequently ceasing photosynthesis. As shown from the results, nitrogen fertilizer application farms in such locations as Nala, Mlali village and Kibakwe pose crop failure risks.

As a way of comparison between the two climate change adaptation strategies, the gains from organic matter strategy outweigh those from nitrogen fertilizer approach. However, it doesn't mean that farms in semi-arid areas do not need N fertilizer, but rather a small N amount may be required, which may not cause excessive vegetative growth which would eventually lead to high transpirational water losses (Mourice et al. 2014b; Aune and Coulibaly 2015).

13.8 Conclusion

This study has demonstrated the modeling approach to evaluate the impacts of projected climate change on maize crop production and adaptation strategies in the Wami River sub basin. The general consensus is that maize yield is bound to decrease in the event of the global change and these changes will have varying degree of change but in a negative direction. Due to an array of advantages associated with the use of soil organic matter, it is evident that the farmers are set to gain more yield if they adopt it in their cropping systems. Organic matter is most important in semi-arid locations of the study area as it would rejuvenate the soils in terms of improved soil water capacity and nutrient retention and release capacity. Nitrogen fertilization, if used as climate change adaptation strategy, may increase crop failure risks in semi-arid locations. However, study is still required to understand the optimum N levels which would reduce climate risks across the study area. What remains to be done is the trade-off analysis of the adaptation strategies presented in this study and more. This would fine tune appropriate climate change-ready farm practices to sustain or surpass current production levels in the future climate.

Acknowledgement Authors of this study wish to acknowledge Agricultural Model Intercomparison Project (AgMIP) and Enhancing Climate Change Adaptation in Agriculture and Water Resources (ECAW) Project through the Soil Water Management Research Programme at Sokoine University of Agriculture (SUA) for technical and financial support for implementing this study.

References

Abraha, M., and M. Savage. 2006. Potential impacts of climate change on the grain yield of maize for the midlands of KwaZulu-Natal, South Africa. *Agriculture, Ecosystems & Environment* 115: 150–160.

Agrell, J., P. Anderson, W. Oleszek, A. Stochmal, and C. Agrell. 2004. Combined effects of elevated CO2 and herbivore damage on alfalfa and cotton. *Journal of Chemical Ecology* 30: 2309–2324.

Ahmed, S.A., N.S. Diffenbaugh, T.W. Hertel, D.B. Lobell, N. Ramankutty, A.R. Rios, and P. Rowhani. 2009. *Climate volatility and poverty vulnerability in Tanzania*, Policy research working paper. Washington, DC: World Bank.

Ahmed, S.A., N.S. Diffenbaugh, T.W. Hertel, D.B. Lobell, N. Ramankutty, A.R. Rios, and P. Rowhani. 2011. Climate volatility and poverty vulnerability in Tanzania. *Global Environmental Change* 21: 46–55.

Ainsworth, E.A., and S.P. Long. 2005. What have we learned from 15 years of free-air CO2 enrichment (FACE)? A meta-analytic review of the responses of photosynthesis, canopy properties and plant production to rising CO2. *New Phytologist* 165: 351–372.

Amuri, N.A. 2015. Enhancing resilience of food production systems under changing climate and soil degradation in semi-arid and highlands of Tanzania. In *Sustainable intensification to advance food security and enhance climate resilience in Africa*. Cham: Springer.

Arndt, C., W. Farmer, K. Strzepek, and J. Thurlow. 2012. Climate change, agriculture and food security in Tanzania. *Review of Development Economics* 16: 378–393.

Aune, J.B., and A. Coulibaly. 2015. Microdosing of mineral fertilizer and conservation agriculture for sustainable agricultural intensification in Sub-Saharan Africa. In *Sustainable intensification to advance food security and enhance climate resilience in Africa*. Cham: Springer.

World Bank. 2012. *The population below poverty line*. http://data.worldbank.org/topic/poverty

Barnabás, B., K. Jäger, and A. Fehér. 2008. The effect of drought and heat stress on reproductive processes in cereals. *Plant, Cell & Environment* 31: 11–38.

Blum, A. 2005. Drought resistance, water-use efficiency, and yield potential – Are they compatible, dissonant, or mutually exclusive? *Crop and Pasture Science* 56: 1159–1168.

Caldwell, C.R., S.J. Britz, and R.M. Mirecki. 2005. Effect of temperature, elevated carbon dioxide, and drought during seed development on the isoflavone content of dwarf soybean [Glycine max (L.) Merrill] grown in controlled environments. *Journal of Agricultural and Food Chemistry* 53: 1125–1129.

Cassman, K.G., A. Dobermann, D.T. Walters, and H. Yang. 2003. Meeting cereal demand while protecting natural resources and improving environmental quality. *Annual Review of Environment and Resources* 28: 315–358.

Enfors, E.I., and L.J. Gordon. 2008. Dealing with drought: The challenge of using water system technologies to break dryland poverty traps. *Global Environmental Change* 18: 607–616.

FAO. 2015. FOASTAT Database. Food and Agriculture Organization, Rome Available from: faostat.fao.org/. Accessed 20 Sept 2015.

Fischer, G., M. Shah, F.N. Tubiello, and H. Van Velhuizen. 2005. Socio-economic and climate change impacts on agriculture: An integrated assessment, 1990–2080. *Philosophical Transactions of the Royal Society, B: Biological Sciences* 360: 2067–2083.

Foley, J.A., R. Defries, G.P. Asner, C. Barford, G. Bonan, S.R. Carpenter, F.S. Chapin, M.T. Coe, G.C. Daily, and H.K. Gibbs. 2005. Global consequences of land use. *Science* 309: 570–574.

Gibbs, H.K., A.S. Ruesch, F. Achard, M.K. Clayton, P. Holmgren, N. Ramankutty, and J.A. Foley. 2010. Tropical forests were the primary sources of new agricultural land in the 1980s and 1990s. *Proceedings of the National Academy of Sciences* 107: 16732–16737.

Griffin, J.J., T.G. Ranney, and D.M. Pharr. 2004. Heat and drought influence photosynthesis, water relations, and soluble carbohydrates of two ecotypes of redbud (Cercis canadensis). *Journal of the American Society for Horticultural Science* 129: 497–502.

Haefele, S.M., Y. Kato, and S. Singh. 2016. Climate ready rice: Augmenting drought tolerance with best management practices. *Field Crops Research* 190: 60–69.

Hunt, L., and K. Boote. 1998. Data for model operation, calibration, and evaluation. In *Understanding options for agricultural production*. Berlin: Springer.

IPCC. 2014a. In *Climate change 2014: Impacts, adaptation, and vulnerability. Part A: Global and sectoral aspects. Contribution of working group ii to the fifth assessment report of the intergovernmental panel on climate change*, eds. C.B. Field, V.R. Barros, D.J. Dokken, K.J. Mach, M.D. Mastrandrea, T.E. Bilir, M. Chatterjee, K.L. Ebi, Y.O. Estrada, R.C. Genova, B. Girma, E.S. Kissel, A.N. Levy, S. Maccracken, P.R. Mastrandrea, and L.L. White. Cambridge/New York: Cambridge University Press.

IPCC. 2014b. Summary for policymakers. In: Climate change 2014: Impacts, adaptation, and vulnerability. Part A: Global and sectoral aspects. Contribution of working group ii to the fifth assessment report of the intergovernmental panel on climate change. In *Climate Change 2014: Impacts, Adaptation, and Vulnerability*, ed. C.B. Field, V.R. Barros, D.J. Dokken, K.J. Mach, M.D. Mastrandrea, T.E. Bilir, M. Chatterjee, K.L. Ebi, Y.O. Estrada, R.C. Genova, B. Girma, E.S. Kissel, A.N. Levy, S. Maccracken, P.R. Mastrandrea, and L.L.W. Cambridge/New York: Cambridge University Pres.

Ito, T., and T. Kurosaki. 2009. Weather risk, wages in kind, and the off-farm labor supply of agricultural households in a developing country. *American Journal of Agricultural Economics* 91: 697–710.

Jat, M.L., J.C. Dagar, T.B. Sapkota, S. Yadvinder, B. Govaerts, S.L. Ridaura, Y.S. Saharawat, R.K. Sharma, J.P. Tetarwal, R.K. Jat, H. Hobbs, and C. Stirling. 2016. Chapter Three – Climate change and agriculture: Adaptation strategies and mitigation opportunities for food security in

South Asia and Latin America. In *Advances in agronomy*, ed. L.S. Donald, 127–235. Academic Press.

Jones, C.A., J.R. Kiniry, and P. Dyke. 1986. *CERES-Maize: A simulation model of maize growth and development*. College Station: Texas A and M University Press.

Kahimba, F.C., K.D. Mutabazi, S.D. Tumbo, K.F. Masuki, and W.B. Mbungu. 2014. Adoption and scaling-up of conservation agriculture in Tanzania: Case of Arusha and Dodoma regions. *Natural Resources* 05: 161–176.

Kang, Y., S. Khan, and X. Ma. 2009. Climate change impacts on crop yield, crop water productivity and food security–A review. *Progress in Natural Science* 19: 1665–1674.

Kimball, B.A. 1983. Carbon dioxide and agricultural yield: An assemblage and analysis of 430 prior observations. *Agronomy Journal* 75: 779–788.

Kimball, B., K. Kobayashi, and M. Bindi. 2002. Responses of agricultural crops to free-air CO 2 enrichment. *Advances in Agronomy* 77: 293–368.

Lal, R. 2009. Challenges and opportunities in soil organic matter research. *European Journal of Soil Science* 60: 158–169.

Leakey, A.D.B. 2009. Rising atmospheric carbon dioxide concentration and the future of C4 crops for food and fuel. *Proceedings of the Royal Society of London B: Biological Sciences* 276: 2333–2343.

Leenaars, J. 2013. Africa soil profiles database, version 1.1. *A compilation of georeferenced and standardised legacy soil profile data for Sub-Saharan Africa (with dataset). ISRIC Report,* 3.

Leyaro, V., and O. Morrissey. 2013. Expanding agricultural production in Tanzania. Scoping study for IGC Tanzania on the National Panel Surveys. International growth Centre (IGC), London School of Economic and Political Science, Houghton Street, London WC2A 2AE.

Lobell, D.B., and C.B. Field. 2007. Global scale climate–crop yield relationships and the impacts of recent warming. *Environmental Research Letters* 2: 014002.

Lobell, D.B., and S.M. Gourdji. 2012. The influence of climate change on global crop productivity. *Plant Physiology* 160: 1686–1697.

Lobell, D.B., M.B. Burke, C. Tebaldi, M.D. Mastrandrea, W.P. Falcon, and R.L. Naylor. 2008. Prioritizing climate change adaptation needs for food security in 2030. *Science* 319: 607–610.

Lokina, R., M. Nerman, and J. Sandefur. 2011. Poverty and productivity: Small-scale farming in Tanzania, 1991–2007. *Preliminary results for International Growth Centre commissioned study.*

Long, S.P., E.A. Ainsworth, A. Rogers, and D.R. Ort. 2004. Rising atmospheric carbon dioxide: Plants FACE the Future*. *Annual Review of Plant Biology* 55: 591–628.

Long, S.P., E.A. Ainsworth, A.D. Leakey, J. Nösberger, and D.R. Ort. 2006. Food for thought: Lower-than-expected crop yield stimulation with rising CO2 concentrations. *Science* 312: 1918–1921.

Lyimo, J.G., and R.Y. Kangalawe. 2010. Vulnerability and adaptive strategies to the impact of climate change and variability. The case of rural households in Semiarid Tanzania. *Environmental Economics* 1: 88–96.

Maestri, E., N. Klueva, C. Perrotta, M. Gulli, H.T. Nguyen, and N. Marmiroli. 2002. Molecular genetics of heat tolerance and heat shock proteins in cereals. *Plant Molecular Biology* 48: 667–681.

Mahoo, H., W. Mbungu, I. Yonah, M. Radeny, P. Kimeli, and J. Kinyangi. 2015. *Integrating indigenous knowledge with scientific seasonal forecasts for climate risk management in Lushoto district in Tanzania*. Copenhagen: CGIAR Research Program on Climate Change, Agriculture and Food Security (CCAFS).

Mbungu, W., V. Ntegeka, F.C. Kahimba, M. Taye, and P. Willems. 2012. Temporal and spatial variations in hydro-climatic extremes in the Lake Victoria basin. *Physics and Chemistry of the Earth, Parts A/B/C* 50–52: 24–33.

Mbungu, W.B., H.F. Mahoo, S.D. Tumbo, F.C. Kahimba, F.B. Rwehumbiza, and B.P. Mbilinyi. 2015. Using climate and crop simulation models for assessing climate change impacts on agronomic practices and productivity. In *Sustainable intensification to advance food security and enhance climate resilience in Africa*. Cham: Springer.

Meinshausen, M., S.J. Smith, K. Calvin, J.S. Daniel, M. Kainuma, J. Lamarque, K. Matsumoto, S. Montzka, S. Raper, and K. Riahi. 2011. The RCP greenhouse gas concentrations and their extensions from 1765 to 2300. *Climatic Change* 109: 213–241.

Mongi, H., A. Majule, and J. LYIMO. 2010. Vulnerability and adaptation of rain fed agriculture to climate change and variability in semi-arid Tanzania. *African Journal of Environmental Science and Technology* 4: 371–381.

Moore, N., G. Alagarswamy, B. Pijanowski, P. Thornton, B. Lofgren, J. Olson, J. Andresen, P. Yanda, and J. Qi. 2012. East African food security as influenced by future climate change and land use change at local to regional scales. *Climatic Change* 110: 823–844.

Morgan, J., D. Pataki, C. Körner, H. Clark, S. Del GROSSO, J. Grünzweig, A. Knapp, A. Mosier, P. Newton, and P.A. Niklaus. 2004. Water relations in grassland and desert ecosystems exposed to elevated atmospheric CO2. *Oecologia* 140: 11–25.

Morton, J.F. 2007. The impact of climate change on smallholder and subsistence agriculture. *Proceedings of the National Academy of Sciences* 104: 19680–19685.

Mourice, S. K., C.L. Rweyemamu, S.D. Tumbo, and N. Amuri. 2014a. Maize cultivar specific parameters for decision support system for agrotechnology transfer (DSSAT) application in Tanzania. *American Journal of Plant Sciences,* 5: 55–65.

Mourice, S., C. Rweyemamu, A. Nyambilila, and S. Tumbo. 2014b. Narrowing maize yield gaps under rain-fed conditions in Tanzania: Effect of small nitrogen dose. *Tanzania Journal of Agricultural Sciences,* 12: 821–833.

Mourice, S.K., S.D. Tumbo, A. Nyambilila, and C.L. Rweyemamu. 2015. Modeling potential rain-fed maize productivity and yield gaps in the Wami River sub-basin, Tanzania. *Acta Agriculturae Scandinavica, Section B—Soil & Plant Science* 65: 132–140.

Mwandosya, M.J., B.S. Nyenzi, and M. Lubanga. 1998. *The assessment of vulnerability and adaptation to climate change impacts in Tanzania*. Dar es Salaam: Centre for Energy, Environment, Science and Technology.

NBS. 2012. National Bureau of Statistics, National panel survey 2009–2010. Available from: www.nbs.go.tz. Accessed on 24 Sept 2015.

Ngana, J., F. Mahay, and K. Cross. 2010. *Wami basin: A situation analysis*. Nairobi: International Union for Conservation of Nature (IUCN).

Nowak, R.S., D.S. Ellsworth, and S.D. Smith. 2004. Functional responses of plants to elevated atmospheric CO2–do photosynthetic and productivity data from FACE experiments support early predictions? *New Phytologist* 162: 253–280.

Paavola, J. 2008. Livelihoods, vulnerability and adaptation to climate change in Morogoro, Tanzania. *Environmental Science & Policy* 11: 642–654.

R_CORE_TEAM. 2013. *R: A language and environment for statistical computing*. R Foundation for Statistical Computing. Vienna. Available at www.R-project.org.

Ray, J.D., R.W. Gesch, T.R. Sinclair, and L.H. Allen. 2002. The effect of vapor pressure deficit on maize transpiration response to a drying soil. *Plant and Soil* 239: 113–121.

Reilly, J., F. Tubiello, B. Mccarl, D. Abler, R. Darwin, K. Fuglie, S. Hollinger, C. Izaurralde, S. Jagtap, and J. Jones. 2003. US agriculture and climate change: New results. *Climatic Change* 57: 43–67.

Riahi, K., S. Rao, V. Krey, C. Cho, V. Chirkov, G. Fischer, G. Kindermann, N. Nakicenovic, and P. Rafaj. 2011. RCP 8.5 – A scenario of comparatively high greenhouse gas emissions. *Climatic Change* 109: 33–57.

Ritchie, J., U. Singh, D. Godwin, and W. Bowen. 1998. Cereal growth, development and yield. In *Understanding options for agricultural production*. Netherlands: Springer.

Rosenzweig, C., and A. Iglesias. 1998. The use of crop models for international climate change impact assessment. In *Understanding options for agricultural production*. Netherlands: Springer.

Rosenzweig, C., J. Elliott, D. Deryng, A.C. Ruane, C. Müller, A. Arneth, K.J. Boote, C. Folberth, M. Glotter, N. Khabarov, K. Neumann, F. Piontek, T.A.M. Pugh, E. Schmid, E. Stehfest, H. Yang, and J.W. Jones. 2014. Assessing agricultural risks of climate change in the 21st cen-

tury in a global gridded crop model intercomparison. *Proceedings of the National Academy of Sciences* 111: 3268–3273.
Rowhani, P., D.B. Lobell, M. Linderman, and N. Ramankutty. 2011. Climate variability and crop production in Tanzania. *Agricultural and Forest Meteorology* 151: 449–460.
Salami, A., A.B. Kamara, and Z. Brixiova. 2010. *Smallholder agriculture in East Africa: Trends, constraints and opportunities*. Tunisia: African Development Bank Tunis.
Sarris, A., S. Savastano, and L. Christiaensen. 2006. *The role of agriculture in reducing poverty in Tanzania: A household perspective from rural Kilimanjaro and Ruvuma, FAO commodity and trade policy research working paper*. Rome: FAO.
Saxton, K.E., and W.J. Rawls. 2009. *Soil water properties calculator*. Version 6.02. Available from: http://hydrolab.arsusda.gov/soilwater/Index.htm.
Schlenker, W., and D.B. Lobell. 2010. Robust negative impacts of climate change on African agriculture. *Environmental Research Letters* 5: 014010.
Schmidhuber, J., and F.N. Tubiello. 2007. Global food security under climate change. *Proceedings of the National Academy of Sciences of the United States of America* 104: 19703–19708.
Shetto, R., and M. Owenya. 2007. *Conservation agriculture as practised in Tanzania: Three case studies: Arumeru district, Karatu district, Mbeya district*. Nairobi: ACT, FAO, CIRAD, RELMA.
Sicher, R.C., and J.Y. Barnaby. 2012. Impact of carbon dioxide enrichment on the responses of maize leaf transcripts and metabolites to water stress. *Physiologia Plantarum* 144: 238–253.
Skidmore, E., J. Layton, D. Armbrust, and M. Hooker. 1986. Soil physical properties as influenced by cropping and residue management. *Soil Science Society of America Journal* 50: 415–419.
Soini, E. 2005. Land use change patterns and livelihood dynamics on the slopes of Mt. Kilimanjaro, Tanzania. *Agricultural Systems* 85: 306–323.
Solomon, D., J. Lehmann, and W. Zech. 2000. Land use effects on soil organic matter properties of chromic luvisols in semi-arid northern Tanzania: Carbon, nitrogen, lignin and carbohydrates. *Agriculture, Ecosystems & Environment* 78: 203–213.
Stone, P. 2001. The effects of heat stress on cereal yield and quality. In *Crop responses and adaptations to temperature stress*, 243–291. New York: Food Products Press.
SUA. 2010. Soil Water Management Research Group (2010). *Economics of climate change for the Agriculture Sector in Tanzania*. Morogoro: Morogoro.
Teyssonneyre, F., C. Picon-Cochard, J.F. Falcimagne, and R. Soussana. 2002. Effects of elevated CO2 and cutting frequency on plant community structure in a temperate grassland. *Global Change Biology* 8: 1034–1046.
Thomas, J., K. Boote, L. Allen, M. Gallo-Meagher, and J. Davis. 2003. Elevated temperature and carbon dioxide effects on soybean seed composition and transcript abundance. *Crop Science* 43: 1548–1557.
Thomson, A.M., K.V. Calvin, S.J. Smith, G.P. Kyle, A. Volke, P. Patel, S. Delgado-Arias, B. Bond-Lamberty, M.A. Wise, and L.E. Clarke. 2011. RCP4. 5: A pathway for stabilization of radiative forcing by 2100. *Climatic Change* 109: 77–94.
Tubiello, F.N., J.-F. Soussana, and S.M. Howden. 2007. Crop and pasture response to climate change. *Proceedings of the National Academy of Sciences* 104: 19686–19690.
Tumbo, S., K. Mutabazi, F. Kahimba, and W. Mbungu. 2011. *Conservation agriculture in Tanzania*. Dar es Salaam: Sokoine University of Agriculture.
Tumbo, S., F. Kahimba, B. Mbilinyi, F. Rwehumbiza, H. Mahoo, W. Mbungu, and E. Enfors. 2012. Impact of projected climate change on agricultural production in semi-arid areas of Tanzania: A case of same district. *African Crop Science Journal* 20: 453–463.
URT. 2002. *National irrigation master plan of Tanzania*. Dar Es Salaam: United Republic of Tanzania.
URT. 2007. United Republic of Tanzania, 2007. National adaptation program of action. Vice.
URT. 2008a. United Republic of Tanzania (URT), 2008. Agriculture sector review and public expenditure review 2008/09. Dar Es Salaam: Ministry of Agriculture Food Security and Cooperatives.

URT. 2008b. United Republic of Tanzania (URT), 2008. The economic survey 2007. Dar-es Salaam: The Ministry of Finance and Economic Affairs.

URT. 2008c. Study on strategies for addressing negative effects of climate change in food insecure areas of Tanzania. In eds. Ministry of Agriculture, F. A. C. Dar Es Salaam: Ministry of Agriculture, Food and Cooperatives.

URT. 2009. *Climate change and agriculture policy brief*. Dar es Salaam: Vice President's Offi ce, Division of Environment.

URT. 2012a. *Priority investments: Irrigation development, water resources and land use management*. Tanzania: United Republic of Tanzania.

URT. 2012b. *Tanzania national climate change strategy*. Dar Es Salaam: Division of Environment, Vice President's Office.

URT. 2012c. *National climate change strategy*. Dar Es Salaam: Vice President's Office, Division of the Environment.

URT. 2014. Ministry of Agriculture, Food Security and Cooperatives (MAFC): Tanzania Agriculture Climate Resilience Plan, 2014–2019. Report 83 pp. Available from: http://www.agriculture.go.tz/publications. Accessed on 25 Sept 2015. Dar Es Salaam: United Republic of Tanzania: Ministry of Agriculture, Food Security and Cooperatives (MAFC).

Van Vuuren, D.P., J. Edmonds, M. Kainuma, K. Riahi, A. Thomson, K. Hibbard, G.C. Hurtt, T. Kram, V. Krey, and J.-F. Lamarque. 2011. The representative concentration pathways: An overview. *Climatic Change* 109: 5–31.

Von Caemmerer, S., and R.T. Furbank. 2003. The C4 pathway: An efficient CO2 pump. *Photosynthesis Research* 77: 191–207.

World_Bank. 2009. *Tanzania: Country brief*. Washington, DC: World Bank.

WRBWO. 2007. *Environmental Flow Assessment (EFA), Wami River Sub-Basin, Tanzania: The Wami hydrology. Volume 1 – general description. Morogoro 2007*. Morogoro: Wami/Ruvu Basin Water Office.

Ziska, L.H., and K. George. 2004. Rising carbon dioxide and invasive, noxious plants: potential threats and consequences. *World Resource Review* 16: 427–447.

Chapter 14
Climatic Variability Impact on Wheat-Based Cropping Systems of South Asia: Adaptation and Mitigation

Amanpreet Kaur, Paramjit Kaur Sraw, and S.S. Kukal

Contents

Abstract Climate change is now a reality, many people prefer to call it as climate variability instead of climate change, which in fact, is much more complex phenomenon from agricultural point of view. The seasons are no longer identifiable as these used to be previously. Sometimes, these prolong or get shortened replacing the other seasons. The increase in climate-extremes in precipitation, temperature, etc. has

A. Kaur • S.S. Kukal (✉)
Department of Soil Science, Punjab Agricultural University, Ludhiana, India
e-mail: sskukal63@rediffmail.com; sskukal1@gmail.com; surindersingh@pau.edu

P.K. Sraw
Department of Agronomy, Punjab Agricultural University, Ludhiana, India

M. Ahmed, C.O. Stockle (eds.), *Quantification of Climate Variability, Adaptation and Mitigation for Agricultural Sustainability*,
DOI 10.1007/978-3-319-32059-5_14

already started taking a tool on the agricultural production as is evident from one of the most productive region of India i.e. Punjab, where the wheat productivity is affected considerably due to elevated temperature conditions over the last few years. Industry and domestic sectors can easily adapt to the changing climatic conditions but for agriculture sector it is not so easy. A lot still needs to be done especially in formulating strategies to mitigate such effects. The plant breeders have to play a big role in redefining the characters of crop plants through genetic engineering so as to develop cultivars to grow efficiently in newer climatic condition. The cultivars for biofuel crops need to be developed as an alternate to fossil fuel energy sources. It is thus a collective effort on the part of plant breeder, soil scientist, agronomist, plant protection scientist and food technologist to work together to develop strategies for mitigating the ill effects of climate change in agricultural sector.

Keywords Climate change • Climate variability • Climate extremes • Precipitation • Temperature • Green house gases • Heat and cold waves • Agriculture and biofuel

14.1 Introduction

Crops such as rice, wheat, maize and barley account for a major proportion of global food grain production and are likely to be influenced by changing climate which in turn is predicted to severely impact the food production under future climate scenarios (IPCC 2013). Crop production can be affected by climate change aspects, such as increased atmospheric CO_2 concentration [CO_2], increased temperature and changed rainfall. In combination, these aspects can either positive or negative effect plant production or in other words, the net effect of climate change on crop yield depends on the interactions between these climatic variables during different crop growth stages. The overall impact of climate change on agriculture is likely to be negative which in turn might result in price hike of agricultural commodities, food supplies and consequently livestock products like meat and milk. Thus, to meet the increasing demand for the food, productivity of wheat-based cropping systems in the region needs to be not only sustained but to be increased under changing climatic conditions. Based on the simulations made with global climate models (GCMs), the projected global average temperature are expected to be between 2 °C and 4.5 °C in the present century (IPCC 2001) and CO_2 concentration to reach 700 ppm at the end of the twenty-first century. These projections made with simulations studies suggest that the increases in CO_2, the major greenhouse gas after water vapour, will cause by the widespread rise in surface air temperatures; alteration in precipitation patterns and the global hydrologic cycle; and increase in the frequency of severe weather events, such as drought spells and flooding (IPCC 1996). In South Asia, the Indian subcontinent, the IPCC has projected 0.5–1.2 °C rise in temperature by 2020, 0.88–3.16 °C by 2050 and 1.56–5.44 °C by 2078, depending upon the scenarios of future development. These environmental changes will further increase pressure on agriculture and scientific community, in addition to ongoing stress of yield stagnation, depleting land and water resources. As per an estimate, an increase in irrigation water requirements of 50 % in developing regions and 16 % in the developed regions is expected between 2000 and 2080. Similarly, higher temperatures tend to reduce yields of many crops and encourage proliferation of weeds and pests. So there will be pressure mainly on agriculture to produce more food from same or even shrinking land and water resources. As agriculture is also a significant contributor to the global climate change through release of green house gases (GHGs) such as methane and nitrous oxide. These gases have 21 and 310 times higher warming potential than that of CO_2.

The primary focus of this review is to present the research findings on climate change, combined effect of different climatic variables and possible adaption and mitigation strategies for climate change impacts on wheat-based cropping systems in South Asia.

14.2 Causes and Evidences of Climate Change in South Asia

Climate change is the long-term shift in weather conditions due to change in different climatic variables like temperature, precipitation and rainfall events, both amount and distribution of rain. Climate change may occur due to natural causes or human-induced causes. Natural causes include changes in volcanic activity and changes in solar radiation, whereas the human activities include fossil fuel burning, change in land use and agriculture. According to U.S. Geological Survey (USGS) reports, human activities are emitting more than 135 times as much CO_2 as volcanoes each year. Fossil fuel burning emits large quantities of CO_2 into the atmosphere. However, apart from fossil fuel combustion, other human activities like industrial processes, change in land use and particularly agriculture, emitting other green house gases like methane from rice fields, enteric fermentation in ruminants and nitrous oxide from application of fertilizers and manures to soils. The concentration of CO_2, methane and nitrous oxide has increased markedly from 280 ppm, 715 ppb and 270 ppb since 1750 AD to 379 ppm, 1774 ppb and 319 ppb, respectively in 2005 (IPCC 2007). This increase in GHG emission has resulted in warming of climate since 1860. The IPCC has also shown through series of observations and modeling that the rate of warming has been much higher in recent decades which further resulted in rise in sea level, decline in glaciers and snow cover, increased frequency of droughts, heavy precipitation, cold days, cold nights and frost has become less severe whereas hot days and hot nights have become more frequent. Yan et al. (2002) in China recorded a gradual decrease in the number of cold days over the twentieth century and an increase in the number of warm days since 1961. Klein et al. (2006) while analyzing data of 159 different meteorological stations from 13 countries of Central and South Asia from period between 1961 and 2000, documented that, 70 % of the stations showed decrease in cold nights and increase

Table 14.1 Projected changes in surface air temperature and precipitation for South Asia under different climate change scenarios

	Temperature (°C)		Precipitation (%)		Temperature (°C)		Precipitation (%)		Temperature (°C)		Precipitation (%)	
Season	A1F1	B1	A1F1	B1	A1F1	B1	A1F1	B1	A1F1	B1	A1F1	B1
Dec–Feb	1.17	1.11	−3	4	3.16	1.97	0	0	5.44	3.93	16	−6
Mar–May	1.18	1.07	7	8	2.97	1.81	26	24	5.22	2.71	31	20
Jun–Aug	0.54	0.55	5	7	1.71	0.88	13	11	3.14	1.56	26	15
Sep–Nov	0.78	0.81	1	3	2.41	1.49	8	6	4.19	2.17	26	10

in warm nights. Similarly Dhorde and Patel (2016) documented the increasing trend in minimum temperatures in Punjab especially during February and March, which is critical grain development phase of wheat crop. They also reported the average of incidence of temperature > 40 °C increased up to 1977–1983 and then decreased till 1991–1997 and thereafter again showed an increasing trend till 2011. The incidence of extreme rainfall events also showed an increasing trend from 2005 to 2011. The global temperature likely to increase in the range of 1.8–4.0 °C by end of this century depending upon different scenarios of future development. The precipitation is also likely to increase except in December-February (Table 14.1). Although impact of climate change is witnessed globally but South Asian countries like India, China are more likely to be affected because of the huge population which largely depend upon agriculture for their livelihood.

14.3 Wheat-Based Cropping Systems of South Asia

The three cereals viz. rice, wheat and maize constitute the major sources of food energy and nutrients for the global population, and together these supply about two-thirds of the total food supply in SA. Other significant staple foods are millets. Rice-wheat (RW) is the predominant cropping system with 12.37 million ha in the subtropical areas of the IGP. The cotton-wheat (CW) system is the second important system after RW in terms of area covering 4.19 million ha dominant in northwest IGP (Pakistan and western India). The maize-wheat (MW) system with 2.84 million ha is less dominant, but maize cultivation is increasing in recent times both in terms of area and production due to higher productivity and profitability and lesser water requirement than winter rice or wheat (Timsina et al. 2010). Other wheat-based cropping systems in South Asia include sugarcane-wheat, groundnut-wheat, rice/maize-potato-wheat, sorghum/pearl millet/pigeon pea-wheat. The RW and MW systems are highly intensive in the northwest and parts of central IGP with a land use intensity of 1.82–1.90 and a liberal and often excessive use of irrigation water and agrochemical inputs to maximize crop yields. Rice is the largest water requiring crop consuming 63 % of the water. Wheat is the second largest crop grown, both under rainfed and irrigated conditions. Total area under wheat cultivation in SA is about 36.1 million ha (73 % of this area is in India) with overall productivity of 2.66 Mg ha^{-1} (Source: FAO 2007 Statistical database http://www.fao-statfacts.org) and the productivity needs to be increased to around 3.8 Mg ha^{-1} from the same area by 2020 due to the population increase and growing urbanization. Sustainability of RW system is now become a major challenging issue due to some of the emerging problems like degradation of natural resources(water, soil and biodiversity), increasing production cost due to high cost of land, labor and chemical inputs. Low input use efficiency (fertilizers, pesticides, labor); environmental pollution, Extreme weather conditions, and fast changing socioeconomic conditions (population increase unemployment, increasing poverty, urbanization, shortage of farm labor, etc.) (Erenstein 2009; Ladha et al. 2009). On the other hand in some parts of IGP like eastern IGP, poor adoption of improved technologies in wheat-based systems resulted in lower productivity and income but land use intensity is high due to multiple cropping system.

14.4 Quantification of Climate Variability Impact on Wheat-Based Cropping System

14.4.1 Overview of Responses to High Temperature, Drought, or [eCO_2]

Different variable of climate change are playing significant role in agricultural crop production in cereal crops. High-temperature stress causes significant reduction in crop production at various growth stages. High temperature stress during vegetative stage limits photosynthesis in maize, wheat, resulting in decreased photo assimilate production and consequently lower shoot biomass accumulation (Wahid et al. 2007; Barnabas et al. 2008). However, reproductive stages of crop development are considered as the most sensitive phases to high-temperature stress across cereals, which alters spikelet production and grain development due to an inadequate assimilate supply, thus resulting in fewer spikelets/grains and decreased sink size (Barnabas et al. 2008). High temperature is also known to advance flowering, induce floral abnormalities such as stamen hypoplasia and pistil hyperplasia, which are hurdles for reproductive success in rice (Takeoka et al. 1991). Moreover, poor anther dehiscence, low pollen germination and decreased pollen viability are identified as major causes of stress-induced sterility in rice and kernel abortion in maize and wheat (Jagadish et al. 2007, 2010; Rang et al. 2011; Barnabas et al. 2008). Post-anthesis high temperature stress induces leaf senescence and decreases overall grain-filling duration. In addition, poor assimilate remobilization and loss of sink activity lead to unused assimilate reserves in the stem and poor grain filling, resulting in low grain weight in wheat and rice (Gooding et al. 2003; Kim et al. 2011; Wang et al. 2012). During grain filling, high temperature stress also affects grain quality by altering protein and starch composition in rice (Lin et al. 2010) and wheat (Hurkman and Wood 2011). High-temperature stress is also known to decrease antioxidants in the rice grains (Britz et al. 2007). Apart from the temperature, altered precipitation (also climatic variability parameter) may affect agricultural productivity. In India intense rainfall events together with reduced number of rainy days has been noted during the latter half of last 50 years and these predominantly affected the rice yield in rainfed area during 1966–2002 (Auffhammer et al. 2012). The drought has been found to have much greater impact than extreme rainfall events. Drought stress also affects different phenological stages of the crop. During early vegetative phase it limits shoot elongation, leaf area, and tillering, possibly by decreased CO_2 assimilation in the leaf and slow nutrient mobilization to growing tissues due to decreased stomatal conductance, transpiration, and low relative water content (Barnabas et al. 2008; Lipiec et al. 2013; Aslam et al. 2013). However, drought stress during the early reproductive stage leads to pollen and spikelet abortion, thus decreasing grain numbers in rice and wheat (Dolferus et al. 2011; Kato et al. 2008). At anthesis, drought stress increases pollen and ovary abortion, induces poor anther dehiscence and restricts panicle exertion due to shortened peduncle length, resulting in higher spikelet sterility in rice, wheat and maize (Rang et al. 2011; Powell et al. 2012; Aslam et al. 2013). The terminal drought at grain filling stage results in early senescence with shorter grain-filling duration and low green flag-leaf area persistence in wheat and barley (Samarah 2005; Foulkes et al. 2007). An increase in protein and/or nitrogen content and change in their composition were documented in wheat under drought stress (Gooding et al. 2003; Ozturk and Aydin 2004). In another study by (Serraj et al. 2011), total plant water use was reduced due to increased CO_2 because stomatal conductance was reduced with increased levels of atmospheric CO_2, thereby reducing the amount of water lost to the atmosphere. Therefore, the elevated atmospheric CO_2 stimulated the root growth and thus increased plant water acquisition and also reduced plant water loss by closing stomata. These two phenomena typically enhance plant water-use efficiency, even under conditions of less-than-optimal soil water content. Besides these physiological traits, [eCO2] promotes leaf area and leaf mass per unit area, thus providing increased surface area for photosynthesis and improves tillering and shoot biomass in wheat (Thilakarathne et al. 2013; Bourgault et al. 2013). In addition, a slight advancement of flowering time in a majority of C3 crops under field conditions has been documented (Craufurd and Wheeler 2009). However, during anthesis stage, [eCO2] has been reported to raise tissue temperature by lowering the critical temperature threshold, resulting in higher spikelet sterility in rice (Matsui et al. 1997). At the grain-filling stage, [eCO2] improves different yield component traits, including spikelet density, 1000- grain weight, panicle density and harvest index in rice and wheat (Shimono et al. 2009; Madan et al. 2012; Högy et al. 2013). The individual effects of these climatic variables but very few studies have looked at the interactions between different aspects of climate change. These climate variables may behave differently in presence of each other.

14.4.2 High Temperature and Drought Interaction Impacts

High temperature often accompanies or interacts with drought stress and as a result of low precipitation the yield of three major cereals (rice, wheat and maize) declined significantly in the northern part of China (Zhang and Huang 2012). Wassmann et al. (2009) documented the occurrence drought and high-temperature stress during the critical developmental stages (flowering and early grain filling) in the major rice-growing areas of Asia. The impact of combined stress has a significantly more detrimental effect on cereal growth and productivity than either high temperature or drought stress applied individually (Shah and Paulsen 2003; Altenbach et al. 2003; Prasad et al. 2011). Hence, drought exacerbated by high-temperature stress or vice versa will have serious implications for future cereal production in arid and semi-arid regions.

14.4.3 Effect on Water Use Efficiency

Photosynthesis (CO_2 uptake) and transpiration (H_2O loss) processes share a common pathway and linkage between these processes is tightly regulated by the extent of stomatal regulation. High-temperature and drought stress both limit the net photosynthetic rate and a decrease in net photosynthesis has been attributed to either stomatal or non-stomatal limitation (Hassan et al. 1998; Yordanov et al. 1999, 2000; Shangguan et al. 1999). In cereals, a few controlled growth chamber studies have quantified the impact of combined drought and high-temperature stress on photosynthesis and transpiration. In wheat, combined stress together resulted in a 66–93 % decrease in photosynthetic rate compared with non-stress conditions in three different experiments and a higher decline was higher than the two stresses imposed independently. Moreover, 24 % increase in water use efficiency (WUE) was observed with drought stress at intermediate/suboptimal, but it decreased by 34 % at higher temperature (Shah and Paulsen 2003; Hassan 2006; Prasad et al. 2011).

14.4.4 Effect on Agronomic Efficiency: Grain Yield and Yield Components

Drought stress causes a decrease in yield primarily by a decrease in grain number either due to a decreased amount of biomass accumulated prior to flowering or direct damage to pollen and ovule viability during the reproductive stage. According to Powell et al. (2012), post-anthesis drought has a strong negative effect on grain filling and grain weight due to an imbalance in grain-filling rate and duration dynamics. The decrease in spike number, more aborted spikes has been observed by El Soda et al. (2010) due to drought stress at various reproductive phases. Similarly high-temperature stress influences the yield and yield attributes by shortening the growth phases. Shortening of grain filling period due to increased temperature stress caused lower grain weight, increased pollen sterility, poor anther dehiscence, which ultimately had negative effect on grain yield reported as by Barnabas et al. (2008) and Powell et al. (2012). These findings were further supported by Samra and Singh (2005) by documenting the decreased wheat production by more than four million tonnes in India because in March the temperatures were higher in IGP by 3–6 °C, which was equivalent to 1 °C/day leading to wheat crop maturity earlier by 10–20 days. According to Prasad et al. (2011), the decreases in leaf photosynthesis was more at elevated temperature as compared to drought alone. However, leaf photosynthetic rate was lowest as result of combined effect of high temperature and drought. Overall decrease of 48–56 % in spikelet fertility, grain numbers and grain yield was observed. Further, they reported that higher temperature decreased in grain weight and the grain yield by 25 % and 56 %, respectively. While the respective decrease due to drought was 48 and 35 % indicated the sensitivity of grain numbers towards higher temperature while the grain weight was more affected by drought stress. The combined effects of higher temperature and drought were greater than individual effects of higher temperature or drought alone for leaf chlorophyll content, grain numbers and harvest index. However, according to van Ittersum et al. (2003), climate change would have positive effects in some regions of the world, especially in Mediterranean environments where lower temperatures is the main plant growth limiting factor. Global warming could potentially have positive effects on crop yield in these regions but on the other hand climate change will have a negative effect on yields of irrigated crops across regions, both due to increase in temperature and changes in water availability. Venkateswarlu and Shanker (2012) predicted that there would be change in rainfall variability and reduction in number of rainy days. Similarly, Changes in rainfall patterns both with respect to amount and distribution of rainfall can have both negative and positive effects on agricultural production. In semi-arid environments, higher rainfall will increase production where lower rainfall is the cause of lower production. However, in high rainfall zones, more rain can increase soil water logging and nutrient leaching which can reduce crop growth. Further, rainfall patterns with a predicted increase in variability are projected to result in increased amounts on fewer rainy days leading to flash floods on the one hand and on and prolonged dry spells on the other, thereby jeopardizing the completion of crop cycles. Drought or water scarcity, owing to climate change has been a recurring challenge in many parts of the world and this is identified as the most important factor limiting plant growth, development, and cereal productivity (Mishra and Singh 2010; Ahuja et al. 2010). Past data showed increase in temperature, decrease in open pan evaporation and irregular trends in rainfall. The projected averaged annual Tmax and Tmin is predicted to increase by 1.1 ± 0.5 °C, 2.5 ± 0.7 °C and 3.5 ± 0.8 °C and1.7 ± 0.5 °C, 3.0 ± 0.4 °C and 4.1 ± 0.6 °C in the year 2020, 2050 and 2080 respectively, as per the GCM studies carried out in the region. If there would be more rainfall in future, the crop yield can be increased because the yield is more sensitive to water availability than temperature. If water availability is reduced in the future, soils of high water holding capacity will be better to reduce the frequency of drought and improve the crop yields. Challinor et al. (2005), Challinor and Wheeler 2008 reported by used PRECIS and the GLAM crop model under present (1961–1990) and future (2071–2100) climate conditions hat extreme temperature has a negative effect on crop yield rather than the mean and high temperatures. Vashisht et al (2013) predicted a decreases in crop yield by 4, 32 and 61 % between the periods from 2021 to 2030,2031–2040 and 2041–2050 respectively by using DSSAT model.

14.5 High Temperature and [eCO_2] Interaction Impacts

14.5.1 Water Use Efficiency

Temperature and CO_2 are two major aspects related to climate change which affect most of the plant processes. Carbon dioxide affects crop production mainly through its direct effect on photosynthesis and stomatal physiology (Uprety et al. 2002; Shimono et al. 2013). Higher rates of photosynthesis and increased WUE with higher [CO2] have been documented by Drake et al. (1997) and Garcia et al. (1998). According to Guo et al. (2010), WUE increased by 40 and 25 % in wheat and maize crops, respectively in the treatment where atmospheric CO_2 concentration reaches nearly 600 ppm as compared to treatments without CO_2 fertilization. However, high temperature alone can negatively impact crop production directly through heat stress and indirectly through higher plant water demand due to increased transpiration (van Herwaarden et al. 1998; Lawlor and Mitchell 2000; Peng et al. 2004). However, higher [CO2] can counteract these negative effects of higher temperatures through decreased stomatal conductance, which reduces transpiration and higher leaf water potential (Garcia et al. 1998; Wall 2001).

14.6 Drought and [eCO_2] Interaction Impacts

14.6.1 Agronomic Efficiency: Grain Yield and Yield Components

Elevated carbon dioxide has a significant role in enhancing grain and biomass yield of crops. Among the cereals, C3 plants like rice and wheat have benefited more than C4 plants like maize and sorghum due to increasing [CO_2] in terms of grain yield, while all these species had a relatively lower decline in grain yield and biomass under combined drought and [eCO_2] conditions than under independent drought stress. The physiological responses, including photosynthesis and overall grain yield have a similar response to combined drought and [eCO_2] compared to independent exposure to drought. The 1000-grain weight increased with [eCO_2] across both wheat and maize, while the ameliorative effect of [eCO_2] was clearly more evident with maize than with wheat with a 19 % increase compared to independent drought stress exposure. The reduction in grain yield with combined drought and [eCO_2] conditions was possibly due to an increased number of panicles per plant in rice and a lesser decrease in grain number per ear or on an area basis in both wheat and maize. Harvest index, however, decreased across both C3 and C4 cereals with [eCO_2], mainly due to a greater proportion of increase in vegetative biomass, and the decline was greater with combined drought and [eCO_2], particularly with sorghum, due to a greater decrease in grain yield. In wheat, duration of grain-filling stage and rate, which are influenced by [eCO_2] and drought are negatively correlated and influence final grain weight (Li et al. 2001). It was observed that under drought stress, the [eCO_2] increased final grain weight in the upper and lower sections of the main stem spike, while under well watered conditions, [eCO_2] increased grain weight in the lower sections of the main stem spike. The increase in grain weight caused by [eCO_2] was attributed to a faster grain-filling rate (Li et al. 2000). Further they reported, grains farther from the rachis or nearest to the rachis were affected proportionately more than those at the center (Li et al. 2001). Also, drought and [eCO_2] had a significant impact on later-formed tillers than on main stem spikes (Li et al. 2000). Similarly, soybean had significantly higher photosynthetic rate (7–16 %) under normal and high soil moisture levels. Total dry matter production was increased significantly from 74.3 to 137.3 % when plants were grown under elevated CO_2 with normal soil moisture level (Madhu and Hatfield 2014). Similarly, open top chamber studies conducted by Kimball (2016) revealed that the biomass and yield were increased by FACE in all C3 species, but not in C4 species except when water was limiting. Yields of C3 grain crops were increased on average about 19%.

14.7 Adaption/Mitigation Strategies to Climate Variability

Agriculture, as also among one of the major divers of climate change is also most likely to be affected which will threaten food security globally. So Adaption and mitigation strategies in this sector are urgently needed. Adaptation of agriculture to climate change has been broadly defined as "any response that improves an outcome" (Reilly and Schimmelpfennig 2000). Adaptation measures also include establishment of disaster risk management plans and risk transfer mechanisms, such as crop insurance and diversified livelihood systems (Reilly and John 1996). Similarly, IPCC (2007) defined mitigation as the "technological changes and substitution that reduce resource inputs and emissions per unit of output". Literature supports that Agriculture and allied fields like forestry, fisheries/aquaculture provide a significant potential for GHG mitigation. The benefits of mitigation activities carried out today will be evidenced in several decades because of the long residence time of GHGs in the atmosphere, whereas the effects of adaptation measures should be apparent immediately or in the near future (Kumar and Parikh 2001). Key adaptation and mitigation strategies are mentioned below.

14.7.1 Genetic Modifications of Crops

Development of crop varieties capable of withstanding high temperature stress, drought stress, pest load imposed by changing climate, needs a greater attention. There will be need to explore several adaptive traits including behavior of different growth regulators under different stress conditions. In addition to these, varieties with high nutrient and radiation use efficiency and also the varieties that can tolerate coastal salinity and salt water inundation are needed. Projected rise in temperature and change in precipitation pattern will reduce nitrogen use efficiency under climate change scenarios through volatilization and leaching losses. So possibly farmers have to apply more nitrogenous crops. Along with this, because of added effect of elevated Carbon dioxide on growth of crops, nitrogen requirement of the crop is likely to increase. So, more efficiency of roots are required for absorption of water and nutrients. Keeping all these points in view, future breeding efforts should be focused on improvement in germplasm of important crops to tolerate heat-stress, drought stress. This would require a great breeding efforts depending upon collection, conservation and distribution of appropriate crop genetic material. In India, considerable progress has been made in the genetic dissection of flowering time, inflorescence architecture, and temperature and drought tolerance in certain model plant system and by comparative genomics in crop plants. Transgenic seeds have several-fold higher germination potential and robust root system in terms of root biomass and length.

14.7.2 Change in Crop Management Practices

Little changes in climatic parameters can be managed by crop management practices like altering the sowing time, plant spacing and supply of inputs. By altering the date of planting, flowering can be avoided to coincide with hottest period (Gadgil 1995). Simulation studies by Jalota et al (2013) also predicted improvement in rice and wheat yield by shifting the transplanting date from June 20 to July 11 in rice and shifting sowing time from Nov 5 to November 26 in wheat. This improvement in rice yield will be 3, 11 and 12 % in midcentury and 3, 14 and 18 % in End century by shifting transplanting date of rice from June 20 to June 27, July 5 and July 11, respectively. The improvement in wheat yield was 69, 126 and 140 % in MC; and 47, 100 and 127 % in EC by shifting planting date of wheat from November 5 to November 12, November 19 and November 26, respectively under A2 and B1 scenarios. The DSSAT model used but Ashish et al 2013 also predicted that Planting of wheat up to November 25 till the years 2030–2031 seems to be helpful to mitigate the climate change effect under Punjab condition. The shift in planting time can be more beneficial in arid and semi-arid regions to decrease the negative effect of climate variability, whereas in the case of rainfed agriculture closer spacing of the crop may help in quick cover of the soil surface and hence reduce the evaporation losses from the soil surface. When the crops are grown primarily on stored soil water, wide rows and low plant populations are highly desirable. A shift from sole crop rotation to a diversified and integrated farming system is also helpful. Horticulture and agroforestry need to be given more encouragement.

14.7.3 Introduction of Legume in the Cropping System

Introduction of legume crop in prevailing crop rotation can also be helpful in future scenarios o climate change. Grain-legume intercrops have many benefits such as better use of resources, weed management, pest and disease reductions, increased protein content of cereals, reduced N leaching as compared to sole cropping systems. This will helpful in cost management and soil fertility management. Establishment of seed banks of major crops would be in highly helpful in major disaster management and extreme climatic events in future changed and unpredictable environments.

14.8 Rain Water Harvesting

Rain water harvesting is simply capturing of rain where it rains and this stored water may be used for many purposes. Climate change will likely to alter the precipitation pattern. There may be more amount of rain in one region and others may face the moderate to severe drought conditions. It is evidenced from the literature that prolonged aridity is associated with human migration to safer and productive areas.

Rain water harvesting has a potential for population stability. In India. Population developed in Thar dessert is an example for this. Indeed, the post -Sarswati society is essentially a rain water harvesting society. As an adaption to climate change, rain water harvesting is more important in present scenario, as reported by Jackson et al (2001).

14.9 Increasing Soil Carbon Storage

Soil carbon sequestration is also one of the most important options to reduce gaseous emission to atmosphere. By increasing carbon concentrations in the soil through better agriculture management practices, reduce the CO_2 emission to atmosphere. Reforestation can also to sequester atmospheric CO_2 into biomass. Therefore, the forest land or land kept for tree plantation should not be used or other purposes. Stubble burning should be stopped. In Indian Punjab, rice residue burning is practiced on large scale because not efficient technique was available for rice residue management. But now "Happy seeder" is recommended by Punjab Agriculture University for Punjab farmers. By using this machine farmers can sow wheat in standing rice stubbles. However for its large scale adoption, high cost of machine and its availability, problem of choking while sowing are obstacle. These problems need to be rectify.

14.10 Reducing GHG Emissions Related to Agriculture

Continuously flooded rice fields are responsible for production, oxidation and transport of methane gas which is major GHG gas. This emission can be reduced by practicing alternate wetting and drying (AWD) instead of continuous flooding. It can be further reduced by irrigating rice field on the basis of soil matric potential. Research data has shown that by following such practices there is no significant reduction in crop production. On the other hand, drainage may lead to increase in emission of nitrous oxide. These management practices have some potential in irrigated areas but it is difficult to follow in low land rainfed rice. Nitrous oxide emissions can be reduced by following appropriate crop management practices which lead to increase in nitrogen use efficiency and yield. The emission of gases can be reduced up to some extent by applying ammonium fertilizer in wet land crop and nitrate fertilizers in aerobic conditions. Nitrification inhibitors can be used. Neem cake is alternate option to mitigate nitrous oxide emission from the soil. Improved management of livestock diet can be helpful in reducing methane emission. Increase in area under agroforestry and biofuel could mitigate GHG emissions.

14.11 Conclusion

Climate change is now a reality, many people prefer to call it as climate uncertainty instead of climate change, which in fact, is much more complex phenomenon from agricultural point of view. The seasons are no longer identifiable as these used to be previously. Sometimes, these prolong or get shortened replacing the other seasons. The increase in climate-extremes in precipitation, temperature, etc has already started taking a toll on the agricultural production as is evident from one of the most productive region in IGPs of India i.e. Punjab, where the wheat productivity is affected considerably due to elevated temperature conditions over the last few years. The GHG emissions are now evident even to a common man, who feels chocked even to respire; the respiratory diseases are on the rise; heat and cold waves taking a toll on the lives of people every year. Most recently the heat wave during 2015 took a heavy toll in Pakistan region when about 1500 people died because of elevated heat conditions.

It is the time now that we must make strategies which include not only mitigation and adaptation measures in agriculture only but in all the sectors like industry, domestic, transport, etc so that GHG emissions are reduced and at the same time technologies are developed to mitigate the ill effects of climate uncertainty on agriculture, industry and domestic sectors. Industry and domestic sectors can easily adapt to the changing climatic conditions but for agriculture sector it is not so easy. A lot still needs to be done especially in formulating strategies to mitigate such effects. For example, the regions for various crops need to be specified depending upon their feasibility. The plant breeders have to play a big role in redefining the characters of crop plants through genetic engineering so as to develop cultivars to grow efficiently in newer climatic conditions. The cultivars for biofuel crops need to be developed as an alternate to fossil fuel energy sources. It is thus a collective effort on the part of plant breeders, soil scientists, agronomists, plant protection scientists and food technologists to work together to develop strategies for mitigating the ill effects of climate change in agricultural sector.

References

Ahuja, I., R.C. de Vos, A.M. Bones, and R.D. Hall. 2010. Plant molecular stress responses face climate change. *Trends in Plant Science* 15: 664–674.

Altenbach, S.B., F.M. Dupont, K.M. Kothari, R. Chan, E.L. Johnson, and D. Lieu. 2003. Temperature, water and fertilizer influence the timing of key events during grain development in a US spring wheat. *Journal of Cereal Science* 37: 9–20.

Aslam, M., M.S.I. Zamir, I. Afzal, M. Yaseen, M. Mubeen, and B.A. Shoai. 2013. Drought stress, its effect on maize production and development of drought tolerance through potassium application. *Cercetari agronomice în Moldova* 46: 99–114.

Auffhammer, M., V. Ramanathan, and J.R. Vincent. 2012. Climate change, the monsoon, and rice yield in India. *Climatic Change* 111: 411–424.

Barnabas, B., K. Jäger, and A. Fehér. 2008. The effect of drought and heat stress on reproductive processes in cereals. *Plant, Cell and Environment* 31: 11–38.

Bourgault, M., M.F. Dreccer, A.T. James, and S.C. Chapman. 2013. Genotypic variability in the response to elevated CO_2 of wheat lines differing in adaptive traits. *Functional Plant Biology* 40: 172–184.

Britz, S.J., P.V.V. Prasad, R.A. Moreau, L.H. Allen, D.F. Kremer, and K.J. Boote. 2007. Influence of growth temperature on the amounts of tocopherols, tocotrienols, and γ-oryzanol in brown rice. *Journal of Agricultural and Food Chemistry* 55: 7559–7565.

Challinor, A.J., and T.R. Wheeler. 2008. Crop yield reduction in the tropics under climate change: Processes and uncertainties. *Agricultural and Forest Meteorology* 148: 343–356.

Challinor, A.J., T.R. Wheeler, P.Q. Craufurd, and J.M. Slingo. 2005. Simulation of the impact of high temperature stress on annual crop yields. *Agricultural and Forest Meteorology* 135: 180–189.

Craufurd, P.Q., and T.R. Wheeler. 2009. Climate change and the flowering time of annual crops. *The Journal of Experimental Botany* 60: 2529–2539.

Dolferus, R., X. Ji, and R.A. Richards. 2011. Abiotic stress and control of grain number in cereals. *Plant Science* 181: 331–341.

Drake, B.J., M.A. Gonzalez-Meler, and S.P. Long. 1997. More efficient plants: A consequence of rising atmospheric CO2? *Annual Review of Plant Physiology and Plant Molecular Biology* 48: 609–639.

Dhorde, A.G., and N.R. Patel. 2016. Spatio-temporal variation in terminal drought over western India using dryness index derived from long-term MODIS data. *Ecological Informatics* 32: 28–38.

El Soda, M., S.S. Nadakuduti, K. Pillen, and R. Uptmoor. 2010. Stability parameter and genotype mean estimates for drought stress effects on root and shoot growth of wild barley pre introgression lines. *Molecular Breeding* 26: 583–593.

Erenstein, O. 2009. *Zero tillage in the rice-wheat systems of the Indo-Gangetic Plains*. IFPRI Discussion Paper 00916. www.ifpri.org/millionsfed.

FAO. 2007. *Conservation agriculture website*. Available online at http://www.fao.org/ag/ca.

Foulkes, M.J., R. Sylvester-Bradley, R. Weightman, and J.W. Snape. 2007. Identifying physiological traits associated with improved drought resistance in winter wheat. *Field Crops Ressearch* 103: 11–24.

Gadgil, D. 1995. Climate change and agriculture: An Indian perspective. *Current Science* 69(8): 649–659.

Garcia, R.L., S.P. Long, G.W. Wall, C.P. Osborne, B.A. Kimball, G.Y. Nie, P.J. Pinter, R.L. Lamorte, and F. Wechsung. 1998. Photosynthesis and conductance of spring-wheat leaves: Field response to continuous free-air atmospheric CO2 enrichment. *Plant, Cell and Environment* 21: 659–669.

Gooding, M.J., R.H. Ellis, P.R. Shewry, and J.D. Schofield. 2003. Effects of restricted water availability and increased temperature on the grain filling, drying and quality of winter wheat. *Journal of Cereal Science* 37: 295–309.

Guo, R., Z. Lin, X. Mo, and C. Yang. 2010. Responses of crop yield and water use efficiency to climate change in the North China Plain. *Agricultural Water Management* 97: 1185–1194.

Hassan, I.A. 2006. Effects of water stress and high temperature on gas exchange and chlorophyll fluorescence in *Triticum aestivum* L. *Brief Communication Photosynthetica* 44: 312–315.

Hassan, I., J. Bendert, and H. Weigel. 1998. Effects of O_3 and water stress on growth and physiology of Egyptian cultivars of tomatoes. *Gartenbauwissenschaft* 76: 122–135.

Högy, P., C. Poll, S. Marhan, E. Kandeler, and A. Fangmeier. 2013. Impacts of temperature increase and change in precipitation pattern on crop yield and yield quality of barley. *Food Chemistry* 136: 1470–1477.

Hurkman, W.J., and D.F. Wood. 2011. High temperature during grain fill alters the morphology of protein and starch deposits in the starchy endosperm cells of developing wheat (*Triticumaestivum* L.) grain. *Journal of Agricultural and Food Chemistry* 59: 4938–4946.

IPCC (Intergovernmental Panel on Climate Change). 1996. Climate change 1995: Summary for policy makers. p. 1–7. In ed. J.T. Houghton et al. Cambridge/New York: Cambridge University Press.

IPCC (Intergovernmental Panel on Climate Change). 2001. *Climate change 2001: The scientific basis—contribution of working group I to the third assessment report of the Intergovernmental Panel of Climate Change*. Cambridge/New York: Cambridge University Press.

IPCC (Intergovernmental Panel on Climate Change). 2007. In: *Climate change 2007: Impacts, adaptation and vulnerability. Contribution of work group II to the fourth assessment report of the Intergovernmental Panel on Climate*. Cambridge: Cambridge University Press.

IPCC (Intergovernmental Panel on Climate Change). 2013. *Working group I contribution to the IPCC fifth assessment report climate change 2013: The physical science basis, Summary for Policymakers*. www.climatechange2013.org/images/uploads/WGIAR5SPM_Approved27Sep2013.pdf.

Jackson, R.B., S.R. Carpenter, C.N. Dahm, D.M. Mcknight, R.J. Naiman, S.L. Postel, and S.W. Running. 2001. Water in a changing world. *Ecological Applications* 11: 1027–1045.

Jagadish, S.V.K., P.Q. Craufurd, and T.R. Wheeler. 2007. High temperature stress and spikelet fertility in rice (*Oryza sativa* L.). *The Journal of Experimental Botany* 58: 1627–1635.

Jagadish, S.V.K., J. Cairns, R. Lafitte, T.R. Wheeler, A.H. Price, and P.Q. Craufurd. 2010. Genetic analysis of heat tolerance at anthesis in rice. *Crop Science* 50: 1633–1641.

Jalota, S.K., H. Kaur, S. Kaur, and B.B. Vashisht. 2013. Impact of climate change scenarios on yield, water and nitrogen –balance and use efficiency of rice – wheat cropping system Agric. *Water Management* 116: 29–38.

Kato, Y., A. Kamoshita, and J. Yamagishi. 2008. Preflowering abortion reduces spikelet number in upland rice under water stress. *Crop Science* 48: 2389–2395.

Kim, J., J. Shon, C.K. Lee, W. Yang, Y. Yoon, W.H. Yang, G.Y. Kim, and B.W. Lee. 2011. Relationship between grain filling duration and leaf senescence of temperate rice under high temperature. *Field Crops Research* 122: 207–213.

Kimball, B.A. 2016. Crop responses to elevated CO2 and interactions with H2O, N, and temperature. *Current Opinion in Plant Biology* 31: 36–43.

Klein, A.M., I. Steffan-Dewenter, and T. Tscharntke. 2006. Rain forest promotes trophic interactions and diversity of trap-nesting hymenoptera in adjacent agroforestry. *The Journal of Animal Ecology* 75: 315–323.

Kumar, K.S.K., and J. Parikh. 2001. Indian agriculture and climate sensitivity. *Global Environmental Change* 11(2): 147–154.

Ladha, J.K., V. Kumar, M.M. Alam, S. Sharma, M. Gathala, P. Chandana, Y.S. Saharawat, and V. Balasubramanian. 2009. Integrating crop and resource management technologies for enhanced productivity, profitability, and sustainability of the rice-wheat system in South Asia. In *Integrated crop and resource management in the rice–Wheat system of South Asia*, ed. J.K. Ladha, Singh Yadvinder, O. Erenstein, and B. Hardy, 69–108. Philippines: International Rice Research Institute.

Lawlor, D.W., and R.A.C. Mitchell. 2000. Crop ecosystem responses to climatic change: Wheat. In *Climate change and global crop productivity*, ed. K.R. Reddy and H.F. Hodges, 57–80. Wallingford: CAB International.

Li, A.G., Y.S. Hou, G.W. Wall, A. Trent, B.A. Kimbal, and P.J. Pinter. 2000. Free air carbon dioxide enrichment and drought stress effects on grain filling rate and grain filling duration in spring wheat. *Crop Science* 40: 1263–1270.

Li, A., Y. Hou, and A. Trent. 2001. Effects of elevated atmospheric CO2 and drought stress on individual grain filling rates and durations of the main stem in spring wheat. *Agricultural and Forest Meteorology* 106: 289–301.

Lin, C.J., C.Y. Li, S.K. Lin, F.H. Yang, J.J. Huang, Y.H. Liu, and H.S. Lur. 2010. Influence of high temperature during grain filling on the accumulation of storage proteins and grainquality in rice (*Oryza sativa* L.). *Journal of Agricultural and Food Chemistry* 58: 10545–10552.

Lipiec, J., C. Doussan, A. Nosalewicz, and K. Kondracka. 2013. Effect of drought and heatstresses on plant growth and yield: A review. *International Agrophysics* 27: 463–477.

Madan, P., S.V.K. Jagadish, P.Q. Craufurd, M. Fitzgerald, T. Lafarge, and T.R. Wheeler. 2012. Effect of elevated CO_2 and high temperature on seed-set and grain quality of rice. *The Journal of Experimental Botany* 63: 3843–3852.

Madhu, M., and J. Hatfield. 2014. Interaction of carbon dioxide enrichment and soil moisture on photosynthesis, transpiration, and water use efficiency of soybean. *Agricultural Sciences* 5: 410–429.

Matsui, T., O.S. Namuco, L.H. Ziska, and T. Horie. 1997. Effects of high temperature and CO_2 concentration on spikelet sterility in indica rice. *Field Crops Research* 51: 213–219.

Mishra, A.K., and V.P. Singh. 2010. A review of drought concepts. *Journal of Hydrology* 391: 202–216.

Ozturk, A., and F. Aydin. 2004. Effect of water stress at various growth stages on some quality characteristics of winter wheat. *Journal of Agronomy and Crop Science* 190: 93–99.

Peng, S., J. Huang, J.E. Sheehy, R.C. Laza, R.M. Visperas, X. Zhong, G.S. Centeno, G.S. Khush, and K.G. Cassman. 2004. Rice yields decline with higher night temperature from global warming. *Proceedings of the National academy of Sciences of the United States of America* 101: 9971–9975.

Powell, N., X. Ji, R. Ravash, J. Edlington, and R. Dolferus. 2012. Yield stability for cereals in a changing climate. *Functional Plant Biology* 39: 539–552.
Prasad, P.V.V., S.R. Pisipati, I. Momčilović, and Z. Ristic. 2011. Independent and combined effects of high temperature and drought stress during grain filling on plant yield and chloroplast EF-Tu expression in spring wheat. *Journal of Agronomy and Crop Science* 197: 430–441.
Rang, Z.W., S.V.K. Jagadish, Q.M. Zhou, P.Q. Craufurd, and S. Heuer. 2011. Effect of high temperature and water stress on pollen germination and spikelet fertility in rice. *Environmental and Experimental Botany* 70: 58–65.
Reilly, Y. 1996. Agriculture in a changing climate: Impacts and adaptations. In *Climate change 1995: Impacts, adaptations, and mitigation of climate change: Scientific- technical analyses*, ed. Robert Watson, Zinyowera Manufu, Moss Richard, and Dokken David, 427–468. Cambridge: Cambridge University Press.
Reilly, J., and D. Schimmelpfennig. 2000. Irreversibility, uncertainty, and learning: Portraits of adaption to long-term climate change. *Climatic Change* 45: 253–278.
Samarah, N.H. 2005. Effects of drought stress on growth and yield of barley. *Agronomy for Sustainable Development* 25: 145–149.
Samra, J.S., and G. Singh. 2005. *Heat wave of March 2004. Impact on agriculture*. Pusa, New Delhi: ICAR – Indian Council of Agricultural Research.
Serraj, R., K.L. Mcnally, I. Slamet-Loedin, A. Kohli, S.M. Haefele, G. Atlin, and A. Kumar. 2011. Drought resistance improvement in rice: An integrated genetic and resource management strategy. *Plant Production Science* 14: 1–14.
Shah, N.H., and G.M. Paulsen. 2003. Interaction of drought and high temperature on photosynthesis and grain-filling of wheat. *Plant and Soil* 257: 219–226.
Shangguan, Z., M. Shao, and J. Dyckmans. 1999. Interaction of osmotic adjustment and photosynthesis in winter wheat under soil drought. *Journal of Plant Physiology* 154: 753–758.
Shimono, H., M. Okada, Y. Yamakawa, H. Nakamura, K. Kobayashi, and T. Hasegawa. 2009. Genotypic variation in rice yield enhancement by elevated CO_2 relates to growth before heading, and not to maturity group. *The Journal of Experimental Botany* 60: 523–532.
Shimono, H., H. Nakamura, T. Hasegawa, and M. Okada. 2013. Lower responsiveness of canopy evapotranspiration rate than of leaf stomatal conductance to open-air CO_2 elevation in rice. *Global Change Biology* 19: 2444–2453.
Takeoka, Y., K. Hiroi, H. Kitano, and T. Wada. 1991. Pistil hyperplasia in rice spikelets as affected by heat-stress. *Sexual Plant Reproduction* 4: 39–43.
Thilakarathne, C.L., S. Tausz-Posch, K. Cane, R.M. Norton, M. Tausz, and S. Seneweera. 2013. Intraspecific variation in growth and yield response to elevated CO_2 in wheat depends on the differences of leaf mass per unit area. *Functional Plant Biology* 40: 185–194.
Timsina, J., M.L. Jat, and K. Majumdar. 2010. Nutrient management research priorities in rice-maizesystems of South Asia. *Better Crops South Asia* 335(1): 65–82.
Uprety, D.C., N. Dwivedi, V. Jain, and R. Mohan. 2002. Effect of elevated carbon dioxide concentration on the stomatal parameters of rice cultivars. *Photosynthetica* 40: 315–319.
Van Herwaarden, A.F., G.D. Farquhar, J.F. Angus, R.A. Richards, and G.N. Howe. 1998. 'Haying-off', the negative grain yield response of dryland wheat to nitrogen fertiliser. I. Biomass, grain yield, and water use. *Australian Journal of Agricultural Research* 49: 1067–1081.
Van Ittersum, M.K., S.M. Howden, and S. Asseng. 2003. Sensitivity of productivity and deep drainage of wheat cropping systems in a Mediterranean environment to changes in CO_2, temperature and precipitation. *Agriculture Ecosystems and Environment* 97: 255–273.
Vashisht, B.B., D.J. Mulla, S.K. Jalota, Samanpreet Kaur, Harsimran Kaur, and Sher Singh. 2013. Productivity of rainfed wheat as affected by climate change scenario in northeastern Punjab, India. *Regional Environmental Change*. doi:10.1007/s10113-013-0412.
Venkateswarlu, B., and A.K., Shanker. 2012. Dryland agriculture: Bringing resilience to crop production under changing climate. In *Crop stress and its management: Perspectives and strategies*, 19–44. Springer Netherlands.

Wahid, A., S. Gelani, M. Ashraf, and M.R. Foolad. 2007. Heat tolerance in plants: An overview. *Environmental and Experimental Botany* 61: 199–223.
Wall, G.W. 2001. Elevated atmospheric CO_2 alleviates drought stress in wheat. *Agriculture, Ecosystems & Environment* 87: 261–271.
Wang, X., J. Cai, F. Liu, M. Jin, H. Yu, D. Jiang, B. Wollenweber, T. Dai, and W. Cao. 2012. Pre-anthesis high temperature acclimation alleviates the negative effects of post-anthesis heat stress on stem stored carbohydrates remobilization and grain starch accumulation in wheat. *Journal of Cereal Science* 55: 331–336.
Wassmann, R., S.V.K. Jagadish, K. Sumfleth, H. Pathak, G. Howell, A. Ismail, R. Serraj, E. Redone, R.K. Singh, and S. Heuer. 2009. Regional vurnerability of climate change impacts on asian Rice production and scope adaptation. *Advance in Agronomy* 102: 91–133.
Yan, Z., P.D. Jones, T.D. Davies, A. Moberg, H. Bergström, D. Camuffo, C. Cocheo, M. Maugeri, G.R. Demarée, T. Verhoeve, E. Thoen, M. Barriendos, R. Rodríguez, J. Martín-Vide, and C. Yang. 2002. Trends of extreme temperatures in Europe and China based on daily observations. In *Improved understanding of past climatic variability from early daily European instrumental sources*, ed. D. Camuffo and P. Jones. Dordrecht: Springer.
Yordanov, I., V. Velikova, and T. Tsonev. 1999. Influence of drought, high temperature, and carbamide cytokinin 4-PU-30 on photosynthetic activity of bean plants. 1. Changes in chlorophyll fluorescence quenching. *Photosynthetica* 37: 447–457.
Yordanov, I., V. Velikova, and T. Tsonev. 2000. Plant responses to drought, acclimation, and stress tolerance. *Photosynthetica* 38: 171–186.
Zhang, T., and Y. Huang. 2012. Impacts of climate and inter annual variability on cereal crops in china from 1980 to 2008. *Journal of the Science of Food and Agriculture* 92: 1643–1652.

Chapter 15
Models to Study Phosphorous Dynamics Under Changing Climate

Waqas Ijaz, Mukhtar Ahmed, Fayyaz-ul-Hassan, Muhammad Asim, and M. Aslam

Contents

Abstract Phosphorus the macronutrient, a component of nucleic acid and helpful for grain developmental phases (reproductive growth) of crops. It is also component of energy carriers like ATP, ADP, NADPH and FADPH which provide energy for

W. Ijaz (✉) • Fayyaz-ul-Hassan
Department of Agronomy, Pir Mehr Ali Shah Arid Agriculture University, Rawalpindi, Pakistan
e-mail: waqas.surani@yahoo.com; drsahi63@gmail.com

M. Ahmed
Department of Agronomy, Pir Mehr Ali Shah Arid Agriculture University, Rawalpindi, Pakistan

Department of Biological Systems Engineering, Washington State University, Pullman, WA, USA
e-mail: ahmadmukhtar@uaar.edu.pk; mukhtar.ahmed@wsu.edu

M. Asim
PARC- Institute of Advanced Studies in Agriculture, National Agricultural Research Centre, Islamabad, Pakistan
e-mail: asimnarc@gmail.com

M. Aslam
Ministry of Food and Agriculture, Islamabad, Pakistan
e-mail: aslamgill@hotmail.com

M. Ahmed, C.O. Stockle (eds.), *Quantification of Climate Variability, Adaptation and Mitigation for Agricultural Sustainability*,
DOI 10.1007/978-3-319-32059-5_15

different physiological processes. Phosphorous plays an important role in the growth, development and yield of crops. However, P causes some environmental problems like eutrophication. The importance of the element necessitate its study through modeling and distribution under changing climate. Since P is present as organic and inorganic form but their fate is different in soils. The inorganic P accounts for 35–70 % while organic form of P accounts for 30–65 % of the total P but it is dominantly available as stabilized forms like diesters. The availability of this P depends upon mineralization processes by soil biota which has dependency upon soil moisture, temperature, physiochemical properties and soil pH. However, the transformation of organic p has strong influence on the availability of P in soil. Therefore, availability of P to crop is extremely complex and its needs to be evaluated using modeling approaches. The Phosphorus Use Efficiency (PUE) for crops might be increased by understanding P-dynamics which may be done by models. The understanding of P dynamics will help to optimized balance use of P. By monitoring P for longer period of time might increase P status of soil. The use of computer models will help to modify fertilizer application which can reduce use of P but will increase PUE. The effects of high temperature, elevated CO_2 and drought on the availability of phosphorous, PUE and its dynamics could be modeled using dynamics models like APSIM, AEP or by using regression modeling approaches.

Keywords Phosphorus • Soil biota • Phosphorus use efficiency • P-Dynamics • APSIM

Abbreviations

ATP	Adenosine Tri-Phosphate
ADP	Adenosine di-phosphate
NADP	Nicotinamide adenine dinucleotide phosphate
FADP	Falvin adenine dinucleotide phosphate
APSIM	Agricultural Production System Simulator
AEP	Agriculture Ecosystem model
PUE	Phosphorus Use Efficiency
IPCC	*Intergovernmental Panel on Climate Change*
NFDC	*National Fertilizer Development Centre*
IFPRI	*International Food Policy Research Institute*
DNA	Deoxyribonucleic acid
RNA	Ribonucleic acid
PAE	P Acquisition Efficiency

15.1 Introduction

Phosphorus the major macronutrient of plant is totally different from nitrogen as it does not come from air and have less solubility in water which leads to slow movement of P to downstream (Tiessen 2008). P is the 11th most abundant element in

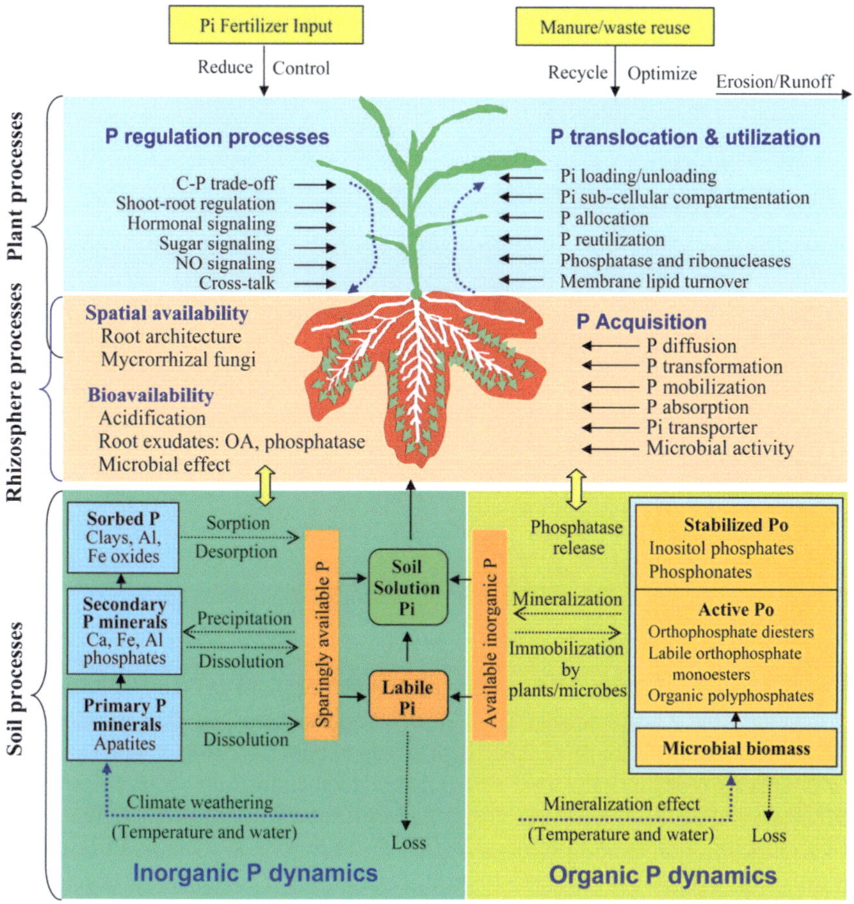

Fig. 15.1 P dynamics in the soil/rhizosphere-plant continuum (Source: Shen et al. 2011)

earth crust and its concentration reaches to 1200 mg P per kg. However, most of the soils contain 200–800 mg per kg while older soils have less compared to younger soils. The binding of P is maximum with Ca or Mg and its solubility in water becomes 0.5 mg P per liter. The decreased solubility than $Ca_3(PO_4)_2$ after weathering is due to Ca leaching which resulted to abundance of Fe and Al. Different physical and chemical reactions (mineralization, immobilizations, dissolution/precipitation and sorption/desorption) controls the availability of P to crops. P fixation is a misnomer as every chemical reaction to some extent is reversible however, since release of P is so negligible therefore, it is considered as non-significant. The organic phosphate esters and inorganic phosphates constitutes 99 % of naturally occurring P. Similarly high negative charge density due to four oxygen per P, $(PO_4)^{3-}$ resulted to its loving attraction to all positively charged ions. P is essential component of heredity material (DNA and RNA) and energy carriers like ATP. The phosphodieaster linkage maintains life on earth which is maintained by P. Since P is present as organic and inorganic form but their fate is different in soils (Turner et al. 2007). The inorganic P accounts for 35–70 % and dynamics of P in soil is further illustrated

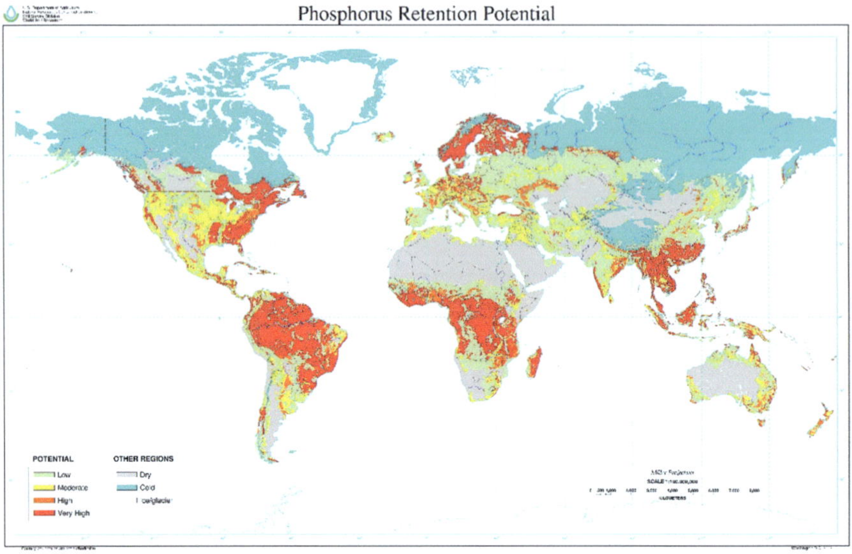

Fig. 15.2 The Phosphorus Retention Potential map (Source: USDA, NRCS 2012)

by Shen et al. (2011) (Fig. 15.1). However, organic form of P accounts for 30–65 % of the total P but it is dominantly available as stabilized forms like diesters (Condron et al. 2005). The availability of this P depends upon mineralization processes by soil biota which has dependency upon soil moisture, temperature, physiochemical properties and soil pH. However, the transformation of organic p has strong influence on the availability of P in soil (Turner et al. 2007). Therefore, availability of P to crop is extremely complex and it needs to be evaluated using modeling approaches since it is associated with P dynamics (s).

The crop productivity over the globe has strong relationship with soil fertility and most of world soils are limited in N and P hence leading to lower soil productivity. However, P deficiency is more often in old weather soils (Lynch and Brown 2011; USDA 2012) (Figs. 15.2 and 15.3). The limited availability of P might be due to several factors like its binding with Fe and Al, human activities (50 % of the agricultural soils in the world) erosion, acidification and mining of nutrients (Hartemink 2003). The developed countries have potential to increase the P status of soil however, in developing countries where P as fertilizer (World Bank 2004) use is limited might result to food security (Lynch 2007). Therefore, use of crops having high phosphorus use efficiency (PUE) and development of cropping systems with greater productivity might tackle the problem of food security (Lynch 2007). P-acquisition is the adoptive value of crop/plant to acquire P and it can be considered as trait called "Phene" (a phenotype is comprised of many distinct phenes). These "Phenes" determines the fitness of crop therefore; it needs to be considered in the domestication of crops. The "phene" which is important for P-acquisition is root hair length (Fig. 15.4). The phenes adaptability is interlinked with surrounding environments like in case of root hairs; length has strong synergism with root hair

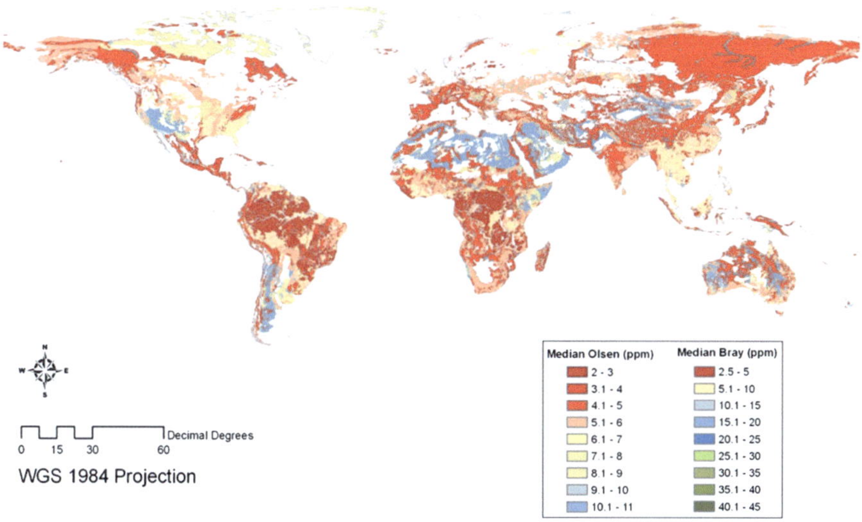

Fig. 15.3 Map of global soil Phosphorus availbility (Source: Lynch 2011)

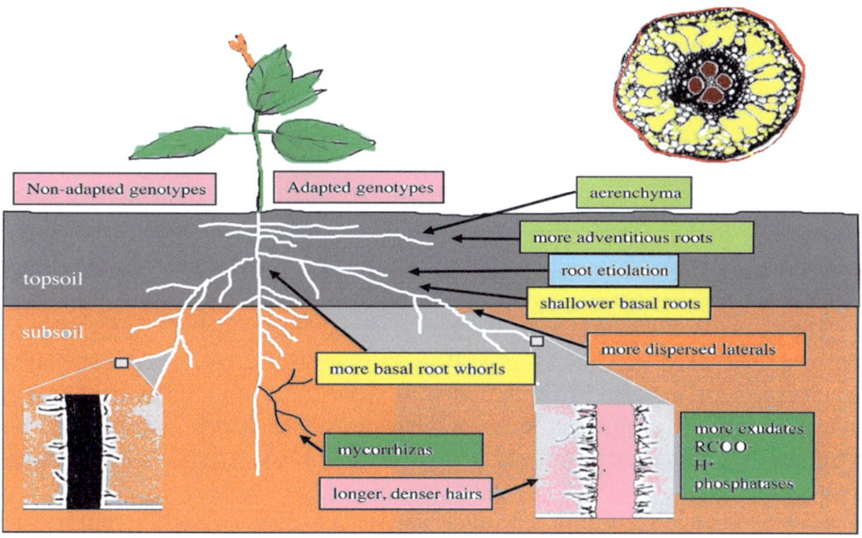

Fig. 15.4 Root phenes associated with genotypic differences in adaptation to low phosphorus (Source: Modified from Lynch 2007)

density (Ma et al. 2001a) for P acquisition. Therefore, this type of interactions needs to be considered for phenes consideration in context with surrounding environment. The adaptation of crops under P limited conditions could be achieved by identification of phenes and understanding of physiological and ecological interactions. This could also be useful for breeding P-efficient crops having high PUE under

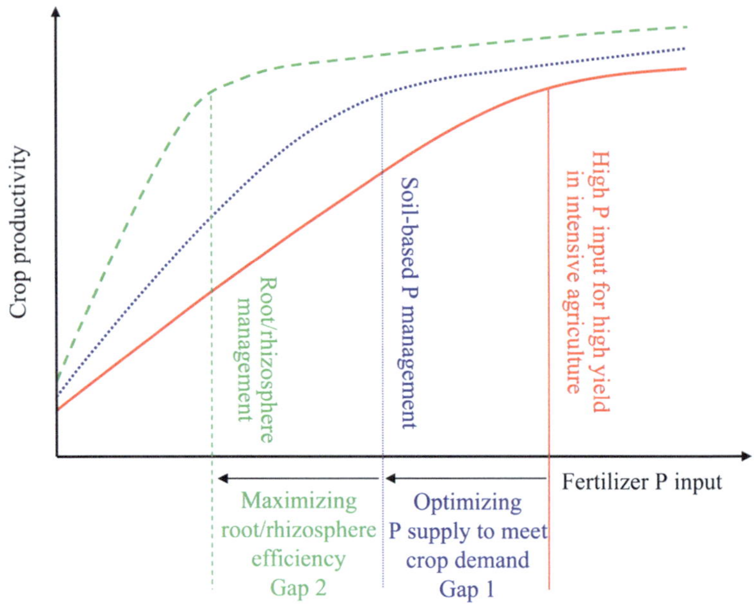

Fig. 15.5 Conceptual model of root/rhizosphere and soil-based nutrient managements for improving P-use efficiency and crop productivity in intensive agriculture (Source: Shen et al. 2011)

limited P. Since P is less mobile therefore, roots display features like mycorrhizal symbioses (Smith and Read 1997), root hair elongation and proliferation (Ma et al. 2001a, b), including rhizosphere modification through secretion of organic acids (Ryan et al. 2001), protons (Hinsinger 2001), and phosphatases (Hayes et al. 1999), and modification of root architecture to maximize P acquisition efficiency (Fig. 15.5, Lynch and Ho 2005).

The increase in R/S ratio (root to shoot) is another feature which plant adapts to compensate P-deficiency (Mollier and Pellerin 1999). Therefore, root growth might be considered to increase PUE (Manske et al. 2000, 2001). In general low P in soil is a constraint and crop adopt different strategies (P acquisition, Topsoil foraging P-solubilizing root exudates, Basal root gravitropism, and Lateral root branching, Reduced metabolic costs of soil exploration, Mycorrhizal symbioses, Phenological and morphological plasticity). The understanding of these strategies might lead to development of such crop genotypes which have high PUE under changing climate. The models might be used to study the role of root architecture in P-acquisition. The earlier results about this aspect using modeling approach indicated that root architecture have significant relationship with PAE (P acquisition efficiency). The change in root architecture resulted to change in PAE. The authors concluded that shallower root architecture might better explore P- rich upper soil layer which might increases the PAE (Yan et al. 2002).

The PUE for crops might be increased by understanding P-dynamics which is possible by using simulation modeling paradigm. The understanding of P dynamics will help to optimized balance use of P. By monitoring P for longer period of time might increase P status of soil by adopting suitable strategies. The use of computer models will help to modify fertilizer application which can reduce use of P but will increase PUE. This will save P by maintaining crop productivity (Fig. 15.4) (Shen et al. 2011). The models like Agriculture Ecosystem (AEP model) could be used to describe P dynamics on farm scale. Cassell et al. (1998) concluded in their work that simulation modeling could be used to understand complex P dynamics and to design management strategies which could achieve the goal of resource sustainability and climate protection.

Phosphorus export from agriculture is major cause of eutrophication (Cassell et al. 1998). P export pattern over time and space is dependent upon human activity, hydrology and physiochemical and biological processes those store and transform P. The ecosystem paradigm was suggested to study P dynamics (Cassell et al. 1998). The AEP model depicted P dynamics with good accuracy and describes P dynamics on farm scale driven by amount of P stored in agricultural soils and systems. Cassell et al. (1998) in their studies concluded that dynamic simulation modeling is valuable tool to study P dynamics for development of policies and management options to achieve climate sustainability. Models are of extreme importance in scientific systems (Frigg and Hartman 2006). A model is a schematic demonstration of the ideology of a system or a set of equations, which shows the performance of a system. Also, a model is "A demonstration of an object, structure or proposal in some form other than that of the individual itself". Models are used to describe and improve the behavior of a system in real and simple form and this simplicity results in the effectiveness of models as it presents the complete explanation of problem (Ahmed 2011; McCown et al. 1996; Murthy 2002). Building, testing, comparing and revising models needs a great deal of time of scientists and the introduction, application and interpretation of these imperative tools needs much general space (Frigg and Hartman 2006). Physiological sub models of complex crop growth includes basic physiological, biophysical and biochemical processes even at cellular or organ level, and it is difficult to conclude a large number of model parameters with which the modelers will have to deal (Mo and Beven 2004). The reaction which occurs within the plants and the interaction of plants with environment is represented in the agricultural models. The agricultural models are the clear picture of reality as it is very difficult and impossible to represent the complete system in arithmetical expressions due to incomplete status of present knowledge and complicacy of the system. In the agricultural sector, universal models do not exist as in the fields of engineering and physics. Complexity level is adopted according to the principle of the model and for different systems; different models can be used (Kumar and Chaeturvedi 2009).

15.2 Statistical and Mechanistic Models

Models are grouped in to different types according to their principle. These include

Statistical Models These models are used to describe the correlation between yield and climate parameters. Crop reaction to the fertilizers can be described by using the simple linear and quadratic models as the scientists used these models as in those researches where the goal is to achieve maximum yield (Willcutts et al. 1998).

Mechanistic Models Models which can describe the correlation between yield and climate parameters along with the function or action of these models. Similarly, "visualization modeling" and "mechanistic" or physiologically based modeling is to describe how a real plant functions using a virtual crop.

Deterministic Models Deterministic models have defined coefficients and predicts accurate yield.

Stochastic Models In these models a possibility factor is involved to every output. These models describe the yield at particular rate.

Dynamic Models Time is taken as variable in these models and both variables remain stable over a known time period.

Static Time is not taken as variable in these models and both variables remain stable over a known time period.

Simulation Models These models are the numerical expression of actual world system and the major objective of crop simulation model is to predict the agricultural yield as a purpose of climate, soil situation as well as crop management.

Descriptive Models These models describe the performance of a system in a simple way and comprises of one or more arithmetical equations.

Explanatory Models These models comprises of quantitative explanation of the functions and processes that causes the performance of the system (Murthy 2002). These models are used to predict the behavior of crop in an area where no crop has been grown earlier and helps in predicting about weather risks and selection of suitable varieties (Van keulen and Wolf 1986).

Process-Based Models These models are used to estimate the influence of fluctuating climate at local and international scales on crops (Rosenzweig and Parry 1994; IFPRI 2009) and the absence of these models for several minor crops confines the opportunity of this technique (Wolf and Van 2003).

Generic Models These models calculate possibly achievable than the real yield. In developing countries where there is a great difference between the expected and the real yield, it is difficult to explain the calculated climatic variations in the real crop yield. These models forecast progressive influence of changing climate on the crop yield than the real possible in fields in regions with great yield differences (Tubiello and Fischer 2007). Future fitness of land for different crops can be compared by using this technique (Stockle et al. 2001). A model can predict the possibilities of production levels in a certain soil type based on rainfall (Kiniry and Bockholt 1998). In short there is a key role of models in the modern science. The role and significance of models in the scientific practice with increasing attention is recognized by the philosophers (Frigg and Hartman 2006).

Phosphorous (P) is the most important yield boosting nutrient for the crops (Schjorring and Nielsen 1987). According to (NFDC 2011), P deficiency in Pakistani soils reaches up to 93 %. These soils have pH >7(alkaline), hence majority of these are Calcareous in nature ($CaCO_3$>3.0 %). A large amount of P fertilizers are precipitated or adsorbed and a very small fraction of it added to soil solution and used by plants (Ahmed et al. 2003). High alkalinity in Pakistani soils decreases the P accessibility to the plants. So the sufficient amount of P fertilizers should be applied to sustain certain level of P in the soil (NFDC 2011). Organic and Inorganic P are two main sources of P (Fig. 15.6; Shen et al. 2011).

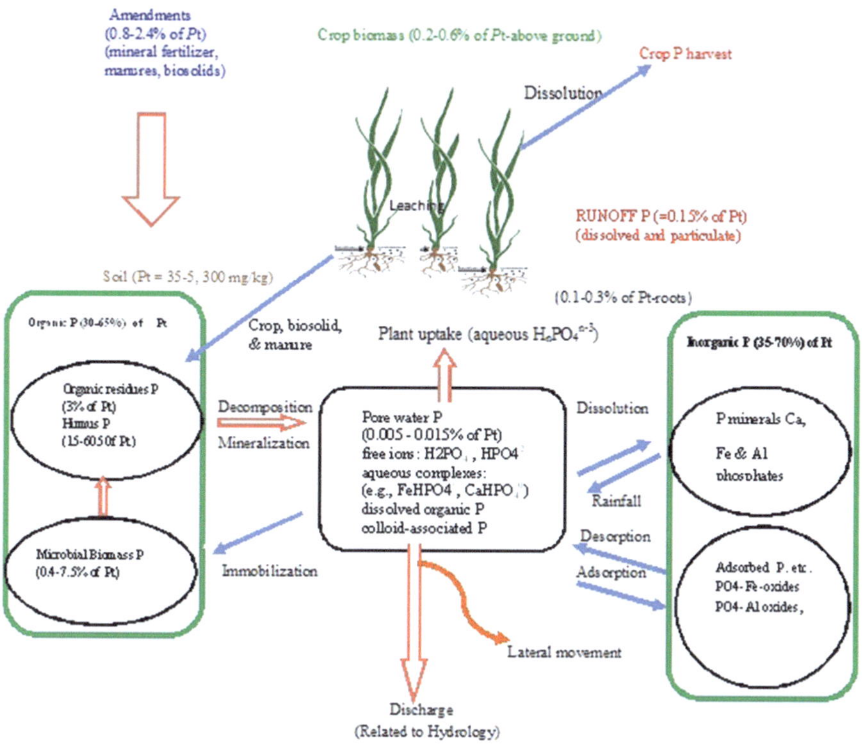

Fig. 15.6 P cycling in crop-soil systems

15.3 Sources of Phosphorus

Organic sources of P includes beef manure, dairy manure, dairy compost, vermi-compost, bone meal, compost having 1, 1, 1, 2, 11–22 and 0.05–2 % of P contents respectively while in poultry manure its concentration is 17 lb/ton (Zublena et al. 1993). Inorganic sources of P include super phosphate, concentrated super phosphate, mono ammonium phosphate, di ammonium phosphate and rock phosphate which contain 21, 45, 49, 47 and 34 % of P contents. Ortho phosphate is better than Poly phosphate (Rehm et al. 2011). Phosphorus plays an important role in plants metabolism, cellular energy transfer, respiration and photosynthesis. It is also the main element of genetic nucleic acid, chromosomes, phosphoproteins and phospholipids. Plant leaf area was reduced to 80 % due to phosphorous deficiency and it also affects the light saturated photosynthesis per unit of leaf area. Similarly, chlorophyll in the leaves of nutrient stressed plants was lowered. Meanwhile highest yield increase was recorded up to 22 % in wheat by the addition of 90 kg P_2O_5 per hectare (Khan et al. 2007). Similarly, sufficient application of P is essential at early stage of crop. P shortage at early crop stage effects the crop growth and it was reported that after this even the amplified application of P cannot recovers crop growth (Bertrand et al. 2003). Agronomically a plant showing good response to the P will be more valuable particularly from the present monetary and ecological point of view. Differences in uptake and use of P resulted in different yield responses of plants (Schjorring and Nielsen 1987). Physiological factors including root morphology (Romer et al. 1988) and Mycorrhizae (Smith and Gianinazzi-Pearson 1988) greatly affect the P uptake efficiency of the plants. Low solubility of P compounds in soil results in the low uptake of P by the plants. Excessive application of P results in leaching of P through macro pores water movement or through transportation in runoff and erosion (Sharpley et al. 2003). Review of modeling of P in current study will enhance our understanding of P availability, application requirement and efficient utilization by each sown crops for monetary as well as environmental points of view.

15.4 Phosphorous Modeling in Soil-Plant Atmosphere

Exact prediction of fertilizer application is becoming more difficult due to growing financial and ecological concerns related to its consumption. The accurate fertilizer rates for the crops can better be predicted by the modified quadratic plateau model than the quadratic plateau model (Bock and Sikora 1990). The correlation of accessible P fertilizer with wheat production can be described better using quadratic and linear plateau models than the quadratic plateau model (Zamune et al. 2005). A significant increase of dry matter production was observed by using linear response and plateau model (Swami and Singh 2008). Spatial variability of soil test P, specific characteristics of soil and production of winter wheat on dry landscapes of central huge plains determines that there are variations of winter wheat production and soil characteristics across landscape. The variables show extreme auto correlation and it helps the Kringing model in exact forecasting of soil characteristics and wheat production (Ortega et al. 1997). The effect of P on individual leaf growth rate and reduced assimilate accessibility for leaf growth in low P situation were testified. The P shortage resulted decline in plants leaf area, grain production and 32 % increase in Phyllochron. Leaf area expansion decreased and photosynthesis per unit leaf area even at higher radiations decreased up to 57 %. It was concluded that the gentle P deficiency at leaf appearance and tillering stage results in 80 % decrease in plant leaf area. The simulation model fails to describe the results due to severe P shortage as well as due to some other effects like direct effects of P on leaf expansion which cannot be explained by model (Rodriguez et al. 1998). The amount of P uptake by wheat plants and its effects on production of different varieties in rain fed as well as irrigated areas having acidic and calcareous soils was determined in an experiment which concludes that production increased significantly in acidic and calcareous soil after the application of P fertilizers. High P in acidic and low P concentration in calcareous soils results in high wheat production (Manske et al. 2001). Integrated effects of N, P, K along with organic matter and water greatly enhance the production which increased with increasing P dosage and 35.6 kg per hectare application of P gives maximum production by using water fertilizer yield model (Duan et al. 2004). P requirement of wheat to get 95 % relative yield by using adsorption isotherm in Freundlich model observed by addition of different amount of P solutions concludes that the highest wheat grain and straw yield was observed by adding 0.15 mg P L^{-1} and 0.50 mg P L^{-1} respectively. Highest P concentration in wheat grain and straw was 0.41 % and 0.16 % respectively. So it is concluded that to get 95 % relative yield 0.2 mg P L^{-1} exterior soil solution and interior P requirement is 0.27 %. The adsorption isotherm in the Freundlich model calculates the P requirement of wheat efficiently (Rehman et al. 2005). It is essential to assess the point to which the nutrient availability decreases the crop yield and the PARJIB model predicts the response of nutrient application to get maximum yield in better way (Reid 2002).

15.4.1 Modeling Phosphorous Availability Under Changing Climate

There will be a major change in production of food due to changing global climate (Rosenzweig and Parry 1994). Nutrient deficits greatly reduce soil productivity. Plants need accessible mineral nutrients in the soil for vigorous development. Shortage of accessible P in soil is a severe problem. P problem is challenging for farmers having small land holdings because of their low accessibility to inputs. The major factors responsible for the shortage of P are: Natural process: High temperature, precipitation, percolation and chemical disintegration are responsible for the nutrient loss. P deficiency is noticed in sandy loam and granite derivative soils.

15.4.2 Human Action

Deficiency of P in the soil is caused due to farm managing techniques, extension of agriculture and further human accomplishments. Reaping of crops, burning of crop remains and the soil destruction directed to the phosphorous loss from the soil. P shortage inhibits growth, cause dark blue green or purple coloration and distresses the root growth (Anonymous 2012). Organic and inorganic phosphorous is produced from plants and bacterial action. These forms of P develop a series of instantly accessible to slowly accessible P which is bonded in an established compound to form a balance between each other. Roots of trees and mycorrhizae cause disintegration which results in phosphorous deficiency (Dijkstra et al. 2003). In an experiment fire showed adverse effects on P accessibility during first year and reduction in available P after 5 years.

15.4.3 Effects of High Temperature

Extremely high incineration temperature reduces the half of P availability to the plants (Thygesen et al. 2011). If the constant rise in temperature cross the specific limit it will also cause decline in crop yield with decreasing soil moisture (Monteith 1981). Effects of P on oat were observed under different temperature to determine P uptake. Soil fertilized with mono and di ammonium phosphate showed high P uptake at 5° C than in 16 and 27° C (Beaton and Read 1965). P concentration in entire plant, leaf, trunk, and roots was highest at 27° C. Increase in P concentration results in maximum phosphorous uptake in leaves and stem. At the temperature of 21° C and 27° C there was maximum P in panicle. There was poor growth at sub optimal temperature which restricts plant growth due to low P contents (Ercoli et al. 1995).

15.4.4 Effects of Elevated CO_2

The global warming and climate change due to increased CO_2 had resulted to the efforts which can reduce the discharge of worldwide CO_2 radiations by intergovernmental panel on climate change (IPCC 2007; Raupach et al. 2007). Supplementary intensification of CO_2 will effect the plant growth considerably, by encouraging the process of photosynthesis directly (Drake et al. 1997) and secondarily by encouraging global warming IPCC (2007). Increasing CO_2 and temperature results in more biomass production and restrictive nutrient accessibility to the plants. Ectoenzymes enhance rhizodeposition due to high atmospheric CO_2 (Lagomarsino et al. 2008) so the demand of P for microbes and plants is also increased (Dijkstra et al. 2003). Sweet gum woodland was treated with high CO_2 to observe the effects of high CO_2 on P availability but no impacts were found on the amount of phosphorous after 2 years of its supplementation. Plantation of popular under higher rate of CO_2 does not deplete P from soil but enhances P in the root zone (Khan et al. 2008). Excessive CO_2 cause 8.6 % decline in P accessibility after the fertilization of soil with P during first year but it shows 69 % rise in above ground trunk development (Webber 1990).

15.4.5 *Effects of Drought*

Availability of P to the plants decreased in the drought. Even a very slight phosphorous scarceness confines the accessibility of phosphorous to the shoots (Turner 1985). Decrease in the mineralization rate of soil is caused by the water stress which results in the reduction of P contents and also causes P accumulation (Sardans and Penuelas 2004). Due to limited presence/supply of water the nutrient availability is also restricted. Decrease in moisture up to 22 % results in 40 % decrease in P uptake by the plants. So the direct impact of low moisture results in the non-availability of P to the plants which will effects the plant growth and consequently there will be a severe effect on the environment (Sardans and Penuelas 2004).

15.5 Conclusion

The present review of P dynamics concluded that it plays an important role in the growth development and all the metabolic processes of the plants thus leaves great impact on the yield of all agronomic crops. Variations in the climatic conditions would result in the limited availability of P to the plants so there is an extreme need of modeling P dynamics under the changing climatic conditions to get optimum yield. The dynamic models like STELLA, APSIM, DSSAT and AEP might be used to understand the P dynamics on farm scale. Since P dynamics was largely driven by short term events like seasons and bio-physiochemical changes therefore, simulation models might be used to understand P dynamics for managerial decisions and policy making. Such decisions and policy making might promote long term environmental, economic and resource sustainability for future generations.

References

Ahmed, M. 2011. *Climatic resilience of wheat (Triticum aestivum) using simulation modeling in Pothwar*. Ph.D., thesis, PMAS-Arid Agriculture University Rawalpindi Pakistan.

Ahmed, N., M. Abid, K. Hussain, M. Akram, and M. Yousaf. 2003. Evaluation of nutrient status in rice growing areas of the Punjab. *Asian Journal of Plant Science* 2(5): 449–453.

Anonymous. 2012. *Phosphorus for agriculture*. www.voguelph.ca/scitale/knowledge/library/effects-of-rising-atmospheric-concentrations-of-carbon-13254108.

Beaton, J.D., and D.W.L. Read. 1965. Effects of temperature and moisture on phosphorous uptake from calcareous Saskatchewan soil treated with several pelleted sources of phosphorous. *Soil Science Society of American Journal* 27(1): 61–65.

Bertrand, I., R.E. Holloway, R.D. Armstrong, and M.J. Mclaughlin. 2003. Chemical characteristics of phosphorous in alkaline soils from southern Australia. *Australian Journal of Agricultural Research* 41: 61–76.

Bock, B.R., and F.J. Sikora. 1990. Modified quadratic plateau model for describing plant response to fertilizer. *Soil Science Society of American Journal* 54: 1784–1789.

Cassell, E.A., J.M. Dorioz, R.L. Kort, J.P. Hoffmann, D.W. Meals, D. Kirschtel, and D.C. Braun. 1998. Modeling posphorus dynamics in ecosystems: Mass balance and dynamic simulation approaches. *Journal of Environmental Quality* 27: 293–298.

Condron, L.M., B.L. Turner, and B. J. Cade-Menun. 2005. Chemistry and dynamics of soil organic phosphorus. In ed. J.T. Sims, A.N. Sharpley. *Phosphorus: Agriculture and the environment*. American Society of Agronomy, Crop Science Society of America, Soil Science Society of America, Inc., Madison, WI, pp. 87–121.

Dijkstra, F.A., N.V. Breemen, A.G. Jongmans, G.R. Davies, and G.E. Likens. 2003. Calcium weathering in forested soils and the effect of different tree species. *Biogeochemistry* 62: 253–275.

Drake, B.G., M.A. Gonzalezmeler, and S.P. Long. 1997. More efficient plants: A consequence of rising atmospheric CO_2? *Annuals Revision of Plant Physiology and Plant Molecular Biology* 48: 609–639.

Duan, Z., H. Xiao, Z. Dong, X. Li, and G. Wang. 2004. Combined effect of nitrogen–phosphorus–potassium fertilizers and water on spring wheat yield in an arid desert region. *Communication in Soil Science and Plant Analysis* 35(1–2): 161–175.

Ercoli, L., M. Mariotti, A. Masoni, and F. Massantini. 1995. Effect of temperature and phosphorous fertilization on phosphorous and nitrogen uptake by sorghum. *Crop Science* 36(2): 348–354.

Frigg, R., and S. Hartman. 2006. Models in science. *The stanford encyclopedia of philosophy* (Fall 2012 Edition), Edward N. Zalta (ed.), forthcoming. http://plato.stanford.edu/archives/fall2012/entries/models-science/.

Hartemink, A.E. 2003. *Soil fertility decline in the tropics*. Wageningen: CABI.

Hayes, J.E., A.E. Richardson, and R.J. Simpson. 1999. Phytase and acid phosphatase activities in extracts from roots of temperate pasture grass and legume seedlings. *Australian Journal of Plant Physiology* 26: 801–809.

Hinsinger, P. 2001. Bioavailability of soil inorganic P in the rhizosphere as affected by root induced chemical changes: A review. *Plant and Soil* 237: 173–195.

International Food Policy Research Institute (IFPRI). 2009. *Climate change impact on agriculture and costs of adaptations*. Food policy report.

IPCC. 2007. *Contribution of working group II to the fourth assessment report of the Intergovernmental Panel on Climate Change*. Cambridge/New York: Cambridge University Press.

Khan, R., A.R. Gurmani, A.H. Gurmani, and M.S. Zia. 2007. Effect of phosphorous application on wheat and rice yield under wheat-rice system. *Sarhad Journal of Agriculture* 23(4): 851–856.

Khan, F.N., M. Lukac, G. Turner, and D.L. Godbold. 2008. Elevated atmospheric CO_2 changes phosphorus fractions in soils under a short rotation poplar plantation (EuroFACE). *Soil Biology and Biochemistry* 40: 1716–1723.

Kiniry, J.R., and A.J. Bockholt. 1998. Maize and sorghum simulation in diverse texas environment. *Agronomy Journal* 90: 682–687.

Kumar, R., and S. Chaeturvedi. 2009. *Crop modeling: A tool for agricultural research*. www.agropedia.iitk.ac.in/?q=content/crop-model.

Lagomarsino, A., M.C. Moscatelli, M.R. Hoosbeek, P. Deangelis, and S. Grego. 2008. Assessment of soil nitrogen and phosphorous availability under elevated CO_2 and N-fertilization in a short rotation poplar plantation. *Plant and Soil* 308: 131–147.

Lynch, J. 2007. Roots of the second green revolution. *Australian Journal of Botany* 55: 493–512.

Lynch, J.P. 2011. Root phenes for enhanced soil exploration and phosphorus acquisition: Tools for future crops. *Plant Physiology* 156(3): 1041–1049.
Lynch, J.P., and K.M. Brown. 2011. Topsoil foraging- an architectural adaptation of plants to low phosphorus availability. *Plant and Soil* 237(2): 225–237.
Lynch, J.P., and M.D. Ho. 2005. Rhizoeconomics: Carbon costs of phosphorus acquisition. *Plant and Soil* 269: 45–56.
Ma, Z., D.G. Bielenberg, K.M. Brown, and J.P. Lynch. 2001a. Regulation of root hair density by phosphorus availability in Arabidopsis thaliana. *Plant, Cell and Environment* 24: 459–467.
Ma, Z., T.C. Walk, A. Marcus, and J.P. Lynch. 2001b. Morphological synergism in root hair length, density, initiation and geometry for phosphorus acquisition in Arabidopsis thaliana: A modeling approach. *Plant and Soil* 236: 221–235.
Manske, G.G.B., J.I. Ortiz-Monasterio, M. Vangrinkel, R. González, S. Rajaram, E. Molina, and P.L.G. Vlek. 2000. Traits associated with improved P-uptake efficiency in CIMMYT's semi dwarf spring bread wheat grown on an acid Andisol in Mexico. *Plant and Soil* 221: 189–204.
Manske, G.G.B., J.I. Ortiz-Monasterio, M. Vanginkel, R.M. Gonzalez, R.A. Fischer, S. Rajaram, and P.L.G. Vlek. 2001. Importance of P uptake efficiency versus P utilization for wheat yield in acid and calcareous soils in Mexico. *European Journal of Argonomy* 14: 261–274.
Mccown, R.L., G.L. Hammer, J.N.G. Hargreaves, D.P. Holzworth, and D.M. Freebairn. 1996. APSIM: A novel software system for model development, model testing and simulation in agricultural systems research. *Agricultural Systems* 50: 255–271.
Mo, X., and K. Beven. 2004. Multi-objective parameter conditioning of three source wheat canopy model. *Agriculture Forest Meteorology* 122: 39–63.
Mollier, A., and S. Pellerin. 1999. Maize root system growth and development as influenced by phosphorus deficiency. *Journal of Experimental Botany* 50: 487–497.
Monteith, J. 1981. Climate variation and the growth of crops. *Journal of Royal Meteorology Society* 107: 749.
Murthy, V. 2002. Crop growth modeling and its applications in agricultural meteorology. *Satellite remote sensing and GIS applications in agricultural meteorology*, 235–261.
NFDC. 2011. http://www.nfdc.gov.pk/public.html.
Ortega, R., D.G. Westfall, and G.A. Peterson. 1997. Variability of phosphorous over landscapes and dryland winter wheat yields. *Better Crops* 81(2): 24–27.
Raupach, M.R., G. Marland, P. Ciais, C. Lequere, J.G. Canadell, G. Klepper, and C.B. Field. 2007. Global and regional drivers of accelerating CO_2 emissions. *Proceedings of the National Academy of Sciences USA* 104: 10288–10293.
Rehm, G., M. Schmitt, J. Lamb, G. Randall, and L. Busman. 2011. *Understanding phosphorous fertilizers, phosphorous in the agricultural environment*. University of Minnesota. www.extension.umn/edu/distribution/cropsystem/dc6288.html-47K.
Rehman, O., A.M. Ranjha, S.M. Mehdi, and A.A. Khan. 2005. Phosphorous requirement of wheat using Modified Freundlich model in Sultanpur (Pakistan) soil series. *International Journal of Agriculture and Biology* 7(1): 74–78.
Reid, J.B. 2002. Yield responses to nutrient supply across a wide range of conditions. 1. Model derivation. *Field Crops Research* 77: 161–171.
Rodriguez, D., W.G. Keltjens, and J. Goudriaan. 1998. Plant leaf area expansion and assimilate production in wheat (*Triticum aestivum* L.) growing under low phosphorous conditions. *Plant and Soil* 200: 227–240.
Romer, W., J. Augustin, and G. Schilling. 1988. The relationship between phosphate absorption and root length in nine wheat cultivars. *Plant and Soil* 11: 199–201.
Rosenzweig, C., and M.L. Parry. 1994. Potential impact of climate change on world food supply. *Nature* 367: 133–138.
Ryan, P.R., E. Delhaize, and D.L. Jones. 2001. Function and mechanism of organic anion exudation from plant roots. *Annuals Revision Plant Physiology and Plant Molecular Biology* 52: 527–560.
Sardans, J., and J. Penuelas. 2004. Increasing drought decrease phosphorous availability in an evergreen Mediterranean forest. *Plant and Soil Journal* 267(1–2): 367–377.

Schjorring, J.K., and N.E. Nielsen. 1987. Root length and phosphorus uptake by four barley cultivars grown under moderate deficiency of phosphorus in field experiments. *Journal of Plant Nutrition* 10: 1289–1295.

Sharpley, A.N., J.L. Weld, D.B. Beegle, P.J.A. Kleiman, W.J. Gburek, P.A. Moore Jr., and G. Mullins. 2003. Development of phosphorus indices for nutrient management planning strategies in the United States. *Journal of Soil and Water Conservation* 58: 137–152.

Shen, J., L. Yuan, J. Zhang, H. Li, Z. Bai, X. Chen, W. Zhang, and F. Zhang. 2011. Phosphorus dynamics: From soil to plant. *Plant Physiology* 156(3): 997–1005.

Smith, S.E., and V. Gianinazzi-Pearson. 1988. Physiological interactions between symbionts in vesicular-arbuscular mycorrhizal plants. *Annuals Revision Plant Physiology* 39: 221–244.

Smith, S.E., and D.J. Read. 1997. *Mycorrhizal symbiosis*, 2nd ed. San Diego: Academic. ISBN 978-0-12-652840-4.

Stockle, C.O., M. Donatelli, and R. Nelson. 2001. CropSyst, a cropping systems simulation model. *European Journal of Agronomy* 18: 289–307.

Swami, S., and M. Singh. 2008. Critical limits of phosphorous for durum wheat in normal and heavy metal polluted soils. *Indian Journal of Agricultural Research* 42(4): 252–259.

Thygesen, L.G., B.J. Hidayat, K.S. Johansen, and C. Felby. 2011. Role of supramolecular cellulose structures in enzymatic hydrolysis of plant cell walls. *Journal of Industrial Microbiology & Biotechnology* 38: 975–983.

Tiessen, H. 2008. Phosphorus in the global environment. In *The ecophysiology of plant-phosphorus interactions*, ed. P.J. White and J.P. Hammond. Dordrecht: Springer.

Tubiello, F.N., and G. Fischer. 2007. Reducing climate change impacts on agriculture: Global and regional effects of mitigation, 2000–2080. *Technol Forecas and Soc Chan* 74: 1030–1056.

Turner, L.B. 1985. Changes in the phosphorous contents of capsicum annum leaves during water stress. *Journal of Plant Physiology* 121: 429–439.

Turner, B.L., A.E. Richardson, and E.J. Mullaney. 2007. *Inositol phosphates: Linking agriculture and the environment*, 304. Wallingford: CAB International.

USDA, NRCS. 2012. http://soils.usda.gov/.

Van Keulen, H., and J. Wolf. 1986. *Modelling of agricultural production: Weather, soils and crops*, Simulation monographs. Wageningen: PUDOC.

Webber, M.G. 1990. Forest soil respiration after cutting and burning in immature aspen ecosystems. *Forest Ecology Management* 31: 1–14.

Willcutts, J.F., A.R. Overman, G.J. Hochmuth, D.J. Cantliffe, and P. Soundy. 1998. A comparison of three mathematical models of response to applied nitrogen: A case study using lettuce. *Horticulture Science* 33(5): 833–836.

Wolf, J., and M. Van Oijen. 2003. Model simulation of effects of changes in climate and atmospheric CO_2 and O_3 on tuber yield potential of potato (cv. Bintje) in the European Union. *Agriculture Ecosystem and Environment* 94: 141–157.

World Bank. 2004. *World development indicators*. Washington, DC: World Bank.

Yan, X., H. Liao, A. Cao, W.J. Horst, M.K. Schenk, A. Bürkert, N. Claassen, H. Flessa, W.B. Frommer, H. Goldbach, H.W. Olfs, V. Römheld, B. Sattelmacher, U. Schmidhalter, S. Schubert, N. Wirén, and L. Wittenmayer. 2002. *The role of root architecture in P acquisition efficiency of different root systems: A case study with common bean and rice Book title: Plant nutrition: Developments in plant and soil sciences*. Dordrecht: Springer. ISBN 978-0-306-47624-2.

Zamune, E., L.I. Picone, and H.E. Echeverria. 2005. Comparison of phosphorous fertilization diagnostic methods for wheat under no tillage. *Soil Tillage Research* 89: 70–77.

Zublena, J., J.C. Barker, and T. A. Carter. 1993. Poultry manure as fertilizer source. *Soil facts*. North Carolina cooperative extension service. www.soil.ncsu.edu/publications/soilfacts/AG-439-05/.

Chapter 16
Studying Impact of Climate Change on Wheat Yield by Using DSSAT and GIS: A Case Study of Pothwar Region

Mahwish Jabeen, Hamza Farooq Gabriel, Mukhtar Ahmed, Muhammad Ahsan Mahboob, and Javed Iqbal

Contents

M. Jabeen (✉) • H.F. Gabriel • M.A. Mahboob • J. Iqbal
Institute of Geographical Information Systems, School of Civil and Environmental Engineering, National University of Sciences & Technology, Islamabad, Pakistan
e-mail: mahwish11@igis.nust.edu.pk; hamza.gabriel@nice.nust.edu.pk

M. Ahmed
Department of Agronomy, Pir Mehr Ali Shah Arid Agriculture University, Rawalpindi, Pakistan

Department of Biological Systems Engineering, Washington State University, Pullman, WA, USA
e-mail: ahmadmukhtar@uaar.edu.pk; mukhtar.ahmed@wsu.edu

M. Ahmed, C.O. Stockle (eds.), *Quantification of Climate Variability, Adaptation and Mitigation for Agricultural Sustainability*,
DOI 10.1007/978-3-319-32059-5_16

Abstract Today global climate change and its impact on crop production is a major issue. Climatic factor such as temperature has been closely linked to agricultural production. According to the international panel on climate change (IPCC) these changes are very rapidly affecting crop productivity. To explore the future climate change impacts on wheat yield in Pakistan, especially the rain fed region of Pothwar which is considered vulnerable to climate change, a wheat crop simulation study was conducted. The specific objectives of the study were to (a) simulate the impact of climate change using DSSAT on wheat yield in the Pothwar region using IPCC climate change scenario for Pakistan and (b) generate spatial maps of wheat yield and correlate with climatic factors. The crop simulation study assessed the impact of rise in maximum and minimum Temperature on the wheat yield. The CERES-Wheat which is a component of the Decision Support System for Agrotechnology Transfer (DSSAT) model was used to input soil, crop management practices and weather data. IPCC Fourth Assessment report B1 scenario for increasing temperature was used in the simulations. The simulated results were imported into ArcGIS to produce regional impact maps for visual assessment and spatial analysis under different climate change scenarios levels through an interface of ArcMap. The DSSAT model simulated results showed that the rise in maximum and minimum temperature decreased the wheat yields across the Pothwar region. The similar methodology could be adopted for different crops in other parts of the country for better mitigation of future food security.

Keywords DSSAT • ArcMap • CERES-wheat • Climate change • Wheat • ArcGIS

16.1 Introduction

Agriculture plays main role in the national economy of Pakistan. Wheat (*Triticum aestivum* L.) is the major worldwide staple food, cash crop and one of the important and older edible cereal crops. Today's climate change is the most emerging issues for the agriculture production in Pakistan and all over the world. Basically change in climate is because of rise in anthropogenic activities and concentration of greenhouse gases (GHGs). Greenhouse gasses trap the solar radiation and increase the earth's temperature. This high temperature might adversely influence the wheat development phases because increase in temperature shortens the crop growing period (Parry et al. 2004).

Greenhouse gases (GHGs) produced due to many anthropogenic activities e.g. (1) Carbon dioxide produced during the discharge of wood, fossil fuels, wastes, and carbon. (2) Methane emitted from industry, agriculture, and waste management activities. (3) Nitrous oxide produced both natural and human sources. Human sources of nitrous oxide are animal manure management, nitric acid production and sewage treatment. Nitrous oxide (N_2O) also produced from biological sources in water and soil, especially microbial action in wet tropical forests. Carbon dioxide (CO_2) concentrations have increased from 220 ppm to 380 ppm because of burning of fossil fuels and deforestation, and so forth. Without any strategy the GHGs outflow, CO_2 would increment from 550 ppm to700 ppm at the mid of the current century and this level of GHGs would result in the rise in temperature from 3 °C to 6 °C (IPCC 2014). Wheat production in all over the world for the period of 2009–2010 decreased up to 0.32 % as compared to earlier wheat crop period (2008–2009) whereas 5.4 % reduction in production of wheat was projected for the period of 2010–2011. This decrease in wheat production was because of several reasons such as poor management, inadequate agronomic practices, and unfavorable climatic situations such as drought, high temperature and salinity. Recent researches on climate change effects predict that temperature and rainfall both have been increased and at the end of the century in south Asia temperature rise will be 3–4 °C (IPCC 2006). Future climate change scenarios depicted global warming might be beneficial in some wheat crop regions, but could decrease the wheat productivity in zones where already temperatures is high (Ortiz et al. 2008).

The high temperature stress is a major factor in limiting the wheat yield. This factor adversely affecting development and growth of wheat and produces low yield in many areas of the world (Buriro et al. 2011). Fluctuation in minimum and maximum temperature from wheat germination to maturity affected the developmental period of wheat grain. "Global warming" and "climate change" both terms have different meanings and have been used from last decades. Both terms are refer to two different physical phenomena and they are frequently used in scientific literature. Basically "Global Warming" is rising trend of average global temperature in long term. And "Climate Change" is the changes in average global temperature, For example, extreme condition of weather, increased prevalence of droughts, changes patterns in precipitation, and other, etc.

Climate changes may result by the anthropogenic and natural phenomena. Variations in the earth's orbit around the sun, volcanic eruptions, and variations in solar output are natural causes of climate change and burning of fossil fuels, industrial pollutants, changes in the earth's albedo due to deforestation of tropical rainforest and warming of average annual temperature due to urbanization are the results of human activities. It is believed that greenhouse gases are main cause of global warming. Earth temperature depends upon the heat energy budget that how much incoming and outgoing heat energy on the earth. Human and anthropogenic activities are the natural causes of changes in balance of energy on earth's surface. Anthropogenic activities hazardly effected the climate by adding carbon dioxide (Janjua et al. 2010) and $CH_{4,}$ in atmosphere in the age of the industrial revolution began in 1750.

Since last few years in Pakistan there are so many factors responsible in decreasing wheat yield (Shakoor et al. 2011). Some land in Pakistan was affected by floods and production also decreased due to excessive rainfall, delayed harvesting of wheat, increase in salinity and water logging conditions, fluctuation in environmental factors, sudden rise in environmental temperature at reproductive stage of crop, large attacked of insects on crop and improper agronomic practices during cropping season. The most significant Green House Gases directly emitted by human activities in the atmosphere includes carbon di oxide, methane and nitrous oxide.

When people of this world first started planting crops thousand years past, the requirement of agriculture was manifest. The question that raised in our forefathers mind are; why do only specific grow in the region? Would there be adequate resources for planting these crops in that area? Perhaps of all agriculture is the first sector where humans realized that there is a strong relation between the agriculture and climate. And over the time human understand and select the appropriate time and condition for the plants growth. They were of course frustrated with the instability of the weather and their impacts on crops, but their coping strategy developed from long time and continue to be used right up to the current day. With change over the time humans change their occupations and start industries with the material that they grow from agriculture sector e.g. sugar industries starting from sugarcane. Increase in industries becoming a major cause of global warming and temperature increased day by day due to ozone depletion. Since the late nineteenth century Global temperature has increased by 0.3–0.6 °C. The warmest period was 1990s from the last 140 years (Jones and Briffa 1992). In Indian context Hingane et al. (1985) stated an increase in average annual temperature by 0.4 °C over 100 years during the twentieth century till 1980s. However, the rise in temperature pattern is not substantial over the whole country; the northeast and northwest South Asia demonstrate some cooling effects. GHG's emissions in Pakistan nearly doubled in the last 16 years (Sayed 2011). Kothawale et al. (2010) reported a rise of 0.5 °C in mean yearly temperature throughout the most recent century. The pattern of spatial and temporal changes in climatic variables because of global warming need more debate and studies are being conducted globally (Herrero et al. 2010).

The study area (Pothwar) lies in the region where the wheat (*Triticum aestivum* L.) is the major food crop. Climate change is the major factor that effect wheat crop in the Pothwar region. Climate effect more severe in the arid region comparatively other region. This study was conducted to assess the effect of temperature change on yield of wheat in connection to future temperature change using DSSAT crop model. This study will help to change the food crop into other valuable crops that can use low water and high temperature for blooming and it might be helpful in earning income and imports. It will also identify the potential adaptation strategies for sustaining the yield of wheat crop. The objectives of the study were to:

- To simulate the climate change and its impact on wheat yield by using DSSAT in Pothwar region using climate change scenario for Pakistan
- To generate spatial maps of wheat yield and correlate with climatic factors.

16.2 Materials and Methods

In an attempt to evaluate the change in climate and its impacts on wheat yield, an analysis of the current climate was made together with an analysis of the between future climate change and wheat crop yield. The analysis involves the development of a number of climate change scenarios and use of DSSAT crop model. The output

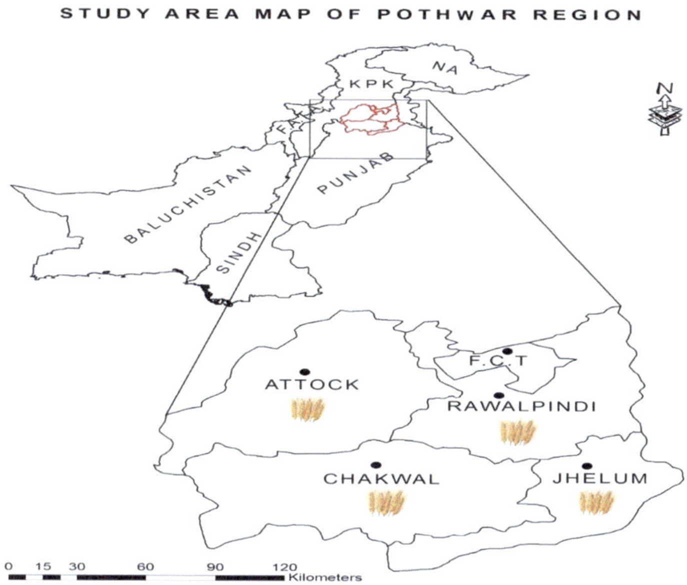

Fig. 16.1 Location map of study area, Pothwar Region, Pakistan

will give likely changes in wheat crop yield as a result of increasing or decreasing temperature and precipitation. This detailed analysis was made in a step-wise manner. First, it involves the description of emissions scenarios, climate model, and selection and evaluation of Global Climate Models. The second step involves the description of crop model and its calibration. Last, the potential impacts of the climate scenarios on wheat crop.

16.2.1 Study Area

The study area include Pothwar region which consist of 4 districts Attock, Chakwal, Jhelum, Rawalpindi and Islamabad Capital Territory (Fig. 16.1). It is located on 32.5° N to 34.0° N *Latitude*, 72° E to 74° E Longitude and altitude 462 m between Indus and Jhelum rivers and surrounded by the Hazara Hills on the north and the Salt Range on the south. This area was selected because of number of reasons: (i) It is among the area where wheat is major crop reported by National Agriculture Research Centre (NARC) Pakistan. (ii) It is among the areas of Pakistan where no irrigation system is developed and the climate change effects could be more severe in future.

The long term climate data (1981–2010) of the study sites were used to calibrate DSSATCERES-Wheat model.

The 90 year projected weather data daily temperature (°C), daily solar radiation (MJ/m^2), and daily total rainfall (mm) was downloaded from MarkSim DSSAT weather file generator (http://gismap.ciat.cgiar.org/MarkSimGCM) using B2 scenario of projected temperature changes in Pakistan over the twenty-first century under IPCC, SRES (Special Report on Emissions Scenarios).

Soil element has been connected as independent variable. This element has been calculated by using the digital soil map of Pothwar Region provided by the Soil Survey of Pakistan. In this map soil classification of the whole area of Pothwar region was performed. This classification considers pedological (soil type) as well as topological (altitude, inclination) components. Soil series data provides information about the soil texture and spatial variability across the field. The data obtained from the Soil Survey of Pakistan digitized was for the further GIS and Statistical analysis. Soil properties data consist of upper and lower horizon depth (cm), percentage of sand, silt and clay, bulk density, organic carbon and pH in water. Crop management data provides information about the field, planting, cultivar, irrigation, fertilizer and harvest was collected from National Agriculture Research Centre (NARC) Pakistan (Fig. 16.2). Historical data of wheat yield was used for the outputs validation of DSSAT model.

16.2.2 DSSAT Calibration

DSSAT calibration is the adjustment of functions and parameters so that simulated data is the same or very close to data obtained from the experimental field. DSSAT model has been validated by a holdout cross-validation. Performing this cross-validation, the information records of the explanatory variables have been divided into a training and into a test information set.

16.2.2.1 Modeling Climate Change Impacts on Agriculture

Due to climate change, conditions of wheat crop will change in the next years. Now the question arises how projected climatic conditions impact on wheat yield. In the past decades many studies estimated the climate change and its impacts on agriculture on the basis of crop development simulation models. DSSAT model simulate crop growth and crop yield levels by means of different input variables (for example, soil characteristics, daily weather parameters, crop characteristics, cropping system management options). Examples of recent crop growth simulation models are DSSAT, CropSyst, CERES and CROPGROW.

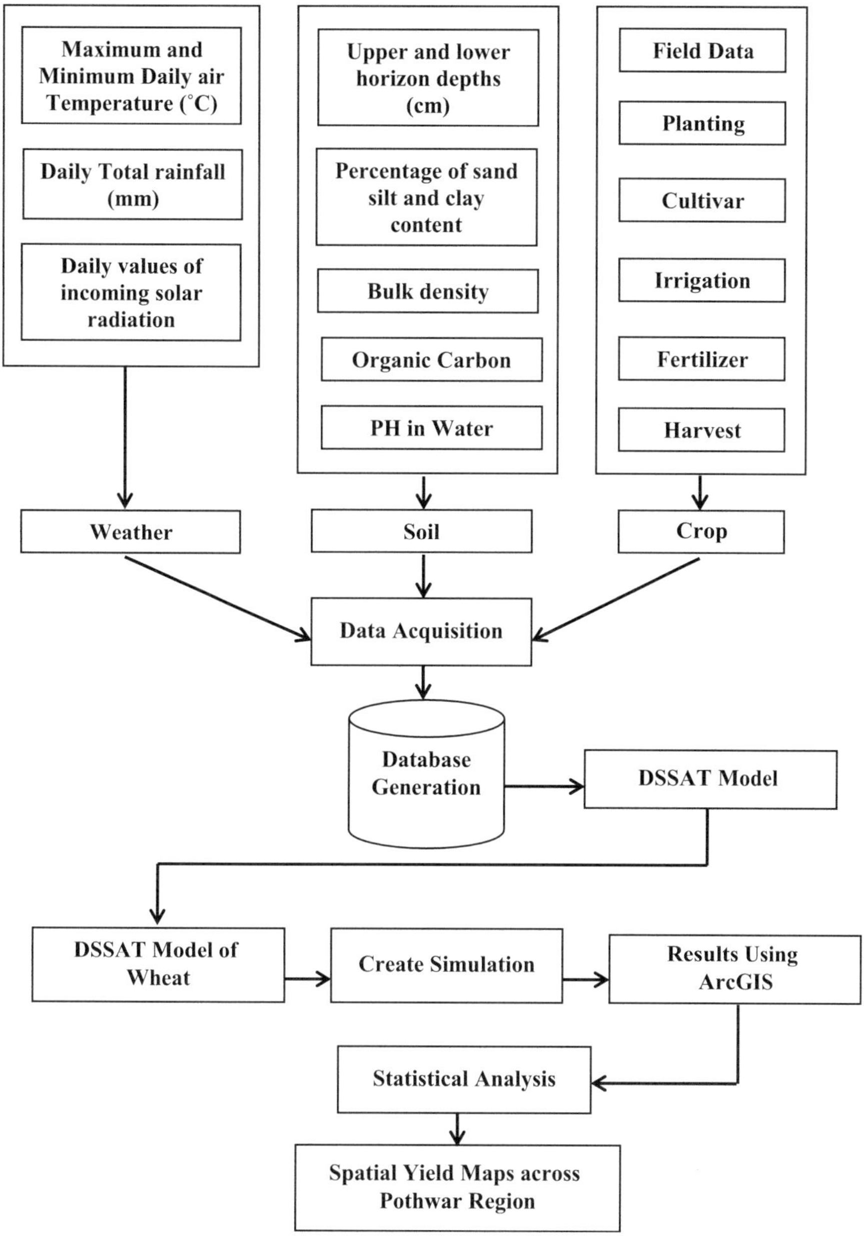

Fig. 16.2 Methodologicl framework

16.2.2.2 Projection of Future Yields

In this research, for the assessment of wheat yield under different conditions of climate, only climate descriptive variables have been altered. Crop management and soil properties variables on the other hand, have been kept constant. Current and historical wheat yield have been simulated by use of the actual weather data, the fixed averages of soil data and management data. Future yield simulation is generated by using future climatic scenarios and constant soil conditions.

16.2.2.3 Soil Type Map

Soil map contained the all types of soil in the Pothwar region. Twelve types of soil are available in the Pothwar region named Calcareous clayey soils, Calcareous loamy and Clayey Soils, Calcareous loamy soils, Calcareous sandy soils, Calcareous silty soils – gullied land complex, Gullied land and bad land, Noncalcareous clayey soils, Noncalcareous loamy soils, Noncalcareous silty soils, Rough broken land, Salt affected soils and Sand dunes and sandy soils. In these soil types 3 or 4 soil types are not suitable for wheat yield but the others are suitable for wheat crop.

16.2.2.4 Projected Climatic Variable Maps

Daily data of Climatic variables for all locations in the Pothwar region were projected by using MarkSim DSSAT weather generator B2 Scenario. Maps of Tmax and Tmin, precipitation and solar radiation created by using Inverse Distance Weighting (IDW) interpolation method because the points are dispersed in the study area so the IDW interpolation method is best for dispersed points.

16.2.2.5 Inverse Distance Weighting (IDW)

The IDW is function used when set of points is dense enough to capture the extent of local surface variation needed for analysis. IDW determines cell values using a linear-weighted combination set of sample points. The weight assigned was a function of the distance of an input point from the output cell location. The greater the distance, the less influence the cell has on the output value. IDW is sensitive to outliers. Furthermore, unevenly distributed data clusters result in introduced errors. The simplest form of IDW interpolation is called, Shepard method and it uses weight function w_i given by:

$$w_i = \frac{h_i}{\sum_{j=0}^{n} h_j} \tag{16.1}$$

Where p is an arbitrary positive real number called the power parameter (typically p=2) and h_j are the distances from the dispersion points to the interpolation point, given by:

$$h_i = \sqrt{(x - x_i)^2 + (y - y_i)^2} \quad (16.2)$$

Where (x, y) are the coordinates of the interpolation point and (x_i, y_i) are the coordinates of each dispersion point. The weight function varies with a value of unity at the dispersion point to a value close to zero as the distance to the dispersion point increase. The weight functions are normalized as a sum of the weights of the unit. Then, the interpolated value of the electric field E(x, y) is given by

$$E(x, y) = \sum_{j\ 0}^{n} w_j\ E(x_j, y_j) \quad (16.3)$$

In order to improve the computational time is possible to set bounds to the dispersion points that contribute to the calculation of the interpolated value, to all those dispersion points within a given search radius centered on the interpolated point. For the particular application developed in this work, it was determined that the most appropriate search radius was 500 m, so that the computation times were manageable.

16.2.3 Model Output Coupled with GIS

After simulation of wheat yield using DSSAT, the next step was to generate wheat yield maps of Pothwar region. To accomplish this objective simulation results were put in ArcMap and generated the projected wheat yield maps for the time period of 2011–2090.

16.3 Results

The DSSAT model was used in simulating wheat yield while ArcGIS was used for spatially displaying the crop yield in Pothwar region from 2010 to 2090. Historical data of wheat yield and other climatic variables was used for calibrating the DSSAT Model. The regression method was used to determine the relationship between wheat yield and different climatic variables.

16.3.1 *Future Projections*

16.3.1.1 Maximum and Minimum Temperature

Global warming resulted from increased concentration of Greenhouse Gases (GHGs) in the atmosphere caused The reported average annual temperature increased was 0.6 °C during 1901–2000 in Pakistan. Annual increase in temperature was 0.24 °C over the period of 1960–2007 as compared to 0.06 °C per decade during 1990–2000 and it has been projected further increase by 1.8–4.0 °C by the end of this century (Sheikh et al. 2011). Table 16.1 depicted the decadal base maximum temperature change in the next century and showed the gradually increasing trend. In Attock district maximum temperature will increase by 2.28 °C from 2011 to 2090. Chakwal district shows maximum temperature increase 2.44 °C from 2011 to 2090. In Islamabad the capital city of Pakistan and hilly area, maximum temperature will increase up to 2.21 °C at the end of twenty-first century. In Jhelum maximum temperature will increase up to 1.98 °C from 2011 to 2090 and Rawalpindi district maximum temperature will increase to 2.07 °C in the next twenty-first century (Table 16.1). Overall maximum temperature increased in all four districts and capital city was in the range of 1.98–2.44 °C while minimum temperature was in the range of 1.87–1.93 (Table 16.1). However, the change in maximum temperature was high in Chakwal district and Jhelum as compared to other districts in the study area. The change in maximum and minimum temperature at the study sites was presented in Figs. 16.3 and 16.4 and Table16.2.

16.3.1.2 Projected Solar Radiation (MJ/m^2)

Solar Radiation will also increase due to change in climate and global warming in the next twenty-first century. In Attock district solar radiation will increase up to 0.49 (MJ/m^2) from 2011 to 2090. In Chakwal district solar radiation will increase 0.19 (MJ/m^2) in the next century from 2011 to 2090 while in Islamabad it will increase up to 0.31 (MJ/m^2) at the end of twenty-first century. In Jhelum solar

Table 16.1 Change in maximum temperature from 2011 to 2090

Years	Attock	Chakwal	Islamabad	Jhelum	Rawalpindi
2011–2090	2.28 °C	2.44 °C	2.21 °C	1.98 °C	2.07 °C
2011–2020	28.31	29.56	27.15	30.32	29.77
2021–2030	28.69	29.87	27.37	30.63	30.05
2031–2040	28.98	29.97	27.62	30.87	30.34
2041–2050	29.27	30.27	27.99	31.17	30.65
2051–2060	29.77	30.76	28.27	31.56	30.95
2061–2070	29.9	31.86	28.3	31.74	31.27
2071–2080	30.43	31.88	28.36	32.02	31.56
2081–2090	30.59	32	29.36	32.3	31.84

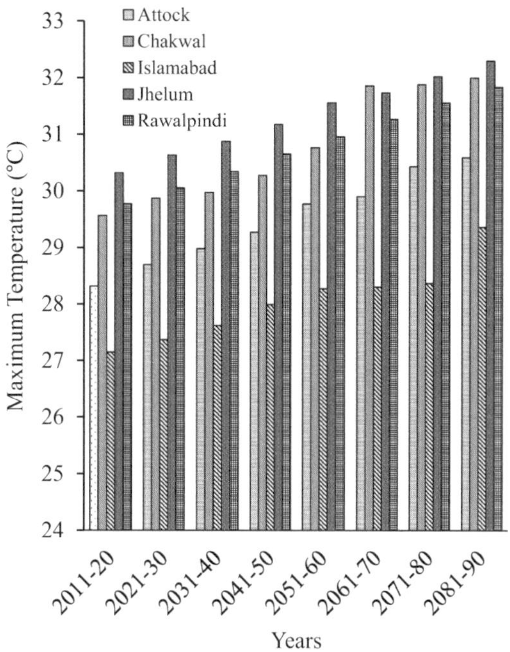

Fig. 16.3 Maximum temperature (°C) of Pothwar region

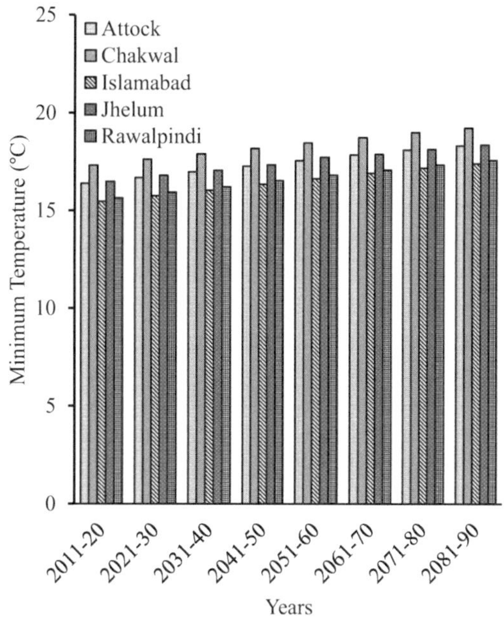

Fig. 16.4 Minimum temperature (°C) of Pothwar region

Table 16.2 Change in minimum temperature (°C) from 2011 to 2090

Projected change in minimum temperature (°C)	Attock	Chakwal	Islamabad	Jhelum	Rawalpindi
2011–2090	1.93 °C	1.9 °C	1.93 °C	1.87 °C	1.93 °C

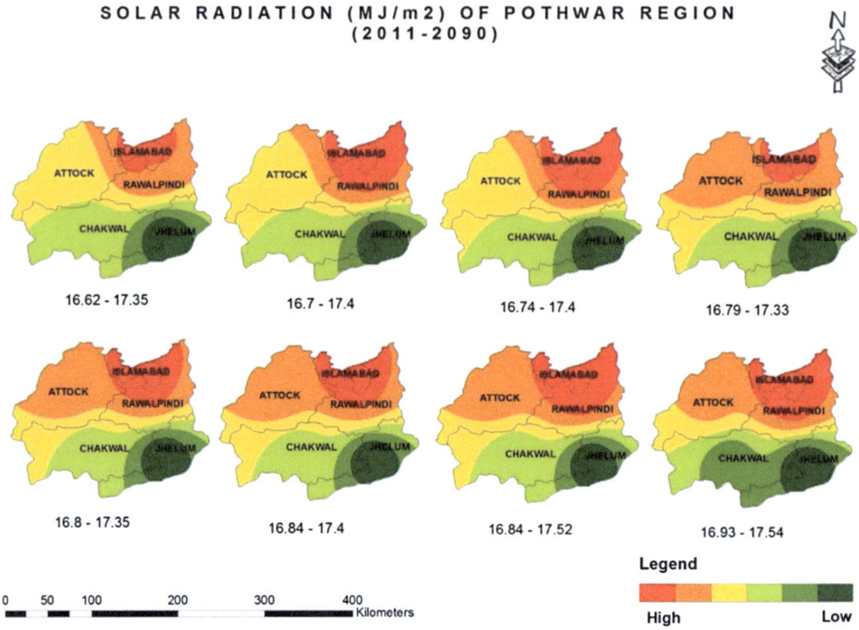

Fig. 16.5 Solar radiation (MJ/m²) from 2011 to 2090

radiation will increase up to 0.25 (MJ/m²) from 2011 to 2090. In Rawalpindi district solar radiation will increase up to 0.34 (MJ/m²) in the next twenty-first century. Overall solar radiation increased in all four districts and capital Pakistan. But the change in solar radiation was high in Attock, Rawalpindi and Islamabad as compared to other districts in the Pothwar region.

Decaded map showed that the solar radiation increase in next twenty-first century (Fig. 16.5). The first map was from 2011 to 2020 in which the projected solar radiation rises up to 17.33 (MJ/m²) in all areas of Pothwar region. The South-East part of the Pothwar region showed the highest rate of solar radiation change and these areas were presented as red color in the map. In next decade 2021–2030 projected solar radiation rise up to 17.35 (MJ/m²) in all areas of Pothwar region.

Similarly, during 2031–2040 projected solar radiation rises up to 17.35 (MJ/m²) in all districts and capital city. While in 2041–2050 projected solar radiation rise up to 17.4 (MJ/m²) in Rawalpindi, Jhelum, Attock and capital city. In general, the average increase in solar radiation from 2011 to 2090 was 17.5 MJ/m².

16.3.1.3 Rainfall

The change in rainfall pattern was observed in all districts of Pothwar region. Rainfall showed both decreasing and increasing trend in different decades from 2011 to 2090 (Fig. 16.6). In Attock district rainfall showed decreasing trend from 2011 to 2090. The change in rainfall was –0.21 mm in Attock district. In Chakwal

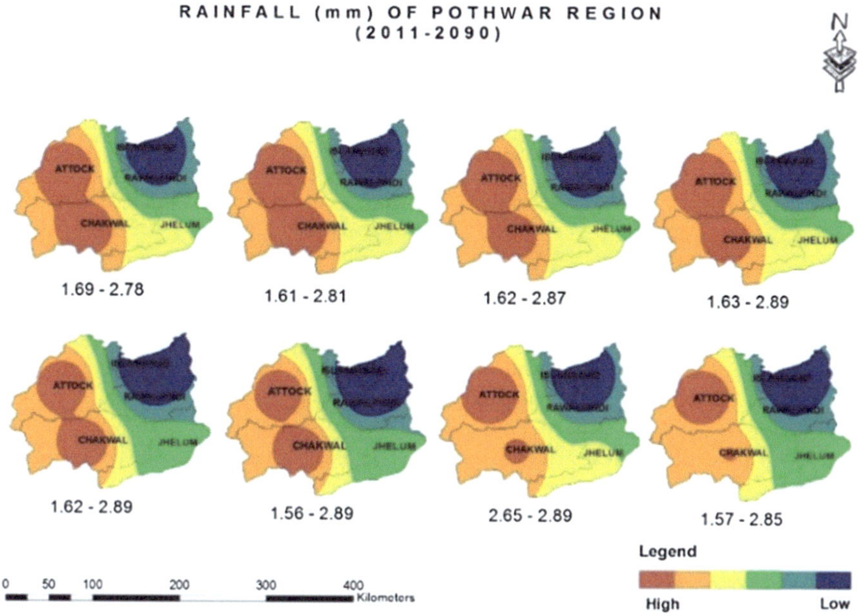

Fig. 16.6 Rainfall (mm) map of Pothwar region from 2011 to 2090

district the behavior of rainfall was increasing during 2041–2050 and 2081–2090 while all other decades showed the decreasing trend. The change in rainfall in Chakwal district was –0.19 mm. Islamabad showed the gradually decreasing behavior in twenty-first century that was –0.26 mm. Jhelum district showed very little change in rainfall from 2011 to 2090 where, the change in rainfall was about–.05 mm. Rawalpindi district also showed the decreasing trend of rainfall and the change in rainfall was about –0.1 mm from 2011 to 2090.

Rainfall maps showed the rainfall behavior in Pothwar region. Some parts of Pothwar region showed the high decreasing trend (Chakwal and Attock) and other districts (Jhelum and Rawalpindi) showed the low or minimum change in rainfall. Islamabad showed the –0.21 mm change in rainfall in the next century and Attock district shows the highest rate of change in rainfall –0.26 mm.

16.3.2 *Simulation Results*

16.3.2.1 DSSAT Yield Estimation

The simulated historical wheat yield data has been presented in Figs. 16.7, 16.8, 16.9, and 16.10 which was used for simulation. Simulation result for all locations of Pothwar region from 1981 to 2010 showed the yield of wheat crop that was compared with the observed values of wheat crop. The model was assessed with

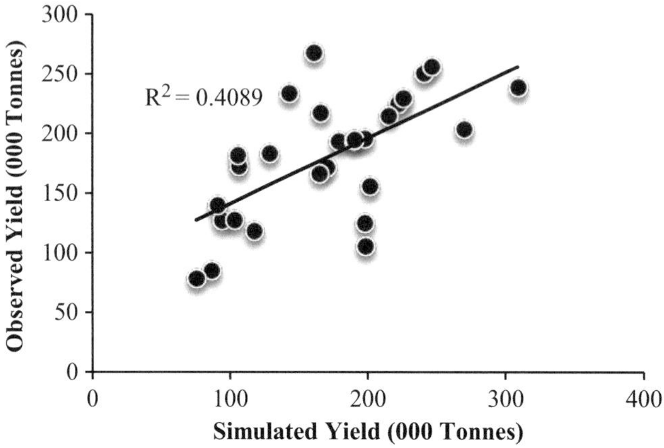

Fig. 16.7 Historical simulated wheat yield (000 tonnes) of Islamabad and Rawalpindi from 1981 to 2010

Fig. 16.8 Historical simulated wheat yield (000 tonnes) of Chakwal district from 1981 to 2010

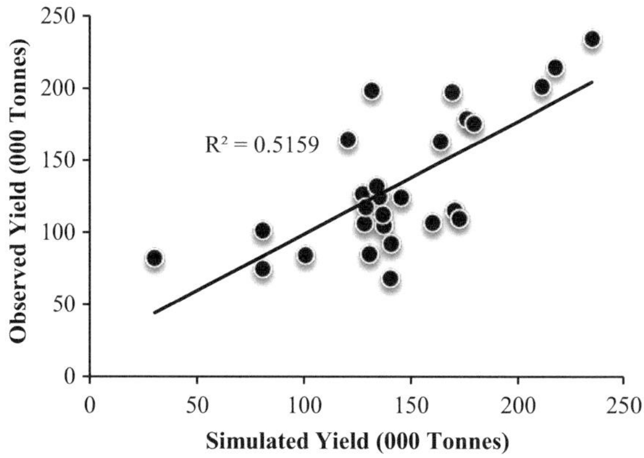

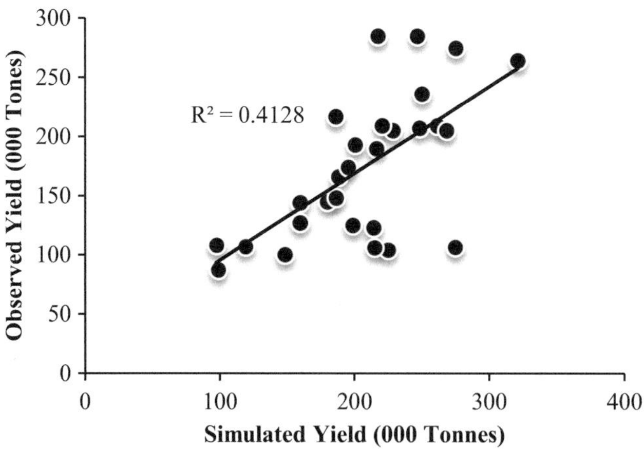

Fig. 16.9 Historical simulated wheat yield (000 tonnes) of Attock from 1981 to 2010

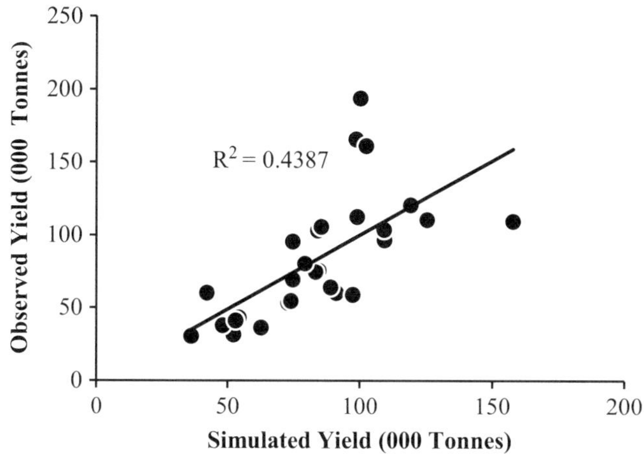

Fig. 16.10 Historical simulated wheat yield (000 tonnes) of Jhelum from 1981 to 2010

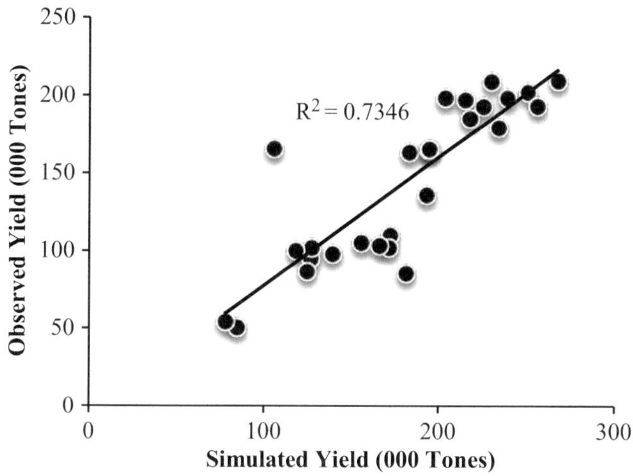

Fig. 16.11 After calibration historical simulation results of Rawalpindi and Islamabad from 1981 to 2010

observed wheat yields for the whole region and per district. Simulated mean wheat yield was larger than statistical mean of yield. The calibrated results were illustrated using Figs. 16.11, 16.12, 16.13, and 16.14 furthermore Figs. 16.15, 16.16, 16.17, 16.18, and 16.19 showed the projected wheat yield from 2011 to 2090. The projected wheat yield show the overall decreasing trend in the next century.

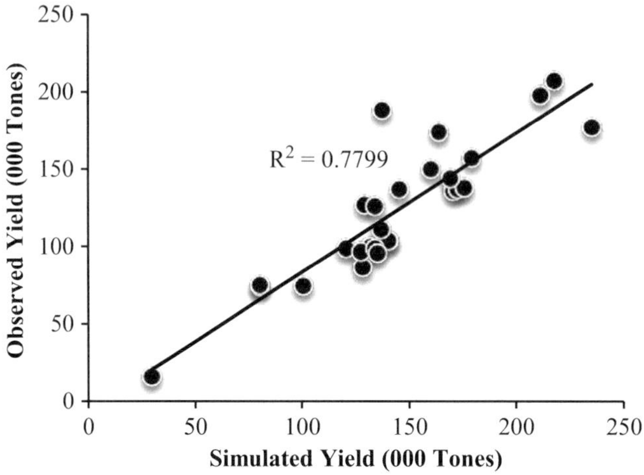

Fig. 16.12 After calibration historical simulation results of Chakwal from 1981 to 2010

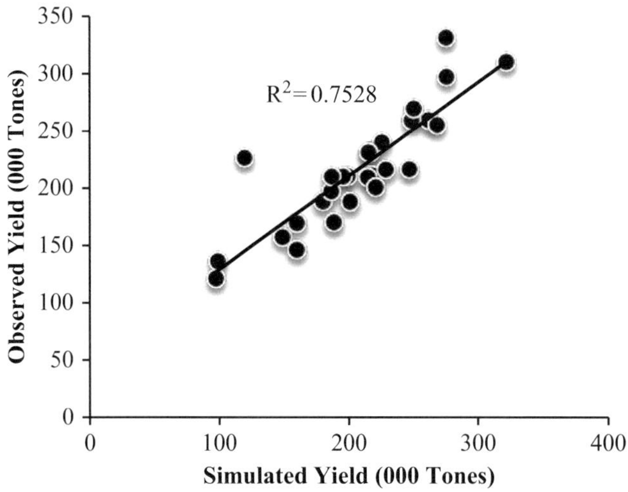

Fig. 16.13 After calibration historical simulation results of Attock from 1981 to 2010

16.3.2.2 Impact of Different Climatic Factors on Wheat Yield

Temperature is one of the important climatic variable that is important for wheat crop. Temperature effects directly and indirectly wheat yield during many developmental phases of grain filling. Different components of yield and yield severely decreased due to increase in temperature (Laghari et al. 2012). The rise in temperature during crop growth result in reduced wheat yield (IPCC 2014). The wheat yield is dependent on different climatic factors and showed the negative correlation between climatic factors and wheat yield. Temperature and other climatic factors influence the development and growth of plant through their effects on physiological processes and stomatal openings (Kayam et al. 2000).

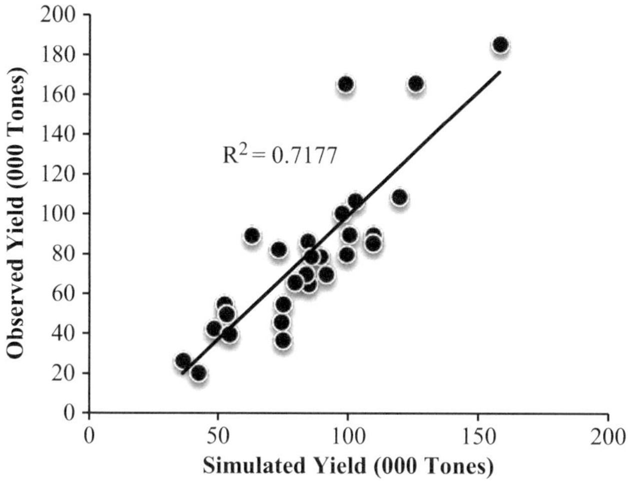

Fig. 16.14 After calibration historical simulation results of Jhelum from 1981 to 2010

Fig. 16.15 Projected wheat yield simulation results of Islamabad from 2011 to 2090

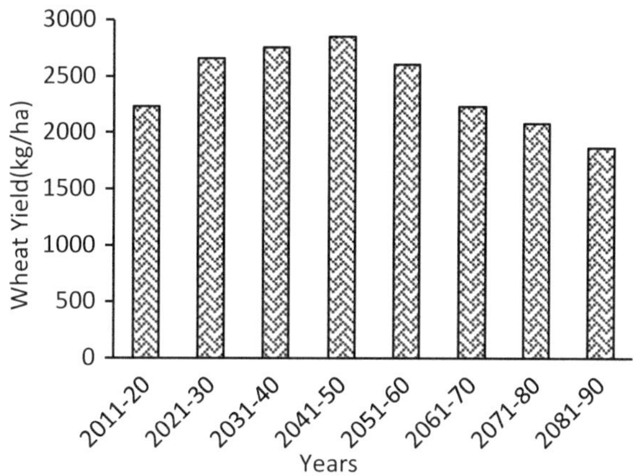

Fig. 16.16 Projected wheat yield simulation results of Rawalpindi from 2011 to 2090

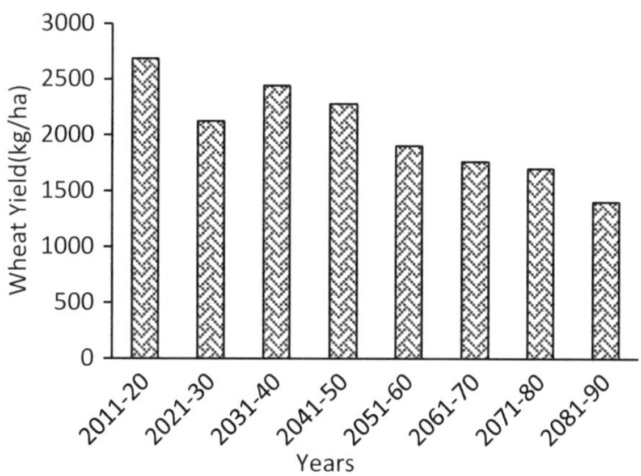

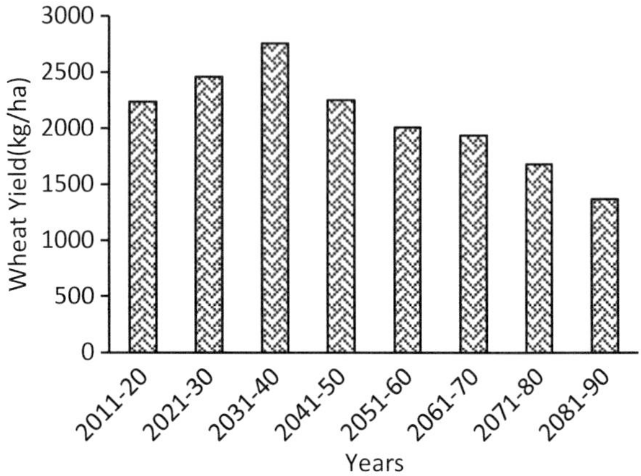

Fig. 16.17 Projected wheat yield simulation results of Attock from 2011 to 2090

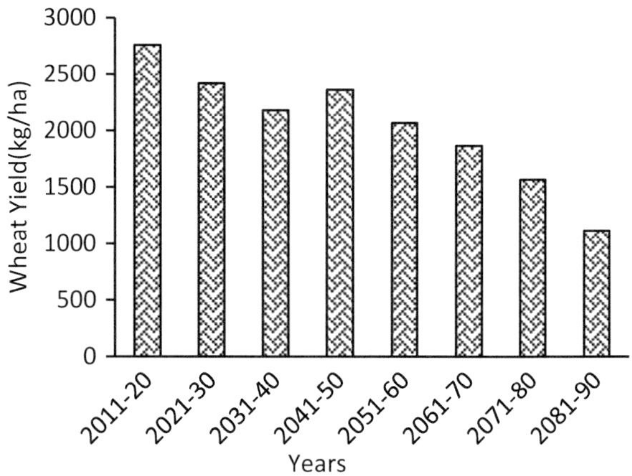

Fig. 16.18 Projected wheat yield simulation results of Jhelum from 2011 to 2090

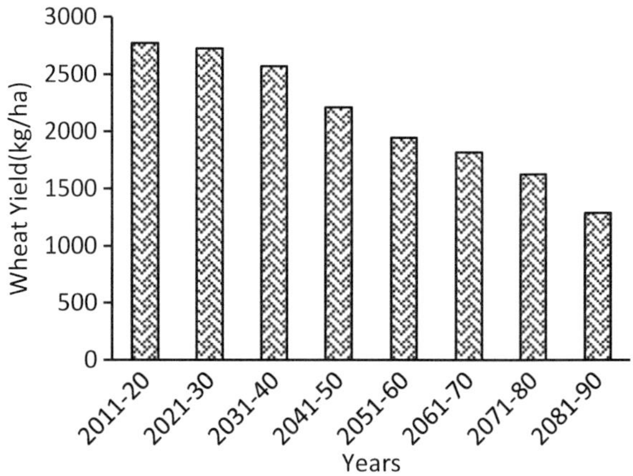

Fig. 16.19 Projected wheat yield simulation results of Chakwal from 2011 to 2090

16.3.2.3 Wheat Yield Maps

The predicted values of yield using DSSAT have been shown spatially by using ArcGIS. The predicted maps of yield was based on soil type and its different climatic variables. Figure 16.20 showed the predicted wheat yield from 2011 to 2020. Part of Jhelum and Chakwal district showed the highest wheat yield that was 2698 kg/ha and lowest yield found in some part of Attock, Rawalpindi and Jhelum areas. Wheat yield from 2021 to 2030 showed the highest value of yield in Pothwar region (2578 kg/ha, Fig. 16.21). Figure 16.22 depicted the wheat yield from 2031 to 2040 where the Jhelum, Attock, and Chakwal has the highest wheat yield (2489 kg/ha)

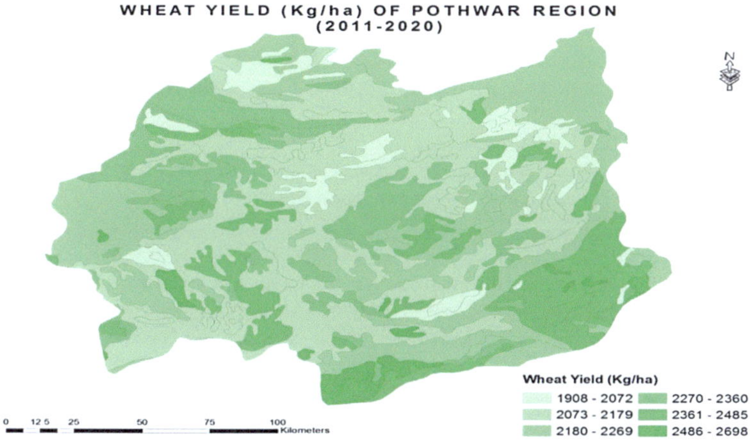

Fig. 16.20 Wheat yield (kg/ha) map of Pothwar region from 2011 to 2020

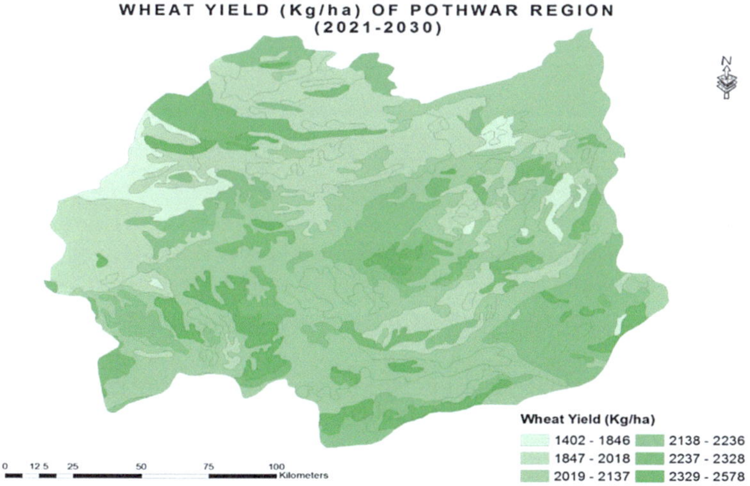

Fig. 16.21 Wheat yield (kg/ha) map of Pothwar region from 2021 to 2030

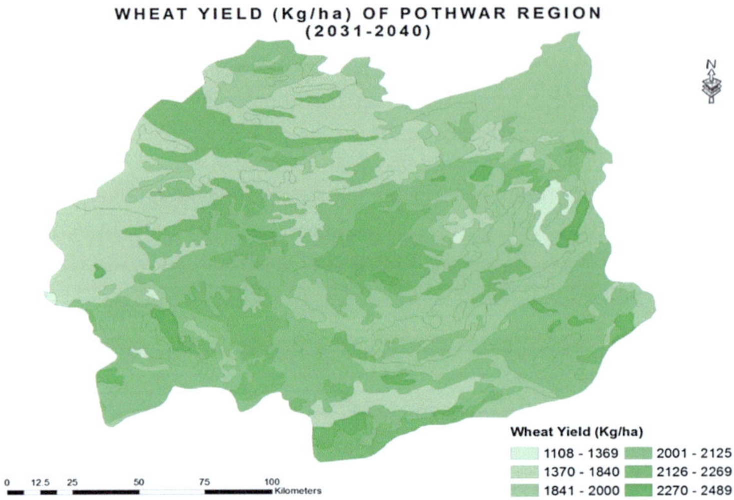

Fig. 16.22 Wheat yield (kg/ha) map of Pothwar region from 2031 to 2040

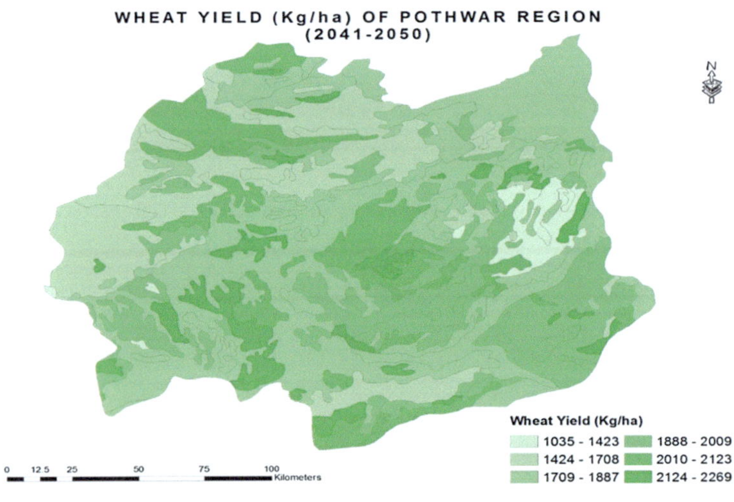

Fig. 16.23 Wheat yield (kg/ha) map of Pothwar region from 2041 to 2050

and the lowest yield in all other parts of the region (1108 kg/ha). The range of wheat yield from 2041 to 2050 as shown in Fig. 16.23 was 1035–2269 kg/ha. The wheat yield in Pothwar region from 2051 to 2060 as shown in Fig. 16.24 was in the range of 1019–2176 kg/ha, while it was 985–2174 kg/ha 2061–2070 (Fig. 16.25). Meanwhile, yield range from 2071 to 2080 was 963–2178 (Fig. 16.26). During last decade (2081–2091) the yield trend remained decreasing (Fig. 16.27).

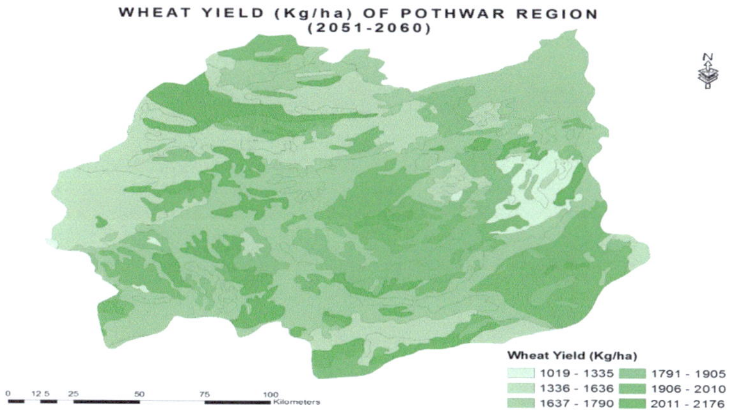

Fig. 16.24 Wheat yield (kg/ha) map of Pothwar region from 2051 to 2060

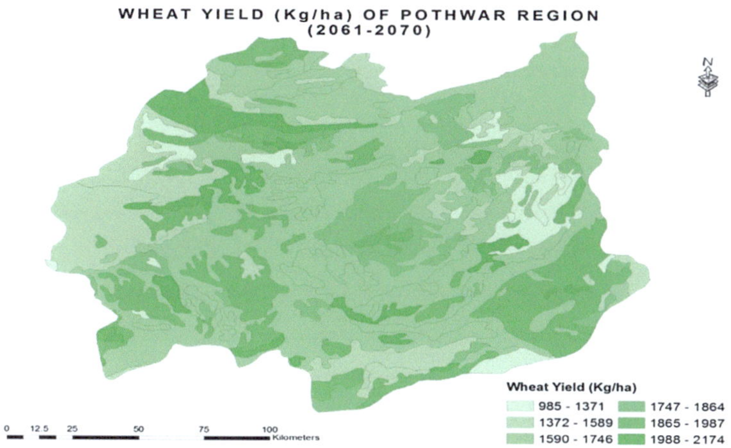

Fig. 16.25 Wheat yield (kg/ha) map of Pothwar region from 2061 to 2070

16.3.3 Statistical Analysis Results

Statistical analysis was performed by using SPSS statistical software. The results showed that the climatic variables were negatively related to wheat yield. Each climatic variable (Tmax, Tmin, Rainfall, and Solar Radiation) affect differently on wheat yield on different locations. At Attock district SPSS results showed that the Tmax, Tmin and Solar radiation effected the wheat yield more as compared to rainfall because increase in temperature shortens the growing period (Table16.3). Slope of the regression equation B showed that Tmax, Tmin, Solar radiation effect

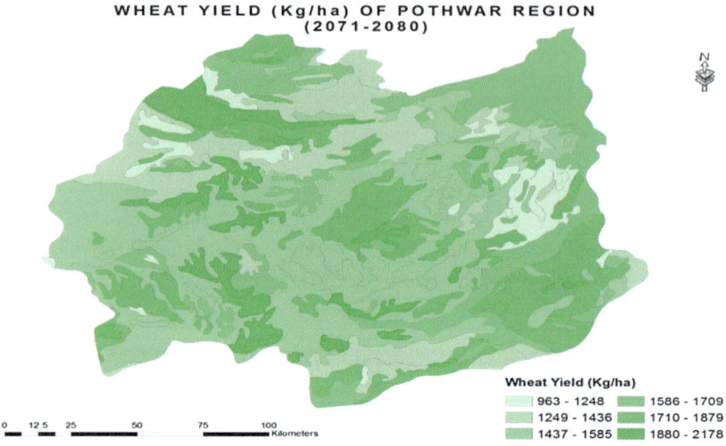

Fig. 16.26 Wheat yield (kg/ha) map of Pothwar region from 2071 to 2080

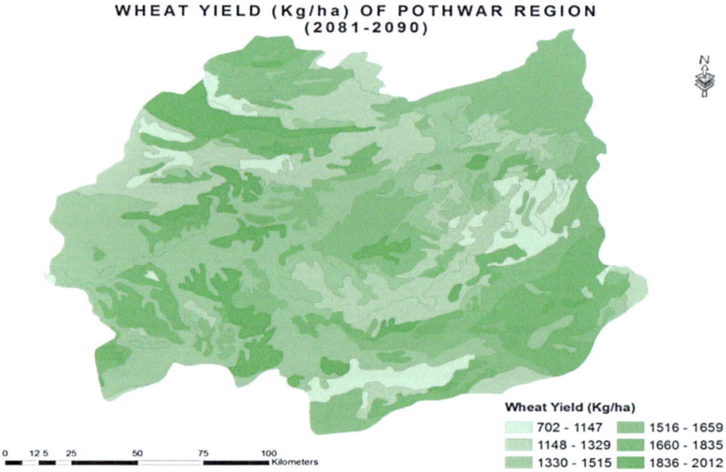

Fig. 16.27 Wheat yield (kg/ha) map of Pothwar region from 2081 to 2090

negatively on wheat yield. Table 16.3 showed Chakwal district has the same behavior like Attock district. Islamabad showed negative affect on wheat. Jhelum district and Rawalpindi districts also showed the negative behavior of climatic variables on wheat yield but rainfall showed the less negative effects as compared to other variables. The results showed that the Tmax and Tmin had negative effect on wheat yield.

Table 16.3 SPSS regression analysis results of wheat yield with different climatic variable

	R^2				B			
Area	Tmax	Tmin	Rainfall	Solar radiation	Tmax	Tmin	Rainfall	Solar radiation
Attock	0.839	0.832	0.571	0.841	−450.569	−533.321	4080.121	−2182.256
Chakwal	0.952	0.987	0.326	0.694	−512.366	−797.642	−2630.904	−4805.907
Islamabad	0.579	0.582	0.690	0.460	−295.850	−301.446	2917.192	−1712.463
Jhelum	0.946	0.937	0.026	0.936	−708.546	−738.018	−579.294	−6234.708
Rawalpindi	0.928	0.927	0.782	0.913	−536.593	−575.650	10879.389	−3619.329

16.4 Conclusion and Recommendations

16.4.1 Conclusion

With the change of climatic conditions, smallholder farmers will find very difficult to operate sustainable agriculture production therefore it is necessary to adopt appropriate strategies in place. Climate model projections using future emissions scenarios showed progressively greater changes in 2050–2090s. The increase in average temperature in Pothwar region by the end of the twenty-first century was between 2 °C and 3 °C. The results of the projections further suggest that, in the medium terms (2050s to 2090s), the abundance of water for agriculture will decline drastically under severe climate conditions. A major implication of precipitation projections is that irrigation and drainage technology are likely to become even more important in the coming decades than they are now. The crop-modeling approach proved to be very useful for evaluating crop water use across a Pothwar region.

Under the IPCC B2 Scenario, different districts have different results. The B2 scenario produces impact of climate, showing the negative effects on districts future wheat yield. The outputs of the DSSAT model confirm that in the future and the medium-term (2050s) under B2 Scenario, the rain fed wheat yield would reduce in all districts. The decrease for the districts located in the south and south east of the Pothwar region is relatively higher as compared to rest of the districts. Generally, the climate change impact on Pothwar region on wheat yield would have more negative impacts. This high sensitivity of crop shows that they should receive more attention. However, some factors, aspects, and uncertainties have not been fully involved in this analysis, like the effects of diseases and pests on crop, future adaptation approaches, which would also modify the results on crop yields. Since the projected climate change impacts on crop yields would be more severe, several strategies are suggested to farmers that they should change according to the new conditions of climate. The simplest one strategy is advance crop sowings taking benefit of the lengthy frost-free time. As the shift in growing season is evidenced from the climate change results, it is worth concluding that shift in planting dates is an effective adaptive measure among many suggested measures.

16.4.2 Recommendations

The current research study recommend the following measures:

- A1B and A2 scenario should be studied for detailed research and predictions in industrial areas
- By altering irrigation specifications proposed methodology can be implemented in irrigated areas
- Similar methodology could be adopted for different crops in other parts of the country for better mitigation of future food security.

References

Buriro, M., Fateh Chand Oad, Muhammad Ibrahim Keerio, Shamsuddin Tunio, A.W. Gandahi, S.W.U. Hassan, and A.S.M. Oad. 2011. Wheat seed germination under the influence of temperature regimes. *Sarhad Journal of Agriculture* 27: 5.

Herrero, M., Claudia Ringler, Jeannette Van DE Steeg, Philip Thornton, Tingju Zhu, Elizabeth Bryan, Abisalom Omolo, Jawoo Koo, and A. Notenbaert. 2010. Climate variability and climate change and their impacts on Kenya's agricultural sector. In *ILRI*, ed. I. L. R. I. Nairobi: International Livestock Research Institute (ILRI).

Hingane, L.S., K. Rupa Kumar, and B.V. Ramana Murty. 1985. Long-term trends of surface air temperature in India. *Journal of Climatology* 5: 521–528.

IPCC. 2006. 2006 IPCC guidelines for national greenhouse gas inventories. Prepared by the National Greenhouse Gas Inventories Programme, ed. H.S. Eggleston, L. Buendia, K. Miwa, T. Ngara, and K. Tanabe. Toyama: IGES.

IPCC. 2014. Climate change 2014: Impacts, adaptation, and vulnerability. Part A: Global and sectoral aspects. Contribution of Working Group II to the Fifth Assessment Report of the Intergovernmental Panel on Climate Change, ed. C.B. Field, V.R. Barros, D.J. Dokken, K.J. Mach, M.D. Mastrandrea, T.E. Bilir, M. Chatterjee, K.L. Ebi, Y.O.Estrada, R.C. Genova, B. Girma, E.S. Kissel, A.N. Levy, S. MacCracken, P.R. Mastrandrea, and L.L. White. New York: Cambridge University Press.

Janjua, P.Z., G. Samad, and N.U. Khan. 2010. Impact of climate change on wheat production: A case study of Pakistan. *The Pakistan Development Review* 49(4): 799–822.

Jones, P.D., and K.R. Briffa. 1992. Global surface air temperature variations during the twentieth century: Part 1, spatial, temporal and seasonal details. *The Holocene* 2: 17.

Kayam, Y., U. Ozsoy, J. Lomas, O. Oden, M. Mandel, and M. Gurbuz. 2000. *The impact of climate change on wheat production of Aegean region in Turkey: The effect of a reduction of rainfall and the increase of temperature on wheat yields*. Available online at http://www.fao.org/sd/climagrimed/pdf/ws01_26.pdf

Kothawale, D.R., J.V. Revadekar, and K. Rupa Kumar. 2010. Recent trends in pre-monsoon daily temperature extremes over India. *Journal of Earth System Science* 119: 51–65.

Laghari, K.A., M.A. Sial', and M.A. Arain. 2012. Effect of high temperature stress on grain yield and yield components of wheat (Triticum aestivum L.). *Science, Technology and Development* 31: 83–90.

Ortiz, R., K.D. Sayre, B. Govaerts, R. Gupta, G.V. Subbarao, T. Ban, D. Hodson, J.M. Dixon, J. Iván Ortiz-Monasterio, and M. Reynolds. 2008. Climate change: Can wheat beat the heat? *Agriculture Ecosystems & Environment* 126: 46–58.

Parry, M.L., C. Rosenzweig, A. Iglesias, M. Livermore, and G. Fischer. 2004. Effects of climate change on global food production under SRES emissions and socio-economic scenarios. *Global Environmental Change* 14: 53–67.

Sayed, A.H. 2011. *Climate change and its realities for Pakistan*. Lahore: Changing Environmental Pattern and its impact with Special Focus on Pakistan.

Shakoor, U., S. Abdul, A. Ikram, and A.Q. Mohsin. 2011. Impact of climate change on agriculture: Empirical evidence from arid region. *Pakistan Journal of Agricultural Sciences* 48: 327–333.

Sheikh, M.M., M.M. Iqbal, G. Ali, and A.M. Khan. 2011. *Global warming in the context of Pakistan: Major concerns and remedial strategies. Changing environmental pattern and its impact with special focus on Pakistan.*

Chapter 17
A Role of Bioinformatics in Agriculture

Zohra Aslam, Jabar Zaman Khan Khattak, Mukhtar Ahmed, and Muhammad Asif

Contents

Abstract Bioinformatics is an interdisciplinary science emerging from interaction of computer, statistics, biology and mathematics to analyze genome arrangement and contents, biological sequence data, predict the structure and function of

Z. Aslam (✉) • J.Z.K. Khattak
Department of Bioinformatics, International Islamic University Islamabad, Islamabad, Pakistan
e-mail: zohra.msbi37@iiu.edu.pk; zzohra.aslam@gmail.com; jabar.khattak@iiu.edu.pk

M. Ahmed
Department of Agronomy, Pir Mehr Ali Shah Arid Agriculture University, Rawalpindi, Pakistan

Department of Biological Systems Engineering, Washington State University, Pullman, WA, USA
e-mail: ahmadmukhtar@uaar.edu.pk; mukhtar.ahmed@wsu.edu

M. Asif
Department of Agronomy, 2004 Throckmorton Plant Sciences Center, Kansas State University, 1712 Claflin Road, Manhattan, KS 66506-0110, USA
e-mail: asifrana@ksu.edu

M. Ahmed, C.O. Stockle (eds.), *Quantification of Climate Variability, Adaptation and Mitigation for Agricultural Sustainability*,
DOI 10.1007/978-3-319-32059-5_17

macromolecules that use in interpreting and decoding plant genome. The broad amounts of data produced in life sciences resulted to the evolution and development of bioinformatics. Omics, bioinformatics and computational tools are very essential to understand genomics and the molecular systems that underlie several plant functions. Various new omic layers such as genome, hormonome, metabalome, interactome, and epigenome analysis have emerged by technological advances. Such integration of information enables and facilitates the identification of expression of gene which helps to interpret the relationship between phenotype and genotype, thus approving from genome to phenome system-wide analysis. Earlier biological research that used laboratories, plant clinics and field is now at *In-silico* or computers level (Computational). Bioinformatics develops software, algorithms, databases and tools of data analysis to make discoveries and infer the information. Application of various bioinformatics tools and databases enable analysis, storage, annotation, visualization and retrieval of outcomes to helps enhanced understanding in living system research. Thus it will help to improve the plant quality based on health care disease diagnosis. In this chapter we describe the bioinformatics approaches (databases and tools) in plant science and implication of next generation sequencing technology (NGS) on crop genetics.

Keywords Bioinformatics • Omics • Agriculture • Sustainable development • Sequencing technology

Abbreviations

NGS	next generation sequencing technology
ETLs	economic trait loci
DNA	deoxyribose nucleic acid
EST	expressed sequence tag
EMBL	European molecular biology laboratory
BLAST	Basic Local Alignment Search Tool

17.1 Introduction

17.1.1 The Bioinformatics Era in twenty-first Century

> The data explosion produced by Human Genome Project has called forth the conception/creation of novel discipline-bioinformatics, whose focus is on the storage, acquisition, analysis, distribution and modeling of various types of information embedded in protein and DNA sequence data.

In the amount of biological information (genomic era) a gigantic explosion have been seen due to advances in genomics and molecular biology. The deluge of genomic information has led to the requirement for databases, specialized tools and computerized methods to organize, store, index, analyze and view the data. Computational biology and bioinformatics are rooted deep in life sciences as well

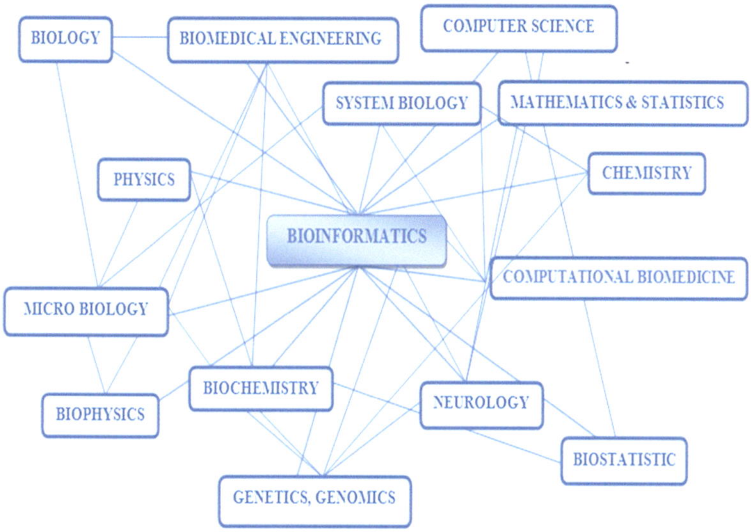

Fig. 17.1 Bioinformatics, an interdisciplinary field

as computer technologies and information sciences. Both approaches maintain close interaction along life sciences to comprehend their full potential.

Bioinformatics could be defined as "enable logical interrogation of structuring of living system information". Bioinformatics is an interdisciplinary field involving physics, mathematics, behavioral sciences, engineering biology, statistics and computer science to analyze and predict structure and function of macromolecules, biological sequence data and genome content as depicted in (Fig. 17.1). The bioinformatics field has emerged as the pressing task involves interpretation and analysis of data related to protein, nucleotide and amino acid sequences and expression of central dogma of cellular biology. This actually (organizing and analyzing data) referred to:

- The implementation and development of tools and software which integrate different type of information related to plant phenomics
- Use of innovative statistics/algorithms to evaluate interactions among large datasets such as to predict protein structure or function and locate gene within a sequence (Vassilev et al. 2005).

Omictechnologies are expanded form of bioinformatics and it sits as an umbrella over biotechnology. Bioinformatics field facing challenges like, to build direct links between genetics (traditional) and across the omic platform complex data integration-through the observed phenotype of plant and the transcriptome, genome, metabalome and proteome. Researchers now demands candidate functions for DNA sequence or for proteins, projected structures. Biotechnology wants smart examining and sieving of diverse, intricate data forms to address particular problems, ranging from wide fields of research to individual knowledge. Meanwhile bioinformatics is increasing and elaborating its applications together with the advancement of postgenomic and new 'omic' technologies, its strength and focus remain in the genomes and DNA sequence analysis (Batley and Edwards 2008, 2009).

This new knowledge of bioinformatics concerned not only with the news handling, facts and information collection but it also deals with the information extraction procedures from the collected data. Bioinformatics have intellectual impacts on different field such as ecology, botany, biology, agronomy, environment and human health.

17.2 Why Bioinformatics?

The community of molecular biology facing the greatest challenge to convert huge amount of data into sensible form which has been produced by sequencing of projects related to genome. The research related to living system such as molecular and genetics level was conducted in the laboratories traditionally but now due to advancement in bioinformatics all information can be generated by the use of computer technologies (Singh et al. 2011).

Bioinformatics emerged as a significant tool to use large amount of data generated by omic technologies in order to deduce logical conclusion about problems. Since use of omic technologies are increasing with great pace with its utilization in different disciplines of sciences. Therefore, it might have vital role in future decision making process under wide spectrum of problems. The use of computer software in the system biology will further help to promote bioinformatics application (Batley and Edwards 2008, 2009). Similarly, by the use of new databases and tools in the molecular biology field the researcher will be able to carry out the exploration and investigation not only genomic level but also at transcriptome, metabalome and proteome. Today the bioinformatics community confronted challenges like efficient and intelligent storage of reliable data generated in vast amount and easy approach to the data. Therefore, such type of computational tools needs to be developed which can extract deep, significant and consequential biological information. The latest example of use and implementation of bioinformatics is in the field of pharma industry to implement bioinformatics tools to depreciate cost and time in the development of molecular markers and drugs (Untergasser et al. 2007).

17.3 Bioinformatics in Agriculture

With the development of modern technologies, bioinformatics came to age with the ability at an ever-decreasing cost to produce sequence information in large amount. Bioinformatics in agriculture (agri-informatics) playing an increasing role. Genome projects for Poaceae family plants like rice and corn plants are in progress and generated genomic data could be used to develop disease insect, pests control programs. The discovery of new genes using computer software were aimed at targeting or improving seed quality, adding micronutrients of plants for human health (nutritional genomics) and engineering plants to deal/cope with metals (phytoremediation). The genetic or genomic data are being shifted to find genes associated with desirable phenotypes called economic trait loci (ETLs), infectious disease resistance and hereditary disorders.

The desired plants for the industrial purposes could be produced by using omics technologies as it can provide guideline from gene to organism level. The fundamental aims of plant genomics research are to have lower cost, safer food, improve food supply raw materials for higher quality and better process ability. Genome-wide research by novel approaches could be used to enhance the effectiveness and efficiency of breeding for plants improvement. The food quality could be increased by the use of biotechnology accelerated by bioinformatics and genomics. High-throughput gene discovery initiated in 1991 by sequencing of expressed sequence tag (EST); set the searchable and large sequence database requirement. Although in many crops the sequencing of EST for gene discovery is still the standard procedure. Reduction in the DNA sequencing cost compelled sequencing of whole-genome (Batley and Edwards 2009). In 2000 by the Arabidopsis Genome Initiative, plant genomics was reformed by the release of Arabidopsis thaliana whole genome sequence. The completion of rice genome sequence was announced by public consortia, 2 years later. Owing to the association/similarities between important crop species and rice (*Oryza sativa*) at genomic level, genome of rice completion had a meaning full influence on both crop bioinformatics and biotechnology (Edwards and Batley 2004).

The genome sequencing of animals and plants will provide tremendous benefits for community of agriculture. Bioinformatics tools and methods could be utilized to find/pursuit specific genes inside the particular genomes which might be beneficial for community related to agriculture and to annotate their functions. The distinct and peculiar genetic information might be later used to have resistant crops related to drought, insects and diseases and to improve the livestock quality making them more productive, healthier and disease resistant.

Plant genomes comparative genetics (genome structure and functions relationship across different strains) revealed that genes organization over evolutionary time has remained more sustained and conserved than was formally believed (Mace et al. 2008; Matthews et al. 2003; Caetano-Anolles 2005). These outcomes recommend that knowledge obtained from crop systems model could be utilized to propose perfections and advancements to food crops such as wheat, rice and maize. *Triticum aestivum* (Wheat), *Oryza sativa* (Rice), *Zea mays* (Maize), and *Arabidopsis thaliana* (water cress) are models of entire ground plant genomes (Paterson et al. 2005; Varshney et al. 2006; Crespi 2013).

The huge amount of sequence data as well as complete genome sequences, is directing on other views on how this data can be interrogated and organized. The great redundancy level in programs of gene innovation is being compressed through reference to complete genome or consensus sequences. Closely related synthetic sequence for a specific crop can be used if complete genome sequence is not available. The ever-growing databases of DNA sequence size remain to accelerate/push competences of bioinformatics, and there is an increasing demand to condense extravagant or redundant data.

The databases development has been supplemented by advancement in data analysis tools, enabling researchers for valuable biological knowledge to mine complex/complicated interacting data and to fully annotate sequences (Edwards and Batley 2004; Batley and Edwards 2008, 2009).

17.3.1 Enhancement for Plant Tolerance Aligned with Stresses (Abiotic and Biotic)

Insect genomics application helps in finding the target sites (novel) and resistance mechanism identification. *Bacillus thuringiensis* genes control severe pests and it has been transferred effectively to potato, maize and cotton. The added characteristics (resist insect attack) resulted to build defensive mechanism in plants means reduced the amount of insecticides that being used.

Roots are the principal organ of plant's which build defense against different type of abiotic stresses. If the plant holding by soil is strong and diverse biologically, the plant might have a greater chance of persisting/enduring stressful situations. Generally the adaptive response of plants to stresses is not so much quick although they are extremely sensitive. Meanwhile adaptive mechanism also differs from species to species. The response of plants to different stresses such as cold/drought/high temperature remained different so every plant have unique trend. Barely any of the feedback or response was found similar. Therefore, plants might become endangered even extinct and their population threatened, where and when abiotic stress is severe specially. The computational and in silico genomics technologies could be used by the scientists to identify gene-enzyme resistance to disease with their transcription factor and promoter region which could help to start resistance mechanism and to augment immunity (Kummerfeld and Teichmann 2006; Pandey and Somssich 2009).

17.3.2 Enhance Nutritional Quality in Depleted Soils

The aim of Nutritional genomics (gene-diet disease interaction) is to provide dietary interventions and study or analysis of susceptible genes of particular species against diseases so that remedy measures can be opted. Recently researchers have succeeded to increase the level of iron, micronutrients and vitamin A into rice by transferring genes, and this could have intellectual impact in reducing anemia and blindness occurrences caused by iron and Vitamin A deficiency respectively. Similarly, technologists of plants have transplanted a gene into tomato from yeast which resulted to increased shelf life of tomato (Singh et al. 2011).

Bioinformatics play a key role to detect metals from contaminated soil from Metagenomics sequencing (Handelsman 2004). The greatest intricate or complex microbial populations lives in soil (arguably houses) and because of complicated sets of interrelating gradients, its ancient history and stable, comparatively resource poor, isolating and protective physical structure. Incredibly this results to different or various set of gene sequence. Advancement has been completed in cultivating and evolving cereal cultivars that have free iron and aluminum toxicities and a greater tolerance for alkalinity of soil (Singh et al. 2011).

17.4 Plant Genomic Databases and Tools of Similarity Searching in Bioinformatics

Crop plants form a vital and large part of global diet; from paper making to pharmaceuticals, their products are highly significant in manufacturing industries. Therefore, studies of plant genomes are of great economic importance. The ever-growing population places increasing demand to improve characteristics of crops such as disease resistance, nutritional value, environmental factors and yield on plant breeders by new technologies such as genetic modifications or by conventional breeding. The key to attain said improvement and advancement lies increasingly in having approach to the latest genomic data (Dicks 2000).

The omic/genomics exponential growth is because of computational tasks of consistently and scientifically storing, gathering, manipulating, consolidating, analyzing and envisioning huge amount of living system information emanate/derive from trials conducted by researchers. Thus, in the comprehensive sense, bioinformatics could be considered as hub of scientific and infrastructure framework in which researchers might take information and convert into knowledge with the help of computers (Singh et al. 2011). Recent development such as new software, standards and tools in bioinformatics have produced better access to the data held within them across the internet as depicted in Fig. 17.2. Since it is fact that bioinformatics discipline is recently organized; there is a remarkable and extraordinary multiplicity of resources of bioinformatics currently available (some are freely available and some are suited to specific tasks ideally).

The rapid growth in the DNA sequence information required the development of databases of particular DNA sequence. In 1986 largest sequence databases emerged from association of GenBank and European molecular biology laboratory (EMBL). This meta-sequence database comprises over 7.4 million DNA sequences of plant and is considered to be standard repository for worldwide public DNA sequences. Based on numerous crops, number of sequencing projects increase in different laboratories, has been contributing progress in databases (Edwards and Batley 2004; Batley and Edwards 2008, 2009).

A database forms a substantial part of any genome project. Database resources can systematize large quantities of data with its hierarchal structures and complex relationships. It is clear that by linking together data sources, much knowledge may be inferred for plant genomes (Dicks 2000). Sequence comparison and assembly

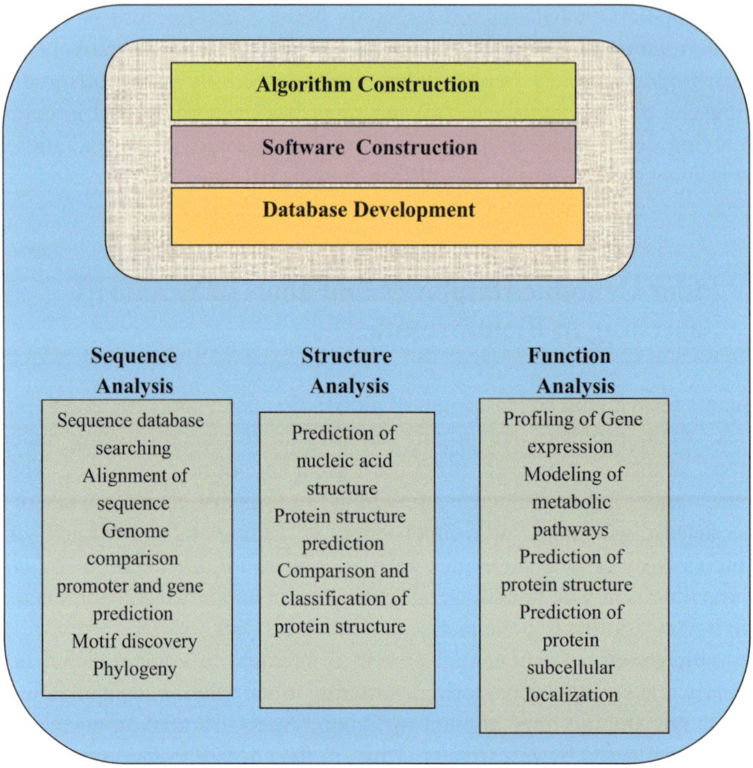

Fig. 17.2 Computation and applications of bioinformatics under sequence, structure and function analysis

tools have been grown resulting to the enlargement of datasets. Tools for sequence assembly and similarity search are basis of various software applications for examining, analyzing and interpreting crop genome information (Edwards and Batley 2004; Duran et al. 2009; Batley and Edwards 2009). Plant genomic databases play a key role in the dissemination and archiving data proceeding from the international genome projects (Dicks 2000). Many of the analysis tools and databases described (Table 17.1), here are available on web as user friendly interfaces.

17.4.1 Sequence and Function Analysis of Protein

In biology the "sequence analysis" term implies subjecting a peptide sequence or DNA to sequence alignment, repeated sequences searches, sequence databases or other methods of bioinformatics on a computer. Sequence alignment in bioinformatics is ordering the sequences of RNA, protein or DNA to find similarity regions that may be a significance of structural, evolutionary or functional relationship

Table 17.1 Plant/crop Bioinformatics online databases

Databases	Description/Organisms	References/URL
BGI Rice Information System	Genome data of rice	Zhao et al. (2004)
Gateway of Brassica Genome	Brassica	Brassica Genome Gateway (2012)
ChloroplastDB	Chloroplast genome data available	Mochida and Shinozaki (2010)
CR-EST	The Crop Expressed Sequence Tag database	Kunne et al. (2005)
Cry-Bt identifier	Database of Cry genes	Singh et al. (2009)
Genomics Database OF Cucurbit	Cucurbitaceae	Cucurbit Genomics Database (2012)
CyanoBase	Cyanobacteria	Mochida and Shinozaki (2010)
DDBJ	DNA Data Bank of Japan	Tateno et al. (2002)
EMBL	European Molecular Biology Laboratory nucleotide sequence database	Lai et al. (2012)
Ensemble Plants	Whole genome data of plant species	EnsemblPlants (2012)
EXPASY	Index to other plant-specific databases	http://www.expasy.org/links.html
		ExPASy (2012)
Genome Database for Rosaceae (GDR)	Rosaceae	Mochida and Shinozaki (2010)
Genevestigator	Gene networks in *Arabidopsis* and rice	https://www.genevestigator.com/gv/
		(Genevestigator (2012))
GRAINGENES	Small-grain crops Genome database	Matthews et al. (2003)
GRAMENE	Cereal genome database	Jaiswal et al. (2006)
GRIN	Plant genetic resources	http://www.ars-grin.gov/
		Germplasm Resources Information Network (2012)
ICIS	International Crop Information System	Fox and Skovmand (1996)
INSDC	International Nucleotide Sequence Database Collaboration	Sugawara et al. (2008)
IRFGC	International Rice Functional Genomics Consortium	Bruskiewich et al. (2006)
IRIS	International Rice Information System	The International Rice Information System (2010)

(continued)

Table 17.1 (continued)

Databases	Description/Organisms	References/URL
KEGG PLANT	Whole genome data of plant species	Mochida and Shinozaki (2010)
KOME database	Oryza Molecular Biological Encyclopedia	Bruskiewich et al. (2006)
LIS	Legume Information System	Lai et al. (2012)
MaizeGDB	Maize	MaizeGDB (2012)
MATDB	*Arabidopsis thaliana*	http://mips.helmholtz-muenchen.de/plant/athal/ (MAtDB (2012))
MOsDB	MIPS rice database	Bruskiewich et al. (2006)
MPSS	Massive parallel signature sequencing gene expression data	Nakano et al. (2006)
NASC	*Arabidopsis thaliana*	http://arabidopsis.info Beale et al. (2002)
NCBI	National Center for Biotechnology Information	Pruitt et al. (2007)
OMAP	Comparative genome physical maps of *Oryza* wild relatives	Ammiraju et al. (2006)
OryGenesDb	Reverse genetics for rice	Droc et al. (2008)
Oryzabase	NIG Oryza genetics database	Integrated Rice Science Database (2012)
OryzaSNP	IRFGC hosted rice single nucleotide polymorphism survey	http://www.oryzasnp.org Clark et al. (2007)
PGD	Plant Genome Data base	Dong et al. (2005)
PGV	Pathogenic Genome Viewer	Marla and Singh (2007)
Phytozome	Plant species (whole genome data available)	Mochida and Shinozaki (2010)
PinTFDB	Plant transcription factor database	Riano-Pachon et al. (2007)
PLACE db	Plant cis-acting regulatory DNA elements database	Higo et al. (1999)
PlantCare	Plant cis-acting regulatory DNA elements database	Higo et al. (1999)
PlexDB	Plant expression data	http://www.plexdb.org/ (Dash et al. (2012))
PRGdb	Platform for plant resistance gene analysis	Sanseverino et al. (2009)

(continued)

Table 17.1 (continued)

Databases	Description/Organisms	References/URL
RAP DB	"Rice Annotation Project" database	Ouyang et al. (2007)
RARGE	Arabidopsis	Mochida and Shinozaki (2010)
RED	Rice expression database	Yazaki et al. (2002)
Rice Array Db	NSF-funded oligo rice gene expression array	Jung et al. (2008)
Rice Blast	*Magnaporthe grisea* genomics	http://www.riceblast.org
Rice Genome Project/IRGSP	International Rice Genome Sequencing Project	Bruskiewich et al. (2006)
Rice Proteome Database	NIAS rice proteome database	Bruskiewich et al. (2006)
RICE-GAAS	Rice Genome Automated Annotation System	Sakata et al. (2002)
RIKEN	*Arabidopsis* and rice functional genomics data	Bruskiewich et al. (2006)
SIGnAL	Arabidopsis	Kayoko et al. (2003)
SOL genomics network	Solanaceae	Sol genomic network (2012)
SoyBase	Soybean	Mochida and Shinozaki (2010)
TAIR	The *Arabidopsis* Information Resource	The Arabidopsis Information Resource (TAIR) (2012)
T-DNA Rice Insertion lines	(Gyn An) Korean T-DNA rice insertion mutants	Plant functional genomics laboratory (2012)
TIGR Rice	rice genome database	Bruskiewich et al. (2006)
Tos17 rice mutants	NIAS rice TOS 17 insertion mutants	Rice *Tos17*Insertion Mutant Database (2012)
TREP	The Triticeae Repeat Sequence Database	(Wicker et al. 2002; Wicker and Buell 2009)
Wheat genome information	Wheat	Lai et al. (2011)
Yale Plant Genomics	Gene expression from rice tissues	Yale Plant Genomics (2012)

among the sequences. Sequence analysis used to assign function on the basis of resemblances between compared sequences to proteins and genes (Stormo 2000). Now a day many techniques and tools are available that could provide sequence alignment comparisons. This could further analyze the alignment product to explain, understand and interpret biological information. BLAST (Basic Local Alignment Search Tool) method is used abundantly for similarity search. It is a heuristic (speed enhancement) amendment of smith-waterman algorithm. Statistical methods in BLAST determine the possibility of specific alignment among sequence regions or sequence that occur by chance and given the composition and size of the databases being searched. The BLAST can be controlled by number of different criteria or parameters which determines the expression of particular characteristics and these demand to be considered carefully (Smith and Waterman 1981). The tools set help us to do more detailed analysis on our concerning or query sequence including mutation identification, compositional biases, CpG islands, hydropathy regions and evolutionary analysis. In bioinformatics sequence alignments are valuable for producing and developing phylogenetic trees, homology models of protein structure and identifying sequence similarity (Singh et al. 2011). In functional studies of protein we contrast the protein succession to derive databases of secondary proteins which have information on protein signatures, domain as well as motifs. Extremely important or indicative hits in opposition to pattern databases can help to estimate the biochemical activity of concerning protein. Profile analysis also acknowledged as motif finding creates global MSA (multiple sequence alignment) which is aimed to arrange conserved undersized sequence motifs amongst the query set sequence. Set of Profile matrices are constructed after the isolation of highly conserved regions. For every conserved constituency, profile matrix is aligned similar to scoring matrix. The incidence count derived at each position from character distribution of conserved region's for each nucleotide or amino acid rather than from empirical distribution. Then other sequences for motif incidence are searched out by profile matrices (Singh et al. 2011). The highly sequential datasets availability sanctions withdrawal of biological characteristics for example, simple sequence repeats molecular markers and single nucleotide polymorphism, which may be used in research of plant biotechnology as in mapping of inherited traits. The accessibility to complete-genome sequence further facilitates mining of dogmatic characteristics such as sequences of novel promoters as well as micro-RNAs. Such tertiary level explanation accommodates or linked to both complex regulatory mechanisms as well as phenotype that response to the environment and governs development (Edwards and Batley 2004; Batley and Edwards 2009).

17.4.2 Insilco Studies of Transcription Factors (TFs) and Cis-acting Elements

DNA binding transcription factors that are sequence-specific are very important molecular switches which influence as well as control several directions including response to climatic variations and development. Genome-wide detection in plants for gene conservation was reported first by transcription factors encoding of Arabidopsis genome (Riechmann et al. 2000). Transcription of eukaryotes is very difficult than prokaryotes. Primarily, DNA or hereditary material and its transcription thus are restricted to nucleus and at this stage nuclear membrane and cytoplasm

Table 17.2 Databases of transcription factor in plants

Databases	Description/Species	References
DRTF	Database of transcription factor	Gao et al. (2006)
DATF	Database of *Arabidopsis* transcription factor	Guo et al. (2005)
PlantTFDB	22 species of plant	Guo et al. (2008)
RARTF	RIKEN *Arabidopsis* transcription factor	Lida et al. (2004)
SoybeanTFDB	Soybean	Mochida et al. (2009)
LegumeTFDB	*Lotus japonicas, Medicago truncatula,* Soybean	Mochida et al. (2010)
AGRIS, AtTFDB	*Arabidopsis*	Palaniswamy et al. (2006)
PinTFDB	20 species of plant	Riano-Pachon et al. (2007)
TOBFAC	Tobacco	Rushton et al. (2008)
DBD	>700 plant species	Wilson et al. (2008)
GRASSIUS, GrassTFDB	Sugarcane, sorghum, maize, rice	Yilmaz et al. (2009)
DPTF	Poplar	Zhu et al. (2007)

are separated from each other. In mitochondria DNA is already exists for transcription, mitochondria use specified RNA polymerase. This concedes for sequential control of gene expression in nucleus through RNA sequestration and allow for RNAs selective transport to site where ribosome reside or the cytoplasm (Singh et al. 2011).

The core promoter of genes encoded protein among eukaryotes is within about 50 bps (base pairs) upstream of transcription initiation site, and contains RNA polymerase II and binding sites. Further UCE's (upstream control elements) is provided transcriptional regulation, generally present in beginning site of upstream within a range of about 200 bps. Sometimes, a TATA box is enclosed for Pol II in core promoter which is extremely preserved sequence of DNA detection for TBP (TATA box binding protein) with its binding at the promoter begins assembly of transcription complex. Combination of enhancers and these UCE's amplified and regulate the basal transcription complex formation. The *cis*-acting elements can regulate gene transcription within the DNA regulatory regions as well as *trans*-acting elements which comprise basal transcription complex and transcription factors (Singh et al. 2011). Presently, various catalogs are used which presents genes statistics putatively in several plant species encoding TFs. These are based on generally computational tools or methods such as hidden Markov model search conserved DNA-binding areas and sequence similarity search (Table 17.2). Recently, integration of TFs-encoding genes data sets has been performed, hence establishing knowledge-based, integrative resource of TFs in terms of comparative genomics of/regarding transcription regulating networks across related plant species (Mochida and Shinozaki 2010).

17.5 Next Generation Sequencing (NGS) Technology

With advancement made in the field of nanotechnology, informatics and microfluidics, alternative technologies have recently emerged to increase the throughput and rapidity of DNA sequencing. The term Next Generation Sequencing (NGS) technology is used to explain techniques collectively other than Sanger sequencing (Service 2006). It is possible to resequence the entire genomes of plants using NGS technology in greater depth, more economically and efficiently than ever before.

The NGS techniques are capable to produce nucleotide sequence data economically at faster pace than that of traditional technology (Sanger Sequencing methodology) (Varshney et al. 2009a; Crespi 2013). The NGS technologies are either available commercially or in development (Shendure et al. 2004; Kling 2005). Commercially available NGS technologies such as Solexa/Illumina, AB SOLiD and Roche/454 have previously exhibited the capability to evade the Sanger sequencing restraining features (Hudson 2008). Currently, Solexa/Illumina, AB SOLiD and Roche/454 are the predominant technologies used in breeding applications and crop genetics. Although Roche/454 is superior in terms of longer sequence reads to Solexa/Illumina and AB SOLiD and highest data output is greater or for both AB SOLiD as well as Solexa/Illumina (Gupta 2008). In terms of sequence data generation or costs per run, Roche/454 is more costly than any AB SOLiD or Solexa/Illumina techniques (Varshney et al. 2009b; Crespi 2013).

17.5.1 Bioinformatics Tools for Analysis of NGS Data

NGS technologies generated DNA sequence reads are diminutive Sanger sequences (traditional method), that make analysis and assembly of NGS data challenging. The NGS technologies can also generate data files of terabyte-sized with each run of instrument, considerably increasing the computer resource sequencing laboratories requirements. Although several algorithms and tools of bioinformatics are presently available (Table 17.3), intentions are in progress to enhance or advance the NGS data arrangement in several laboratories (Varshney et al. 2009a; Crespi 2013).

Most of the software and tools accommodate limited analysis and assembly. However, because Next Generation Sequencing technologies are suited particularly for variation discovery and for resequencing for SNP (single nucleotide polymorphism), available software is biased against this application. However, the gap is open; but still there is a thrust for the advancement of better platforms and bioinformatics tools to assist the sequence analysis of next generation sequencing data in a trustworthy, accessible and efficient way (Varshney et al. 2009a; Crespi 2013).

Table 17.3 Bioinformatics tools for analysis of NGS data

Sequence Variant Discovery Tools	
SNPsniffer	Tool for discovery of SNP designed specifically for Roche/454 sequences (Varshney et al. 2009a).
SeqMap:	Tool to detect indels and multiple substitutions and map short sequences to reference genome (Jiang and Wong 2008).
ssahaSNP:	Tool to detect indels and homozygous SNPs (Varshney et al. 2009b).
Atlas-SNP:	Tool for invention of SNPs as well as indel through genome resequencing by means of NGS technologies (Wheeler et al. 2008).
Assembly, Alignment and visualization Tools	
Velvet:	Tool for short reads as well as paired assembly of de novo (Zerbino and Velvet 2008).
GMAP:	Program for alignment of map using minimal memory plus time with genome sequence as well as cDNA sequences (Wu and Watanabe 2005).
EULER:	To produce short-read assembly and assist Sanger sequencing assembly of combined reads as well as NGS (Chaisson and Pevzne 2008).
RMAP	Tool for aligning of short reads with reference genome (Varshney et al. 2009b).
MOSAIK	Tool used for pair wise configuration of NGS data (Varshney et al. 2009b).
SOAP	Program for alignment of gapped as well as ungapped short reads to assist smRNA discovery, Pair-end resequencing, reference sequences and mRNA tag sequence mapping (Li et al. 2008).
SHARCGS	Device for short reads assembly of de novo (Varshney et al. 2009a).
Zoom	Tool to carry out post-analysis and millions of short reads mapping with reference genomes (Varshney et al. 2009a, b).
VCAKE	Tool with robust error detection for de novo assembly of short reads (Jeck et al. 2007).
EagleView	Display tool to inspect the excellence of genome assembly visually (Varshney et al. 2009b).
Integrated Tools	
PanGEA	Tool for mapping of NGS data to entire genomes, with the identification of SNP (Kofler et al. 2009).
CLCbio Genomics Workbench	Tool for Sanger sequence data, NGA reference assembly and de novo (Varshney et al. 2009b).
SeqMan genome analyzer	Software with capability of detection of SNPs and aligning of NGS and Sanger data (Varshney et al. 2009b).
MAQ	Short reads mapping and assembly program. It may also use simple assembly visualizer to report SNPs as well as indel (Maqview) (Varshney et al. 2009b).
NextGENe™	Software to investigate NGS data for transcriptome analysis, de novo assembly, SNP and indel detection (Varshney et al. 2009b).

17.5.2 Uses of NGS Technology

NGS technology has been applied on a wider scale. For example, number of plant species developing SNP based markers, both where reference genome is accessible (Medicago and Arabidopsis) and where it is not (Eucalyptus and maize). NGS technologies used for draft sequencing where reference genome sequences are not available through methods including pool of BACs (Bacterial artificial chromosomes) clones that can facilitate or promote quick genome assembly. Interestingly, the Next generation sequencing technologies for orphan or so-called minor crop species are proved valuable for efficient as well as rapid improvement in genomic possessions (Varshney et al. 2009b; Crespi 2013). NGS technologies are fast and becoming excellent or valuable techniques for analysis of gene expression particularly for species where reference genome sequences are available already (Weber et al. 2007; Cheung et al. 2006). Intentions are underway to avail/use NGS technologies for mapping, population biology, alien and wide crosses introgression, epigenetic modification and association mapping (Varshney et al. 2009a; Crespi 2013).

17.5.3 Prospects of NGS for Crop Improvement

The initial goal of NGS technologies was resequencing. They are presently being used in various crop species to explore/investigate de novo genome sequencing, including common bean, pigeon pea and wheat. NGS technologies have significant suggestions/implications for breeding and crop genetics. The development of genomic resources at large scale, including sequence data and transcript, physical and genetic maps and molecular markers along with further potential applications is significant. Genome sequencing (both de novo and resequencing) and transcriptome is increasing for crop plants using NGS technology. In the amount of available genomic data, NGS technique use becomes a quantum leap for crops such as pigeon pea and chicken pea for which many genomic resources were not available (Varshney et al. 2009b).

An important use of NGS technologies is in studies of gene appearance, for which NGS technologies has capability to substitute trials of microarray in coming years; in contradiction to further approaches of gene demonstration like real-time PCR along with microarray. NGS applications render insights into temporal and spatial gene expression control due to their capability to recognize all records of RNA developed at a particular occasion (Varshney et al. 2009a). NGS technologies may also hasten the conversion skill development for crop plants as it will be effortless to alter/modify genes due to convenience of increasing genomic statistics. While at present, the analysis of wider-scale NGS technologies has a challenge, advancement is being made for this task both in developing new approaches and improving existing tools (Varshney et al. 2009a).

17.6 Future Horizons

Organisms include bacteria, fungi, viruses, nematodes, oomycetes, viroids, phytoplasmas, virus-like organisms, parasitic plants and protozoa cause infectious diseases. Plant pathology concerns the study of origin, nature, course of disease, pathogen identification, syndrome cycle, resistance of plant diseases, epidemiology of plant diseases, plant disease management, pathosystem genetics and how diseases of plant affect animals and humans. Genome progression from parasitic plants, viruses, virus-like organism, fungi, bacteria, viroids, oomycetes, nematodes, protozoa along with phytoplasmas gives opportunities to analyze and understand interaction of plant-pathogens which helps to disease diagnose, make transgenic plant (disease resistance) and management (Kim et al. 1997). Gene-to-gene assumption articulate that plants holds foremost resistance single R genes that contain complementary avirulence genes that recognize pathogens specifically. Virulence genes may be described as genes that program protein production in pathogens which is recognized temporarily by plants (that contain complementary R genes) indirectly or directly.

Plants should protect themselves to survive from several pathogens. Few ramparts are activated by pathogen recognition and some are very essential (like a variety of anti-microbial agents). The recognition process comprises the semi-dominant and dominant product of resistance gene (R) exists in the plant as well as *Avr* (leading avirulence aspect) derived or programmed by pathogen. *Avr* factor detection by host plant simulates and activates plant's defenses by starts signal transduction pathways thus compromising the pathogen ability to colonize plant. The interaction between pathogens and plants are dynamic, simple and complex.

Several strategies have been recently envisaged for characterization, functional analysis and identification of plant genes involved in signaling, responses to abiotic and biotic stresses and triggering. *In-silico* biology, Bioinformatics and computational techniques play key function to identify pathogen plant interface at genome with genes of pathogens as well as plants (Wan et al. 2002).

17.7 Conclusion

Bioinformatics is a rapidly evolving and expanding field. Understanding and interpreting specific functions of plants that arise in particular plant species is necessary to discover useful genes to enhance and promote plant functions. The integration of bioinformatics and omics data sets from diverse plant species is then important and essential to promote/enhance translational research (to engineer plant systems) in an outcome or response to the emanating demands of mankind. We are also confronting the various global issues such as food, water, climate changes and global warming. Globally, for computational biology the international society serves as a global community of experiment or practice in the field. Meanwhile for bioinformatics in Asia, the Asia Pacific Bioinformatics Network (apbionet.org) is a good regional source.

References

Ammiraju, J.S.S., M.Z. Luo, J.L. Goicoechea, W.M. Wang, D. Kudrna, C. Mueller, J. Talag, H. Kim, N.B. Sisneros, B. Blackmon, E. Fang, J.B. Tomkins, D. Brar, D. Mackill, S. Mccouch, N. Kurata, G. Lambert, D.W. Galbraith, K. Arumuganathan, K.R. Rao, J.G. Walling, and N. Gill. 2006. The Oryza bacterial artificial chromosome library resource: Construction and analysis of 12 deep-coverage large-insert BAC libraries that represent the 10 genome types of the genus Oryza. *Genome Research* 16: 140–147.

Batley, J., and D. Edwards. 2008. Bioinformatics: Fundamentals and applications in plant genetics and breeding. In *Principles and practices of plant molecular mapping and breeding*, ed. C. Kole and A.G. Abbott, 269–302. Enfield: Science Publishers, Inc.

Batley, J., and D. Edwards. 2009. Mining for Single Nucleotide Polymorphism (SNP) and Simple Sequence Repeat (SSR) molecular genetic markers. In *Bioinformatics for DNA sequence analysis*, ed. D. Posada, 303–322. New York: Humana Press.

Beale, M., P. Dupree, K. Lilley, J. Beynon, M. Trick, J. Clarke, M. Bevan, I. Bancroft, J. Jones, S. May, S.K. de Van, and O. Letser. 2002. GARNet, the Genomic Arabidopsis Resource Network. *Trends in Plant Science* 7(4): 145–147.

Brassica Genome Gateway. 2012. http://brassica.bbsrc.ac.uk/. Accessed 29 Oct 2012.

Bruskiewich, R., T. Metz, and G. Mclaren. 2006. Bioinformatics and crop information systems in rice research. *IRRN* 31(1): 5–12.

Caetano-Anolles. 2005. Gene-interleaving patterns of synteny in the Saccharomyces cerevisiae genome: Are they proof of an ancient genome duplication event? *Crop Science* 45: 1809–1816.

Chaisson, M.J., and P.A. Pevzne. 2008. Short read fragment assembly of bacterial genomes. *Genome Research* 18: 324–330.

Cheung, F., B.J. Haas, S.M.D. Goldberg, G.D. May, Y. Xiao, and C.D. Tow. 2006. Sequencing Medicago truncatula expressed sequenced tags using 454 life sciences technology. *BMC Genomics* 7: 272–282.

Clark, R.M., G. Schweikert, C. Toomajian, S. Ossowski, G. Zeller, P. Shinn, N. Warthmann, T.T. Hu, G. Fu, D.A. Hinds, H. Chen, K.A. Frazer, D.H. Huson, B. Schölkopf, M. Nordborg, G. Rätsch, J.R. Ecker, and D. Weigel. 2007. Common sequence polymorphisms shaping genetic diversity in *Arabidopsisthaliana. Science* 317(5836): 338–342.

Crespi, M. 2013. *Root genomics and soil interactions*, 296. Ames: Wiley.

Cucurbit Genomics Database 2012. http://www.icugi.org/. Accessed 29 Oct 2012.

Dash, S., J. Van Hemert, L. Hong, R.P. Wise, and J.A. Dickerson. 2012. PLEXdb: Gene expression resources for plants and plant pathogens. *Nucleic Acids Research* 40(1): 1194–1201.

Dicks, J. 2000. Plant genome databases: From references to inferences tools. *Briefing in Bioinformatics* 2: 138–150.

Dong, Q., C.J. Lawrence, S.D. Schlueter, M.D. Wilkerson, S. Kurtz, C. Lushbough, and V. Brendel. 2005. Comparative plant genomics resources at PlantGDB. *Plant Physiology* 139: 610–618.

Droc, G., M. Ruiz, P. Larmande, A. Pereira, P. Piffanelli, J.B. Morel, A. Dievart, B. Courtois, E. Guiderdoni, and C. Perin. 2008. OryGenesDB: A database for rice reverse genetics. *Nucleic Acids Research* 34: 736–740.

Duran, C., D. Edwards, and J. Batley. 2009. Molecular marker discovery and genetic map visualisation. In *Applied bioinformatics*, ed. D. Edwards, D. Hanson, and J. Stajich, 165–189. New York: Springer.

Edwards, D., and J. Batley. 2004. Plant bioinformatics: From genome to phenome. *Trends in Biotechnology* 22(5): 232–237.

Ensembl Plants. 2012. http://plants.ensembl.org/index.html. Accessed 29 Oct 2012.

Expasy. 2012. http://www.expasy.org/links.html. Accessed 30 Oct 2012.

Fox, P.N., and B. Skovmand. 1996. The International Crop Information System (ICIS) – connects gene bank to breeder to farmer's field. In *Plant adaptation and crop improvement*, ed. M. Cooper and G.L. Hammer, 317–326. Wallingford/Oxon: CAB Inter.

Gao, G., Y. Zhong, A. Guo, Q. Zhu, W. Tang, W. Zheng, X. Gu, L. Wei, and J. Luo. 2006. DRTF: A database of rice transcription factors. *Bioinformatics* 22: 1286–1287.

Genevestigator. 2012. https://www.genevestigator.com/gv/. Accessed 29 Oct 2012.

Germplasm Resources Information Network. 2012. http://www.ars-grin.gov/. Accessed 10 Oct 2012.

Guo, A., K. He, D. Liu, S. Bai, S. Gu, L. Wei, and J. Luo. 2005. DATF: A database of Arabidopsis transcription factors. *Bioinformatics* 21(10): 2568–2569.

Guo, A.Y., X. Chen, G. Gao, H. Zhang, Q.H. Zhu, X.C. Liu, Y.F. Zhong, X. Gu, K. He, and J. Luo. 2008. PlantTFDB: A comprehensive plant transcription factor database. *Nucleic Acids Research* 36: 966–969.

Gupta, P.K. 2008. Ultrafast and low-cost DNA sequencing methods for applied genomics research. *Proceedings of the National Academy of Sciences, India* 78: 91–102.

Handelsman, J. 2004. Metagenomics: Application of genomics to uncultured microorganisms. *Microbiology and Molecular Biology Reviews* 68: 669–685.

Higo, K., Y. Ugawa, M. Iwamoto, and T. Korenaga. 1999. Plant cis-acting regulatory DNA elements (PLACE) database:1999. *Nucleic Acids Research* 27(1): 297–300.

Hudson, M. 2008. Sequencing breakthroughs for genomic ecology and evolutionary biology. *Molecular Ecology Resources* 8: 3–17.

Integrated Rice Science Database. 2012. http://www.shigen.nig.ac.jp/rice/oryzabase/top/top.jsp.

Jaiswal, P., J. Ni, I. Yap, D. Ware, W. Spooner, K. Youens-Clark, L. Ren, C. Liang, W. Zhao, K. Ratnapu, B. Faga, C. Payan, M. Fogleman, C. Hebbard, S. Avraham, S. Schmidt, T.M. Casstevens, E.S. Buckler, L. Stein, and S. Mccouch. 2006. Gramene: A bird's eye view of cereal genomes. *Nucleic Acids Research* 34: 717–723.

Jeck, W.R., J.A. Reinhardt, D.A. Baltrus, M.T. Hickenbotham, V. Magrini, E.R. Mardis, J.L. Dangl, and C.D. Jones. 2007. Extending assembly of short DNA sequences to handle error. *Bioinformatics* 23(21): 2942–2944.

Jiang, H., and W.H. Wong. 2008. SeqMap: Mapping massive amount of oligonucleotides to the genome. *Bioinformatics* 24: 2395–2396.

Jung, K.H., C. Dardick, L.E. Bartley, P. Cao, J. Phetsom, P. Canlas, Y.S. Seo, M.S. Shultz, Q. Ouyang, B.C. Yuan, E. Frank, L. Ly, Y. Zheng, A.P. Jia, K. Hsia, H.H. An, D. Chou, G.C. Rocke, P.S. Lee, G. Schnable, C.R. Buell An, and P.C. Ronald. 2008. Refinement of light-responsive transcript lists using rice oligonucleotide arrays: evaluation of gene-redundancy. *PLoS ONE* 3(10): 3337.

Kayoko, Y., J. Lim, J.M. Dale, H. Chen, P. Shinn, C.J. Palm, A.M. Southwick, H.C. Wu, C. Kim, M. Nguyen, P. Pham, R. Cheuk, G. Karlin-Newmann, S.X. Liu, B. Lam, H. Sakano, T. Wu, G. Yu, M. Miranda, H.L. Quach, M. Tripp, C.H. Chang, J.M. Lee, M. Toriumi, M.M.H. Chan, C.C. Tang, C.S. Onodera, J.M. Deng, K. Akiyama, Y. Ansari, T. Arakawa, J. Banh, F. Banno, L. Bowser, S. Brooks, P. Carninci, Q. Chao, N. Choy, A. Enju, A.D. Goldsmith, M. Gurjal, N.F. Hansen, Y. Hayashizaki, C. Johnson-Hopson, V.W. Hsuan, Kei Iida, M. Karnes, S. Khan, E. Koesema, J. ISHIDA, P.X. JIANG, T. Jones, J. Kawai, A. Kamiya, C. Meyers, M. Nakajima, M. Narusaka, M. Seki, T. Sakurai, M. Satou, R. Tamse, M. Vaysberg, E.K. Wallender, C. WONG, Y. Yamamura, S. Yuan, K. Shinozaki, R.W. Davis, A. Theologis, and J.R. Ecker. 2003. Empirical analysis of transcriptional activity in the Arabidopsis genome. *Science* 302: 842–846.

Kim, E., H. Kosack, and D.G.J. Jonathan. 1997. Plant disease resistance genes. *Annual Reviews of Plant Physiology and Plant Molecular Biology* 48: 575–607.

Kling, J. 2005. The search for a sequencing thoroughbred. *Natoliya Biotechnology* 23: 333–1335.

Kofler, R., T.T. Torres, T. Lelley, and C. Schlötterer. 2009. PanGEA: Identification of allele specific gene expression using the 454 technology. *BMC Bioinformatics* 10: 143.

Kummerfeld, S.K., and S.A. Teichmann. 2006. DBD: A transcription factor prediction database. *Nucleic Acids Research* 34: 74–81.

Kunne, C., M. Lange, T. Funke, H. Miehe, T. Thiel, I. Grosse, and U. Scholz. 2005. CR-EST: A resource for crop ESTs. *Nucleic Acids Research* 33(Database issue): D619–D621.

Lai, K., P.J. Berkman, M.T. Lorenc, C. Duran, L. Smits, S. Manoli, J. Stiller, and D. Edwards. 2011. Wheat genome.info: An integrated database and portal for wheat genome information. *Plant Cell Physiology* 53(2): e2.
Lai, K., M.T. Lorenc, and D. Edwards. 2012. Genomic databases for crop improvement. *Agronomie* 2(1): 62–73.
Li, R., Y. Li, K. Kristiansen, and J. Wang. 2008. SOAP: Short oligonucleotide alignment program. *Bioinformatics* 24: 713–714.
Lida, K., M. Seki, T. Sakurai, M. Satou, K. Akiyama, T. Toyoda, A. Konagaya, and K. Shinozaki. 2004. Genome wide analysis of alternative pre-mRNA splicing in *Arabidopsis thaliana* based on full length cDNA sequences. *Nucleic Acid Research* 32(17): 5096–5107.
Mace, E.S., R. Varshney, V. Mahalakshmi, K. Seetha, A. Gafoor, Y. Leeladevi, and J.H. Crouch. 2008. In silico development of simple sequence repeat markers within the aeschynomenoid/dalbergoid and genistoid clades of the Leguminosae family and their transferability to Arachishypogaea, groundnut. *Plant Science* 174(1): 51–60.
MaizeGDB. 2012. http://www.maizegdb.org/.
Marla, S., and V.K. Singh. 2007. PGV – Prokaryotic Genome Viewer. *Silico Biology* 7(4–5): 543–555.
MATDB. 2012. http://mips.helmholtz-muenchen.de/plant/athal/.
Matthews, D.E., V.L. Carollo, G.R. Lazo, and O.D. Anderson. 2003. Grain genes, the genome database for small-grain crops. *Nucleic Acids Research* 31: 183–186.
Mochida, K., and K. Shinozaki. 2010. Genomics and bioinformatics resources for crop improvement. *Plant Cell Physiology* 51(4): 497–523.
Mochida, K., T. Yoshida, T. Sakurai, K. Yamaguchi-Shinozaki, K. Shinozaki, and L.S. Tran. 2009. In silico analysis of transcription factor repertoire and prediction of stress responsive transcription factors in soybean. *DNA Research* 16: 353–369.
Mochida, K., T. Yoshida, T. Sakurai, K. Yamaguchi-Shinozaki, K. Shinozaki, and L.S. Tran. 2010. Legume TFDB: An integrative database of Glycine max, Lotus japonicus and Medicago truncatula transcription factors. *Bioinformatics* 26: 290–291.
Nakano, M., K. Nobuta, K. Vemaraju, S.S. Tej, J.W. Skogen, and B.C. Meyers. 2006. Plant MPSS databases: Signature-based transcriptional resources for analyses of mRNA and small RNA. *Nucleic Acids Research* 34: 731–735.
Ouyang, S., W. Zhu, J. Hamilton, H. Lin, M. Campbell, K. Childs, F. Thibaud-NISSEN, R.L. Malek, Y. Lee, L. Zheng, J. Orvis, B. Haas, J. Wortman, and C.R. Buell. 2007. The TIGR rice genome annotation resource: Improvements and new features. *Nucleic Acids Research* 35: 883–887.
Palaniswamy, S.K., S. James, H. Sun, R.S. Lamb, R.V. Davuluri, and E. Grotewold. 2006. AGRIS and AtRegNet. A platform to link cis-regulatory elements and transcription factors into regulatory networks. *Plant Physiology* 140: 818–829.
Pandey, S.P., and I. Somssich. 2009. The role of WRKY transcription factors in plant immunity. *Plant Physiology* 150(4): 1648–1655.
Paterson, A.H., M. Freeling, and T. Sasaki. 2005. Grains of knowledge: Genomics of model cereals. *Genome Research* 15: 1643–1650.
Plant Functional Genomics Laboratory. 2012. http://www.postech.ac.kr/life/pfg/.
Pruitt, K.D., T. Tatusova, and D.R. Maglott. 2007. NCBI reference sequences (RefSeq): A curated non-redundant sequence database of genomes, transcripts and proteins. *Nucleic Acids Research* 35: 61–65.
Riano-Pachon, D.M., S. Ruzicic, I. Dreyer, and B. Mueller-Roeber. 2007. PLNTFDB: An integrative plant transcription factor database. *BMC Bioinformatics* 8: 42–53.
Rice *Tos17* Insertion Mutant Database. 2012. http://tos.nias.affrc.go.jp/.
Riechmann, J.L., J. Heard, G. Martin, L. Reuber, C. Jiang, J. Keddie, L. Adam, O. Pineda, O.J. Ratcliffe, R.R. Samaha, R. Creelman, M. Pilgrim, P. Broun, J.Z. Zhang, D. Ghandehari, B.K. Sherman, and G. Yu. 2000. Arabidopsis transcription factors: Genome-wide comparative analysis among eukaryotes. *Science* 290(5499): 2105–2110.

Rushton, P.J., M.T. Bokowiec, T.W. Laudeman, J.F. Brannock, X. Chen, and M.P. Timko. 2008. TOBFAC: The database of tobacco transcription factors. *BMC Bioinformatics* 9: 53–55.

Sakata, K., Y. Nagamura, H. Numa, B.A. Antonio, H. Nagasaki, A. Idonuma, W. Watanabe, Y. Shimizu, I. Horiuchi, T. Matsumoto, T. Sasaki, and K. Higo. 2002. RiceGAAS: An automated annotation system and database for rice genome sequence. *Nucleic Acids Research* 30: 98–102.

Sanseverino, W., G. Roma, M.D. Simone, M. Faino, S. Melito, E. Stupka, L. Frusciante, and M.R. Ercolano. 2009. PRGdb: A bioinformatics platform for plant resistance gene analysis. *Nucleic Acids Research* 38: 814–821.

Service, R.F. 2006. The race for the $1000 genome. *Science* 311(5767): 1544–1546.

Shendure, J., R.D. Mitra, C. Varma, and G.M. Church. 2004. Advanced sequencing technologies: Methods and goals. *Natural Genetics* 5: 335–344.

Singh, V.K., S. Ambwani, S. Marla, and Akumar. 2009. Cry-Bt identifier: A biological database for PCR detection of Cry genes present in transgenic plants. *Bioinformatics* 4: 182–183.

Singh, V.K., A.K. Singh, R. Chand, and C. Kushwaha. 2011. Role of bioinformatics in agriculture and sustainable development. *International Journal of Bioinformatics Research* 3: 221–226.

Smith, T.F., and M.S. Waterman. 1981. Identification of common molecular subsequences. *Journal of Molecular Biology* 147: 195–197.

Sol Genomic Network. 2012. http://solgenomics.net/.

Stormo, G.D. 2000. DNA binding sites: Representation and discovery. *Bioinformatics* 16(1): 16–23.

Sugawara, H., O. Ogasawara, K. Okubo, T. Gojobori, and Y. Tateno. 2008. DDBJ with new system and face. *Nucleic Acids Research* 36: 22–24.

Tateno, Y., T. Imanishi, S. Miyazaki, K. Fukami-Kobayashi, N. Saitou, H. Sugawara, and T. Gojobori. 2002. DNA Data Bank of Japan (DDBJ) for genome scale research in life science. *Nucleic Acids Research* 30: 27–30.

The Arabidopsis Information Resource (TAIR). 2012. http://www.arabidopsis.org/index.jsp.

The International Rice Information System. 2010. http://www.iris.irri.org.

Untergasser, A., H. Nijveen, X. Rao, T. Bisseling, R. Geurts, and J.A.M. Leunissen. 2007. Primer3Plus, an enhanced web interface to Primer3. *Nucleic Acids Research* 35: 71–74.

Varshney, R.K., A.D. Hoisington, and K.A. Tyagi. 2006. Advances in cereals genomics and applications in crop breeding. *Trends in Biotechnology* 24: 1–10.

Varshney, R.K., S.N. Nayak, G.D. May, and S.A. Jackson. 2009a. Next-generation sequencing technologies and their implications for crop genetics and breeding. *Trends in Biotechnology* 27(9): 522–530.

Varshney, R.K., T.J. Close, N.K. Singh, D.A. Hoisington, and D.R. Cook. 2009b. Orphan legume crops enter the genomics era. *Current Opinion of Plant Biology* 12: 202–210.

Vassilev, D., J. Leunissen, A. Atanassov, A. Nenov, and G. Dimov. 2005. Application of bioinformatics in plant breeding. *Biotechnology* 19: 139–152.

Wan, J., F.M. Dunning, and A.F. Bent. 2002. Probing plant-pathogen interactions and downstream defense signaling using DNA microarrays. *Functional and Integrative Genomics* 2(6): 259–273.

Weber, A.P.M., K.L. Weber, K. Carr, C. Wilkerson, and J.B. Ohlrogge. 2007. Sampling the Arabidopsis transcriptome with massively parallel pyrosequencing. *Plant Physiology* 144(1): 32–42.

Wheeler, D.A., M. Srinivasan, M. Egholm, Y. Shen, L. Chen, A. Mcguire, W. He, Y.J. Chen, V. Makhijani, G.T. Roth, X. Gomes, K. Tartaro, F. Niazi, C.L. Turcotte, G.P. Irzyk, J.R. Lupski, C. Chinault, X.Z. Song, Y. Liu, Y. Yuan, L. Nazareth, X. Qin, D.M. Muzny, M. Margulies, G.M. Weinstock, R.A. Gibbs, and J.M. Rothberg. 2008. The complete genome of an individual by massively parallel DNA sequencing. *Nature* 452: 872–876.

Wicker, T., and C.R. Buell. 2009. Gene and repetitive sequence annotation in the Triticeae. *Plant Genetics Genomics Crop Model* 7: 407–425.

Wicker, T., D.E. Matthews, and B. Keller. 2002. TREP: A database for Triticeae repetitive elements. *Trends in Plant Science* 7: 561–562.

Wilson, D., V. Charoensawan, S.K. Kummerfeld, and S.A. Teichmann. 2008. DBD-taxonomically broad transcription factor predictions: New content and functionality. *Nucleic Acids Research* 36: 88–92.

Wu, T.D., and C.K. Watanabe. 2005. GMAP: A genomic mapping and alignment program for mRNA and EST sequences. *Bioinformatics* 21: 1859–1875.

Yale Plant Genomics. 2012. http://plantgenomics.biology.yale.edu/.

Yazaki, J., N. Kishimoto, M. Ishikawa, D. Endo, K. Kojima, Microarray Center, and S. Kikuchi. 2002. The Rice Expression Database (RED): Gateway to rice functional genomics. *Trends in Plant Science* 12: 563–564.

Yilmaz, A., M.Y. Nishiyama Jr., B.G. Fuentes, G.M. Souza, D. Janies, J. Gray, and E. Grotewold. 2009. GRASSIUS: A platform for comparative regulatory genomics across the grasses. *Plant Physiology* 149(1): 171–180.

Zerbino, D.R., and E.B. Velvet. 2008. Algorithms for de novo short read assembly using de Bruijn graphs. *Genome Research* 18: 821–829.

Zhao, W., J. Wang, X. He, X. Huang, Y. Jiao, M. Dai, S. Wei, J. Fu, Y. Chen, X. Ren, Y. Zhang, P. Ni, J. Zhang, S. Li, J. Wang, G.K. Wong, H. Zhao, J. Yu, H. Yang, and J. Wang. 2004. BGI-RIS: an integrated information resource and comparative analysis workbench for rice genomics. *Nucleic Acids Research* 132: D377–D382.

Zhu, Q.H., A.Y. Guo, G. Gao, Y.F. Zhong, M. Xu, M. Huang, and J. LUO. 2007. DPTF: A database of poplar transcription factors. *Bioinformatics* 23(10): 1307–1308.

Index

M. Ahmed, C.O. Stockle (eds.), *Quantification of Climate Variability, Adaptation and Mitigation for Agricultural Sustainability*,
DOI 10.1007/978-3-319-32059-5

MIX
Papier aus verantwortungsvollen Quellen
Paper from responsible sources
FSC® C105338

If you have any concerns about our products,
you can contact us on
ProductSafety@springernature.com

In case Publisher is established outside the EU,
the EU authorized representative is:
Springer Nature Customer Service Center GmbH
Europaplatz 3, 69115 Heidelberg, Germany

Printed by Libri Plureos GmbH
in Hamburg, Germany